AF615932

PENETRATION TESTING

EUROPEAN
SYMPOSIUM
ON
PENETRATION
TESTING
amsterdam 1982

PROCEEDINGS OF THE SECOND EUROPEAN SYMPOSIUM ON PENETRATION TESTING / ESOPT II / AMSTERDAM / 24-27 MAY 1982

Penetration Testing

Edited by
A.VERRUIJT
Delft University of Technology
F.L.BERINGEN
Fugro B.V., Leidschendam
E.H.DE LEEUW
Delft Soil Mechanics Laboratory

VOLUME ONE:
Standard penetration test
Dynamic probing / Swedish weight sounding

A.A.BALKEMA / ROTTERDAM / 1982

The texts of the various papers in this volume were set individually by typists under the supervision of each of the authors concerned.

For the complete set of two volumes, ISBN 90 6191 250 4
For volume 1, ISBN 90 6191 251 2
For volume 2, ISBN 90 6191 252 0

Distributed in USA & Canada by: MBS, 99 Main Street, Salem, NH 03079, USA

Printed in the Netherlands

Table of contents

Dynamic probing / Swedish weight sounding

Cone penetration test

Introduction

The first European Symposium on Penetration Testing – ESOPT I – was held in Stockholm, Sweden in 1974. During its Brighton meeting in September 1979 the Sub-Committee on Standardization of Penetration Testing in Europe decided to arrange a second international symposium. The Netherlands National Society for Soil Mechanics & Foundation Engineering was honoured with the request to organize this symposium.

The theme of the second European Symposium on Penetration Testing – ESOPT II – was set to be the interpretation of penetration tests:

- Cone Penetration Test (CPT),
- Standard Penetration Test (SPT),
- Weight Sounding Test (WST) and
- Dynamic Probing Test (DPA and DPB).

Recommended standards for these tests were presented by the Sub-Committee on Standardization of Penetration Testing in Europe to the Executive Committee of the International Society for Soil Mechanics & Foundation Engineering during its Tokyo meeting in July 1977. The recommendations were adopted and have been included in the minutes of the Executive Committee meeting and published in volume 3, page 95–152, of the proceedings of the Tokyo Conference. A further report with a recommended standard for the Light Dynamic Probing Test (DPL) was presented to the Executive Committee at its Stockholm meeting in June 1981 and will be published in the last volume of the proceedings of the Stockholm Conference.

Following a call for papers the Scientifc Committee of ESOPT II was agreeably surprised to receive not less than 168 summaries. 142 of these materialized into symposium papers written by authors from more than 30 different countries all over the world. Roughly 2/3 of these papers are on Cone Penetration Testing, while the remaining 1/3 are equally distributed over Standard Penetration Testing and Weight Sounding Testing plus the Dynamic Probing Tests. About half of all written papers have been selected for short oral presentation at the symposium.

The proceedings of this symposium are divided in two volumes. Volume 1 contains the state-of-the-art reports plus the papers on SPT, WST, DPA, DPB and DPL, while volume 2 contains the state-of-the-art report plus all papers on CPT. The proceedings have been edited by the Scientific Committee of ESOPT II.

The Scientific Committee has had comprehensive discussions about the most efficient way of presenting the written material. Several classifications attempts were made and rejected. Finally it was decided to simply arrange the papers according to the type of test, although it was realized that this method could not yield a unique result since some papers treat more than one test. Within their category the papers are arranged in alphabetic order by the name of the (first) author.

One of the lists the Scientific Committee itself used was considered to be so helpful in finding papers of a certain kind that it was decided to include this list in the proceedings. The list clearly demonstrates that the reader interested in only one of the four types of penetration tests, may also find relevant material in one of the sections of the proceedings devoted to another type. The editors hope that the reader will find this list useful.

The interest shown in ESOPT II and the large response as expressed in the generated number of papers, clearly shows that the importance and elegance of penetration testing is recognized now by a large portion of the geotechnical profession and that geotechnical engineers all over the world are actively working to try and improve and better understand these effective and relatively simple tools. At the Stockholm meeting of the Executive Committee the President of ISSMFE suggested that the terms of reference of the Sub-Committee should be altered so as to cover the topic world wide. Subsequently the Sub-Committee's title was changed to Sub-Committee on Penetration Testing. The editors sincerely hope that ESOPT II will help and stimulate the Sub-Committee and its chairman Prof. Bengt B.Broms in the succesful continuation of its important but difficult task.

Although the quality of the papers in these proceedings show quite a large variation, most papers present interesting material and together they contain a wealth of information which will take some time to digest. The Scientific Committee expresses the hope and the expectation that the two proceedings-volumes of ESOPT II presented here, will prove to be an important reference manual for the interpretation of penetration tests for a long time to come and that they will contribute to the success of ESOPT II.

Finally the Scientific Committee wishes to thank the publishers Messrs. Balkema, the authors and all those who have assisted and contributed in editing and producing these proceedings.

The Editors

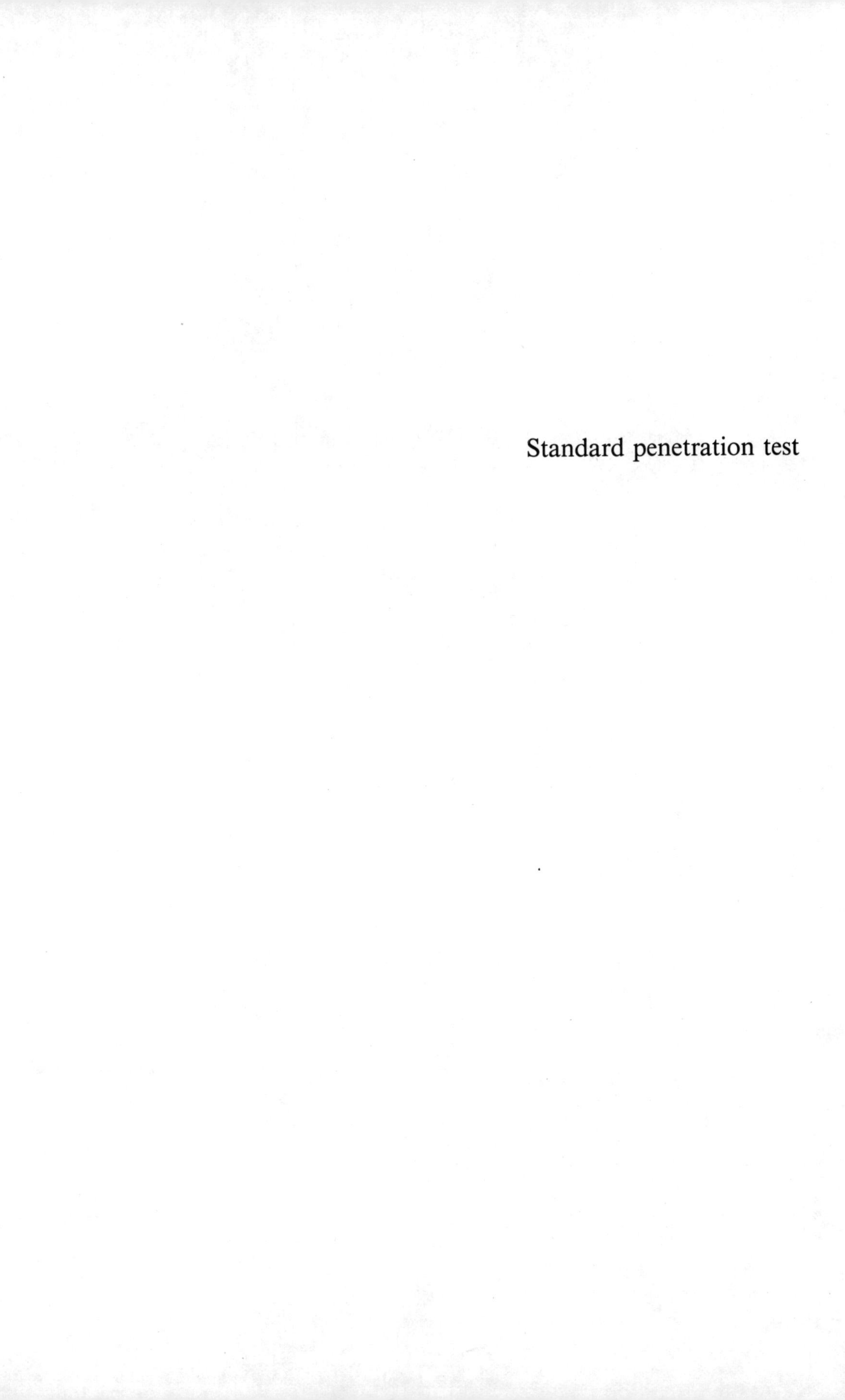

Standard penetration test

Proceedings of the Second European Symposium on Penetration Testing / Amsterdam / 24-27 May 1982

Standard penetration test
State-of-the-art report

IVAN K.NIXON
Engineering Laboratory Equipment Ltd., Hemel Hempstead, UK

SYNOPSIS

Important developments have taken place since ESOPT 1974 both with respect to the test method as well as the interpretation of the results. The considerable amount of research into the method being undertaken in the USA and Japan is reviewed against the new European Standard for the test. Japan has raised an objection to the specified rod size. Driving techniques have been closely examined in the search for a more precise standard in the USA. In reviewing the correlations with the various soil parameters particular attention is given to large scale laboratory experiments on the relationship between SPT N value, overburden pressure and relative density in sands that has emphasised the need for consideration of the other factors involved. Outline descriptions are given of eleven new methods for estimating settlement in granular soils and the results of a comparative study between eighteen methods. The test is important for estimating the danger of liquefaction and statistical methods are being suggested. Examples are included for compaction control by SPT. Besides its application in soils the test is widely used in Britain when investigating weak rocks and techniques have been proposed to aid pile design.

1 HISTORICAL BACKGROUND OF SPT

1.1 Scope of report

This review of the Standard Penetration Test (SPT) is intended to present the principal developments, both in the method of execution and its interpretation, that have taken place over the past 8 - 10 years, set against the earlier background. The datum has been arbitrarily taken for convenience as an undefined combination of three milestones in the subject; the very comprehensive state-of-the-art report on the SPT (de Mello 1971), the first edition of the standard reference book on the penetrometer (Sanglerat 1972) and the Proceedings of the First European Symposium on Penetration Testing (1974).

1.2 Origin and growth of the SPT

The Standard Penetration Test - originated about 1927 (Fletcher 1965), developed by Raymond Concrete Pile Company (Mohr 1943) and publicized by Terzaghi and Peck (1948) - has been in use for some 55 years. It is practised worldwide and to a greater extent than any other soil test. Horn (1979) has reported that 'the SPT has been and is likely to remain a keystone in soil exploration practice in North America'. It was used on 40 out of 49 US nuclear power plant site investigations made between 1954 and 1975 that he reviewed, and far exceeded in use all other samplers. Mori (1979) has reported that to his knowledge '... in Japan more than 90% of borings during the preliminary investigation phase are carried out with SPT', where it has been in use since 1953.

Despite continual criticism of the crudity of the test the number of papers published on it has flowed unabated over the years and since about 1972 the rate has increased with additional research into the dynamics of the test arising partly from the need to evaluate the liquefaction potential when siting major onshore and offshore structures using SPT data (Schmertmann 1978).

1.3 Advantages of the SPT

For a balanced viewpoint on the SPT it is appropriate, before starting to consider in some depth the weaknesses of the test, just to record its several advantages:

- Equipment is relatively simple and rugged
- Procedure is easy to execute and permits frequent tests
- Sampling facility is included
- Suits practically all soils and weak rocks
- Covenient both above and below the water table

No other insitu test combines this range of flexibility.

2 SPT METHOD OF TEST

2.1 The need for Standardisation

We continue to mislead both ourselves and those for whom we work by all of us using the same title for this test, thereby suggesting that it is executed in the same manner universally when, in fact, there are very significant variations between countries and sometimes within a country. 'If the testing procedure was better controlled, more precise correlations based on the N value might offer better design methods and answer some of the criticisms at the use of the test' (Ladd et al 1977). Moreover continuing research would have a better yardstick by which to judge its results. There need be no worries that standardisation would stifle further development on the testing method where sufficient strength of conviction existed, as is evident from the considerable improvements that have taken place over the years in laboratory testing despite the existence of many national standards.

Summarizing the situation on the method of test during the period under review; whilst we in Europe have devoted a considerable effort towards the establishment of a recommended Standard particulary for the region and possibly internationally, that was formally recorded during the International Conference at Tokyo (Anon. 1977), other countries outside Europe, notably the USA and Japan, have been carrying out detailed research on individual aspects of the procedure, and sometimes without as yet coming to any significant conclusion, by way of the revision of a National Standard.

Such is the need in my view for standardisation internationally that in the following paragraphs an attempt is made to explain briefly the principles behind each part of the new European Standard (reproduced in full in Appendix A) and then to refer to the detailed research.

2.2 Split-barrel sampler

Fundamentally the design adopted for the new Standard is the metric equivalent of the well established thick walled split tube of external diameter 2 in. (now 51 mm ± 1 mm), as originally proposed by the Raymond Concrete Pile Company, with a minimum length of 18 in. (now 457 mm), carrying a driving shoe at one end having the same external diameter as the barrel, and a coupling (with ball valve) at the other to connect with the sampler rods. A solid steel 60° cone is also specified that may be used to replace the driving shoe when in gravelly soils, as was originally suggested by Palmer and Stuart (1957) which gives results of the same order as for the shoe, or possibly only slightly higher (Rodin 1961, Gawad 1976).

The remaining critical dimension is the internal diameter which has again been fixed to correspond with the established figure of $1^3/_8$ in. (now 35 mm ± 1 mm).

2.2.1 Omission of liner. The USA Standard designation D 1586-67 (reapproved 1974) permits the internal diameter of split barrel (not the driving shoe), to be enlarged to $1^1/_2$ in. (38. 1 mm) 'provided it contains a liner of 16 gage (1.5 mm) wall thickness'. 'Almost all SPT samplers used in the United States have an enlarged inside diameter ... but drillers almost always use them without the liners' (Schmertmann 1979). In this paper, dealt with in more detail later under Correlations, Schmertmann believed that removing the liner from the split-tube section reduces internal wall friction in the split section to zero, and concluded as a result that sample recovery was greatly increased (in one case from 11.9 in. with liners to 17.8 in. without), removal was much easier but, on the other hand, because the tendency to 'plug' was reduced so was end bearing resistance, as well as the value of N when measured with a liner. According to his theory in apportioning wall and end bearing resistance, as applied to his field test results, this indicated that the reduction in N values, when liners are omitted, might rise to 50% in loose/weak soils (N ≃ 2 at 3 m, 10 at 30 m) and reach 30% in strong/dense soils (N ≃ 10 at 3 m, 60 at 30 m). The percentage reduction in N increases with decreasing N in any type of soil.

2.3 Sampler rods

The basic rule laid down in the new European Standard is that the stiffness shall be equal to or greater than type 'AW' drill rods (43.7 mm O.D., 34.1 mm I.D., and approximately 6 kg/m weight), with steadies at 3 m intervals for holes deeper than 15 m, otherwise 'BW' drill rods (54.0 mm O.D., 44.4 mm I.D., and approximately 8 kg/m weight) should be used. By recognising only the 'W' series of drill rods a difficulty has arisen in that the older and slightly smaller 'A' size drill rods (41.2 mm O.D.) are no longer acceptable although they are still widely used in many countries and acceptable in current Standards, e.g. ASTM D 1586-67 and JIS (Japanese Industrial Standard) A-1219-1976.

The Japanese Society for Soil Mechanics and Foundation Engineering (1981), after detailed research, has gone so far as to oppose the proposal in this matter of limiting rod diameter to the 'W' series, having regard to the submission for Internaional Standardisation of the SPT made at Tokyo in 1977.

2.3.1 Physical characteristics of rods. Evidence that it would be reasonable to accept the 'A' size drill rod has been obtained in Japan and the USA. Koreeda et al (1981) compared SPT results using standard JIS 40.5 mm O.D. rod and a 50 mm O.D. rod in two types of ground, one of which was to a depth of 45 m in a saturated sand and silt deposit and found no significant differences except for slightly higher N values with the 'N' rods in hard clay at depths below 15 m. Fletcher (1965) as a result of tests with 'N' and 'A' rods found the former gave slightly higher values and questioned their use. Brown (1977) in the USA compared SPT results in 6 boreholes using 'A' rod (40 mm O.D., 5.8 kg/m) and 'N' rod (60 mm O.D., 7.5 kg/m) without intermediate guides, to a depth of 34 m in loose becoming very dense granular deposits. He also found no significant difference overall, including the standard deviation for either size of rod, or by soil type.

Clearly a lower limit needs to be retained for rod diameter, particularly with respect to rod length. An example of inadequate stiffness is suggested by significant differences that were found in N values between different closely spaced boreholes at an overwater site in Brazil (Bogossian et al 1981 and Bogossian 1981) where one size of rod was used with an O.D. of 33 mm, I.D. of 23 mm and weighing only 2.75 kg/m that was thought to have given rise to whipping (Mori 1981).

Turning more directly to the subject of rod length the general consensus for a long time has been that the effect on the N value is negligible, although the evidence for this was meagre and some doubt was raised by de Mello (1971). Possibly as a result of this questioning and other factors a considerable amount of research has taken place in recent years on the dynamics of the test, particularly in the USA and Japan, that concern a number of aspects of which the rod length is but one.

2.3.2 Energy considerations in rods. Use of the wave equation as applied to pile driving analysis was made in a preliminary manner by Palmer and Stuart (1957) to evaluate the effect on the rods, from which they deduced that the rod weight in relation to the impact forces was small except in very weak ground, thereby concluding that the rods generally had little influence on the N value. The significance being limited to a long rod length in soils of low resistance was also found by McLean et al (1975) by means of a more refined theoretical use of the wave equation and taking advantage of computer facilities, which indicated that the variations with 'A' rod between 10 ft. and 200 ft. were generally only about 3 to 4 blows per 6 in. (152.4 mm) of sampler penetration but higher for weak soils.

A long series of valuable experiments to measure the dynamic forces has been taking place at the University of Florida under the direction of Professor Schmertmann. These consisted (Schmertmann 1978) of dynamic force-time measurements from just beneath the anvil and just above the SPT sampler using waterproofed strain gauge hollow load cells that screwed directly into the string of SPT rods; the data being displayed on an oscilloscope. Separate tests showed the load cells to have negligible effect on the results. It was found that 'for rod lengths exceeding 6 m (20 ft.) 90+ percent of the compression wave energy has already entered the rods before the hammer senses any effect from the soil around the sampler. Soil resistance at the sampler, and therefore the N value, has no effect on determining the energy input from the hammer beneath the anvil'. 'Rods less than 6 m (20 ft.) cause progressively significant reductions in hammer energy input because of the progressively earlier separation of rods from the hammer'.

More recently (Schmertmann and Palacios 1979), as a result of further field measurements and wave equation modelling on a computer, it has been concluded that rod whip or other effects with 'A' rods to a depth of at least 21 m (70 ft.) has a negligible effect on wave energy transmission. Moreover, varying slack of the rod joints had only a very small effect on N values.

Experiments on the rods in Japan have taken two forms, Uto and Fuyuki (1981) carried out tests on rods suspended horizontally with strain gauges near each end, up to lengths of 120 m, and found loss of wave energy due to an end impact was small. 'For practical reasons, a formula was proposed for correcting N values when rods of more than 20 m are used:

$$N = N_m (1.06 - 0.003\ell) \qquad (1)$$

where N is the corrected value, N_m as measured and ℓ the length of rods in metres'.

Matsumoto and Matsubara (1981) reported tests on rods in boreholes using one set of JIS Standard 40.5 mm O.D. to a depth of 46 m, and two non-standard rods of 50 mm and 60 mm O.D. to a depth of 56 m. Strain gauges were fitted at 10 m intervals. The measured stress waves were similar for all sizes of rod with little variation with increasing depth, except for time of arrival, and it was concluded that it is unnecessary to have intermediate lateral support or to use a larger diameter of rod than the 'A' size.

Evidence contrary to the foregoing has been reported (SPT ASTM Workshop 1979) of some tests in the Port of Los Angeles Testing Laboratory using rods in the horizontal position, as did the Japanese, but from which it was concluded 'that the energy delivered to a resisting medium is universally proportional to rod length, and that length is a significant factor with respect to the energy delivered'.

2.4 Driving technique

The new Standard has again adopted the metric equivalent of the well-established hammer mass of 140 lb (now 63.5 kg ± 0.5 kg) falling freely 30 in. (now 0.76 m ± 0.02 m) with a guide, on to a steel anvil rigidly fixed into the rod string. A self-tripping hammer is recommended to be used without causing additional forces on the rods. Special precautions are also to be taken to ensure the guide does not affect the hammer efficiency.

This part of the equipment and its use is undoubtedly the major factor causing variability in the performance of the test. Progress to obtain better correlations between one site and another is not possible until there is an improvement in this matter. The considerable amount of research that has been undertaken in the period under review is therefore not only a very encouraging sign for those interested in the future of this test, but it is an important reminder of the unsatisfactory bases of the existing correlations that rely upon results drawn from different sources.

2.4.1 Rope and cathead hoist. By way of introducing the research, all of which has centred around a 140 lb hammer being used in one form or another reference may be made to the contribution by Frydman (1970) to the discussion on the paper by Ireland et al

(1970) in which he stated that after a series of field tests in granular materials it was concluded that lifting and dropping the hammer by use of a rope on a cathead hoist with two turns gave a penetration resistance of the order of 1.4 times that obtained using the trigger mechanism. This was generally confirmed by Serota and Lowther (1973), except that their laboratory test based on Gibbs and Holtz (1957) work with dry dense sand, gave one turn of the rope as equivalent to the trip hammer they tested, and two turns an increased resistance in the ratio 1:1.4. These two results regarding the efficiency of the drop together with much experience by others on the advantage of the trip or triggered hammer to secure better reliability in the distance of the drop, compared with techniques depending upon visual control combined with manual dexterity, has led to trip hammers being specified by many civil engineers.

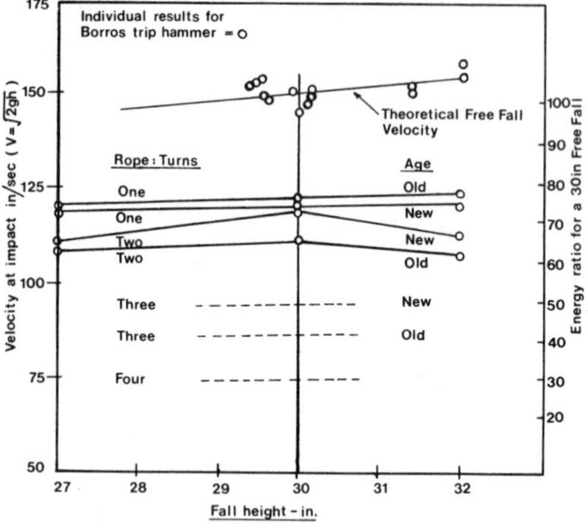

Fig. 1 Velocity at impact v. drop for several driving methods. (After Kovacs et al 1978 and Kovacs 1979).

Nonetheless visual control of the drop with rope and cathead operation is still widely used, particularly in USA and Japan, and a series of studies led by Kovacs has provided valuable insight into the actual variations that occur and their consequences. Measurements using lightbeam oscillograph and a reflective target have been made of the velocity of the hammer on contact with the anvil-rod system (see Fig. 1). Kovacs et al (1977) disclosed that the impact velocity was significantly influenced by the number of turns of the rope around the cathead, the age of the rope, its moisture content, the method of releasing the rope and the distance the hammer fell. They found that even one turn when released did not give a completely free fall, confirming that Serota and Lowther's triggered hammer was not free fall as they themselves showed by the N value being lowest for an unguided hammer. Overall Kovacs et al concluded that the N value may vary by a factor of two or more, all other factors being the same. On the basis of the variability in applied energy, field tests showed an inverse linear relationship to exist between the N value and the delivered energy for a given insitu condition. The foregoing results that included tests with new safety hammers were extended by Kovacs et al (1978) in trials with a special hoisting device designed to reproduce mechanically the manual test with an experienced driller using two turns of the rope on a cathead. The use of the rope 'is not especially safe', having caused injury and even fatalities. It was explained that one turn of the rope was rarely used in the USA because of the extra effort involved when lifting the hammer prior to applying the next blow and normally two or even three turns are used. Many operators, its seems, also vary the number of turns throughout the day in an attempt to compensate for the changing characteristics of the rope with variations in temperature and humidity. A study was undertaken of 37 different rigs (Kovacs 1980) as a result of which it was found that cathead on different rigs rotated in both directions in about equal numbers, and this also naturally affected the length of contact of the rope on the drum, as well as its position with respect to the ground. Subsequently the tests have included a triggered hammer whose speed at impact was within one percent of the free fall velocity. (Note this is different to results by Serota and Lowther (1973) where their trip hammer corresponded to 1 turn of the rope). Figure 1 shows the results of various measurements.

The rate at which blows are delivered with triggered hammers affects their performance, depending upon the design. Too fast a speed in one case (Kovacs 1979) increased the dropping distance by 2 or 3 in. (51 - 76 mm).

2.4.2 Hammer and anvil interaction. Another series of research studies has been in progress at the University of Florida by Professor Schmertmann to which reference has already been made concerning the rods. Use of the load cell just below the anvil (Schmertmann 1978) showed that hammer falls impeded by rope-cathead friction and other causes gave an impact anywhere between 30 to 80 percent of its true free fall energy, confirming work by Kovacs et al (1978). A detailed exposition of the energy dynamics of the overall test system has been made by Schmertmann and Palacios (1979) in which the results from the load cells at each end of the rods were analysed using the wave equation; and probably for the first time due account was taken of the effect of the resistance of the ground. The theory was reported to reflect closely the experimental results. Fig. 2 is a particular case in this analysis. Fig. 2b shows the decaying series of pulses synchronous with the stress wave traverse up and down the rods, while Fig. 2c shows the penetration that occurs in response to the force pulses. In the case shown the first pulse produced 31% of the total penetration and it took 5 wave cycles and 26 msec from impact to achieve 90% penetration. Fig. 3 shows results of actual measurements of the vertical displacement of the top of the rods by means of a highspeed camera by Yamada et al (1981) that shows a similar pattern to the US results in Fig. 2.

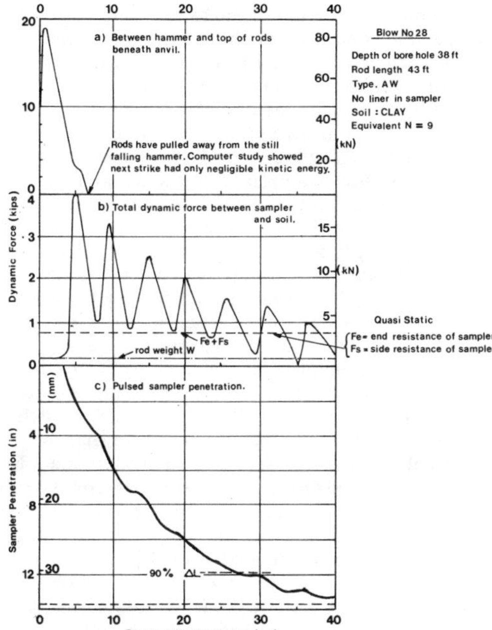

Fig. 2 Wave equation model results (after Schmertmann and Palacios 1979)

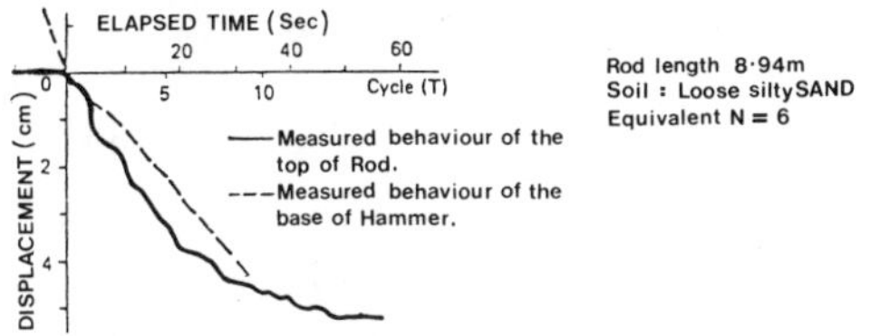

Fig. 3 Dynamic behaviour at top of rod observed with highspeed camera (after Yamada et al 1981).

However although research in Japan on the dynamic behaviour of the driving system has been in progress for some years and they have confirmed the increased resistance with the rope and cathead method as compared to trip hammers, it appears that agreement has not been reached and it is considered that more work is required (Shioi et al 1981).

In Britain the present BS 1377:1975 that describes the test method specifies a hammer weight of 65 kg, i.e. 1.5 kg more than the agreed European Standard.

Summarising the situation from this survey there are strong feelings that the driving technique needs to be more closely defined, less susceptible to the human factor and not restricted to the hammer impact but to include the whole system. It has been suggested that the energy reaching the sampler is standardised but not necessarily related to the theoretical 0.76 m (30 in.) free drop of the standard hammer, but adjusted e.g. by a reduced drop, to give the same energy as has been used in the past, in order not to upset existing correlations. A 50% efficiency has been applied in one case for calibration purposes (Schmertmann et al 1978). Other refinements that have been mentioned are wave traps in the rods and corrections for the potential energy component (SPT ASTM Workshop 1979).

2.5 Boring and test techniques

Guidance on this aspect in the new European Standard contains all the well known precautions including the use of drilling mud to minimise disturbance prior to the test. No boring technique is excluded.

The diameter of the borehole should be between about 60 and 200 mm. After a 'seating drive' of 0.15 m (6 in.) (which shall include the initial 'own weight' penetration) the number of blows required (rate not to exceed 30 per min.) for the further penetration of 0.30 m (12 in.) shall be recorded, also the number for the intermediate distance of 0.15 m (6 in.). Precautions are included for hard ground.

Assuming careful use of the boring tool, whatever its form, then the greater the hole diameter the more liability there is for a deeper bulb of disturbed ground. The literature in general would suggest an upper limit of about 100 mm which is to some extent confirmed by Lake (1974) who found reduction in SPT in granular materials of the order of 25 - 50% in moving from 125 mm to 200 mm using shell and auger borings.

The value of maintaining the water level at or slightly above that of the groundwater is limited to the avoidance of hydraulic disturbance. The use of drilling mud is one practical solution if an attempt is to be made to replace the overburden pressure (Schmertmann 1978).

A warning is required with the increasing use of screw augers where withdrawal must produce excessive suction and consequent disturbance. Continuous flight augers with a hollow stem from which the tests are made appear to offer advantages, at least above the watertable, but below it the method has been found unreliable (Schmertmann 1975).

3 CORRELATIONS OF SPT DATA WITH OTHER SOIL PARAMETERS

3.1 The role of empiricism

This chapter deals with the more commonly used parameters for which correlations have been proposed with the SPT. Some suggestions have been made that it would be preferable when applying the results of empirical tests to engineering problems to seek a direct relationship thereby eliminating intermediate discrepancies. On the other hand there can be advantages by working via interpreted soil parameters that adjustments can be made that would not otherwise be possible, such as to allow for insitu stress changes in the case of foundations, and for the local conditions as explained in the relative density section.

3.2 Cone penetration test (CPT)

Schmertmann (1979) has presented a static theory for SPT in terms of CPT data and hence provides a means of quantitatively estimating one from the other. Earlier suggestions had been made by Schmertmann (1971) with which Sanglerat (1972) generally concurred.

From vertical force equilibrium, the following equation is developed that expresses the additional force F to be added to the gravitational force on the sampler, W (potential energy, that is required for penetration (see Fig. 4)).

$$F + W = C_1 q_c A_e + C_2 f_c (d_i + d_o) \pi L \qquad (2)$$

where it is assumed

(a) SPT end bearing resistance is a constant C_1 times static cone end bearing q_c, times area A_e (SPT and CPT end areas similar, 10.7:10 cm^2).

(b) SPT wall frictional resistance is a constant C_2 times local friction on the CPT sleeve f_c, times the area inside and outside.
d_i = inside diameter
d_o = outside diameter

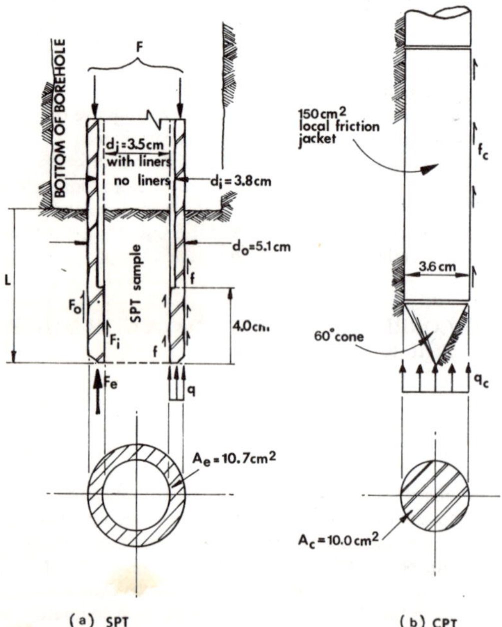

Fig. 4 Comparison of SPT and CPT components of penetration resistance (after Schmertmann 1979)

By dividing a test over three equal increments of penetration and assuming the resistance remains directly proportional to the added energy; also that q_c and f_c do not change, then it is possible to derive 3 equations expressing the ratio of forces for each increment in comparison with the final one, from which N can be obtained from q_c and f_c or vice versa, given the proportionality constants C_1 and C_2.

In order to evaluate these constants a series of field tests was made consisting of a close pattern of four CPT soundings surrounding two SPT borings. USA standard split barrel samplers with and without liners were used and from the results it was deduced that C_1 was approximately unity from 16 comparisons in several different types of soil, but not weak clays. The parameter f_c was tentatively evaluated as 0.7 for all soils when using the Begemann friction-cone tip which reduces in diameter above the cone and hence it was suspected to include some bearing resistance at the bevel at the bottom of the friction sleeve close to the tip. For the parallel sided sleeve $C_2 = 1.0$ was suggested. Other comparisons were made using the WES research work on relative density (Marcuson and Bieganousky 1977a and 1977b): One other implicit assumption was that there was no significantly different effective stress effect between the dynamic and quasi-static penetration. Some results of this work are given in Fig. 5.

A recent comparison in field test results between SPT and CPT data has been obtained from a study in Poland (Borowczyk and Frankowski 1981) where 32 different sites were investigated by the probing techniques recommended for the European Standards.

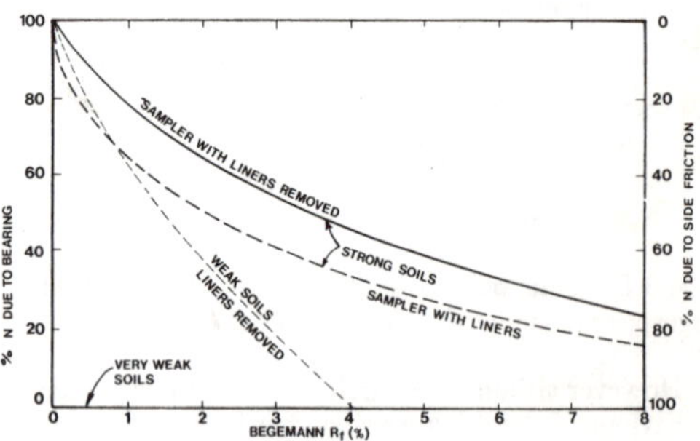

Fig. 5 Percentages of N due to end bearing and side friction for various CPT friction ratios (after Schmertmann 1979).

The soils were all non-cohesive Quarternary materials with grain sizes from 0.5 to 15 mm. Density index was derived empirically from the results. The relation between SPT and CPT in Fig. 6 should be viewed with caution as it shows a variation in the q_c/N ratio according to the N value whereas usually the variation is related to grain size, and further examination may show this is reflected in the results.

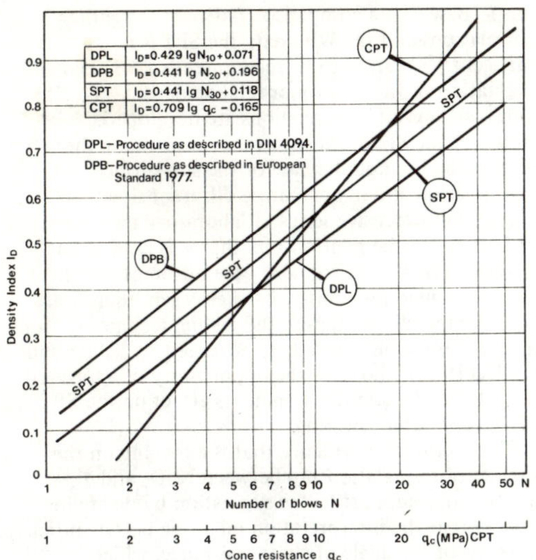

Fig. 6 Relation between density index I_D and number of blows N or cone resistance q_c (after Borowczyk and Frankowski 1981).

A number of results has been published relating the cone resistance q_c, normally in kg/cm^2, to the N value (e.g. Meigh and Nixon 1961, Rodin 1961 and Sutherland 1974). More recently additional proposals have been made (Meigh and Hobbs 1975, Aoki and De Alencar Velloso 1975, Sanglerat 1979) that have been summarised to give typical results as shown in Table 1

Table 1 Typical values of the ratio q_c/N

Soil description	Ratio q_c/N
Silty or sandy clay	2
Sandy silt	3
Fine sand	4
Fine to medium sand	5
Medium to coarse sand	8
Coarse sand	10
Gravelly sand	8 - 18
Sandy gravel	12 - 18

3.3 Relative density of sands

The initial use of relative density to provide a simple means of classifying the degree of compaction of cohesionless soils into broad bands according to the results of the SPT is universally accepted. Wide recognition has also been given to the adjustment for the overburden effect proposed by Gibbs and Holtz (1957) as a result of their laboratory tests. At the same time the parameter was made more precise and this gave it added importance which in many respects continues to this day. Other researchers (Schultze and Melzer 1965, Bazaraa 1967) carrying out similar studies but in different ways, however, arrived at different correlations between N value, relative density (D_r) and the effective overburden pressure (σ'_v). This naturally led to a close examination (e.g. Tavenas and Rochelle 1972) of all aspects of this approach in the interpretation of SPT data.

Having regard to developments since ESOPT 1974, it is relevant to appreciate the principal factors involved:

(i) The generally adopted definition for relative density, or more correctly 'relative voids ratio' (Selig and Ladd 1973), involves the ratio of two small differences between relatively large numbers and is therefore very sensitive to errors in each of the three densities concerned. Errors up to 20% in D_r have been suggested by Tavenas et al (1973).

(ii) Methods of measuring density vary with different results, and are not always defined.

(iii) Methods of executing SPT also vary with consequences already described.

(iv) Placement techniques in the laboratory tests vary between laboratories and for achieving different densities. Water content has varied from zero to submerged.

(v) Surcharge techniques combined with tank design and its geometry all significantly influence the stress distribution before and during the test.

A recent review of the uncertainty in measuring D_r by Halder and Tang (1979) suggested that the Gibbs and Holtz relationship could contain a systematic bias and be unconservative.

The application of SPT to predict insitu D_r for determining liquefaction potential increased the need for validation and as a result an important series of carefully executed large scale laboratory tests has been done by the Waterways Experimental Station (WES) Vicksburg, USA on four different types of sand to investigate the reliability of the SPT to determine relative density. The tests were made (Marcuson and Bieganousky 1977a) in a stacked ring soil container 2 m (6 ft.) deep by 1.2 m (4 ft.) I.D. consisting of alternating layers of steel rings 25 mm (1 in.) square section grooved top and bottom to retain 4.8 mm ($^3/_{16}$ in.) thick rubber spacing rings; an arrangement which has been found to overcome almost totally the transfer of a significant proportion of the soil stresses to the tank wall when solid, from the deadweight of sand and the surcharge loads. (Gibbs and Holtz used a solid tank). A number of different raining techniques had to be employed to place the sands in the tank because of difficulties when trying to use one method to obtain uniform conditions for each of the selected densities. Overburden loads were applied via a waterbag to ensure free strain loading and in all cases except one the sandbed was submerged. (Gibbs and Holtz used plane strain). SPT were made at four levels with drilling mud through each of four sleeves in the tank top using a conventional USA pattern split spoon sampler, but without the liner, and a triggered hammer to ASTM D 1586 - 67, equivalent to the European Standard. (Gibbs and Holtz employed a cathead and 2 turns of the rope).

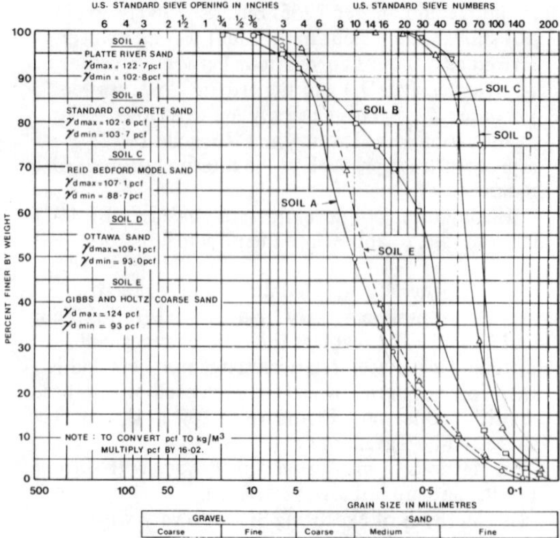

Fig. 7 Gradings and densities for the 4 WES and 1 USBR sands used in laboratory tests for determining relative density with SPT (after Marcuson 1978)

Gradings and limiting densities for the four sands are shown in Fig. 7 together with those for the coarse sand used in the classic earlier US Bureau of Reclamation related tests (Gibbs and Holtz 1957).

The 32 SPT results with the four different sands are summarised in Fig. 8 where it will be seen that there was a significant spread for the two fine sands 'C' and 'D', as well as variations with previous tests. WES was satisfied that SPT results in a given test specimen were reproducible.

A few tests were made in the fine sands 'C' and 'D' with an over-consolidation ratio (OCR) of 3 which indicated that although the results were slightly greater than those for normally consolidated specimens 'it would be difficult to establish the stress history of a site from field N values'(Marcuson and Bieganousky 1977a). The measurements would not appear to bear out the suggestion by Schmertmann (1975) that the insitu horizontal stress has at least twice the proportional effect of the vertical stress.

The principal conclusions (Marcuson 1978) were:

(i) A relationship existed between N value, the effective overburden pressure (σ'_v) and the relative density (D_r), but the spread of the WES data suggested that a unique family of curves for all sands under all conditions was not valid. Generally while there is fair convergence at low values of D_r, divergence is significant at high values of D_r.

(ii) Differences in testing techniques, both during the WES work and more significantly with those used by Gibbs and Holtz (1957), were thought to explain some of the variances.

(iii) Sand type, e.g. grain size, grading and particle shape also influenced the results (re size see Ishihara and Watanabe 1976, re particle shape see Holubec and D'Appolonia 1973).

(iv) 'Based on a comparison between the earlier correlations and the WES data the SPT is not sufficently accurate to be recommended for final evaluation of the density or relative density at a site, unless site-specific correlations are developed. However the SPT does have value in planning the undisturbed sampling phase of the sub-surface investigation and in comparing different sites'.

Another extensive series of laboratory tests on sands to investigate the relationship, that was carried out in Egypt (Gawad 1976), also gave different results to those obtained previously, but here again there were considerable differences in the testing technique. More inconsistent results have been reported by Kilker and Lucks (1981) who used the opportunity of a lowered watertable at a site to examine its effect on the SPT before and after lowering.

The conflicting evidence that is emerging on this subject of the relationship between N, D_r and σ'_v lends strong support to the suggestion by de Mello (1971) that the interpretation might be better suited in terms of the angle of shear resistance, which is already acknowledged to combine the several factors concerning the soil and ground conditions, that are now being suggested to explain some of the variances. If the relative density parameter is to be retained and improved then more test standardisation and a better adherence to it is essential (Holtz 1973).

Utilisation of the existing data on the relationship N, σ'_v and D_r has lately been given renewed value by the application of probabilistic and statistical methods. Fardis and Veneziano (1981a) have developed a reference model particularly suitable for computer use, based upon the WES laboratory and Gibbs and Holtz fine sand test data, and allowing for estimated density testing errors, all related to normally consolidated material. This is then combined with a few related measurements on site of D_r and from careful sampling and the corresponding N value, in order to derive a site-specific model to infer D_r from N at individual locations, thereby taking account of the local insitu structure, consolidation history, soil type and the manner in which the SPT is performed.

3.4 Shear strength

In clays a better quality sample should be taken for shear strength measurement than is obtainable with the SPT. That said, its use for this parameter is related solely to the undrained condition. In weak and sensitive clays remoulding due to penetration of the thick walled sampler must reduce strength and hence the N value in relation to the undisturbed state. A number of correlations have been given by Schmertmann (1975). Arising from his more recent work on the dynamics and statics of the SPT he has suggested that N should relate more closley to the cyclic undrained strength from considerations of the side friction factor and stress wave penetration (Schmertmann 1979).

In sands the preference by de Mello (1971) for an N - ϕ correlation rather than N - D_r, included in his report a collection of the meagre data available, and it is unfortunate that more such measurements where not given

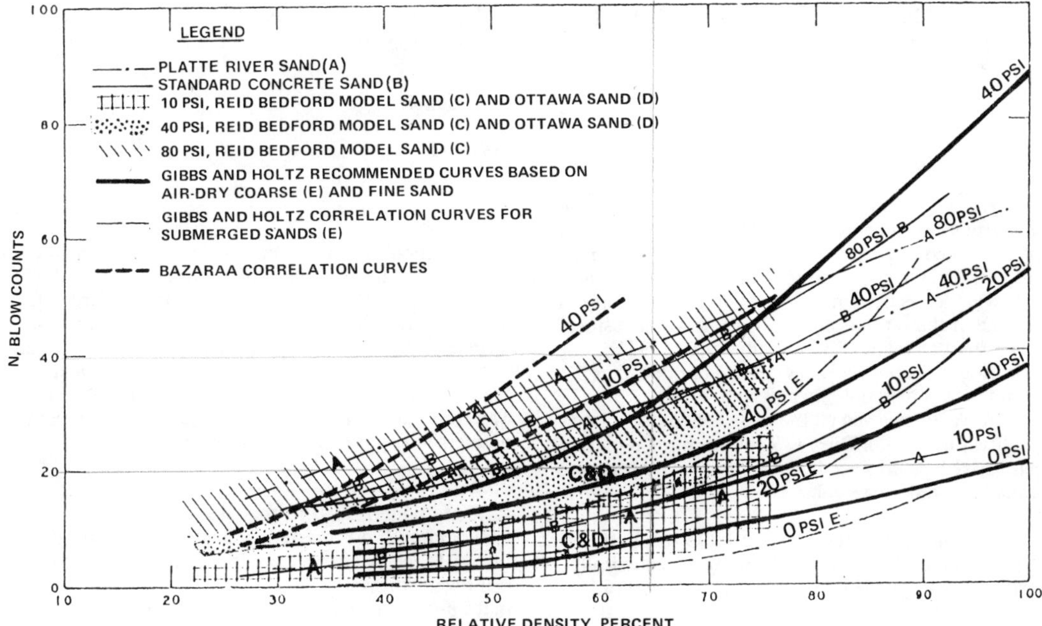

Fig. 8 Summarized WES results of N value v. D_r compared with correlations proposed by Gibbs and Holtz and by Bazaraa. (after Marcuson 1978)

for the sands used for the WES research projects referred to earlier.

3.5 Deformability

Six different proposals, suggested between 1965 and 1974, for predicting deformability from N values, together with the relevant soil type and basis of the method, have been listed by Mitchell and Gardner (1975) in their state-of-the-art report on insitu measurement of volume change characteristics. The methods were found to be widely divergent.

A review of the existing correlations has also been made by Natarajan and Tolia (1979) who concluded that relative density and the overburden effect should both be taken into account. Working at the Central Road Research Institute, New Delhi, it is understood that tests are planned in sandy soils to relate the compressibility with various confining pressures at different values of relative density.

Schultze and Biedermann (1977) have described the results of a series of comparative tests between various penetration tests and two types of pressuremeter over a range of fine grained soils in several countries. Analysis suggested the overburden pressure had no influence on the results. The pressuremeter modulus (E_p) with which the N values were compared is more closely related to a shear modulus, corresponding to a deviatoric stress field, which is customarily used to estimate settlement.

Difficulties frequently occur when sampling glacial tills (including 'boulder clays'), consisting typically of varying amounts of coarse particles from sand size to boulders, often with some silt and clay. An opportunity was therefore taken to compare available oedometer tests on 100 mm diameter samples with SPT N values from 17 different sites (Stroud and Butler 1975) and it was found that a simple correlation appeared to exist for the coefficient of volume compressibility of the form

$$m_v = \frac{1}{fN}, \text{ where N is SPT value} \quad (3)$$

Values of .f ranged from about 450 kN/m^2 for materials of medium plasticity, to over 600 kN/m^2 for materials with a plasticity index less than 20. They also considered that for overconsolidated clays and boulder clays the final settlement could be adequatley predicted using N values by means of quasi-elastic analysis and the vertical drained elastic modulus.

4 ENGINEERING APPLICATIONS OF SPT IN SOILS

4.1 Settlement in granular soils

A very thorough survey of the application of N values to predict settlements in granular soils was made by Sutherland (1974) for the Cambridge UK Conference on 'Settlement of Structures' (COSOS). He concluded that it was now generally agreed that the 1948 Terzaghi and Peck 'prescription' led to over-conservative results but owing to insufficient evidence on the reliability of any one of the various modifications which had since been proposed (about 13 in number up to 1971 according to Talbot 1981) he considered that they should be regarded only as 'aids to design'.

Sutherland reviewed all the main proposals as far as the direct method of Parry (1971). Additional methods that I have located since that date are briefly reviewed in the following paragraphs.

4.1.1 Method of Schultze and Sherif (1973). Settlement observations of 48 footings on sand were statistically correlated with the N values directly measured or deduced, using linear elastic theory, to arrive at the following expression for mean settlement.

$$s\,(cm) = \frac{q\,Bf}{1.71\,N^{0.87}\,(B/B_1)^{0.5}(1 + 0.4\,D/B)} \qquad (4)$$

where q kg/cm^2 = mean contact pressure without reduction for excavation
B(cm) = breadth of footing
B_1(cm) = 1 cm
f = Influence factor depending upon width:length ratio and thickness of compressible stratum
N = Arithmetical mean of SPT N values; if necessary by layers
D(cm) = Depth of embedment

Reapplication to the 48 cases gave predictions generally within the range of ± 40% of the measured values of settlement.

4.1.2 Method of Peck, Hanson and Thornburn (1974). A new set of design charts was offered from those suggested by Terzaghi and Peck (1948). The widely used adjustment in submerged low permeability sands, when N is greater than 15, was dropped in favour of a conservative interpretation. Adjustments were included for the overburden pressure, but more conservative than Gibbs and Holtz (1957), as well as for a high watertable.

4.1.3 Method of Meigh and Hobbs (1975). They proposed for footings an initial calculation of settlement (s_i) using the given net bearing pressure and the charts of Terzaghi and Peck (1948), but ignoring the effect of the watertable. The result was then to be corrected to compensate for grain size and grading, utilising if need be typical values in Table 2, as follows:

$$s_{corrected} = s_i \frac{4N}{q_c} \qquad (5)$$

For example,

In medium coarse sand $q_c = 8$ N
Hence $s_{corr} = \frac{1}{2} s_i$

4.1.4 Method of Tomlinson (1975). An adjustment of N, as measured on the basis of Gibbs and Holtz (1957), was proposed to allow for the effect of overburden to the extent that N adjusted became 4 N at the ground surface. This adjustment factor has however since been considered optimistic and there is preference for the slightly lower factor suggested by Thorburn (1963). A watertable adjustment and the reduction of N in dense submerged low permeability sands were retained.

4.1.5 Method of Burland et al (1977). A completely fresh approach was offered by analysing the available case histories where site conditions and actual settlement were known, as shown in Fig. 9, which revealed that upper limits could be assessed by broadly classifying the ground as loose, medium dense and dense.

It was suggested that the method was sufficiently accurate for routine design. Probable settlement of medium dense or dense ground might be taken as half the upper limit values in which case the maximum settlement would not normally exceed about 1.5 times this value. The uppermost bound marked L was to be treated as only tentative.

The analysis did show that for footing widths greater than 3 m the scatter in actual observed settlement was less than the range of prediction by the other methods.

4.1.6 Method of Parry (1977). By extending his earlier work (Parry 1971) for estimating bearing capacity in sand from N values, and using an assumed relationship between D_r and $\emptyset$, he arrived at a very simple expression for ultimate bearing capacity (q_f) where footing breadth exceeds its depth and is fully buried

$$q_f = 0.24\,N\,MN/m^2 \qquad (6)$$

As measured N values were considered to reflect the presence of groundwater no adjustment was required unless a change in level was anticipated.

Utilising case histories given by Simons and Menzies (1977) a new expression was also given for estimating settlement

$$s = 300\,q\,{}^{B}\!/_{N} \qquad (7)$$

It was concluded that bearing capacity is likely to dictate the design for shallow foundations only up to a width of about 2 metres.

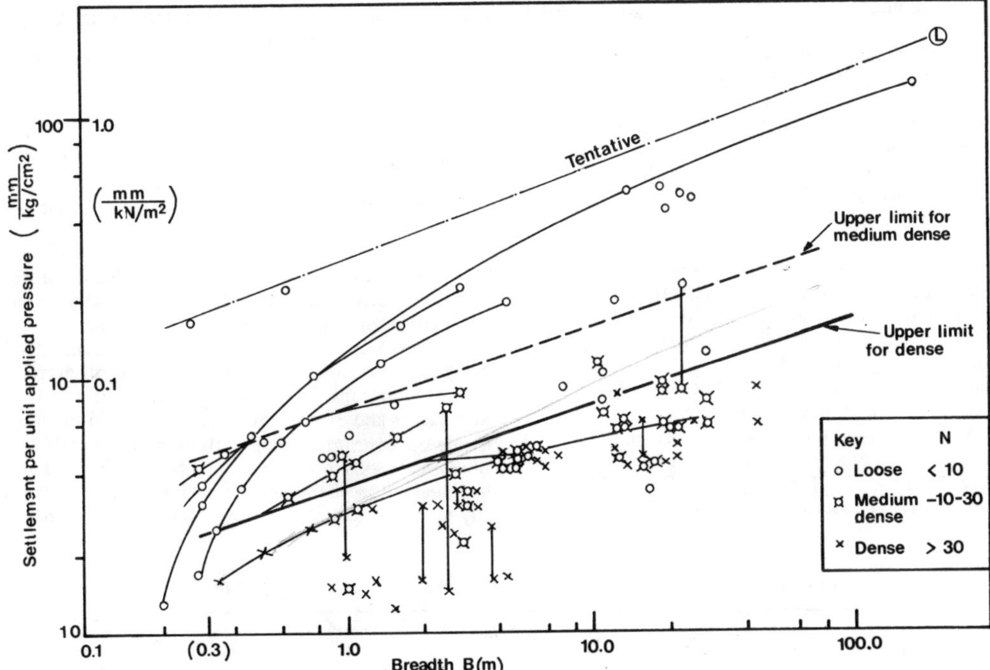

Fig. 9 Observed settlement of footings on sand of various relative densities (after Burland et al 1977)

4.1.7 Modification by Christian and Carrier (1978). According to Dikran (1980) improved charts for influence factors were proposed for computing settlement using elastic theory (e.g. D'Appolonia et al 1970), based upon proposals by Burland (1970) who studied the effect of embedment of foundations using finite element methods, and Giroud (1972) who considered the average settlement of a flexible loaded area.

4.1.8 Method of Parry (1978). In continuing to support the use of the plate bearing test (but preferably larger than 0.3 m diameter) in sand, he advocated that 'full and proper account is taken of the variation of soil stiffness with depth' and proposed the expression for the foundation settlement

$$s_f = s_p \times \frac{B_f}{B_p} \times \frac{N_p}{N_f} \qquad (8)$$

where s_f = Settlement foundation
s_p = Settlement plate
B_f = Breadth foundation
B_p = Breadth plate
N_f = N value beneath foundation
N_p = N value beneath plate

It was pointed out naturally that due account would have to be taken of any subsequent change in the ground on the N values.

4.1.9 Method of Louw (1977). A method was proposed according to Talbot (1981) that was highly complicated and used the CBR test as a model footing. A number of objections has been raised at this method which includes many assumptions (Dikran 1980).

4.1.10 Method of Oweis (1979). This was a method based upon an interpretation of 51 plate bearing tests where SPT data was known. The analytical model for each layer at a site incorporates an expression for the secant modulus of deformation in terms of N corrected according to Bazaraa (1967) for the overburden pressure, the mean effective normal stress compatible with the foundation loads, and the vertical strains induced by the foundation load. Settlements are then calculated from the equivalent model using linear elastic theory. Test cases for three sites were examined and the predictions were all within 30% of the observed settlement.

4.1.11 Method of Arnold (1980). A method is proposed involving the establishment of the relative density from which strains below a footing are deduced at any horizontal plane by means of an empirical stress-strain relationship. Utilising the Boussinesq distribution of vertical stress and the effect of increased modulus with depth, according to the breadth of the footing, the individual strains per layer are summed to give the total settlement.

A formula based on the Gibbs and Holtz (1957) curves is proposed for determining the relative density from the N values. The stress-strain relationship is that given by Terzaghi and Peck (1948) for a 0.305 m sq. plate test on sand.

86% of the 94 case histories analysed with this method were within ± 100% of the measured settlement value.

4.1.12 Performance of methods to predict settlement. Practical difficulties continue to persist to enable satisfactory samples of granular soil to be recovered for laboratory tests to provide an alternative approach for estimating settlement. A review is therefore apposite on the general performance of and between SPT and CPT methods. Talbot (1981) has collected 360 published comparisons between measured and predicted load-induced settlements for 75 different foundations, in a review of the use of insitu penetration test data. The summarised results are given in Table 2.

Table 2 Performance of settlement predictions by SPT and CPT methods (after Talbot 1981)

Bases of prediction	No. of methods	No. of predictions	Ratio of $\frac{\text{Predicted s}}{\text{observed s}}(R_s)$			
			maximum	minimum	average	standard deviation
SPT	18	213	19.79	0.12	2.01	1.07
CPT	7	147	23.79	0.11	1.71	1.09
combined	25	360	23.79	0.11	1.92	1.09

Talbot found that of the 40 methods he located to assess settlement, 15 had not been applied to cases reported in the literature. Subsequent application of a method by others generally yielded wider variations in the ratio R_s, predicted to observed, than was obtained by the originator when demonstrating its validity. In referring to the more obvious reasons for differences between methods such as the non-standard SPT, he noted that authors often do not specify which settlement is being calculated and suggested use of the 'characteristic point' as defined in DIN 4019 (1959). The settlement at this point is the same for either a flexible or rigid raft.

The established SPT methods, i.e. those used in over a dozen published predictions, that Talbot found to be most consistently reliable, were those proposed by D'Appolonia et al (1970) and Parry (1971). It is of interest to note that both are based on direct correlations with SPT (Jordan 1977), an approach that has been strongly advocated for some time by de Mello (1971 and 1979).

4.1.13 Creep. It has been noted by Hobbs (1981) that there appears to be no guidance in the published SPT methods for estimating settlement, whether or not they take account of creep. Schmertmann (1970) considered creep did occur and was a separate entity for which he gave guidance when using CPT data, and presumably this would also apply in sands tested by SPT. This would seem reasonable if the elastic theory is involved as is so often the case.

4.2 Estimating liquefaction potential

This subject dominates the application of the SPT in the study of soil dynamics. The literature on it is extensive particularly in the USA and Japan, arising from the concern for the safety of nuclear power plants.

4.2.1 The nature of liquefaction. The mechanics of flow were first described by Terzaghi and two papers presented at our second International Conference held at Rotterdam (Koppejan et al 1948, Peck and Kaun 1948), showed liquefaction could occur as a result of several types of vibrations, including the passage of trains, earthquakes and explosions. Major research in the last decade has concentrated upon its effect in saturated cohesionless soils under undrained conditions. Besides the liquefaction phenomenon which normally concerns only loose fine sands and arises from the development of high pore pressures at constant volume leading to a complete loss of shear strength; it is now commonly linked with another phenomenon termed cyclic mobility, also due to high pore pressures, but which can occur in loose or dense sands and entails large deformations but not loss of shear strength or flow.

4.2.2 Applications of SPT to predict liquefaction potential. The use of the SPT to predict liquefaction potential came to the fore with the classic study of the Niigata earthquake of 1964 when extensive liquefaction occurred in level sandy ground, and comparisons were able to be made of N values before and after the earthquake in conjunction with the degree of damage to reinforced concrete structures (Yoshimi 1977). A simple empirical rule was suggested by Ohsaki (1970) after studying the Tokachioki earthquake that liquefaction is not a problem if the N value exceeds twice the depth of the test measured in metres. Use of this was validated by a subsequent earthquake.

By combining the field observations of different earthquakes it was possible to relate their recorded average horizontal shear stresses with the determination locally of SPT v. depth or equivalent relative density, and thereby to distinguish at level sites between where liquefaction had occurred and where it had not (e.g. Seed and Idriss 1971). The next step has been to adjust the average N value for the sand layer in question, from its actual effective

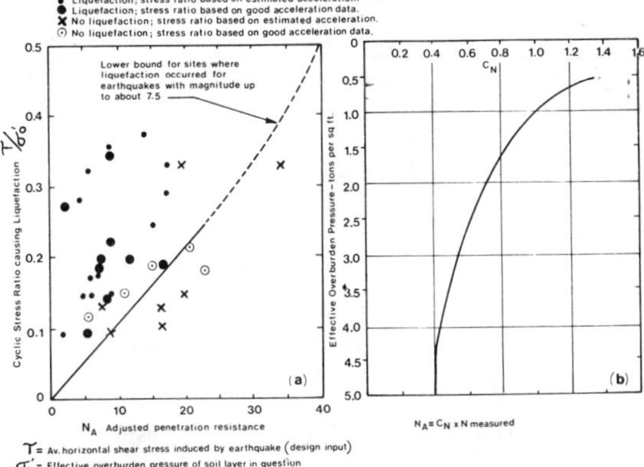

Fig. 10 (a) Correlation between stress ratio causing liquefaction in the field and adjusted penetration resistance (N_A) for magnitudes 7.5.
Fig. 10 (b) Relationship to obtain N_A based on average of relationships N,σ_v by Peck et al 1974, Gibbs and Holtz 1957 and Marcuson and Bieganousky 1977. (all after Seed et al 1977).

overburden pressure σ_v back to a standardised σ_v, (1ton/sq. ft. (107 kN/m^2) Seed et al 1977, 1.5 kgs/sq. ft. (71 kN/m^2) Castro 1975) using normally the Gibbs and Holtz (1957) relationship for N and σ_v. More details of Seed et al (1977) approach is shown in Fig. 10.

More recently independent correlations, from liquefaction studies in mainland China, have given 'a high degree of agreement' (Seed 1979) between the critical boundary for the Chinese criteria and the lower bound as marked in Fig. 10. For earthquakes of magnitudes less than 7.5 the correlation is considered conservative. Suggestions on the positions of lower bounds for earthquakes of different magnitudes have been proposed by Valera and Donovan (1977) who considered that agreement with Seed et al (1977) was reasonably good up to magnitudes of 6.5, but above this there was a significant difference.

In Japan use continues to be made of relative density in one form or another derived from N values for predicting liquefaction potential, and in so doing it has been found desirable to take account of grain size otherwise the results can be conservative for fine or silty sands (Ishihara and Watanabe 1976, and Tasuoka et al 1980).

Statistical analysis has also now begun to be applied for treating field data to obtain empirical evaluations of liquefaction potential (e.g. Christian and Swiger 1975, Christian 1980, Fardis and Veneziano 1981b). Broadly the methods involve combining laboratory test results, and their uncertainties, in conjunction with the N value to derive a model for the prediction.

A separate use common in Japan of the N value with regard to earthquake engineering is for the prediction of the shear stress developed during the earthquake which involves an esitmation of S wave velocity (Japanese Society 1981). Schmertmann (1975) has also considered the possiblity and tentatively concluded that a correlation may exist. Hoar and Stokoe (1981) experimenting for a similar purpose in natural clay soil used the SPT sampler as a crosshole source but apparently did not measure N values.

The reliance placed upon the SPT N value for predicting liquefaction in the light of the considerable variations in the method of test now being investigated (see chapter 2) is indeed remarkable. The sooner there is improved international standardisation the better. In the meantime it would be helpful in records of experience, such as embodied in Fig. 10a, to add more details of the test method, e.g. boring technique, hole diameter, use of mud, driving technique and whether liner was omitted. It then becomes possible to take steps when using the experience to adjust N values obtained by other methods and thereby to obtain better comparisons.

Notwithstanding the difficulty of variability in the SPT method the empirical evaluation of liquefaction potential via the N value is very widely used and Peck (1979) has suggested that it is unjustified to state that cyclic laboratory testing is any better with its many complications.

4.2.3 Factors common to SPT and liquefaction potential. Besides the fact that the N value has now been directly related to the phenomena, thereby eliminating the uncertainties involved in assessing relative density; it has been noted that a large number of the factors which significantly affect the possibility of liquefaction in sands, as listed in Table 3, are also affected in the same way with the penetration resistance (Seed 1979).

4.3 Compaction control

Compaction depth, e.g. by vibro or dynamic compaction techniques, particularly with its advantages when placing fill overwater, is most conveniently monitored by some form of insitu test. It needs to be sufficiently sensitive, not only to distinguish the change due to compaction, but also to register adequately the final relative density.

The suitability of the SPT for this purpose has been illustrated by tests at two sites where papers have been published since ESOPT 1974. At an overwater reclamation site on Ohgishima Island near Tokyo (Saito 1977 and Ladd et al 1977) sand was placed initially by barges and then hydraulically. Many N values from 90 borings were analysed in detail statistically by Saito and compared with density tests on fixed piston samples. Several different vibratory compaction techniques were used and compared. Generally, however, N values were increased 2 or 3 times those before compaction but the improvement was noted to be much less where fines (< 74 μm) were present and became quite negligible at about 20% fines content, as found elsewhere. Density measurements on samples taken from lower levels with the piston sampler were unable to detect so large an increase in the relative density as

Table 3 Factors affecting the liquefaction potential and penetration resistance (after Seed 1979)

Factor (1)	Effect on stress ratio required to cause liquefaction (2)	Effect on penetration resistance (3)
Increased relative density	Increases stress ratio for cyclic mobility or liquefaction	Increases penetration resistance
Increased stability of structure	Increases stress ratio for cyclic mobility or liquefaction	Increases penetration resistance
Increase in time under pressure	Increases stress ratio for cyclic mobility or liquefaction	Probably increases penetration resistance
Increase in K_o	Increases stress ratio for cyclic mobility or liquefaction	Increases penetration resistance
Prior seismic strains	Increases stress ratio for cyclic mobility or liquefaction	Probably increases penetration resistance

indicated by the N values and a pressuremeter was used to show that the higher N values were due to increased horizontal stresses. The maximum density tests on the sand samples I believe to be conservative insofar as 120 cc specimens of dry sand were not vibrated or saturated but only tapped under a fixed surcharge of 1 kg/cm^2.

At another overwater site at Singapore where dynamic compaction of sand fill was used, SPT, CPT and Menard pressuremeter tests (PMT) were all employed (Choa et al 1979). The increase in the N values was again 2 or 3 times that before compaction which was the same as for the CPT and both had a similar amount of scatter. The N values also served to select the grid spacing for the dropping weight.

A detailed comparison of the SPT, CPT, PMT and laboratory tests in a compacted fly ash fill in West Virginia, USA, has been described by Seals et at (1977) but no measurements were made prior to compaction.

4.4 Bearing capacity of piles

In researching for this report very little new data was encountered on this subject with a direct reference to use of SPT data in soils.

Two methods have been in use for many years for calculating the bearing capacity of isolated piles directly from the N value (Meyerhof 1956 and Japanese Society 1981). Experience in Japan is that the Meyerhof formula requires a safety factor around 4 to 7, whereas by taking account of the type of pile and method of installation a factor of 3 is found sufficient for the Japanese formula, as used in the specification for Highway Bridges. Other established methods involve empirical correlations between N and either a basic soil parameter or the CPT.

A recent survey in the UK of piling in glacial tills, including 'boulder clays' (Weltman and Healy 1978) has suggested, on limited information, that cohesion in kN/m^2 can be taken as 2.5 N (for average value) or 3.5 N (for lower bound) for calculating end resistance. Unit shaft resistance is suggested as 2 N kN/m^2 for driven piles from Meyerhof (1956) up to a limit of 100 kN/m^2 but in the case of bored piles in granular deposits a reduction factor is advised to be applied to N before assessing ϕ; unity at N = 10 with a linear reduction to 0.25 at N = 30 for impermeable gravels and 0.25 at N = 50 for permeable gravels (Weltman and Healy 1978).

5 USE OF SPT IN WEAK ROCKS

Weak rocks is another group of materials difficult to sample with consequent attempts to apply the SPT to interpret their engineering characteristics. The group comprises soft and weathered rocks, including chalk, marl, shale and poorly cemented sandstones (Meigh and Wolski 1979).

A straightforward example of discriminating between soil and weak rock with SPT equipment has been described by Kotzias and Stamatopoulos (1980) within Tertiary deposits in Greece where the 'penetrability', defined as the penetration in mm caused by 60 standard blows, was used to identify soils if penetrability exceeded 300 mm per 60 blows, and rock if less than 120 mm per 60 blows. Between these values it was classed as an intermediate material.

The extensive use in Britain of the SPT for assessing the qualities of weak rocks was referred to by Rodin et al (1974) at ESPOT 1. Its use in chalk, however, has drawn criticism from several quarters. Dennehy (1975) showed that simple grading according to N values (e.g. Wakeling 1970) did not take account of the age of the chalk and others have found the correlation inconclusive in Upper Chalk (e.g. Lord and Smith 1976). After concluding that SPT values were unreliable for a geological classification of chalk, as would be expected, Clayton (1978) suggested the N value was being mainly influenced by the density. Searle (1979) has taken the matter a step further by describing chalk grades in terms of both density and the degree of cementation, with extremes of either being possible. Hence he concluded that although a high SPT value is likely to represent less weathered rock, and a low value a highly weathered rock, a single value of N must straddle more than one grade. 'It would seem that greater caution is required in inferring chalk grades from SPT results even when sites are very thoroughly investigated' (Hobbs 1977). However in conjunction with a knowledge of the detailed geology SPT data is useful to differentiate between weak and weathered chalk where N values are less than 20, and stronger material if the values are greater than 20. Use of the solid cone in weathered chalk tends to give greater values than when using a cutting shoe.

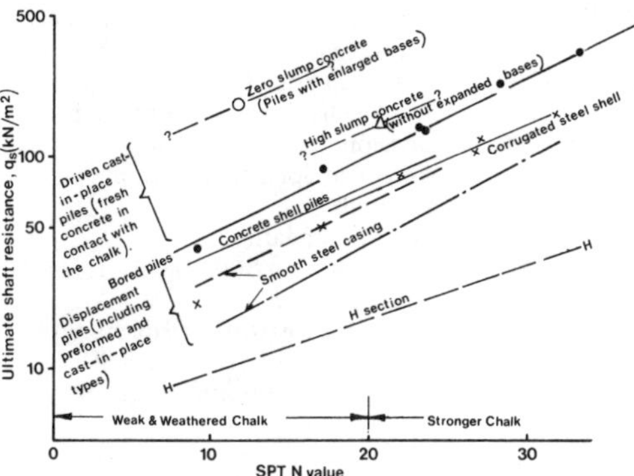

Fig. 11 Ultimate shaft resistance against N values for piles in chalk. (After Hobbs and Healy 1979).

Regarding correlations with the physical properties of weak rocks, information continues to be collected. Meigh and Wolski (1979) have plotted unconfined compression strengths for three different rocks with the N value, Hobbs (1974) gave some deduced field modulus results for chalk, while Stroud (1974) gave values for some marls and overconsolidated clays.

Concerning direct engineering applications of SPT in weak rocks, its main use is for pile design. Most contributions to the Geotechnique Symposium in print on piles in weak rocks (1976) included SPT data. Significant use has been made of it for the design of piles in chalk by Hobbs and Healy (1979) who analysed 346 piles of 9 different types at 97 UK sites. In their report that includes several interesting correlations, e.g. Fig. 11, it is concluded that boring and sampling with SPT 'offer the best readily available means of assessing the bearing capacity of piles, both bored and driven in chalk, particularly in flinty chalk'. Naturally the need for a sufficient number of tests is emphasised.

Another example of the use of SPT data for piling has been given by Leach and Thompson (1979) who used N values to obtain shear strength and deformation modulii of weak mudstone which they confirmed from pile tests was a satisfactory means for the prediction of ultimate capacity and load settlement behaviour of large diameter bored piles. Hagenaar and Van den Berg (1981) have described an application to piling in carbonate sediments and rocks in the Red Sea.

6 CONCLUSIONS

6.1 Standardisation

The most important outstanding matter regarding the SPT is the lack of standardisation of the test method at an international level. The trend appears in favour of an improvement.

The new European Standard has been agreed and published. Is this to be more widely accepted, possibly with some adjustments to accommodate minor objections such as raised by the Japanese Society, or will it be challenged by something more precise such as the standardisation of the driving energy at the sampler head, that is being discussed in the USA?

Before this can proceed, however, its value must be decided and how it is to be reliably achieved. An integrated research programme would be needed involving all the variables, including the performance of different trip hammers.

6.2 Liners

The value of the split spoon sampler for sampling is limited at best to classification tests. Therefore facilities to incorporate a liner, as offered in the USA, is not necessary; and by keeping the bore parallel sided throughout at 35 mm, this would eliminate influence on the SPT N value when liners are omitted as revealed by Schmertmann (1979).

6.3 Relative density

The objections raised by de Mello (1971) to the interpretation of the SPT N value in terms of the relative density have not been dispelled but rather they have increased.

Researchers in the future undertaking laboratory experiments should examine more closely the effects of the many variables involved in order to provide better quality information for statistical analysis. At the same time more complete information should always be given of the material being tested to include its basic physical properties, e.g. shear strength, compressibility; besides the results of the standard classification and index tests.

6.4 Prediction of settlement on granular soils

The SPT is an acceptable test in routine cases for evaluating the settlement of structures founded on

Table 4 SPT N values and sand properties (after Mitchell and Katti 1981)

	Very loose	Loose	Medium dense	Dense	Very dense
SPT N value (blows/0.3 m)*	<4	4 - 10	10 - 30	30 - 50	>50
CPT cone resistance (kg/cm^2)*	<50	50 - 100	100 - 150	150 - 200	>200
Equivalent relative density (%)**	<15	15 - 35	35 - 65	65 - 85	85 - 100
Dry unit weight (kN/m^3)	<14	14 - 16	16 - 18	18 - 20	>20
Friction angle (^{o})	<30	30 - 32	32 - 35	35 - 38	>38
Cyclic stress ratio causing liquefaction (τ / σ_o')***	<0.04	0.04 - 0.10	0.10 - 0.35	>0.35	–

* At an effective vertical overburden pressure of 100 kPa
** Freshly deposited, normally consolidated sand
*** From Seed (1979), Fig. 6a

granular soils.

Direct methods such as proposed by D'Appolonia et al (1970), Parry (1971) and Burland et al (1977) are favoured, although a cross-check with more than one method is strongly recommended.

6.5 Liquefaction

The SPT has become an important test for use in conjunction with the geomorphological data, anticipated changes in the soil and groundwater regime, in order to assess whether or not liquefaction is likely to occur. No other field test is so widely used and at present it may be preferable to laboratory tests.

Statistical models have been proposed for analysis but the use of direct correlations based on experience appear better to suit the method.

6.6 Weak rocks

The SPT is of value to interpret weak rocks and glacial till for geotechnical purposes but its limitations must be acknowledged and the results analysed in conjunction with the detailed geological information.

Empirical formulae have been proposed using SPT N values for the design of piles in chalk and marls.

6.7 Correlations of SPT N values with other soil parameters

In support of the use of the SPT as a guide Mitchell and Katti (1981) have proposed approximate correlations as set out in Table 4. The study of case histories using this table would be a promising area of research.

7 ACKNOWLEDGEMENTS

The writer wishes to express his appreciation for the critical review of the original draft by Dr. B.K. Menzies, Mr. N.B. Hobbs and Dr. B.O. Skipp. He also wishes to thank Mrs. M. Letch, Librarian of Soil Mechanics Ltd., for valuable assistance in locating references.

8 REFERENCES

Abbreviations:

ASCE GED	American Society of Civil Engineers Geotechnical Engineering Division
ASCE SMFD	American Society of Civil Engineers Soil Mechanics and Foundations Division
ASTM	American Society for Testing and Materials
BS	British Standards
COSOS	Conference on Settlement of Structures, UK
DIN	Deutsches Institut für Normung
ECSMFE	European Conference on Soil Mechanics and Foundation Engineering
ESOPT	European Symposium on Penetration Testing
ICSMFE	International Conference on Soil Mechanics and Foundation Engineering

Anonymous 1977, Proposed European Standard of penetration testing, IX ICSMFE, Tokyo 3:95-120

Aoki, N. and De Alencar Velloso 1975, An approximate method to estimate the bearing capacity of piles, 5th PanAM CSMFE, Buenos Aires 1:367-376

Arnold, M. 1980, Prediction of footing settlements on sand, Ground Engineering 13:2:40-49

ASTM 1586:1967, Standard method for penetration test and split barrel sampling of soils, ASTM Book of Standards Part 19

Bazaraa, A.R.S.S. 1967, Use of the standard penetration test for estimating settlement of shallow foundations on sand, PhD thesis, University of Illinois

Bogossian, F. 1981, Private communication

Bogossian, F. and C.F. Dias Machado 1981, Energy dissipation on the SPT rods, X ICSMFE, Stockholm, 2:449-450

Borowczyk, M. and Z.B. Frankowski 1981, Dynamic and static sounding results interpretation, X ICSMFE, Stockholm, 2:451-454

Brown, R.E. 1977, Drill rod influence on standard penetration test, ASCE J.GED 103:GT11:1332-1336

BS 1377:1975, Methods of test for soils for civil engineering purposes, Test 18 Determination of the penetration resistance using the split barrel sampler, BSI London, pp.103-104

Burland, J.B. 1970, Discussion of Session A, Proc. Conf. on Insitu investigations in soils and rocks, London pp. 61-62

Burland, J.B., B.B.Broms and V.F.B. de Mello 1977, Behaviour of foundations and structures, State-of-the-Art Review, IX ICSMFE, Tokyo, 3:495-546

Castro, G.1975, Liquefaction and cyclic mobility of saturated sands, ASCE J.GED 101:GT6:551-569

Choa, V., G.P. Karunaratue, S.D. Ramaswamy, A. Vijiaratnam and S.L.Lee 1979, Compaction of sandfill at Changi Airport, VI Asian Reg. Conf. SMFE, Singapore, 1:137-140

Christian, J.T. and W.F. Swiger 1975, Statistics of liquefaction and SPT results. ASCE J.GED 101:GT11 1135-1150, also closure 102:GT12:1279-1281

Christian, J.T. and W.D.Carrier 1978, Janbu, Bjerrum and Kjaernsli's chart reinterpretated, Can. Geotech. J. 15:123-138

Christian, J.T. 1980, Probabilistic soil dynamics: State-of-the-Art, ASCE J.GED 106:GT4:385-397

Clayton, C.R.I. 1978, A note on the effects of density on the results of standard penetration test in chalk, Géotechnique 29:1:119-123

D'Appolonia, D.J.,E.D'Appolonia and R.F. Brissette 1970, Discussion on settlement of spread footings on sand, ASCE J.SMFD 96:SM2:754-762

Dennehy, J.P. 1975, Correlating the SPT N value with chalk grade for some zones of the Upper Chalk, Géotechnique 25:3:610-614

Dikran, S. 1980, Prediction of settlement of structures in sands, using results of penetration testing, Geotech. Eng. Div., University of Surrey, UK

DIN 4019 1959. Flächengründungen und Fundamentsetzungen, Erlaüterungen Deutsches Institut für Normung, Berlin

Fardis, M.N. and D.Veneziano 1981a, Estimation of SPT-N and relative density, ASCE J.GED, 107: GT10:1345-1359

Fardis, M.N. and D.Veneziano 1981b, Statistical analysis of sand liquefaction, ASCE J.GED 107: GT10:1361-1377

Fletcher, G.F.A. 1965, Standard Penetration test, its uses and abuses, ASCE J.SMFD 91:SM4:67-75

Frydman, S. 1970, Discussion, Géotechnique 20:4: 454-455

Gawad, T.E.A. 1976, Standard penetration resistance in cohesionless soils, Soils and Foundations 16:4: 47-60

Gibbs, H.J. and W.G.Holtz 1957, Research on determining the density of sands by spoon penetration testing, IV ICSMFE,London 1:35-39

Giroud, J.P. 1972, Settlement of rectangular foundations on soil layers, ASCE J.SMFD 98:SMI:149-154

Hagenaar, J. and J. Van den Berg 1981, Installation of piles for marine structures in the Red Sea, X ICSMFE, Stockholm, 2:727-734

Halder, A. and W.H. Tang 1979, Uncertainty analysis of relative density, ASCE J.GED 105:GT7:899-904

Hoar, R.J. and K.H.Stokoe 1981, Crosshole measurement and analysis of shear waves,X ICSMFE, Stockholm, 3:223-226

Hobbs, N.B. 1974, Rocks - A review paper, COSOS, Cambridge, UK, pp. 579-739

Hobbs, N.B. 1977, Behaviour and design of piles in chalk - an introduction to the discussion of the papers on chalk. Piles in weak rocks, Thomas Telford Ltd., London, pp. 149-175

Hobbs, N.B. and P.R. Healy 1979, Piling in chalk, Construction Industry Research and Information Association, UK, Report PG6

Hobbs, N.B. 1981, Private communication

Holtz, W.G. 1973, The relative density approach - uses, testing requirements, reliability and shortcomings, ASTM Spec. Sym. Evaluation of relative density and its role in geotechnical projects involving cohesionless soils, pp. 5 - 17

Holubec, I. and E.D'Appolonia 1973, Effect of particle shape on the engineering properties of granular soils, ASTM Spec. Sym. Evaluation of relative density and its role in geotechnical projects involving cohesionless soils, pp.304 - 318

Horn, H.M. 1979, North American experience in sampling and laboratory dynamic testing ASTM Geotech. Testing J. 2:2:84-97

Ireland, H.O., O. Moretto and M. Vargas 1970, The dynamic penetration tests; a Standard that is not standardised, Géotechnique 20:2:185-192

Ishihara, K. and T.Watanabe 1976, Sand liquefaction through volume decrease potential, Soils and Foundations, 16:4:61-70

Japanese Society SMFE 1981, Present state and future trend of penetration testing in Japan, Separate report at X ICSMFE, Stockholm

Jorden, E.E. 1977, Settlement in sand - methods of calculating and factors affecting, Ground Engineering 10:1:30-37

Kilker, W.E. and A.S. Lucks 1981, Effect of change in effective stress on SPT-N values, X ICSMFE, Stockholm, 2:497-600

Koppejan, A.W., B.M. van Wamelen and L.J.H. Weinberg 1948, Coastal flow slides in the Dutch Province of Zeeland, II ICSMFE, Rotterdam, 5:89-96

Koreeda, K., T.Yoshihashi and T. Muromach 1981, Present state and future trend of penetration testing in Japan, Japanese Soc. SMFE, p.13

Kotzias, P.C. and A.C. Stamatopoulos 1980, Penetrability for discriminating in soil and rock, ASCE J.GED 106:GT2:193-198

Kovacs, W.D., J.C. Evans and A.H. Griffith 1977, Towards a more standardised SPT, IX ICSMFE, Tokyo, 2:269-276

Kovacs, W.D., A.H. Griffith and J.C. Evans 1978, An alternative to the cathead and rope for the standard penetration test, ASTM Geotech. Testing J. 1:2:72-81

Kovacs, W.D. 1979, Velocity measurement of free-fall SPT hammer, ASCE J.GED 105:GT1:1-10

Kovacs, W.D. 1980, What constitutes a turn ASTM Geotech, Testing J. 3:3:127-130

Ladd, C.C., R. Foott, K. Ishihara, F. Schlosser and H.G. Poulos 1977, Stress-deformation and strength characteristics, IX ICSMFE, Tokyo, 2:421-480

Lake, L.M. 1974, Discussion on Session 1, Granular soils, COSOS, Cambridge, UK, p. 663

Leach, B.A. and R.P. Thompson 1979, The design and performance of large diameter bored piles in weak mudstone rocks, VII ECSMFE, Brighton, 3:101-108

Lord, E.R.F. and W.E. Smith 1976, The misuse of SPT N value correlations with Upper Chalk grades, Géotechnique 26:1:217-220

Louw, J.M. 1977, Estimating settlements on cohesionless soils from SPT data, Civil Eng. S.Africa, 19:12:275-284

Marcuson, W.F. and W.A. Bieganousky 1977a, Laboratory standard penetration tests on fine sands, ASCE J.GED 103:GT6:565-588

Marcuson, W.F. and W.A. Bieganousky 1977b, SPT and relative density in coarse sands, ASCE J.GED 103:GT11:1295-1309

Marcuson, W.F. 1978, Determination of insitu density of sands, Dynamic geotechnical testing, Denver, ASTM STP 654:318-340

Matsumoto, K. and M. Matsubara 1981, Present state and future trend of penetration testing in Japan, Japanese Soc, SMFE, p.15

McLean, F.G., A.G. Franklin and T.K. Dahlstrand 1975, Influence of mechanical variables on the SPT, ASCE Spec. Conf. GED, Insitu measurement of soil properties, Raleigh, 1:287-318

Meigh, A.C. and I.K. Nixon 1961, Comparison of insitu tests of granular soils, V ICSMFE, Paris, 1:499-507

Meigh, A.C. and N.B. Hobbs 1975, Soil Mechanics, Section 8. Civil Engineer's Reference Book, 3rd Ed., Newnes-Butterworth, London

Meigh, A.C. and W. Wolski 1979, Design parameters for weak rocks, VII ECSMFE, Brighton, 5:59-79

de Mello, V.F.B. 1971, The penetration test, 4th PanAm Conf. SMFE, Puerto Rico, 1:1-86

de Mello, V.F.B. 1979, Discussion on design parameters for granular soils, VII ECSMFE, Brighton, UK, 4:253

Meyerhof, G.G. 1956, Penetration tests and bearing capacity of cohesionless soils, ASCE J.SMFD, 82: SM1:1-19

Mitchell, J.K. and W.S. Gardner 1975, Insitu measurement of volume change characteristics, ASCE Spec. Conf. GED Insitu measurement of soil properties, Raleigh, 2:279-345

Mitchell, J.K. and R.K. Katti 1981, Soil improvement - State-of-art report (preliminary), X ICSMFE, Stockholm, General reports, p.264

Mohr, H.A. 1943, Exploration of soil conditions and sampling operations, Harvard Univ. SM series 21

Mori, H. 1979, Review of Japanese Sub-surface investigation techniques, J. S.E. Asian Soc. Geotech. Eng. 10:219-242

Mori, H. 1981, Soil exploration and sampling, X ICSMFE, Stockholm, Gen. Reports pp. 97-112

Natarajan, T.K. and D.S. Tolia 1979, Modulus of elasticity of sandy soils by sounding methods, 6th Asian Reg. Conf. SMFE, Singapore, 1:51-54

Ohsaki, Y. 1970, Effects of sand compaction on liquefaction during the Tokachioki earthquake, Soils and Foundations 10:2:112-128

Oweis, I.S. 1979, Equivalent linear model for predicting settlements of sand bases, ASCE J.GED 105:GT12: 1525-1544

Palmer, D.J. and J.G. Stuart 1957, Some observations on the standard penetration test and the correlation of the test insitu with a new penetrometer, IV ICSMFE, London, 1:231-236

Parry, R.H.G. 1971, A direct method of estimating settlements in sand from SPT values, Proc. Symp. Interaction of Structures and Foundations, Midlands SMFE Soc., Birmingham, UK, pp.29-37

Parry, R.H.G. 1977, Estimating bearing capacity in sand from SPT values, ASCE J.SMFD 103:GT9: 1014-1019

Parry, R.H.G. 1978, Estimating foundation settlements in sand from plate bearing tests, Géotechnique 28:1:105-118

Peck, R.B. and W.V. Kaun 1948, Descriptions of flow slides in loose sand, II ICSMFE, Rotterdam, 2:31-33

Peck, R.B., W.E. Hanson and T.H. Thornburn 1974, Foundation Engineering, J.Wiley & Sons New York, 2nd Ed.

Peck, R.B. 1979, Liquefaction potential: Science versus practice, ASCE J.GED 105:GT3:393-398

Rodin, S. 1961, Experiences with penetrometer with particular reference to the standard penetration test, V ICSMFE, Paris, 1:517

Rodin, S., B.O. Corbett, D.E. Sherwood and S. Thorburn 1974, Penetration testing in the UK, State-of-art report, ESOPT I, Stockholm, 1:140-146

Saito, A. 1977, Characteristics of penetration resistance of a reclaimed sandy deposit and their change through vibratory compaction, Soils and Foundations 17:4:31-43

Sanglerat, G. 1972, The penetrometer and soil exploration, lst Ed., Elsevier Publishing Co., Amsterdam

Sanglerat, G. 1979, The penetrometer and soil exploration, 2nd Ed., Elsevier Publishing Co., Amsterdam

Schmertmann, J.H. 1970, Static cone to compute static settlement over sand, ASCE J.SMFD 96: SM3:1011-1043

Schmertmann, J.H. 1971, The importance of side friction and lateral stress to the SPT N value, 4th PanAm Conf. SMFE, Puerto Rico, Vol. 3

Schmertmann, J.H. 1975, Measurement of insitu shear strength, State-of-art report, ASCE Spec. Conf. GED Insitu measurement of soil properties, Raleigh, 2:57-138

Schmertmann, J.H. 1978, Use of the SPT to measure dynamic soil properties? - Yes, but....! Dynamic Geotech. Testing, Denver, ASTM STP 654:341-355

Schmertmann, J.H., T.V. Smith and R.Ho 1978, Example of an energy calibration report on a standard penetration test (ASTM Standard D1586-67) drill rig, ASTM Geotech. Testing J. 1:1:57-61

Schmertmann, J.H. 1979, Statics of SPT, ASCE J.GED 105:GT5:655-670

Schmertmann, J.H. and A. Palacios 1979, Energy dynamics of SPT, ASCE, J.GED 105:GT8:909-926

Schultze, E. and K.J. Melzer 1965, The determination of the density and the modulus of compressiblity of non-cohesive soils by soundings, VI ICSMFE, Montreal, 1:354-358

Schultze, E. and G. Sherif 1973, Prediction of settlements from evaluated settlement observations for sand, VIII ICSMFE, Moscow, 1.3:225-230

Schultze, E. and B. Biederman 1977, Pressuremeter penetrometer and oedometer tests,IX ICSMFE, Tokyo, 1:271-276

Seals, R.K., L.K. Moulton and D.L. Kinder 1977, Insitu testing of a compacted fly ash fill, ASCE Spec. Conf. Geotechnical practice for disposal of solid waste materials, pp.493-516

Searle, I.W. 1979, The Interpretation of Begemann friction jacket cone results to give soil types and design parameters, VII ECSMFE, 2:265-270

Seed, H.B. and I.M. Idriss 1971, Simplified procedure for evaluating soil liquefaction potential, ASCE J.SMFD, 97:SM9:1249-1273

Seed, H.B., K. Mori and C.K. Chan 1977, Influence of seismic history on liquefaction of sands, ASCE J.GED 103:GT4:246-270

Seed, H.B. 1979, Soil liquefaction and cyclic mobility evaluation for level ground during earthquakes, State-of-the-Art ASCE J.GED 105 GT2:201-255

Selig, E.T. and R.S. Ladd, 1973 Evaluation of relative density measurements and applications, ASTM Spec. Sym. Evaluation of relative density and its role in geotechnical projects involving cohesionless soils, Los Angeles, pp. 487-504

Serota, S. and G. Lowther 1973, SPT practice meets critical review, Ground Engineering, UK, 6:1:20-22

Shioi, Y., K. Uto, M. Fuyuki and T. Iwasaki 1981, Present state and future trend of penetration testing in Japan, Japanese Soc. SMFE, pp.5-19

Simons, N.E. and B.K. Menzies 1977, A short course in foundation engineering, Butterworth & Co. (Publishers) Ltd., London, 1st Ed.

SPT ASTM Workshop 1979, A technical report on the SPT during June 1979 committee week, ASTM Geotech. Testing J. 2/3:176-180

Stroud, M.A. 1974, The standard penetration test in insensitive clays and soft rocks, ESOPT I, Stockholm, 2.2:367-375

Stroud, M.A. and F.G. Butler 1975, The standard penetration test and the engineering properties of glacial materials, Conf. on Engineering behaviour of glacial materials, University of Birmingham, UK, pp.124-135

Sutherland, H.B. 1974, Granular materials: Review paper session 1, COSOS, Cambridge, UK, pp;473-499

Talbot, J.C.S. 1981, The prediction of settlements using insitu penetration test data, MSc Dissertation, University of Surrey, UK

Tatsuoka, F., T. Iwasaki, K. Tokida, S. Yasuda, M.. Hirose, T. Imai and M. Komno 1980, Standard penetration tests and soil liquefaction potential evaluation, Soils and Foundations, 20:4:95-111

Tavenas, F.A. and P. La Rochelle 1972, Accuracy of relative density measurements, Géotechnique 22:4:549-562

Tavenas, F.A., R.S. Ladd and P. La Rochelle 1973, Accuracy of relative density measurements: results of a comparative test program, ASTM Spec. Sym. Evaulation of relative density and its role in geotechnical projects involving cohesionless soils, Los Angeles, pp.18-60

Terzaghi, K. and R.B. Peck 1948, Soil mechanics in engineering practice, John Wiley & Sons, New York

Thorburn, S. 1963, Tentative correction chart for the standard penetration test in non-cohesive soils, Civ. Eng. Pub. Wks. Rev. 58:6:752-753

Tomlinson, M.J. 1975, Foundation design and construction, Pitman Publishing Ltd., London, 3rd Ed.

Uto, K. and M. Fuyuki 1981, Present state and future trend of penetration testing in Japan, Japanese Soc. SMFE, p.14

Valera, J.E. and N.C. Donovan 1977, Soil liquefaction procedures - A review, ASCE J.GED 103:GT6:607-625

Wakeling, T.R.M. 1970, A comparison of the results of standard site investigation methods against the results of a detailed geotechnical investigation in the Middle Chalk at Mundford, Norfolk, Conf. Insitu investigation in soils and rocks, BGS London, pp.17-22

Weltman, A.J. and P.R. Healy 1978, Piling in 'boulder clay' and other glacial tills, Construction Industry Research and Information Association, UK - Report PG5

Yamada, J., M. Fuyuki and K. Uto 1981, Present state and future trend of penetration testing in Japan, Japanese Soc. SMFE, p.12

Yoshimi, Y. 1977, Liquefaction and cyclic deformation of soils under undrained conditions, Session 4 Review, IX ICSMFE, Tokyo, 2:613-623

APPENDIX A

RECOMMENDED STANDARD FOR THE SPT TEST

by The Subcommittee on Standardisation of Penetration Testing in Europe

1 SCOPE

1.1 This method describes a procedure for determining the resistance of soils to the penetration of a split-tube sampler and obtaining disturbed samples of soil in a borehole for identification purposes. The test provides information on soil variability and stiffness.

The test is made by dropping a free falling hammer weighing 63.5 kg onto the drill rods from a height of 0.76 m. The number of blows N necessary to achieve a penetration of 0.30 m (below the seating drive) is regarded as the penetration resistance. (This test was developed in the USA and has been widely known as the 'Standard Penetration Test'.)

2 APPARATUS

2.1 Boring equipment

2.1.1 The boring equipment shall be capable of providing a reasonably clean hole to ensure that the penetration test is performed on relatively undisturbed soil.

2.1.2 When wash boring, a side-discharge drilling bit should be used but not a bottom-discharge drilling bit. Jetting through an open-tube sampler with water and then testing when the desired depth is reached shall not be permitted.

2.1.3 The process of jetting through an open-tube sampler with drilling mud and then testing when the desired depth is reached may be used, provided the flow and pressure of the drilling mud does not disturb the soil at the depth of the test drive (see clause 3.2.2).

2.1.4 When shell and auger boring, the drilling tool shall have a diameter which is not more than 90% of the internal diameter of the casing or of the borehole if no casing is used.

2.1.5 When drilling in soil that will not allow a hole to stay open, casing or drilling mud shall be used.

2.1.6 The diameter of the borehole should be between 60 and 200 mm approximately.

2.2 Split–barrel sampler

2.2.1 The sampler shall have the dimensions shown in Fig. 1.

2.2.2 The drive shoe shall be of hardened steel. It shall be replaced when it becomes significantly damaged or distorted. (See note 1 clause 5).

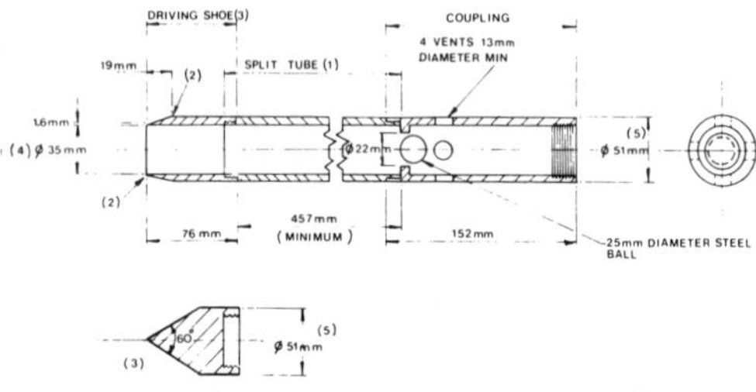

Fig. 1 Cross section of the SPT sampler

2.2.3 The central section of the sampler shall be of steel and of split-tube construction to allow examination and easy removal of the sample.

2.2.4 The sampler head shall have four 13 mm (minimum) diameter vent ports and shall contain a 25 mm steel ball check valve seated in an orifice of not less than 22 mm diameter which is located below the vent ports to improve sample recovery when there is water in the borehole. The ball and its seat shall be constructed and maintained so as to give a watertight seal when the sampler is withdrawn. (See note 2 clause 5).

2.3 Sampler rods

2.3.1 The rods used for driving the sampler should have a stiffness equal to or greater than type AW drill rods (43.7 mm O.D., 34.1 mm I.D. and approximately 6 kg/m weight). For holes deeper than 15 m steadies shall be used at intervals of 3 m, or alternatively rods with a stiffness equal to or greater than type BW drill rods (54.0 mm O.D., 44.4 mm I.D. and approximately 8 kg/m weight). (See note 3 clause 5).

2.3.2 Tolerance on straightness: when measured over the whole length of the rod by rolling against a straightedge, the maximum deviation shall not be greater than 1 in 1000.

2.3.3 The rods should be tightly coupled.

2.4 Drive weight assembly

2.4.1 The drive weight assembly shall comprise:
(a) A steel driving head or anvil screwed to the top of the sampling rod. (See note 4 clause 5).
(b) A steel hammer of 63.5 kg (±0.5 kg) mass
(c) A guiding assembly which will ensure that the hammer has a free fall of 0.76 m (±0.02 m).

Special precautions shall be taken to ensure that the energy of the falling weight is not reduced by friction between the drive weight and the guide. (See note 5 clause 5).

3. TEST PROCEDURE

3.1 Preparing the borehole

3.1.1 The borehole shall be carefully cleaned out to the test elevation using equipment that will ensure that the soil to be tested is not disturbed by the operation.

3.1.2 The water level in the boring shall at all times be maintained at or slightly above the ground water level. (See note 6 clause 5).

3.1.3 The drilling tool shall be withdrawn slowly to prevent loosening of the soil in the test section.

3.1.4 Where casing is used, it shall not be driven below the level at which the test is to commence.

3.2 Penetration test

3.2.1 The sampler shall be lowered to the bottom of the borehole and the following information recorded:

(a) Size and depth of casing
(b) Depth to the bottom of the borehole
(c) Water level (or mud where used) in the borehole
(d) If a solid steel cone is used in place of the driving shoe, this should be stated and referred to as the SPT (cone)
(e) Type of rods
(f) Amount of penetration of the sampler into the soil under the combined weight of sampler and rods
(g) Type of hammer

3.2.2 The sampler shall be driven in two stages, as follows:

Seating drive: A penetration of 0.15 m (which includes the initial penetration of the sampler under its own weight). If the 0.15 m penetration cannot be achieved in 50 blows, the latter shall be taken as the seating drive.

Test drive: A further penetration of 0.30 m. The number of blows required for this 0.30 m penetration is termed the penetration resistance N. If the 0.30 m penetration cannot be achieved in 50 blows (or 100 blows if a solid cone is used), the test drive shall be terminated.

The rate of application of hammer blows should not exceed 30 blows/minute. The number of blows required to effect each 0.15 m of penetration shall be recorded. If the seating or test drive is terminated before the full penetration, the record should state the amount of penetration for the corresponding 50 blows.

3.3 Removal of sample and labelling

3.3.1 The sampler shall be raised to the surface and opened. Place representative sample or samples of the soil into air-tight containers. (See note 7 clause 5).

3.3.2 Lables shall be fixed to the container with the following information:

(a) Site
(b) Borehole number
(c) Sample number
(d) Depth of penetration
(e) Length of recovery
(f) Date of sampling

4. REPORTING OF RESULTS

4.1 The following information shall be reported (Fig. 2):

(a) Penetration record (as described in clause 3.2.2)
(b) The depths between which penetration resistance was measured
(c) Information on the ground water level and the water level in the borehole at the start of each test.
(d) The soil type and description as identified from the sample (with a soil profile of the borehole if the data permits this).

The following information shall also be given with the report:

(e) Date of boring
(f) Borehole number
(g) Boring method and size of casing used
(h) Size and weight of rods used for the penetration test
(j) Type of hammer and anvil

BORING METHOD SHELL AND AUGER	LOCATION COORDINATES SITE E J097 N	BOREHOLE E2
BORING DIAMETER (mm) 150		SHEET OF
CASING DIAMETER (mm) 160		GROUND LEVEL (m) 3·20
BORING EQUIPMENT		DATE COMMENCED 28 MAY 74

	SAMPLES AND IN SITU TESTS		SPT		CASING	WATER	DATE AND	DESCRIPTION OF STRATA	LEVEL	LEGEND
	DEPTH (m)	TYPE	BLOWS 150mm	N	DEPTH (m)	DEPTH (m)	DEPTH (m)		(M)	
						0·70	28/6	LOOSE BECOMING MEDIUM DENSE GREY FINE TO MEDIUM SAND.		
1	1.00/1·30	S	3/4	7	0·85					
2	2·00/2·30	S	4/5	9	1·85					
3	3·00/3·30	S	10/13	23	2·85					
4	4·00/4·30	S	11/14	25	3·85					
5	5·00/5·30	S	8/11	19	4·86		5·50		−2·30	
6	5·80/6·10	S	14/17	31	5·65			DENSE TO VERY DENSE FINE TO MEDIUM SAND.		
7	7·00/7·23	S	23/(27)	(50)	6·86		7·60		−4·40	
8	7·90/8·20	SC	20/24	44	7·75		8·20	DENSE GREY COARSE SAND AND GRAVEL	−5·00	
9										
10									FIG.	

Fig. 2 Example of reporting SPT test results.
Key: S = SPT, SC = SPT (cone)

5 EXPLANATORY NOTES

Note 1. Clause 2.2.2

The drive shoe of the SPT sampler is not designed to provide inside clearance with the sampling tube. Hence any significant inward distortion of the cutting edge shall not be permitted.

Note 2, Clause 2.2.4

Alternative designs of check valves are permitted provided they give equal or better performance.

Note 3; Clause 2.3.1

The stiffness of the rod used during testing is believed to affect the penetration resistance, especially because a light rod 'whips' under the hammer blows.

Note 4. Clause 2.4.1 (a)

A loose anvil shall not be permitted. It is an advantage for the striking face of the anvil to be domed 3 mm in 100 mm (approximately) to inhibit glancing blows between the drop weight and anvil.

Note 5. Clause 2.4.1 (c)

A hammer incorporating a self-tripping mechanism will overcome the effect of any friction in the lifting tackle and its use is recommended. A hammer with a free fall will give more reproducible results and also smaller N values than a hammer operated by a friction winch with manilla rope to lift and drop the hammer.

The self-tripping hammer should be raised slowly to ensure that the inertia of the hammer does not carry it above the prescribed level in excess of the prescribed free fall of 0.76 m (±0.02 m). Also, the pick-up assembly should be lowered slowly to avoid significant impact on the hammer.

A down-the-hole drive weight assembly acting directly on the sampler and mounted in a watertight chamber is permissible.

Note 6. Clause 3.1.2

To avoid hydraulic disturbance when boring in sand, the water pressure at the bottom of the borehole should correspond to the piezometric pressure in the surrounding ground at the test level. In artesian conditions, this will not be the same as the pressure corresponding to the standing water table.

Information on the water level in the layer(s) tested shall be recorded. Particular attention should be given where artesian conditions are encountered, as is sometimes found when penetrating through an impervious layer into a pervious layer below water level. Any operation that gives an opportunity for an upward flow of water to loosen the soil should also be recorded.

Note 7. Clause 3.3.1

The sample obtained with the SPT split-tube sampler is used for identification. It must be regarded as disturbed from the point of view of determining deformation or strength properties.

Proceedings of the Second European Symposium on Penetration Testing / Amsterdam / 24-27 May 1982

SPT and the compressibility of cohesionless soils

A.G.ANAGNOSTOPOULOS & B.P.PAPADOPOULOS
National Technical University, Athens, Greece

1 INTRODUCTION

The first quantitive prediction of settlements using N values has been given by Terzaghi and Peck (1948). This method was widely accepted but now it is recognised that the results are rather conservative. From the contemporary methods the one suggested by Meyerhof (1965) and Peck, Hanson, Thornburn (1974) are considered being the more reliable for engineering applications. The last two methods recommend allowable bearing pressures of the order of about 50% greater than Terzaghi and Peck.

Besides the methods of direct settlement estimates from SPT results there are other ones which introduce the use of modulus of compressibility (E_s), estimated from SPT tests along with the applications of the theory of elasticity, as also the method suggested by Parry (1971) which has the form of an elastic equation where E is replaced by N.

In the present paper a new method for the settlement estimation is proposed in the form of an elastic equation by means of a direct use of SPT results, which takes into account the linear relationship between the modulus of compressibility (E_s, from oedometer tests) and the stress level. For the latter an extended laboratory investigation has been done on cohessionless soils of the Quaternary in the Greek area, both on undisturbed and semi-disturbed samples in conjuction to SPT results performed at depths nearby the sampling. Furthermore laboratory tests have been carried out on "artificial" samples from sand, and the effect of the stress level and preconsolidation have been investigated in a wide pressure range.

2 THE COMPRESSIBILITY OF COHESSIONLESS SOILS

2.1 Results from oedometer tests on undisturbed and semi-disturbed samples of normally consolidated fine sands (SP) with a silt content up to 10%, silty sands (SM) and on non-plastic sandy silts (ML) have been correlated with the SPT results. For pressures up to 600 kPa the modulus of compressibility E_s has been found practically increasing linearly with the effective pressure (σ'), according to the relationship $E_s = E_{so} + \lambda\sigma'$ (1), for all the above samples.

From oedometer tests on "artificial" samples of sands with various grain size distributions (fine sands and fine to medium sands), of various initial densities (achieved by a preconsolidation), has been found that a linear relationship between E_s and the effective pressure (σ') holds, according to equation (1) up to the pressure of 800 kPa (Anagnostopoulos, 1979). The influence of the preconsolidation on the coefficient λ for preconsolidations ranging between 25 and 3200 kPa has been also considered.

Figure 1a illustrates the correlation of the coefficient λ against the effective overburden pressure. Figure 1b shows the correlation of the coefficient λ versus the preconsolidation pressure for the "artificial" samples of sands. From the afore mentioned the following are concluded : a) The coefficient λ is practically independent of the effective overburden pressures for the case of normally consolidated cohessionless materials. b) The coefficient λ is also practically independent

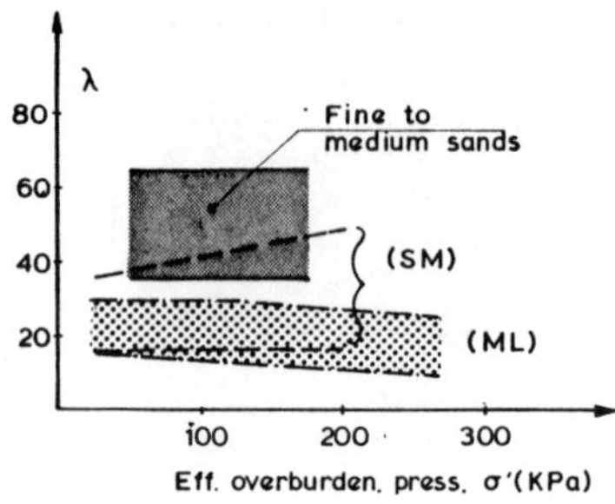

Fig. 1a

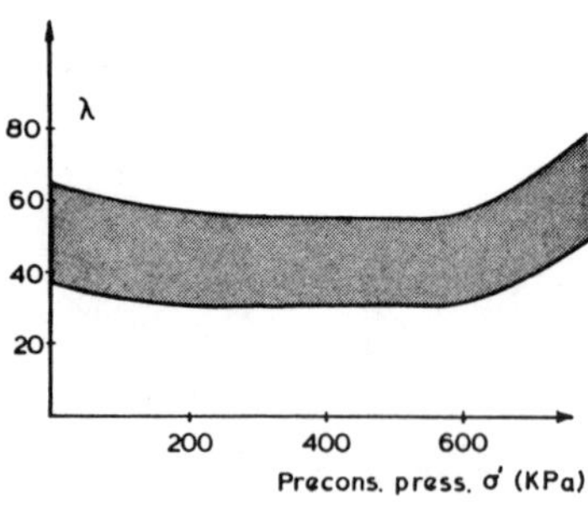

Fig. 1b

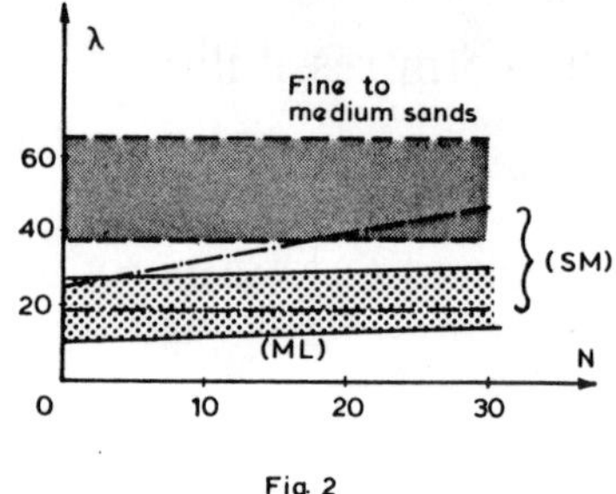

Fig. 2

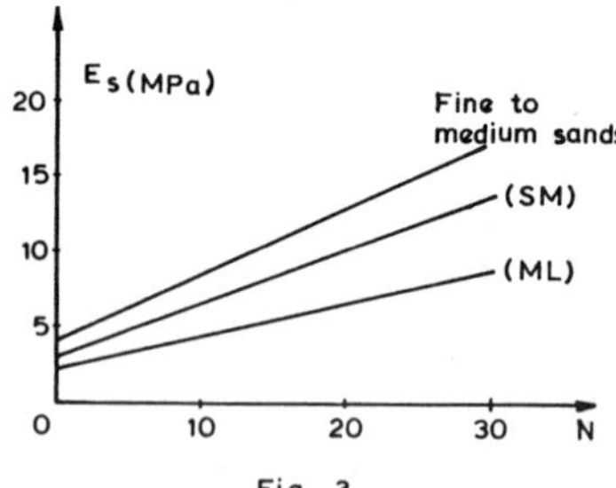

Fig. 3

of the preconsolidation pressure for sands tested. c) From all the above tests it is evident that the coefficient λ is closely dependent on the soil type and varies in relatively narrow limits for pressures up to 600 kPa (Fig.2).

2.2 From the tests on undisturbed and semi-disturbed samples -mentioned in paragraph 2.1- E_S values for pressures equal to the effective overburden have been correlated to SPT results. It has been observed the trend of a linear relation between the number of blows N and E_S, expressed by the formula $E_S=C_1+C_2N$ (2). The same exactly has been previously proposed by Schultze and Menzenbach (1961).

It can be stated that the coefficient C_1 has been found to be of the same order for all the three types of the investigated soils, whereas C_2 is more sensitive with the soil type, taking lower values for the ML samples (Fig.3). A linear relationship between E_S and N in the form of equation (2) is acceptable by many researchers, and consequently can be applied in settlement calculations by the help of equations in elastic form.

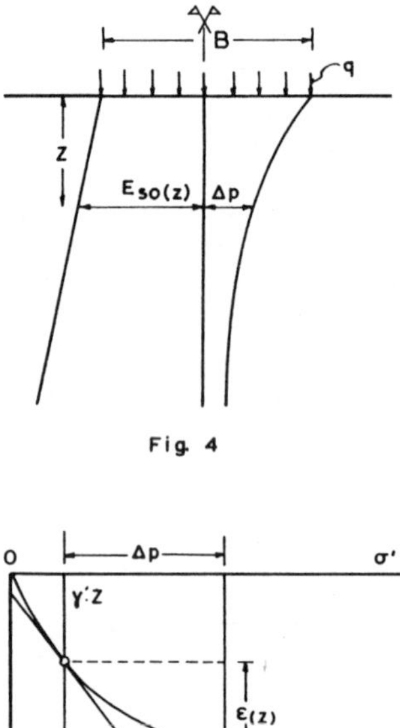

3 PROPOSED METHOD

A simple method of settlements estimation,

Fig. 4

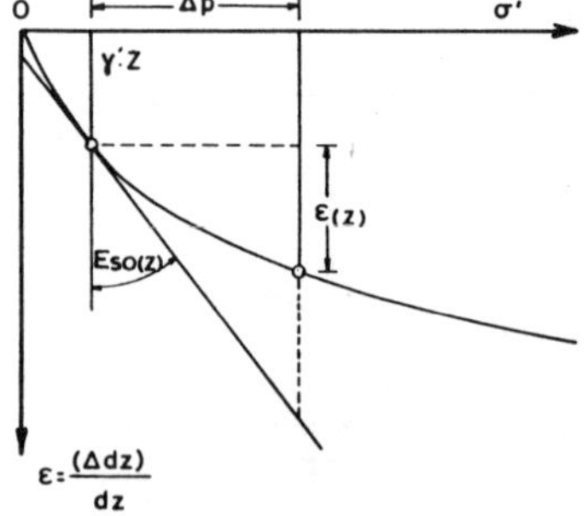

Fig. 5

in cohessionless soils, leading to an elastic formula is based on the integration of the vertical strains over a compressible depth H. The simple assumption of stress distribution 2:1 has been considered herein. The results of this latter assumption are very close to the one for an elastic solution and compatible to the direct use of the modulus of compressibility E_s for the settlement calculation.

Figure 4 illustrates the stress distribution according to the before mentioned assumption as well as the linear variation of $E_{so(z)}$ against the depth z, which is a result of the increasing effective overburden pressure. In every depth z, the $E_{so(z)}$ is the tangent modulus of compressibility in the stress level of the overburden. In the soil region immediately under the foundation the strains are considerably smaller than those resulting from the ratio $\Delta p/E_{so(z)}$, (where Δp is the additional foundation pressure). This is attributed to the fact that the additional pressures Δp are generally much higher than the overburden pressures γ'z, for which the moduli $E_{so(z)}$ have been considered (Fig.5).

If E_{so} is the modulus of compressibility for σ'=0, according to equation (1), then $E_{so(z)}$ can be expressed as :

$$E_{so(z)}=E_{so}+\lambda\gamma'z \qquad (3)$$

For an additional pressure Δp, due to a foundation loading, the strain at the depth z equals :

$$\frac{\Delta dz}{dz}=\frac{1}{\lambda}\ln\frac{E_{so(z)}+\lambda\Delta p}{E_{so(z)}}=\frac{1}{\lambda}\ln\frac{E_{so}+\lambda\gamma'z+\lambda\Delta p}{E_{so}+\lambda\gamma'z} \qquad (4)$$

This is a consequence of the fact that strains and stresses are related to the equation :

$$\varepsilon=\frac{1}{\lambda}\ln\frac{E_{so}+\lambda\sigma'}{E_{so}} \qquad (5)$$

i.e.

$$E_s=d\sigma'/d\varepsilon=E_{so}+\lambda\sigma'$$

The settlement of a compressible layer H is given by the following equation :

$$S=\int_0^H\frac{1}{\lambda}\ln\frac{E_{so}+\lambda(\gamma'z+\Delta p)}{E_{so}+\lambda\gamma'z}\,dz$$

or

$$S=\frac{1}{\lambda}\int_0^H\ln\left[1+\frac{\lambda\Delta p}{E_{so}+\lambda\gamma'z}\right]dz \qquad (6)$$

The settlement S is obtained by the numerical integration of equation (6), as a function of the geometry of the foundation, the modulus of compressibility of the layer and its rate of increase with the effective stresses. It is common practice to consider a compressible layer H=B÷2B, where B is the foundation width.

The settlement S can also be expressed under the form : $S=f_c\cdot S_1$ (7)

where f_c : correction (reduction) factor

S_1 : settlement, $=f_1\cdot\frac{qB}{E_{s(m)}}$

f_1 : influence factor given from the diagram of Figure 6

q : foundation pressure

B : foundation width

$E_{s(m)}$: average modulus of compressibility over the depth B

N_m : average blow count over the depth B

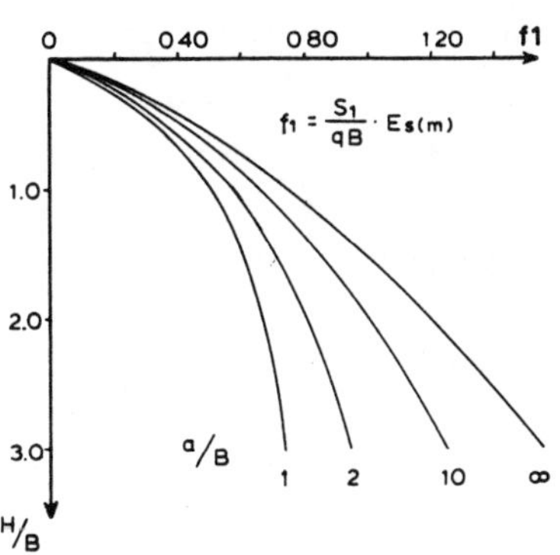

Fig. 6

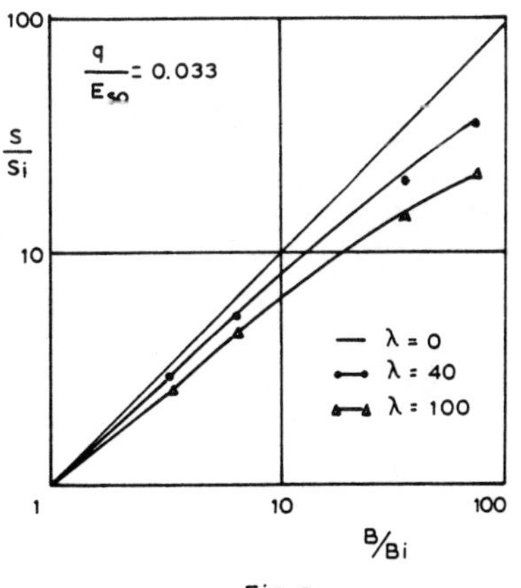

Fig. 7

Depth B: part of the compressible layer where the most significant strains occur.

Thus equation (7) can be written under the form :

$$S=f_c\ f_1 \frac{qB}{E_{s(m)}} = f_c\ f_1 \frac{qB}{C_1+C_2N_m} \quad (8)$$

According to the above, for the evaluation of the settlement S_1, the modulus of compressibility $E_{s(m)}$ can be assumed as constant over the compressible layer. (This of course should be considered as a first approximation). The correction factor f_c depends mainly on the ratio a/B (dimensions of the foundation), the ratio q/E_{so} and the coefficient λ. This factor, f_c, is a reduction factor due to the linearly increasing modulus of compressibility with the effective stresses.

If S_i is the settlement of a standard plate of 0.3 m in width and S the settlement (according to equ. 8) of a foundation with the same bearing pressure and a width B in metres, Figure 7 illustrates the settlement ratio S/S_i as a function of the width B and the coefficient λ, which expresses the increase of the soil stiffness with depth.

For a square foundation with a width B or a circular one of diameter D=B, resting on a comrpessible stratum H=2B with an average blow count N_m (over a depth B), the settlement S_1 is given by :

$$S_1 \cong 0.7 \frac{qB}{C_1+C_2N_m} \text{ and hence } S=f_c \cdot 0.7\frac{qB}{C_1+C_2N_m}$$

In the case of medium dense sands, where λ varies between 40-60, for q/E_{so} values between 0.01 up to 0.05 the correction factor f_c takes values 0.65-0.75, and consequently :

$$S \simeq 0.5 \frac{qB}{C_1+ C_2N_m}$$

In the case of dense sands where λ is of the order of 100 the correction factor f_c is always lower than 0.6 and thus the settlement S of the order :

$$S \simeq 0.4 \frac{qB}{C_1+C_2N_m}$$

For the loose silty sands or sandy silts, where λ is ranging between 10-15, the modulus of compressibility $E_{s(m)}$ can be considered as constant over the compressible layer i.e. f_c=1 and the settlement equals :

$$S \simeq 0.7 \frac{qB}{C_1+C_2N_m}$$

4 CONCLUSIONS

1 From laboratory tests in cohessionless soil samples, has been found that the modulus of compressibility E_s is related to the effective pressure linearly for pressures up to 600 kPa, as follows :

$$E_s = E_{so} + \lambda\sigma'$$

The coefficient λ, mainly depends on the soil type and especially for sands does not depend on the preloading.

2 Comparison of the tangent modulus (at stresses equal to the overburden pressure) to the blow count N from SPT, have verified relationships of the form :

$$E_s = C_1 + C_2N$$

3 By means of SPT results and an appropriate selection of the coefficient λ the evaluation of the settlement is possible through the equation :

$$S= f_c f_1 \frac{qB}{C_1+C_2N_m}$$

This equation is especially useful in cases of low count number N.

5 REFERENCES

Anagnostopoulos, A.G. 1979, Discussion: Design parameters for granular soils. Proc.7th Eur.C.S.M.F.E. Brighton, Vol 4:265.

Meyerhof, G.G. 1965, Shallow Foundations. ASCE JSMFD. 91. SM-2:21-31.

Parry, R.H.G. 1971, A direct method of estimating settlements in sands from SPT values. Proc.Symp.Interaction of Structures & Found. Birmingham:29-32.

Peck, R.B., W.E. Hanson. T.H. Thornburn, (2 ed.) 1974. Foundation Engineering. N.Y., London, Sydney, Toronto. J.Wiley and Sons, Inc.

Schultze E., Menzenbach E. 1961, Standard Penetration Test and Compressibility of Soils. Proc. 5th ICSMFE. Paris, Vol 1: 527-532.

Terzaghi K., R.B.Peck 1948, Soil Mechanics in Engineering Practice, N.Y., J.Wiley and Sons, Inc.

Proceedings of the Second European Symposium on Penetration Testing / Amsterdam / 24-27 May 1982

Prediction of the bearing capacity of piles based exclusively on N values of the SPT

LUCIANO DÉCOURT
Luciano Décourt, Engenheiros Consultores Ltda., São Paulo, Brazil

1 INTRODUCTION

A paper was presented at the VI Brazilian Conference on Soil Mechanics and Foundation Engineering (Rio de Janeiro-1978), by the author and Ing. Arthur R. Quaresma, in which the author's experience in estimating the bearing capacity of piles based solely on data from Standard Penetration Tests (SPT) was confronted with data from load tests, most of them belonging to Ing. Quaresma's file.

For that study, N values greater than 15 along the pile shaft were not considered since traditional piles are not usually driven through soil strata of high consistency and/or density.

Recently the world has witnessed the increased use of bored piles of large dimensions and great bearing capacity, executed with the aid of bentonite slurry and mechanical excavation which enables them to be built through highly resistent soils with high N values. Interest has, then, arisen towards investigating the validity of the method originally developed by the author for traditional piles, i.e. precast concrete piles, Franki piles etc, as applied to piles of high bearing capacity, especially those going through soil with high N values.

2 BEARING CAPACITY OF TRADITIONAL PILES (precast, Franki, etc)

In the author's previous paper, the bearing capacity of piles was estimated with basis only on the data supplied by the traditional borings, i.e., tactual-visual classification of soil layers and N values of the Standard Penetration Test which in Brazil is carried out every meter of depth.

The conclusions of that study were as follows.

2.1 The pile ultimate bearing capacity (Q_u) is given by

$Q_u = Q_s + Q_p$ where Q_s is the ultimate load capacity resulting from skin friction resistance along the pile shaft, and Q_p is the load capacity resulting from the ultimate point resistance of the pile.

2.2 The load capacity resulting from the skin friction resistance is given by:

$$Q_s = p \quad L \quad q_s$$

where

p = pile perimeter
L = pile length
q_s = adhesion along the shaft

Let $\bar{N}$ be the average value of N along the shaft, then

$$q_s = \frac{\bar{N}}{3} + 1 \ (t/m^2)$$

This expression is to be valid for all soils.

In the mentioned paper, the authors did not considier N values smaller than 3 or greater than 15. That was due to the fact that in the Brazilian experiments with very soft soils (N=0) adhesion values smaller than 2t/m² (19,61Pa) were very unusual. Also, when dealing with medium dense or stiff soils, adhesion values higher than 6t/m² (58,84Pa) rarely ever accoured.

The above expression applies for estimating the skin friction resistance of the various concrete piles such as precast piles, Franki piles, Strauss piles, as well as the skin friction resistance of the shaft of open air caissons.

2.3 The point bearing capacity of displacement piles is given by:

$$Q_{\bar{p}} = A_p \quad N \quad K$$

where

A_p: area of the point the pile
K : coeficient depending on soil type at the point of the pile. Four types of soil were considered, as shown on table 1 below.

Table 1 - K values

soil type	K (t/m²)
clays	12
clayey silts*	20
sandy silts *	25
sands	40

*residual soils

2.4 The allowable load of the pile was given by adopting a global safety factor of 2

$$Q_{all} = Q_{u/2}$$

3 A PROPOSITION FOR A METHOD TO DETERMINE THE BEARING CAPACITY OF DISPLACEMENT PILES

We consider that the method presented in 2 coud be generalized for all soils including dense sands and stiff clays.

The only difference would be the limit of N that is proposed to be 50 instead of 15 as formely advised for estimating the shaft resistance of the piles.

N values bigger than 50 must be computed as 50 since such values are characteristic of soil strata that practically cannot be penetrated by the sampler, and it makes no sense considering a penetration test where penetration is practically impossible.

4 SOME CONSIDERATION ON THE PILE-SOIL LOAD TRANSMISSION MECHANISM OF "BENTONITE PILES"

4.1 Skin friction and point behavior

There seems to be a consensus that the bearing capacity of "bentonite piles" derives mostly from the high values of skin friction developed between the shaft and the surrounding soil. The notion that the film of bentonite slurry left between the concrete and the soil was a hindrance to the development of such high shaft resistance has been completely over come.

Actually, nowadays, the notion is just the the opposite. Quoting, for instance, Burland (Burland, 1975)"...although the investigation is not yet complete, preliminary results suggest that in some cases bentonite may inhibit the stress relief and consequent softening of silty clays during boring and thereby actually give rise to higher shaft resistance than for similar piles bored in the dry".

In fact, the presence of the bentonite slurry seems to minimize the relief of stresses along the shaft excavation, thus preserving the high resistance of the soil along the concrete-soil interface.

On the other hand, the skin friction resistance is totally mobilized with deformations of about 10mm or at the most 15mm, regardless of shaft dimensions. This observation of ours seems to clash with the opinions of other authors who seek to relate the deformation necessary for full skin friction mobilization with the shaft diameter. Furthermore, we've observed that once the curve skin friction resistance x deformation reaches its peak, there seems to be no further changes in resistance, its value remaining constant and equal to the peak value or slightly smaller than it.

The development of point resistance, on the other hand, requires deformations of considerable magnitude. A clear example of such situation can be found in an interesting paper presented in the Stockholm 1981 Conference by W.Prodinger and CH. Veder (Prodinger and Veder, 1981).

Figure 1 shows a Load-Settlement Diagram taken from that paper, which shows clearly that the maximum value of skin friction resistance occoured for a total settlement of 12mm. For such deformation the load applied in the point of the pile was practically nill. Only for deformation of several centimeters did the point resistance reach values as high as those of the skin friction resistance. This fact is highly important and anyone designing foundations of this type should bear it in mind.

In the following items the author presents a proposition for a method to determine the allowable load of high capacity piles excavated and cast under bentonite slurry, considering separately the usual cases where settlements should be limited to a few milimiters, as well as certain special cases where settlements of several centimeters are to be allowed.

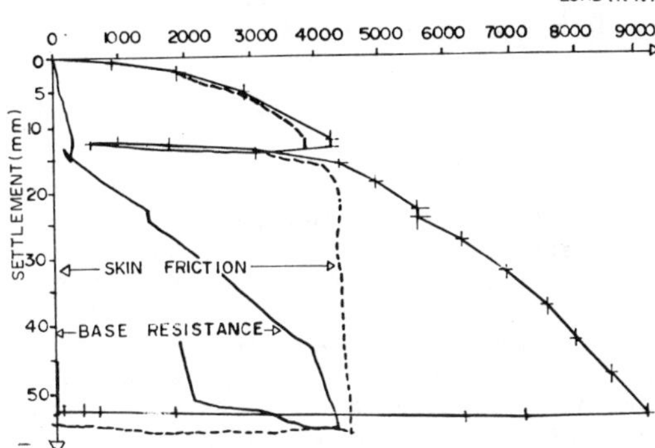

Figure 1 - W. Prodinger and C.H. Veder Load Settlement Diagram for a diaphragm wall ellement 0,5 x 1,5m, L= 24,4m

4.2 Execution problems

As we've seen, there is a consensus among authors that the skin friction resistance is highly trustworthy, while doubts have arisen as to the point resistance.

As a matter of fact, the point resistance besides requiring great settlements for its total mobilization may also be affected by an inadequate scraping of the bottom of the excavation. Quoting once more Burland: "I am less confident about relying on end bearing resistance for piles cast under bentonite as it is difficult to check that the base is clean and free of remoulded material" (Burland, 1975).

In our jobs we've sought to check the cleaness of the base as well as the soil resistance by performing a penetration test (SPT) imediatly before casting the pile.

The SPT apparatus is lowered through the tremie-pipe. The SPT sampler is then opened and examined by our inspecting engineer who observes the thickness of remoulded soil as well as the contamination of the bentonite slurry into the natural soil. Furthermore, a careful record of blow count for each 15cm is kept, which allows us to evaluate the eventual loss of resistance in the remoulded and/or contaminated zone.

5 A PROPOSITION FOR A METHOD TO DETERMINE THE ALLOWABLE LOAD OF HIGH CAPACITY PILES EXCAVATED AND CAST UNDER BENTONITE

5.1 Usual cases

What we call usual cases, are foundations designed to bear small differential settlements - of about 5mm - which is usually accomplished by limiting total settlements to about 10mm.

For such cases, let it be made clear that only the lateral skin friction resistance may be considered.

The ultimate bearing capacity derived from skin friction along the shaft shall than be determined by using the afore mentioned expression:

$$Q_s = p \cdot L \cdot q_s$$

where

$$q_s = \frac{\overline{N}}{3} + 1 \ (t/m^2)$$

N values smaller than 3 must be taken as 3 and N values greater than 50 must be taken as 50 for the reasons exposed before.

As to safety factors, these will be dealt with on ítem 7.

In the present case, the point bearing capacity is not taken into account, due to the fact that settlements of several centimeters are required for its mobilization. Any point resistance developed will be taken as an additional safety factor.

Data from load tests so far analysed by us seem to demonstrate that foundations thus designed are bound to suffer total settlements of 10mm at the most.

5.2 Cases in which bigger total settlements are allowed

In such cases, besides the skin friction resistance (Q_s) calculated according to ítem 5.1., an additional resistance may be considered as derived from the point allowable bearing capacity ($Q_{p.all}$).

In engineering practice an analogy with caissons with base on clay can be made, and the allowable point pressure can be obtained in an expedite way by dividing a safe estimate (see ítem 6) of the average point N value (average of N at point level, one meter above and one meter below) by 3. The value thus reached is the allowable pressure in kg/cm². To change this into kN/m² a multiplying factor of 98.07 must be used that is $Q_p = 32,7$ N (KN/m²).

In founfations designed under such criteria we've observed (Décourt, L. 1978) that the settlement (in cm) to be expected is 2/3 of the base diameter (in meters)

$$\rho_p(cm) = \frac{2}{3} D(m)$$

An additional settlement due to the deformation of the contaminated and/or remoulded soil must also be taken into account (ρ_c). Such settlement may be estimated by using simple elastic theory expressions such as

$$\rho_c = \frac{\sigma . L}{E}, \text{ where}$$

σ: point pressure
L: thickness of remoulded and/or contaminated layer
E: may be estimated through correlations with N values, such as the ones we've suggested in a former paper (Décourt L, 1978).

E= 15N (kg/cm²) for clays
E= 30N (kg/cm²) for sands

Calling ρ_s the settlement necessary for mobilization of the shaft resistance the total settlement would then be given by:

$$\rho = \rho_s + \rho_p + \rho_c$$

6 SAFETY FACTORS

The global safety factor should be reached by the product of several safety factors.

6.1 Soil heterogeneity and the fortuitous errors in SPT execution (F_p)

Due to, either the natural heterogeneity existing even in homogeneous soils, and/or to the various errors inherent to the SPT, an individual safe N value can only be reached if the original N value is divided by 1.5. In fact, we've observed that if X is the value of N at a certain depth in a homogeneous soil, a new boring performed very close to the first one, at the same depth, will display an N value between 1.5 X and X/1.5. When average values, such as those along the pile shaft, are dealt with, the dispersion of values is obviously quite smaller. We've observed that in such case, a division by 1.1 instead of 1.5 is more than enough for stablishing a safe average value.

As to the average value for the point, once only three N values are involved, the factor will be between 1.1 an 1.5, In this case we suggest 1.35 be used.

6.2 Formulation used (F_f)

In practical terms, the problem to be solved is that of evaluating the ultimate bearing capacity on basis of a certain theory or method. Since any theory or method is subject to its own limitations, a certain safety margin needs to be adopted.

In the case of the author's proposed formulae, once the partial safety factors proposed on ítem 5.1 are considered, the value of F_f to be adopted will be 1, as it can be easyly understood by analysing Figure 2.

6.3 Excessive deformation (F_d)

All theories seek to determine the ultimate loads, which are only reached with big deformations. This applies even when we consider loads somewhat smaller than the ultimate load.

It is, then necessary to bear in mind the difference between the deformation necessary for the total development of skin friction, and that necessary to mobilize the point resistance.

Due to the deformations necessary for the development of these resistances, we think that no safety factor for deformation needs to be used for the skin friction resistance while for the point resistance this factor should be at least 2, and in several cases 3, varying basically according to the point dimension.

6.4 Working Loads (F_w)

No engineers when designing a foundation shall actually load a pile with its ultimate bearing capacity, even if it has been safely estimated. Nevertheless, if the safety factors proposed on ítens 5.1 through 5.5 have been considered, it seems enough to consider a safety factor of 1.2 as related to the working load of the pile.

6.5 A new concept of safety factor for piles

The safety factor F shall, then be given by:

$F = F_p \, F_f \, F_d \, F_w$, where

F_p: safety factor related to soil parameters

F_p= 1.1 - skin friction

F_p= 1.35- point bearing capacity

F_f: safety factor related to the formulation adopted
If the formula proposed in this paper is adopted the author recomends F_f= 1,0. In fact, as it can be seen in Figure 2, once the safety factors F_p recomended for the soil parameter are used no other is necessary for a safe assessement of the ultimate bearing capacity of the pile.

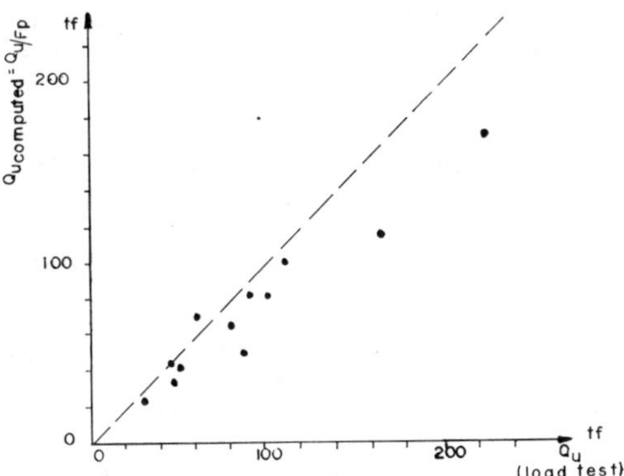

Figure 2 - Computed versus load test values of the ultimate bearing capacity of piles (Q_u)

F_d: safety factor related to excessive deformation
Doubtlessly $F_d = 1$ for skin friction
As for the point $F_d = 2$ to 3 depending on the pile diameter
For the usual cases 2,5 is suggested

F_w: safety factor related to working load. The author recomends 1.2 provided that the other partial safety factors are used.

Thus, we'll have

- for skin friction

$$F_s = 1.1 \times 1,0 \times 1,0 \times 1,2 = 1,32 \cong 1.3$$

- for the point

$$F_p = 1,35 \times 1,0 \times 2,5 \times 1,2 = 4,05 \cong 4.0$$

The allowable bearing capacity will then be given by:

$$Q_{all} = \frac{Q_s}{1.3} + \frac{Q_p}{4}$$

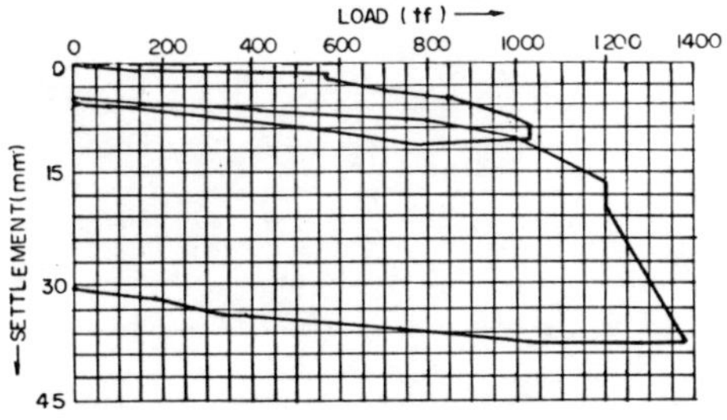

Figure 3 - Load test on a "Bentonite Pile"

7 ANALYSIS OF A "BENTONITE PILE" WITH DIAMETER D= 1,50m AND LENGTH L= 23,20M

Figure 3 shows the results of the load test to which the pile was submitted.

The average value of N along the pile shaft was $\bar{N}$ 29,17, the adhesion being thus given by

$$q_s = \frac{\bar{N}}{3} + 1 = 10,7 t/m^2$$

7.1 Skin friction resistance

The skin friction ultimate resistance was estimated by applying the adhesion value above, as follows

$$Q'_s = q_s \quad D \quad \pi \quad L$$

$$Q'_s = 1.169,2t$$

If the soil parameters were to be safely estimated, a safety factor should by applied (see 5.1), and Q'_s would then become Q_s

$$Q_s = 1062,7t$$

The load test showed that for loads of 1050t and 1200t, total settlements were 10 and 20mm respectively. Somewhere between those two values the ultimate skin friction resistance (Q_s) was reached.

Therefore, it seems quite reasonable to admit the load of 1050t, to which total settlement of 10mm was reported, to be the the ultimate skin friction resistance mobilized in the load test.

This value (1.050t) comes quite close to the estimated skin friction resistance to which the safety factor related to soil parameters (1.1) was applied (1062.7t)

8 ANALISYS OF TWO PRECAST CONCRETE PILES DRIVEN THROUGH SOILS WITH HIGH N VALUES

Meyerhof and others presented a paper in the X International Conference on Soil Mechanics and Foundation Engineering (Meyerhof, Bronw and Mooland, 1981) in which the bearing capacities of two hexagonal concrete piles driven through highly consistent soil (N values from 20 to 52) were evaluated with basis on several field tests (SPT, deep sounding, pressuremeter, screw plate test) as well as laboratory tests (triaxial, unconfined compression tests), by using formulae which take into account either total or effective pressures. The results were then compared with results of load tests cárried out to failure.

It is, then of interetest to compare the results which would be obtained if the author's method were to be applied. A single modification was introduced here to the original 1978 paper: the top limit of 15 to the N values was extended, to 50 according to what has already been proposed on ítem 3.

Table 2 shows that the results obtained are quite close to the best achieved by Meyerhof and constitute a precise evaluation of the failure loads given by the load test.

TABLE 2 - RESULT COMPARISON

ESTIMATE	MEYERHOF'S		AUTHOR'S	
Pile Type	420	800	420	800
Point (1)	85	165	250	490
Shaft adhesion (2)	700	1060	947	1521
Shaft friction (3)	1210	1850	-	-
Total (1)+(2)	785	1225	1197	2012
Total (1)+(3)	1295	2015	-	-
Measured ultimate load (KN)	1160	1780	1160	1780

9 CONCLUSION

The presente state of knowledge on the matter leads us to believe that the method proposed constitutes an expedite way for determining the allowable bearing capacity of the various types of piles.

Our initial recomendation that the skin friction resistance be limited to q_s=6,0t/m² or N=15 should still be kept for "Strauss" piles and open air caissons.

For displacement piles and piles excavated and cast under bentonite this limit should be q_s= 17t/m² or N=50.

As to safety factors, the concept presented in this paper seems to constitute an evolution over the traditional adoption of a single factor related to the ultimate load.

10 REFERENCES

Burland, J.B. 1975, Contribuitions to the Paper Bentonite Piles in Durban by J. Everett and C,M. Mcmillan. Soil Mechanics and Foundation Engineering Sixth Regional Conference for Africa - Durban. Vol. Two.- A.A. Balkema

Décourt, L, Quaresma A.R. 1978, Capacidade de Carga de Estacas a partir de Valores de SPT. VI Congresso Brasileiro de Mecânica dos Solos e Engenharia de Fundações - Rio de Janeiro - ABMS.

Décourt, L. 1978, Resistência e Conpressibilidade de uma argila de São Paulo a partir de Ensaios de Campo e de laboratório. VI Congresso Brasileiro de Mecânica dos Solos e Engenharia de Fundações - Rio de Janeiro - ABMS.

Décourt, L. 1978 Written Discussion VI Congresso Brasileiro de Mecânica dos Solos e Engenharia de Fundações - Rio de Janeiro - ABMS

Meyerhof G.G., Brown J.D., Mooland G.D., 1981, Prediction of Friction Pile Capacity in a till, X ICSMFE - Stockholm- A.A. Balkema.

Prodinger W., Veder C.H. 1981, Bearing Capacity of Floating Groups of diaphragm walls X ICSMEF - Stockholm - A.A. Balkema

Proceedings of the Second European Symposium on Penetration Testing / Amsterdam / 24-27 May 1982

Modulus of elasticity for sand determined by SPT and CPT

H.DENVER
Danish Geotechnical Institute, Lyngby, Denmark

1 INTRODUCTION

This paper deals with the determination of the modulus of elasticity for sand based on standard penetration tests (SPT) and cone penetration tests (CPT).

Various algorithms have been proposed to calculate Young's modulus (E) in the geotechnical literature. In order to select a suitable method, a test programme has been carried out on three selected locations where the soil profile mainly consists of homogeneous sand layers.

The moduli of elasticity for the sand profiles are determined by pressuremeter tests and screw-plate tests and are then compared with the results from the SPT tests and CPT tests.

2 TEST SITES

Three different test sites, where the soil profile mainly consists of fine (silty) sand to a depth of 15 metres, have been selected in order to perform a laboratory and an in-situ testing programme on the sand.

A water tower is actually built at each location and settlement records of the water towers are compared with calculated settlements. However, this paper deals only with the problem of how to determine E on the basis of in-situ tests - the rest of the above-mentioned subjects are described elsewhere (Denver 1980, Denver 1981).

The three locations, at Birkerød, Søllerød and Lyngby, are situated in the northern vicinity of Copenhagen and are, from now on, referred to as A, B and C.

The grain size distribution is plotted as mean curves for three samples from each location on Fig.1. The sand from all three locations is described as fine sand. The in-situ void ratios are measured to $e = 0.68$, 0.74, 0.68 for A, B and C, respectively, and the density is characterized as medium to dense for all three test sites.

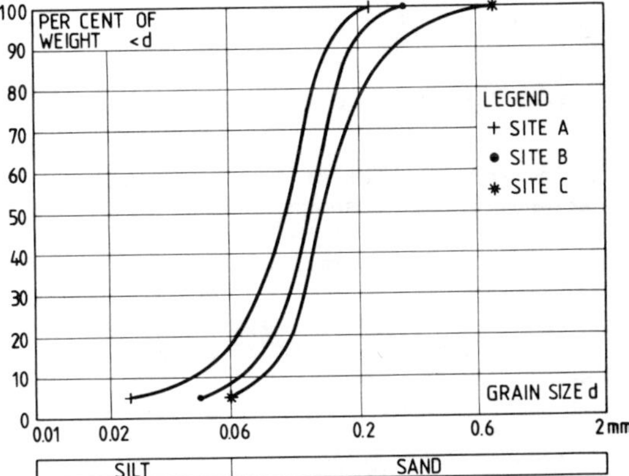

Fig.1: Grain size distribution for the three test sites. Each curve represents a mean from three samples taken from representative depths.

No ground water table is observed at the locations at the depths where the in-situ tests are performed.

Apart from the first metre, the entire profile consists of the sand described above (of diluvial or glacial nature) to the operating depths. The only exception is site C, where the upper 3 m consist of glacial clay (boulder clay).

3 IN-SITU TEST PROGRAMME

The in-situ tests are situated in two or three groups around the water tower to facilitate:

(i) a settlement calculation for the water tower (as mentioned in paragraph 2), and

(ii) a comparison of the different testing methods.

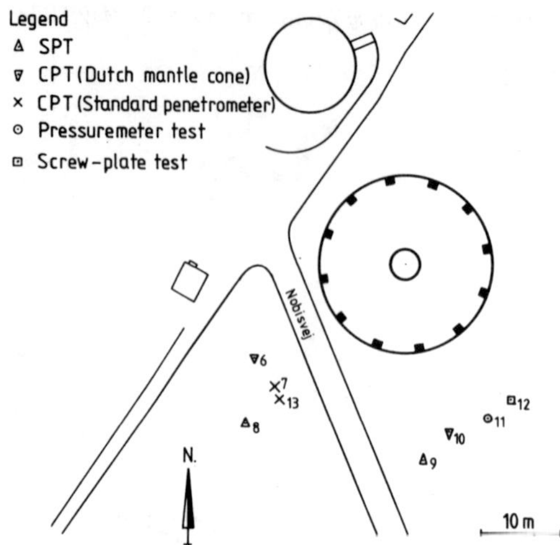

Fig.2: In-situ test arrangement - site A.

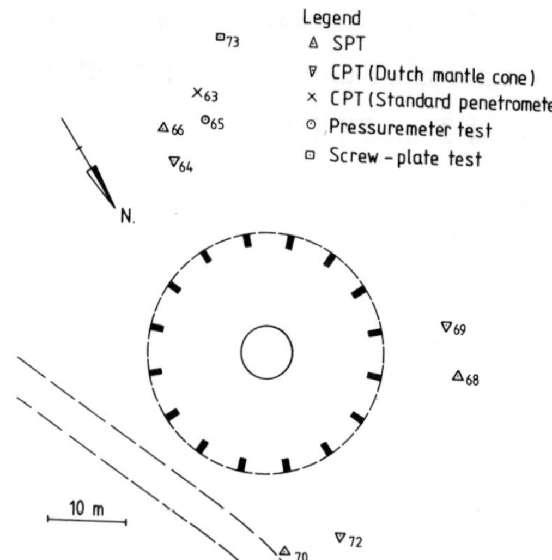

Fig.3: In-situ test arrangement - site B.

The total number of tests is 26, and the arrangements are shown in Figs.2 - 4.

3.1 Standard penetration tests

Standard penetration tests are perfomed according to the Recommended Standard for the SPT Test.

3.2 Dutch cone penetration tests

Dutch cone penetration tests are performed by means of the Dutch mantle cone penetrometer tip (Type M1) following the Recommended Standard for the Cone Penetration Test.

3.3 Standard cone penetration tests

The CPT tests are performed with the standard penetrometer tip (electric cone - Fugro type) following Recommended Standard for the Cone Penetration Test (CPT).

The testing is not perfectly continous - the pushing is halted each metre to allow a new rod to be mounted.

3.4 Pressuremeter tests

Pressuremeter tests are routine tests at the Danish Geotechnical Institute, and the equipment used is the Ménard pressuremeter.

It was not necessary to make special arrangements due to the fact that the soil is sand. As mentioned before, no water table has been observed, and the sand was moist in a manner that made it possible, by careful manual dry drilling, to obtain a bored hole with a perfectly undisturbed wall in the lowest 1.5 m. The upper part of the hole was protected by a casing.

When a test in one depth was completed the casing was pushed down to the bottom of the hole and the drilling was continued to a greater depth.

3.5 Screw-plate tests

It was not possible to screw the standard equipment (the Field Compressiometer from "Geonor") down to the desired depths, and a modification was made to apply the screw-plate to the bottom of a 150 mm cased boring.

Furthermore, to avoid densification of the sand due to the massive central core, the screw-plate was mounted on a hollow pipe to allow the sand to penetrate the pipe when the plate was screwed down.

When a certain depth is reached, the sand will stop penetrating into the pipe. If the plate is screwed any further, the result is a densification of the soil. If, on the other hand, a load test is performed at this depth, both the central sand core and the plate are active during the test.

Laboratory tests have been performed to determine this depth and precautions are made to be able to screw the plate down to a depth where the influence of the unloading caused by the bored hole is negligible. These subjects are described elsewhere (Denver 1980) and shall not be treated further here.

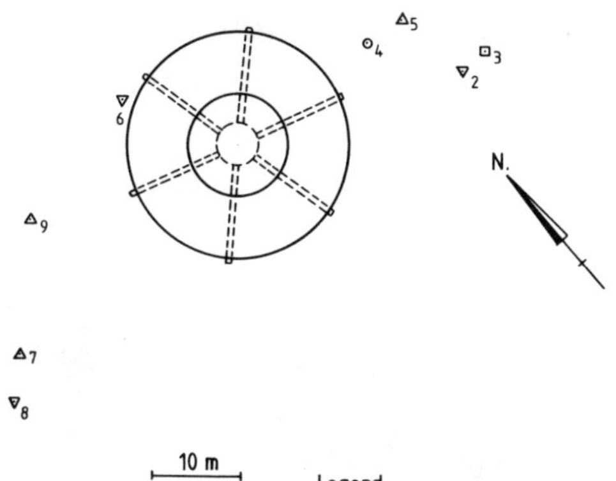

Fig.4: In-situ test arrangement - site C.

4 DETERMINATION OF E FOR THE TEST SITES

Young's modulus of elasticity is determined for one group of the in-situ tests by means of the pressuremeter tests and the screw-plate tests. The value of Young's modulus is chosen as the average of the values found by the tests, neglecting the big differences between these tests (e.g. the effect of anisotropy).

4.1 Pressuremeter tests

E is normally deduced from pressuremeter test by the equation

$$E_p = 2(1+\mu)\,\frac{dp}{dv}\,v_0 \qquad (1)$$

where E_p denotes the pressuremeter modulus, μ is Poisson's ratio, v_0 the initial volume of the measuring cell, dp/dv is the increment of pressure over the corresponding increment of volume.

Equation (1) is based on the assumption of an axisymmetric expansion of an infintely long cylindrical opening in a perfectly elastic medium. dp/dv is taken from the linear part of the pressuremeter curve (i.e. v plotted as a function of p). This interval of p is often called the "pseudo elastic" region.

In this paper we assume:

$$E = E_p \qquad (2)$$

and $\mu = 0.33$.

4.2 Screw-plate tests

A certain loosening of the sand over the plate will occur when the plate is screwed down.

If a full bonded contact between a rigid circular plate and the surrounding soil is assumed, and the plate is embedded completely at a great distance from the surface, then the dimensionless settlement (λ_{rc}) is referred by Selvadurai & Nicholas, 1979, to be:

$$\lambda_{rc} = \frac{\delta E}{r_0 p} = \frac{\pi(3-4\mu)(1+\mu)}{16(1-\mu)} \qquad (3)$$

where δ is the settlement, r_0 is the radius of the plate, and p is the average stress applied on the plate area.

If, on the other hand, it is assumed that the stiffness is described as if the plate was placed on the bottom of a deep cylindrical hole, then the dimensionless settlement is

$$\lambda_{rb} = \frac{\delta E}{r_0 p} \approx f\,\lambda_{rt} \qquad (4)$$

where $\lambda_{rt} = 0.5\pi(1-\mu^2)$ is the dimensionless settlement of a rigid, smooth plate resting on the surface of a semi-infinite elastic medium. f is a reduction factor, used when the plate is resting on the bottom of a cylindrical hole. For $\mu = 0.33$ Burland, 1970, found $f = 0.82$ for a circular load for $\mu = 0.33$.

The mean value of λ from equation (3) and (4), for $\mu = 0.33$, is used to determine E from the screw-plate tests:

$$E = 0.90\; r_0 p/\delta \qquad (5)$$

The settlement δ is not a linear function of p. Nevertheless, if only an interval of $p = [400|1000]$ kPa is observed, the function is quite linear, and p/δ is estimated as $\Delta p/\Delta\delta$ from this interval.

Again, this interval coincides more or less with that of the pseudo-elastic region of the pressuremeter tests, and is also relevant to the settlement calculations for normally loaded spread footings.

4.3 Standard penetration tests

Different correlations between the number of blows N and Young's modulus E are in use to-day. The most commonly used correlations are a linear relationship as expressed in equation (6)

$$E = s_1 N + s_2 \qquad (6)$$

where s_1 and s_2 are constants. Table 1 contains a selection of the constants s,

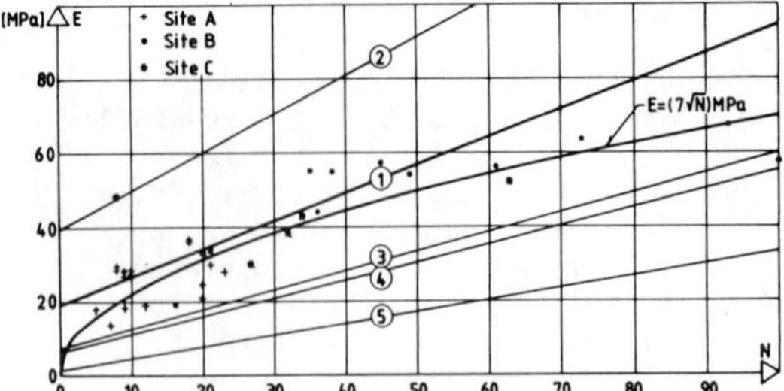

Fig.5: Young's modulus E as function of N.

and the respective functions are plotted on Fig.5.

It should be emphasized that very different procedures are used to determine E by the different authors, and some of the procedures are closely connected to a specific method of settlement calculation.

Table 1. Values of s_1 and s_2 to equation (6).

Number of curve on Fig.5	s_1 [MPa]	s_2 [MPa]	Remark	Reference
1	0.756	18.75	Normally loaded sand and gravel	D'Appolonia et al, 1970
2	1.043	36.79	Preloaded sand	-"-
3	0.517	7.46		Schultze & Menzenbach, 1961
4	0.478	7.17	Sand - saturated	Webb,1970
5	0.316	1.58	Clay & sand	-"-

If the "true" E for the soil profile at the test sites is determined as the average of the value from the pressuremeter tests (2) and the screw-plate tests (5), it is then possible to plot E v N on Fig.5.

A parabolic relation:

$$E = B\sqrt{N} \tag{7}$$

where $B = 7$ MPa, is seen to yield a better correlation to the measured relations than the linear functions.

4.4 Dutch cone penetration tests

Similar to the standard penetration tests, different linear correlations between q_c and Young's modulus E have been proposed:

$$E = h_1 q_c + h_2 \tag{8}$$

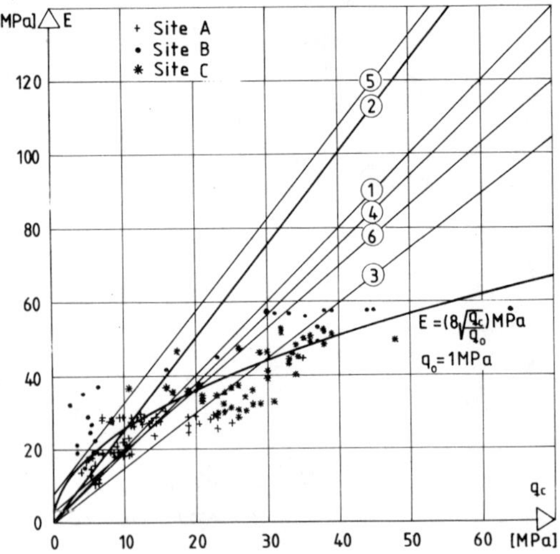

Fig.6: Young's modulus E as function of q_c.

where h_1 and h_2 are constants. Some values of h are shown in Table 2 and the functions are plotted in Fig.6. As for the standard penetration tests, very different procedures are applied by the authors to obtain E.

Table 2. Values of h_1 and h_2 to equation (8)

Number of curve on Fig.6	h_1	h_2 [MPa]	Remark	Reference
1	2	0		Schmertmann,1970
2	2.5	0	Modified method	Schmertmann,1978
3	1.5	0		De Beer & Martens, 1957
4	1.9	0		Meyerhof, 1965
5	2.5	7.17	Sand - saturated	Webb,1970
6	1.67	2.40	Clayey sand - saturated	-"-

Young's modulus E for the soil profile is determined by means of the procedure described in section 4.3, and E v q_c is shown in Fig.6.

Fig.6 also includes a parabolic relation:

$$E = A\sqrt{q_c/q_0} \tag{9}$$

where $A = 8$ MPa and $q_0 = 1$ MPa. Obviously, equation (9) yields a better correlation with the test points than the linear relationship.

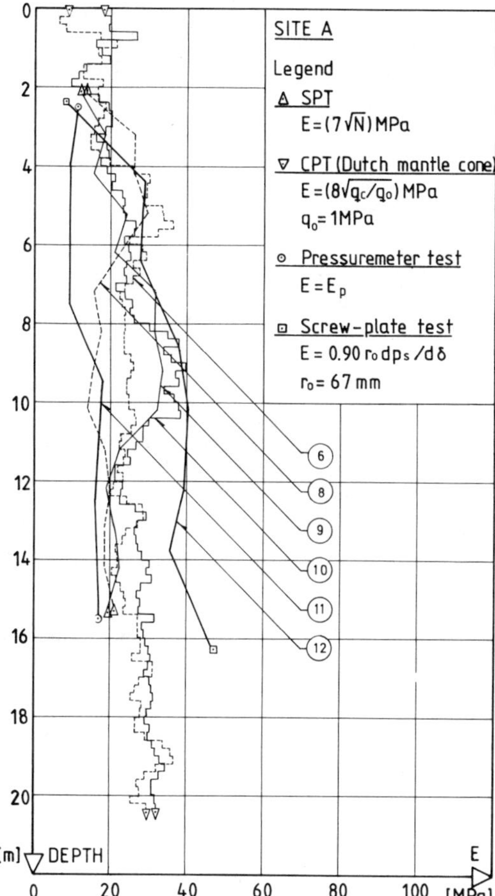

Fig.7: The E profile - Site A.

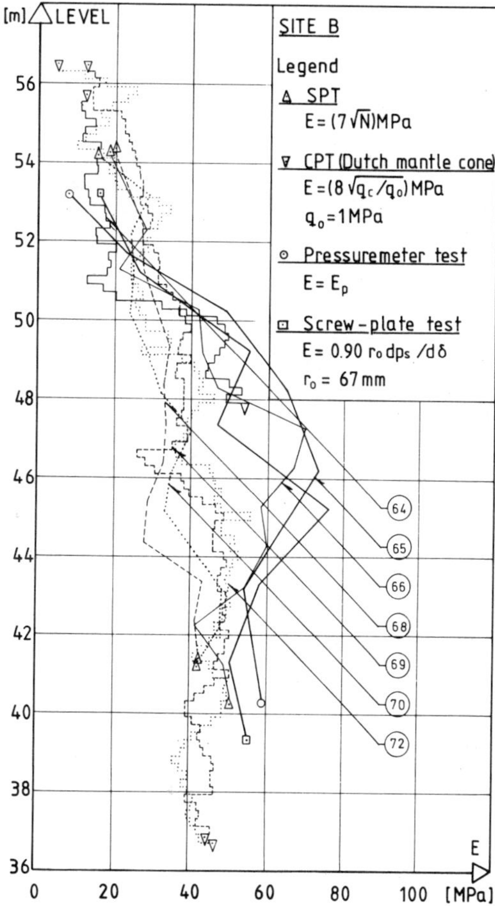

Fig.8: The E profile - Site B.

4.5 Standard cone penetration tests

Very few standard cone penetration tests are carried out, and the cone resistance (q_{cs}) is related to the cone resistance of the Dutch cone penetration tests (q_c), and not directly to Young's modulus.

Different authors have proposed a constant value of q_{cs}/q_c for sand, ranging from 1.0 to 1.3.

The Danish tests yield the following mean value and standard deviation of the ratio of corresponding values of q_{cs} and q_c: $m(q_{cs}/q_c) = 1.4$, $s(q_{cs}/q_c) = 0.7$.

5 DISCUSSION OF THE RESULTS

The calculation of E using the above-mentioned procedures are shown in Fig.7 to Fig.9 for the three test sites. The tests belonging to the same test group on the site are drawn with the same line pattern. The group including the pressuremeter test and the screw-plate test is drawn with solid lines (only curves with the same line pattern should be compared).

No systematic divergence is observed between E from pressuremeter tests and screw-plate tests. Furthermore, although showing considerable scatter, no systematic divergence between curves with equal line pattern is observed.

Apparently, the parabolic functions fulfil their purpose and yield an easy relation between q_c and N by eliminating E in equation (7) and (9):

$$q_c/N = 0.77 \text{ MPa} \tag{10}$$

6 CONCLUSIONS

For the soil investigated (fine sand over the ground water table) the following conclusions are drawn:

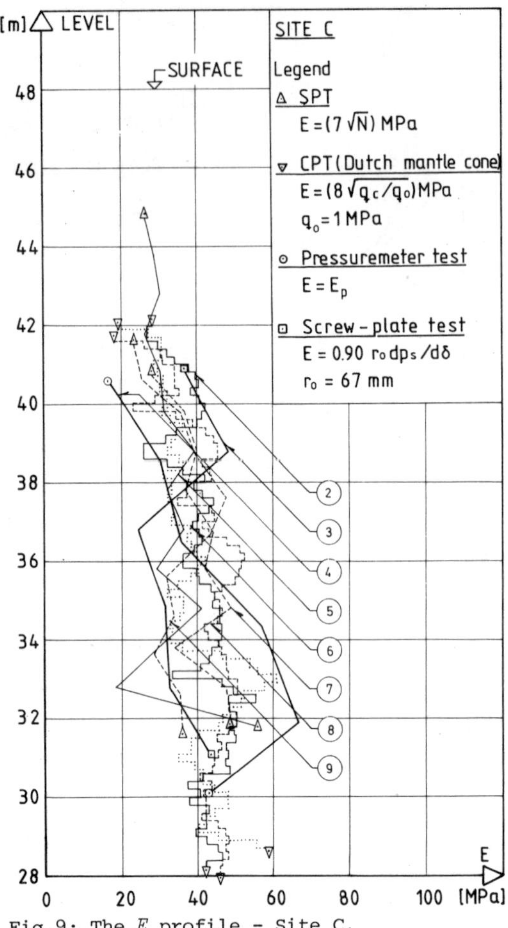

Fig.9: The E profile - Site C.

(i) Young's moduli can be calculated by applying pressuremeter tests, screw-plate tests, standard penetration tests, and Dutch cone penetration tests by means of the equations (2), (5), (7), and (9), respectively.

(ii) By elimination of E it is possible to predict the other tests by means of results from one of the tests mentioned in (i).

(iii) Taking the findings of other authors into account, it is proposed to use the relation

$$q_{cs}/q_c = 1.3 \quad (11)$$

to calculate the cone resistance for the Dutch mantle cone (q_c) by means of results from standard penetrometer tests (q_{cs}).

7 ACKNOWLEDGEMENTS

Acknowledgements are made to the National Agency of Technology and the National Research Administration, from whom I have received grants for this study.

The author would also like to thank the Danish Geotechnical Institute for sponsoring the preparation of this paper.

8 REFERENCES

Burland, J.B. 1970, Disc., Session A.Proc. Conf.on In Situ Investigations in Soils and Rocks, Brit.Geotechn.Soc., London.

D'Appolonia, D.J., et al. 1970, Discussion on settlement of spread Footings on sand, J.Soil Mech. and Found.Div.ASCE, Vol.96, No.SM2.

De Beer, E., A.Martens 1957, Method of computation of an upper limit for the influence of heterogeneity of sand layers on the settlements of bridges, Proc.4th Int. Conf.on Soil Mech.and Found.Engn., London.

Denver, H. 1980, Calculation of settlements of Footings on sand, Ph.D.Thesis, Technical University of Denmark (in Danish with English summary).

Denver, H. 1981, A method of settlement calculation, Proc.10th Int.Conf.on Soil Mech. and Found.Eng., Vol.2, Stockholm.

Meyerhof, G.G. 1965, Shallow foundations, J. Soil Mech.and Found.Div.Proc.ASCE, Vol.91, No.SM2.

Recommended standard for the cone penetration test (CPT), Proc.9th Int.Conf.on Soil Mech.and Found.Eng., Appendix A, Vol.3, Tokyo 1977.

Recommended standard for the SPT test, Proc. 9th Int.Conf.on Soil Mech.and Found. Eng., Appendix C, Vol.3, Tokyo, 1977.

Schmertmann, J.H. 1970, Static cone to compute static settlement over sand, J.Soil Mech.and Found.Div., Proc.ASCE, Vol.96, No.SM3.

Schmertmann, J.H. 1978, Improved strain influence factor diagrams, J.Geotechn.Eng. Div., Proc.ASCE, Vol.104, No.GT 8.

Schultze, E., E.Menzenback 1961, Standard penetration and compressibility of soils, 5th Int.Conf.on Soil Mech.and Found.Engn., Paris.

Selvadurai, A.P.S., T.J.Nicholas 1979, A theoretical assessment of the screw-plate test, 3rd Int.Conf.on Numerical Methods in Geomechanics, Aachen.

Webb, D.L. 1970, Settlements of structures on deep alluvial sandy sediments in Durban, South Africa, In-situ Investigations in Soils and Rocks, Brit.Geotechn.Soc., London.

Proceedings of the Second European Symposium on Penetration Testing / Amsterdam / 24-27 May 1982

SPT blowcount variability correlated to the CPT

B.J.DOUGLAS
Ertec Western Inc., Long Beach, California, USA

1 INTRODUCTION

The blowcount measured in the Standard Penetration Test (SPT) depends heavily upon test equipment which influences magnitude of incident sampling energy. However, knowledge of incident energy does not directly allow scaling of blowcounts to represent those of some other energy-level hammer. That is, the ratio of blowcounts obtained with different constant energy hammers is not constant, but depends upon soil strength and gradation. In addition, the SPT blowcount is procedure dependent, thus even if incident energy is standardized and soil indices known, uncertainties will still exist.

There are two primary needs for SPT standardization. One is to allow provision of a higher degree of certainty in present practice, and the other is to allow assessment of the comparison of new SPT measurements with measurements taken in the past. Both of these needs are critical to the continued use of the SPT for liquefaction potential assessments using the simplified SPT blowcount methods. Both of these needs are met by the Cone Penetrometer Test.

The procedures and equipment of the quasi-static electric Cone Penetrometer Test (CPT) are easily standarized. In addition, the test directly provides continuous measurements of the soil indices needed to assess the effects of variation in incident SPT hammer energy on blowcounts. Further, the CPT can be used to accurately predict any specific constant energy SPT blowcounts. Thus the CPT method can be directly used for assessment of liquefaction potential.

2 FIELD INVESTIGATIONS

Standard Penetration and Cone Penetrometer Tests were conducted at three sites, subsequently termed SD, SS, and DU. The SD site comprises recent hydraulically-filled sands and silts. The SS site comprises interbedded fluvial silty sands, gravels and clays of Holocene Age. The DU site comprises sands, silts and clays of eolian, fluvial, and lacustrine origin ranging in age up to 1 million years.

Standard Penetration Tests were conducted using a variety of equipment and procedures. At the SD site, all borings were rotary wash using a baffled bit. SPT hammers were of two types: an automatic-release trip hammer with cylinder weight, and a so-called standard 2-rope-turn-around-the-cat-head cylinder hammer. Both hammers were of similar weight and geometry. These systems are subsequently identified as R.W.Trip;SD, and R.W.Standard;SD.

These same pieces of equipment were used at the SS and DU sites with different operators. Designations for those sites follow the same pattern as of the SD site. In addition, data is presented from a third system at the DU site. This system consisted of a hollow-stem auger boring performed without drilling mud and using a steel block with guide pin for the hammer. The system is designated Auger, Pin;DU. Finally, data obtained by Youd and Bennet (1981) at the SS site using a water-filled hollow stem auger and Mobil safety hammer are also presented and designated Auger,Safety;SS.

All soundings were conducted following ASTM D-3441 using an electric cone penetrometer. All SPTs were performed following ASTM D-1586 except that liner samplers were used without liners.

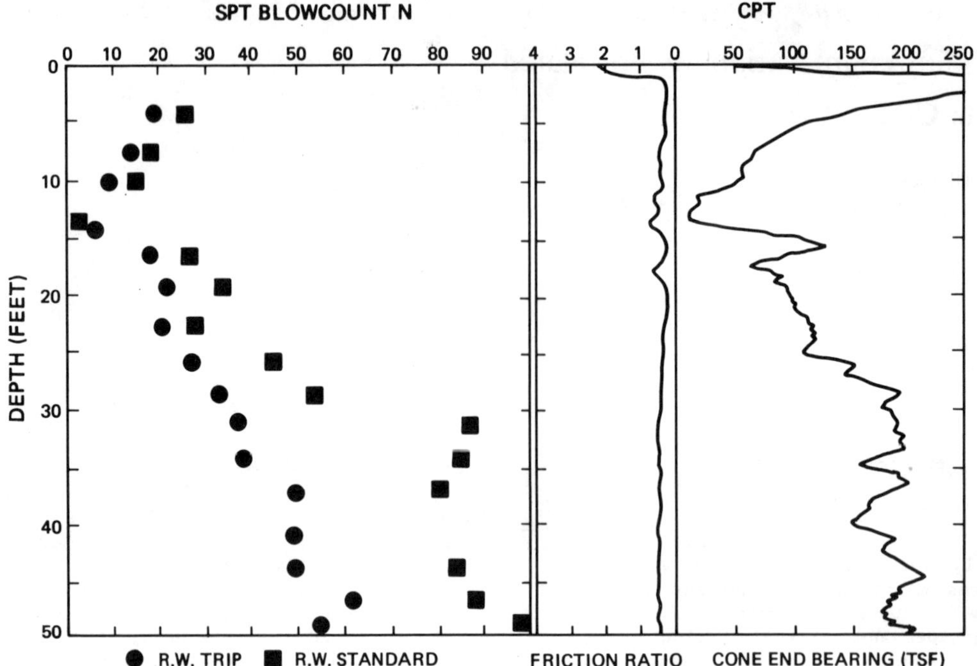

FIGURE 1: AVERAGE CONE PENETROMETER AND STANDARD PENETRATION TEST DATA; SAN DIEGO SITE

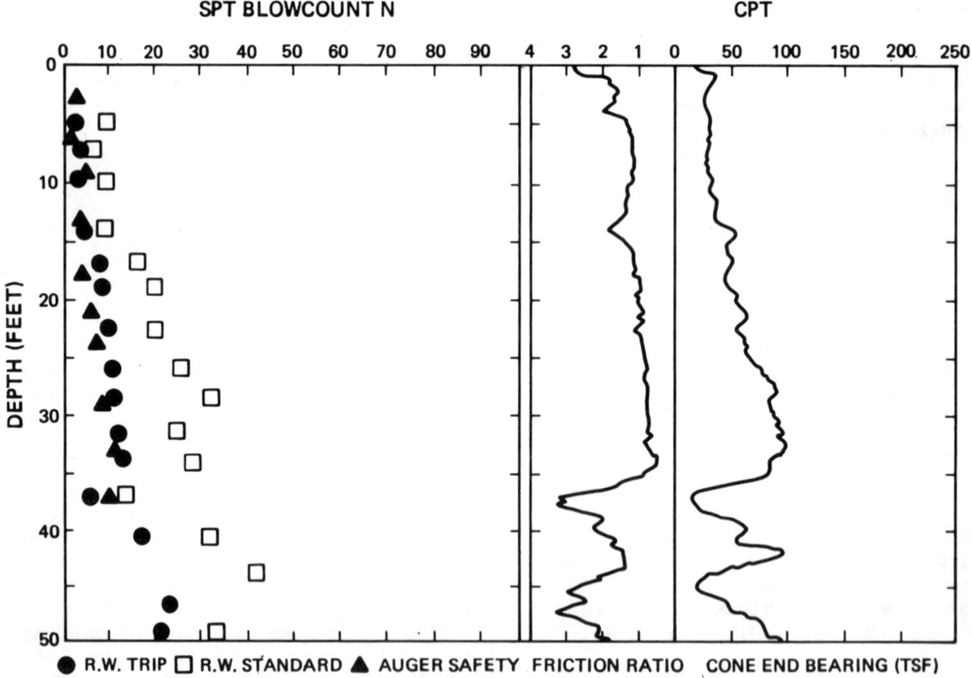

FIGURE 2: AVERAGE CONE PENETROMETER AND STANDARD PENETRATION TEST DATA; SALINAS SITE

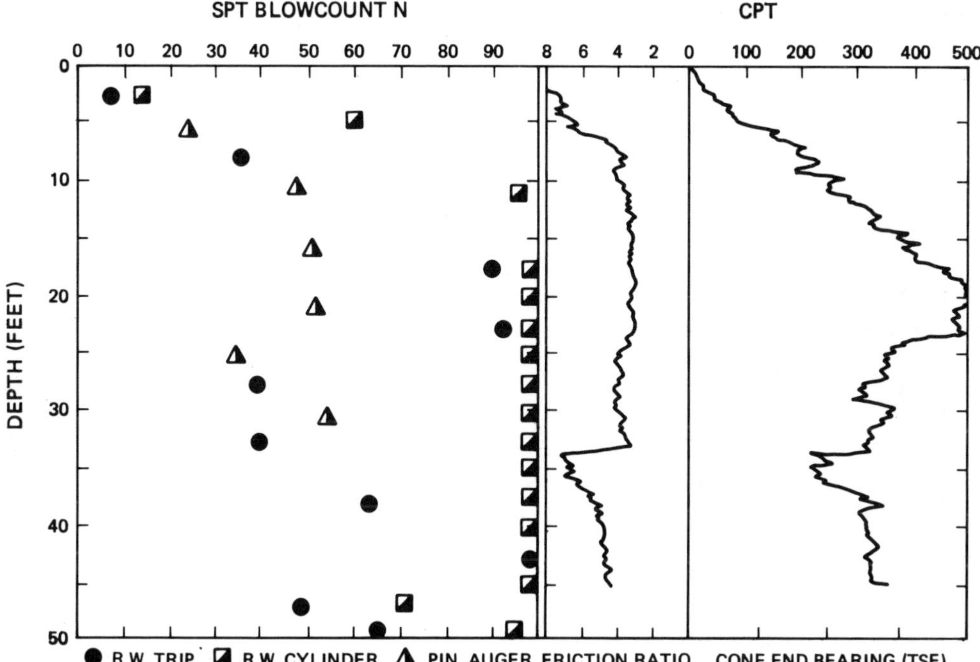

FIGURE 3: AVERAGE CONE PENETROMETER AND STANDARD PENETRATION TEST DATA; D.U. SITE

3 DATA ANALYSES

The results of the SPT and CPT investigations at each site are summarized in Figures 1 through 3. All data in those figures represent site averages. There is a large range in blow counts obtained using different systems, with the Auger methods producing the lowest and the trip hammer next lowest and the standard hammers the highest. Next, the ratios of blowcounts obtained with the different hammers is not a constant, but tends to increase with increasing penetration resistance or with increasing CPT friction ratio. Finally, the higher energy SPT data show a smoothing of resistances over the depth of investigation, while the lower energy SPT data show less smoothing, and the CPT data show the least.

All standard and trip hammer measurements were conducted using controlled procedures but different crews. The variability in ratios of standard to trip hammer blowcounts must then result from differences in operators, soil type, or soil resistance. To investigate these differences the ratios for each site were compared against the soil gradation and resistance. Gradation and resistance were quantified using the CPT data.

The SPT blow count ratios for standard and trip hammer data are plotted on the CPT Soil Behavior Type Classification Chart shown in Figure 4 (Douglas and Olsen 1981; Schmertmann, 1978). It is seen that although the blow count ratios are variable, some trends do exist. First, an increase in cone end bearing or friction ratio corresponds to an increase in blowcount ratio. Considering that the blowcount increases with increasing end bearing or friction ratio (Martin and Douglas, 1981; Schmertmann, 1976) it follows that the blowcount ratio nonlinearly increases with increasing blowcount. That is, a lower-energy hammer shows blowcounts increasing at a faster rate than higher energy-hammers in increasing resistance soils.

Blowcount ratios from different sites obtained by different operators were not constant even when performed in similar soils of similar resistance at similar depths, as seen in Figure 4. The differences may result from variations in procedure. For example, Kovacs (1981) shows that SPT operators provide variability in hammer drop height on a daily basis, and that the effort expended in trying to "throw" the lifting rope off of the cat head also influences the energy transmitted. Likewise, the drilling

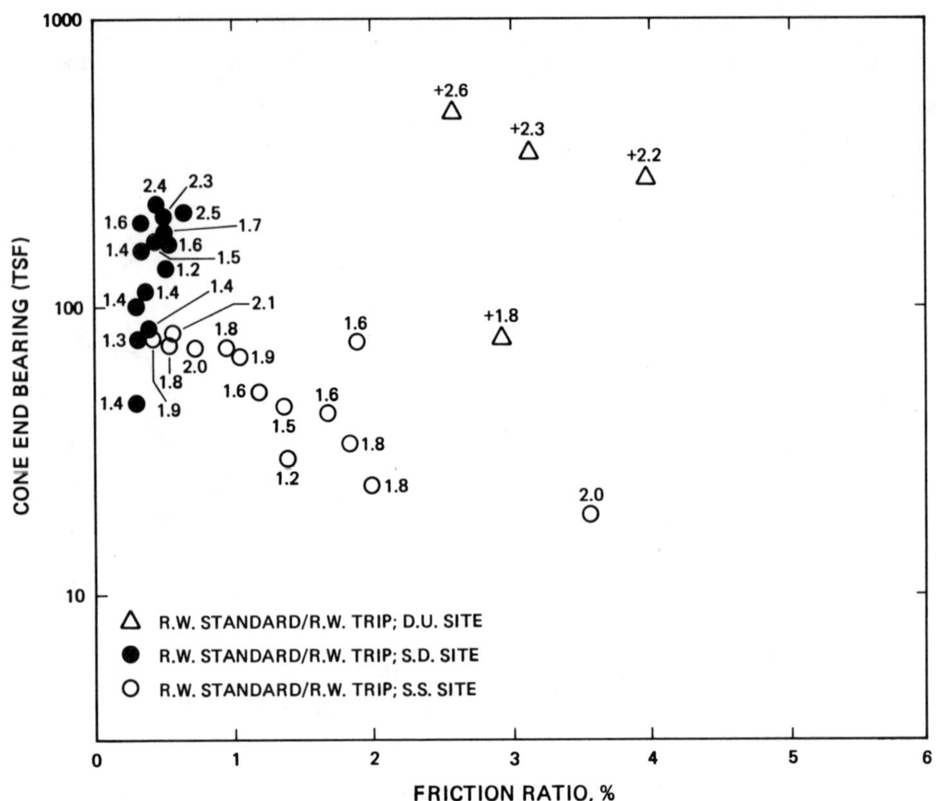

FIGURE 4: SPT BLOWCOUNT RATIOS ON CPT SOIL BEHAVIOR TYPE CLASSIFICATION CHART

procedure prior to performance of the SPT can influence the effective stress and degree of soil disturbance at borehole bottom (Schmertmann, 1977; Douglas et al, 1981) and therefore blowcount.

Standardization of the SPT through adjustment of weight or fall distance will not allow assessment of previous blowcounts because of the importance of drilling procedure. Even with procedure standardization, it would be necessary to assess the soil types and resistance prior to performing any energy-ratio scaling.

The CPT is easily standarized and can be used to assess previous SPT data. Much research into the correspondence of the CPT and SPT has been performed (Martin and Douglas, 1981; Schmertmann, 1976; Sanglerat, 1972, deMello, 1971). The modeling of the SPT sampler penetration using CPT-measured soil stresses (Schmertmann, 1976) provides the best correlations between SPT N value and CPT end bearing. Such a procedure previously developed (Douglas et al, 1981) to allow routine use of CPT measurements for SPT type liquefaction potential assessments was used as a basis to compare the different SPT data presented in this paper.

Comparison of the averaged CPT and SPT data from each site and each hammer system using the CPT-sampling-energy procedure is shown in Figure 5. The CPT-SPT relations are fairly constant for a specific hammer from site to site, but the ratio of blowcounts predicted for different hammers increases with increasing blowcount or CPT sampling energy dissipation function. Thus the blowcount ratios can be predicted for different zones of the CPT Soil Behavior Type Classification Chart using the CPT-SPT correlation method and the data in Figure 5.

The predicted ratios for the trip and standard hammers are shown in Figure 6. The dependence of the ratio on soil type and resistance is easily seen. The predicted ratios agree well with the measured ratios shown in Figure 4, although the data of Figure 6 represents averages of the correlations for all sites.

The data in Figure 6 reveals the same

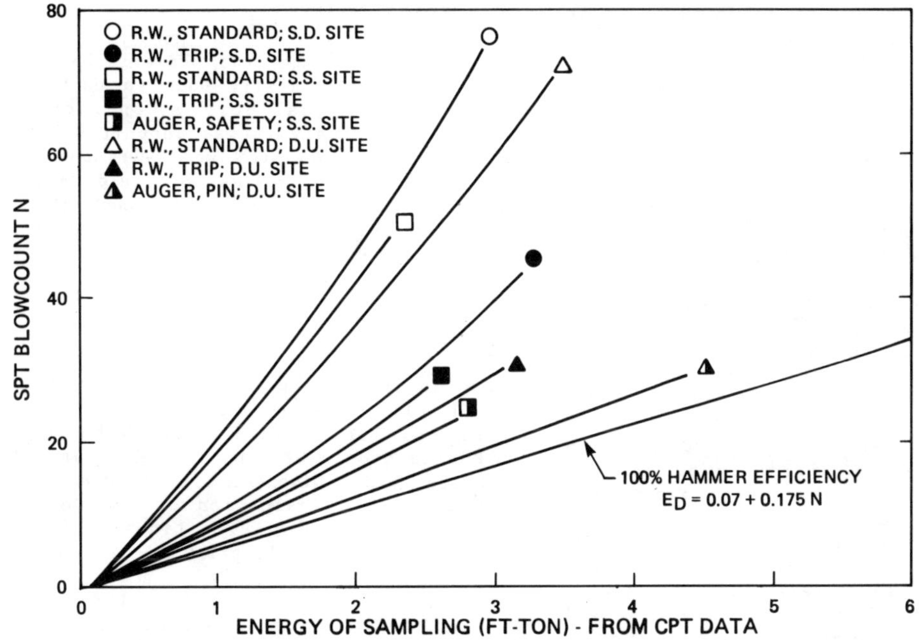

FIGURE 5: CALCULATED SAMPLING ENERGY VS MEASURED BLOWCOUNTS

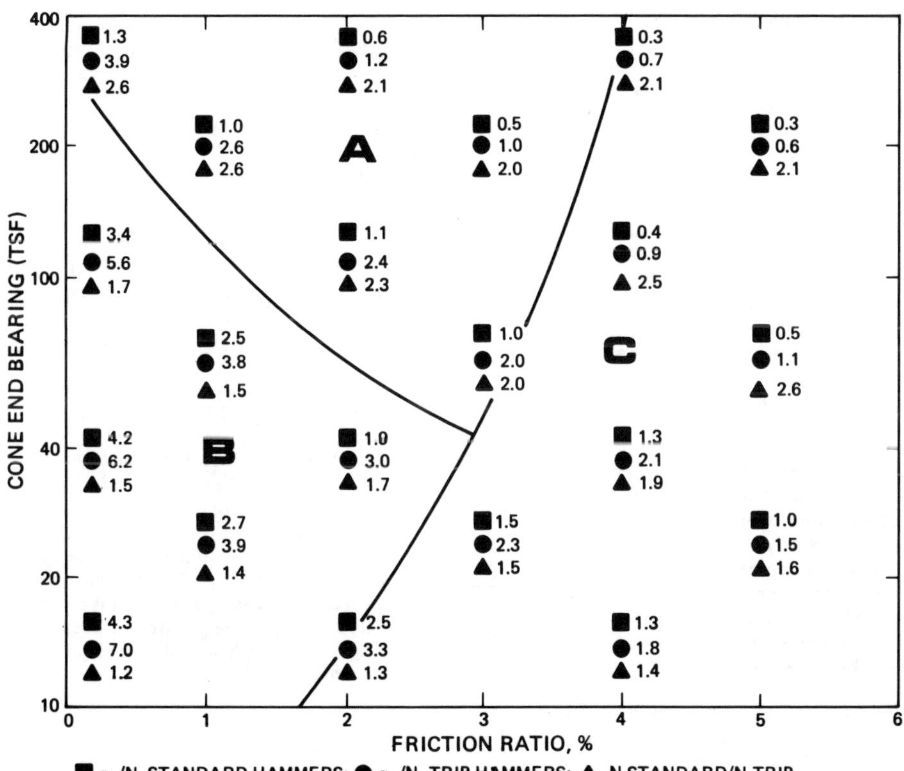

FIGURE 6: q_c/N RATIOS AND BLOWCOUNT RATIOS

range of q_c/N ratios as found by other researchers (Martin and Douglas, 1981). Comparing the different hammer N values predicted for each zone of the classification chart with the cone end bearing at that point gives the q_c/N ratio data shown in Figure 6. It is seen that q_c/N ratios of from 0.5 to 7.0 result as a function of hammer type, soil resistance, and soil gradation.

The data shown in Figures 5 and 6 can be used to assess incident energy of previous SPT data or to convert CPT to SPT data for specific materials or applications. For example, considering liquefaction potential assessments, the susceptible soils can be defined as sands and silty sands having corrected blowcounts less than about 35. Examination of q_c/N ratios of Figure 6, and the CPT-soil classifications (Douglas and Olsen, 1981) allows designation of Zone B as being representative of liquefaction susceptible soils. Within this zone, the q_c/N ratio can conservatively be treated as a constant of 4.0 for the standard hammer. The hammer ratios can likewise be held at a constant of 1.4.

The assumption of a constant q_c/N ratio of 4.0 will result in under prediction of actual blowcounts in silty soils. However, for these soils the ratio can be changed to 2.5, or to the actual value representative of each material type zone.

4 CONCLUSIONS

Based on the data presented in this paper, several conclusions can be drawn regarding the variability of blowcounts in the SPT, the reliability and use of incident energy standarization as a means to provide repeatable SPT data or to reassess historical SPT data and the use of the CPT for blowcount type assessments.

- o The blowcount ratio between different constant incident energy hammers is not constant but depends upon soil density and gradation;
- o Because blowcount ratio is not constant, incident energy scaling of historical blowcounts is only feasible if soil indices are considered;
- o The CPT method can be used to assess the system energy efficiency represented in historical blowcount data;
- o For liquefaction assessments in sands, the CPT end bearing can be related to the SPT using a constant ratio of q_c/N equal to 4.0

5 ACKNOWLEDGEMENTS

The data upon which this paper is based was developed primarily under a United States Geological Survey Contract. Much of the analysis was supported by Ertec Western Research and Development Grants. Finally, volumes of field data were processed and reduced through the conscientous efforts of Messrs. George Edmonds and Mike Leue.

6 REFERENCES

Douglas, B.J., and Olsen, R.S., 1981, Soil Classification Using Electric Cone Penetrometer, ASCE Special Technical Publication, St. Louis, Missouri.

Douglas, B.J., Olsen, R.S., and Martin, G.R., 1981, Evaluation of the Cone Penetrometer for SPT - Liquefaction Assessment, ASCE Preprint 81-544, St. Louis, Missouri, October.

Kovacs, W.D., Salomone, L.A. and Yokel, F.Y., 1981, Energy Measurement in the Standard Penetration Test, NBS Building Science Series 135, National Bureau of Standards, Washington, D.C., USA.

Martin, G.R. and Douglas, B.J., 1981, Evaluation of the Cone Penetrometer for Liquefaction Hazard Assessment, U.S. Geological Survey Open File Report 81-234.

de Mello, V.F.B., 1971, The Standard Penetration Test, 4th Pan American Conference on Soil Mechanics and Foundation Engineering, San Juan, Puerto Rico, Vol. 1.

Sanglerat, G., 1972, The Penetrometer and Soil Exploration, Elsevier Publishing Company, New York.

Schmertmann, J.H., 1976, Predicting the q_c/N Ratio, Final Report D-636, Engineering and Industrial Experiment Station, Department of Civil Engineering, University of Florida, Gainesville, Floria, USA.

Schmertmann, J.H., 1977, Use the SPT to Measure Dynamic Soil Properties? Yes, But...!, ASTM Symposium on Dynamic Field and Laboratory Testing of Soil and Rock, Denver, Colorado, June.

Schmertmann, J.H., 1978, Guidelines for Cone Penetration Test Performance and Design, U.S. Department of Transportation, Federal Highway Administration Report No. FHWA-TS-78-209.

Proceedings of the Second European Symposium on Penetration Testing / Amsterdam / 24-27 May 1982

New analytical correlations between SPT, overburden pressure and relative density

F.GIULIANI
Centro de Ingeniería Océanica, Buenos Aires, Argentina

F.L.GIULIANI NICOLL
Empresa de Estado Hidronor, Argentina

1 INTRODUCTION

The use of the SPT associated to empirical relationships between blowcount number and relative density of cohesionless soils for the determination of shear strength, bearing capacity and liquefaction potential, is continued in many countries, mainly owing to the low cost of the equipment involved in field testing, notwithstanding many objections and discussions about the reliability of the method.

The influence of the relative density in shear strength, bearing capacity and liquefaction potential is nevertheless well established, notwithstanding some evidences of large dispersions in its determination, both in laboratory and field work.

On the other hand there are important contributions showing the influence of the overburden pressure in the relationship between blowcount and relative density, although perhaps so obvious as the relationship between cone penetration resistance and angle of shear strength.

A somewhat undecided controversy about the relationships between blowcount in the SPT, overburden pressure and relative density, as given by Gibbs and Holtz's chart and Bazaraa's formula, deserved interesting comments by V.F.B. de Mello in his State of the Art Paper on the SPT, among them that values of relative density estimates by the chart are likely to be greater than in the field, so agreeing with Bazaraa and Peck, but objecting the discontinuity in Bazaraa's formula as well as a common fallacy in pretending to include all classes of sands in one single relationship of properties.

This paper presents tables and graphs based in those controversial relationships looking for some clarification of the controversy and leading to a new and more satisfactory analytical expression.

2 ANALYTICAL EXPRESSION SUBSTITUTING GIBBS AND HOLTZ'S CHART. BAZARAA'S FORMULA

Taking from Gibbs and Holtz's chart the coordinates of a number of points representative of its complete domain, a regression analysis led to the following expression:

$$RD/100 = 1.5 \left(\frac{N}{F}\right)^{0.222} - 0.6 \qquad (1)$$

where,

$$F = 0.0065\, \bar{\sigma}^2 + 1.68\, \bar{\sigma} + 14$$

The accuracy of this expression has been verified in every part of the chart. On the other hand we have Bazaraa's formula:

$$RD/100 = 0.2236 \sqrt{\frac{N}{a + b\,\bar{\sigma}}} \qquad (2)$$

where,

for $\bar{\sigma} \leqq 15$, a = 1, b = 0.2

" $\bar{\sigma} > 15$, a = 3.25, b = 0.05

In these expressions and others, and in the tables in this paper, the overburden pressure $\bar{\sigma}$ is given in metric tons per square meter (t/m^2), its approximate equivalent of newtons per square centimeter or decibars.

3 NUMERICAL COMPARISON

Values of relative densities from formulas (1) and (2) were computed for blowcount numbers between 2 and 50 and overburden pressures between 0 and 50 t/m^2. These values, presented in table 1, show that those from Gibbs and Holtz's chart are under Bazaraa's only for very low blowcount (under 5) and high overburden pressure (over 20 t/m^2) i.e. inside the boundaries in solid lines at the corner of the table.

On the other hand, G & H's values are higher than Bazaraa's everywhere outside those boundaries, but with very large deviations only inside the boundaries in dash

Table 1 Values of relative densities computed from G & H's and Bazaraa's relationships

$\bar{\sigma}$	N	2	5	10	20	30	40	50
0	GH	37	59	79	>100			
	B	32	50	71	100			
5	GH	27	47	65	86	>100		
	B	22	35	50	71	87	100	
10	GH	21	40	56	76	88	98	>100
	B	18	29	41	58	71	82	91
20	GH	(13)	30	45	62	74	83	90
	B	16	24	34	48	59	69	77
30	GH	(8)	23	37	53	64	72	79
	B	15	23	32	46	56	65	73
40	GH	(4)	19	32	47	57	65	71
	B	14	22	31	44	53	62	69
50	GH	—	15	27	42	51	59	65
	B	13	21	29	42	51	59	66

Table 2 Averages and percentages of deviation from values compared in table 1

$\bar{\sigma}$	N	2	5	10	20	30	40	50
0	Av	34.5	54.5	75	>100			
	±D	7.2	8.2	5.3				
5	Av	24.5	41	57.5	78.5	94	>100	
	±D	10.2	14.6	13.0	9.5	7.4		
10	Av	19.5	34.5	48.5	66	78.5	90	>100
	±D	7.7	15.9	15.5	13.6	10.8	8.9	
20	Av	16	27	39.5	55	66.5	76	83.5
	±D	—	11.1	13.9	12.7	11.3	9.2	7.8
30	Av	15	23	34.5	49.5	60	68.5	76
	±D	—	0	7.2	7.1	6.7	5.1	3.9
40	Av	14	20.5	31.5	45.5	55	63.5	70
	±D	—	7.3	1.6	0.3	3.6	2.4	1.4
50	Av	13	18	28	42	51	59	65.5
	±D	—	6	3.4	0	0	0	0.7

lines at blowcounts less than 30 amd overburden pressures less than 20 t/m^2.

Consequently there seems to be better agreement in about 70 % of the whole domain, as shown more clearly in table 2 with averages and percentages of deviation.

4 GRAPHIC COMPARISON

Considering that Bazaraa's formula suggests an almost linear relationship between RD and the square root of the blowcount, it was decided to compare charts on this basis, as shown in figs. 1 and 2, drawing a system of curves at constant overburden pressures.

In fig. 1 the curves represent G and H's values, having been added for comparison in dash lines the curves for **0 and** 50 t/m^2 representing Bazaraa's values. As all curves should tend to the origin, it seems evident the presence of some source of gross errors in measurements by Gibbs and Holtz at high pressures (over 20 t/m^2) and low blowcounts (under 10).

In fig, 2 the curves represent Bazaraa's values, having been added for comparison in dash lines the curves for 0 and 50 t/m^2 representing G & H's values. It is clear here that all the curves in solid lines fall inside the confines of those in dash lines, but there is also evidence of some source of error in measurements by Bazaraa, producing large separation of curves at low pressures and close proximity of those at high pressures.

Nevertheless the differences are not too large, in fact less than ± 10 % in about 70 % of the whole domain of practical application, as shown before in table 1.

5 NEW CORRELATIONS

A reasonable agreement of both relationships in the largest part of their common domain as shown by tables 1 and 2, as well as the approximate linearity of RD vs. $\sqrt{N}$ shown in figs. 1 and 2, leaving out only the relatively small field where the Gibbs and Holtz curves are unreliable because they deviate upwards instead of converging to the origin, i.e. only for blowcounts under 10 and overburden pressures over 20 t/m^2, led to the drawing of curves of averages vs. square root of blowcount given in fig. 3.

A somewhat surprising result was an almost perfect linearity, except for slight deviations in curves of 30 to 50 t/m^2 caused by anomalies around blowcounts 5 and 10. Therefore linear regression analyses were performed, including all the averages defining each curve, to obtain the slopes of straight lines converging to the origin, corresponding to overburden pressures from 0 to 50 t/m^2 respectively, as shown in table 3.

Further regression analyses were then tentatively performed with the reciprocals of those values of slope, one adjusted to a linear semilogarithmic function and the other to a second degree equation, to obtain the following expressions of the slope:

$$S1 = \frac{1}{4.188 + 0.639\,\bar{\sigma}^{0.606}} \qquad (3)$$

$$S1 = \frac{1}{3.723 + \sqrt{0.216 + 0.994\,\bar{\sigma}}} \qquad (4)$$

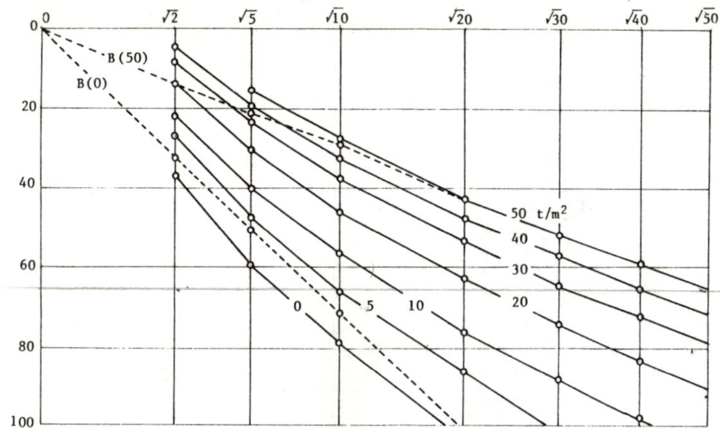

Fig.1 Curves of RD vs.$\sqrt{N}$ from Gibbs and Holtz's values compared with Bazaraa's confines.

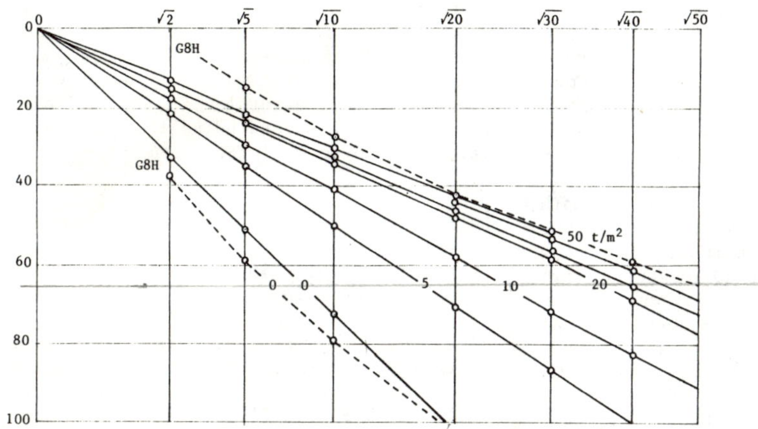

Fig.2 Curves of RD vs.$\sqrt{N}$ from Bazaraa's values compared with Gibbs and Holtz's confines.

Values from these formulas and their percentages of deviation are compared to those used as data, as shown in table 3. As the deviations are smaller for equation (3), then the best approximation of relative density will be obtained with the following:

$$RD/100 = \frac{\sqrt{N}}{4.188 + 0.639\ \bar{\sigma}^{0.606}} \qquad (5)$$

This is obviously a continuous function, convergent to the origin and therefore also valid for very low blowcount numbers as well as down to zero pressure.

6 RELATIVE DENSITY AND ITS RELATIONSHIP TO OTHER PROPERTIES

The need of varied relationships between RD and SPT according to differences in soil characteristics, as suggested by V. de Mello, is an interesting proposition, although perhaps not practically feasible.

In fact one of the authors participated a few years ago in laboratory work with a similar purpose, using sands obtained from widely different areas and selecting thirteen grain size distributions for triaxial testing at different relative densities, but while no influence of grain size distribution or mineral composition could be detected, roughly approximate relationships between RD and angle of friction as well as between RD, dry unit weight and coefficient of uniformity were verified.

A recent reevaluation of these experimental data led to the following approximate expressions, with unit weight in t/m^3:

$$tg\ \phi = 0.575 + 0.361\ RD^{0.866} \qquad (6)$$

$$\gamma a = 1.33\ (1 + 0.23\ RD)\ U^{0.1} \qquad (7)$$

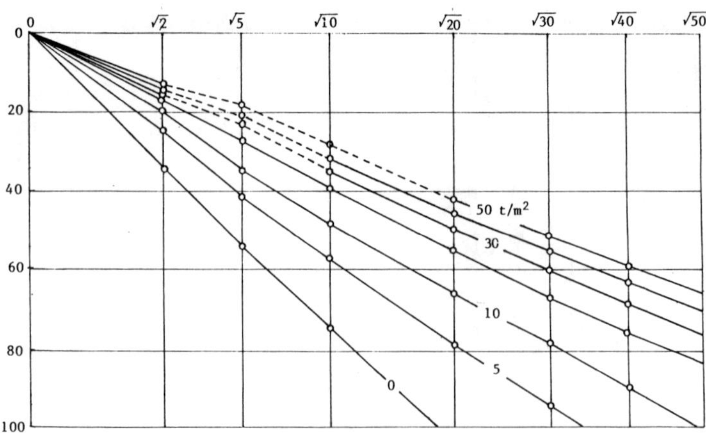

Fig.3 Curves of average RD from Gibbs and Holtz's and Bazaraa's values

While not assuming strict and general validity of these expressions, they are shown here only as evidence of the extreme complexity of the problem, even not considering mineral composition of the sands.

Another aspect of the authors' personal experience is that in large areas of the western plains of their country along the Andes, there are fine silty sands to a depth of few meters, with the water table close to the surface, where notwithstanding blow count less than one and RD negative, two story buildings of brick and concrete on continuous footings are well supported at allowable pressure of 0.5 kg/cm^2. Consequently all known relationships between SPT, RD and bearing capacity are not valid in cases like these, but on the other hand those buildings are probably bound to collapse if affected by earthquakes of not very high intensity.

Table 3 Regression values of the ratio $\sqrt{N}$/RD vs. overburden pressure

Rg	$\bar{\sigma}$ S1	0 .2388	5 .1737	10 .1434	20 .1199	30 .1080	40 .0996	50 .0934
(3)	Rg	.2388	.1700	.1478	.1232	.1085	.0984	.0907
	Dif%	0	-2.2	+3.3	+2.7	+0.5	-1.2	-2.9
(4)	Rg	.2388	.1666	.1447	.1216	.1086	.0995	.0927
	Dif%	0	-4.0	+0.9	+1.4	+1.5	-0.1	-0.8

7 CONCLUSIONS

A new analytical correlation between SPT, overburden pressure and relative density has been obtained by statistic analysis of available data. While seemingly more consistent than former relationships, its application to foundation engineering problems can not be independent of sound judgement, since it represents only average conditions for most natural sands, not including some very fine saturated silty sands with very low blowcount in SPT.

REFERENCES

Begemann, H.K.S. 1965, The Friction Jacket Cone in Determining Soil Profile. P. 6th Int. Conf. on S.M. and F.E.

Gibbs, H.J., Holtz, W.H. 1957, Research on Density of Sands by Spoon Penetration Testing. P.4th Int. Conf. on S.M.and F.E.

Mello de, V.F.B. 1971, The Standard Penetration Test. P.4th Panam. Conf.on S.M. and F.E.

Meyerhof, G.G. 1956, Penetration Test and Bearing Capacity of Cohesionless Soils.J. ASCE.

Peck, R.B., Bazaraa, A.R.S. 1968, Discussion of paper by D'Appolonia, D.G. & E, Brissette, R., Settlement of Footing on Sand. J. ASCE.

Seed, Bolton H., Idriss, I.M. 1971, Simplified Procedure for Evaluating Soil Liquefaction Potential. J. ASCE.

Suarez, R. et alii, 1962, Resistencia al corte de arenas. Proc. 1ª Reunión Argentina de M. de S.

Terzaghi, K., Peck, R.B. 1948, 1973, Soil Mechanics in Engineering Practice. J.Wiley and Sons.

Tschebotarioff, G.P.1953, 1973, Foundations, Retaining Walls and Earth Structures. Mc Graw-Hill.

Proceedings of the Second European Symposium on Penetration Testing / Amsterdam / 24-27 May 1982

The use and interpretation of SPT results for the determination of axial bearing capacities of piles driven into carbonate soils and coral

J.HAGENAAR
Frederic R.Harris, The Hague, Netherlands

1 INTRODUCTION

Recent publications by various authors (Angemeer et al 1973, McClelland 1974, Stevenson and Thompson 1978) have established that piles driven into carbonate deposits and coral formations have encountered low driving resistance.

Conditions favouring the deposition of such carbonate sediments have been shown to exist between the latitudes 30 degrees N and 30 degrees S.

The engineering properties of the carbonate sediments required for the analysis and design of pile foundations vary significantly from those of silica sands, silts and clays.

A series of loading tests on steel pipe piles driven into such sediments and coral was carried out in connection with various offshore projects along the Red Sea Coast of Saudi Arabia.

This paper presents the results of a study made to use the data obtained from the load tests for establishing bearing capacity factors. These factors are correlated with standard penetration resistances (SPT N values). Special attention was given to depth of penetration, diameter, plugged and unplugged condition of the driven steel pipe piles.

2 GENERAL SUB-SURFACE CONDITIONS

The Red Sea shelf bordering the coastal plain of Saudi Arabia is covered with numerous coral reefs (Hagenaar and v.d. Berg 1981). The reefs extend generally in long strips parallel with the coast, and are called fringing reefs when attached to the coast, or barrier reefs when detached from the coast. Coral debris from breakage by wave action falls into the voids and cavities of the reefs and forms blankets of skeletal sands down their slopes.

The sea floor between the coral heads and reefs is covered with marine sediments. These consist of loose to medium dense carbonate sands, silts and clays with layers of coral.

From the inland mountains, material has eroded to form the coastal plain, which consists of gravel, sand and silts. These alluvial deposits extend from the shoreline out to underly and intermix with the reef deposits and marine sediments. The alluvial deposits were encountered in offshore boreholes as brown, dense to very dense sands and gravels with cobbles. The gravels and cobbles are from igneous or metamorphic origin.

The fringing reef forms a barrier that prevents the carbonate detritus and sediments, as well as the land derived alluvial deposits, from being transported into the deeper waters of the Red Sea. This backreef lagoon area consists of fine-grained carbonate sand, silt, and clay, which vary in density from very loose to medium dense. The depth of the loose to very loose sediments encountered in boreholes drilled reached maximum depths of 10 to 15 meters.

The subsurface conditions existing at the test pile locations are shown in figure 1. Test locations A and B were located offshore a barrier reef, C and D in a backreef lagoon area, and test location E on a fringing reef.

3 STANDARD PENETRATION RESISTANCES (SPT N VALUES)

The SPT N values measured in the coral formations ranged generally from 15 to 50 blows per foot. However, due to the presence of cavities, N values of less than 4 have been recorded. Layers of strong

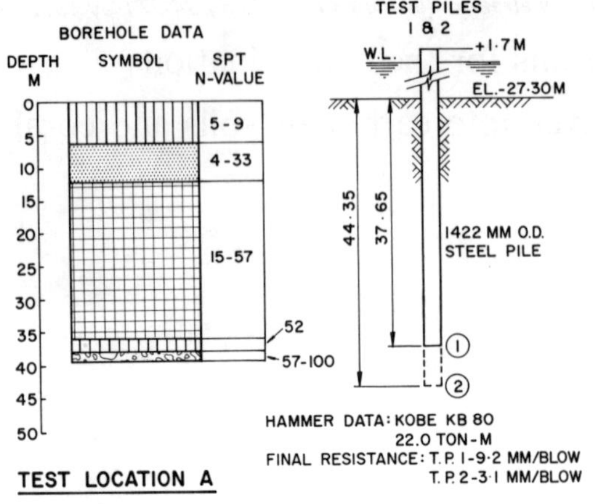

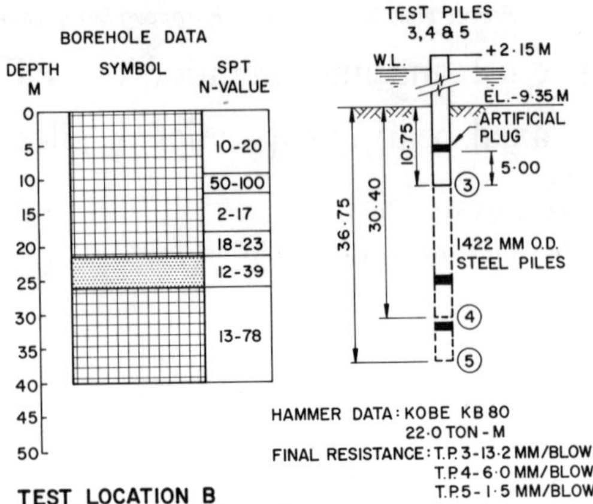

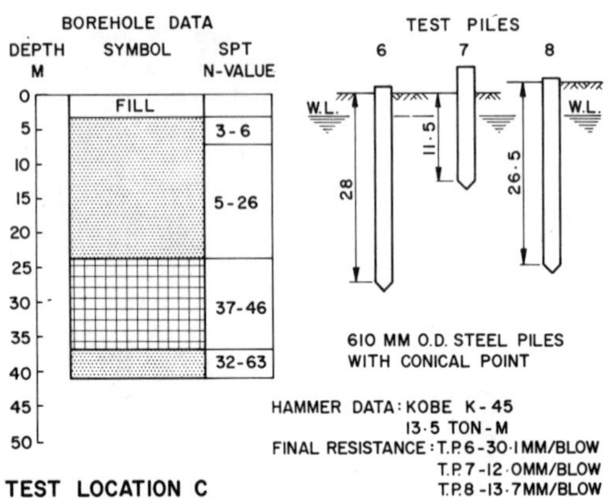

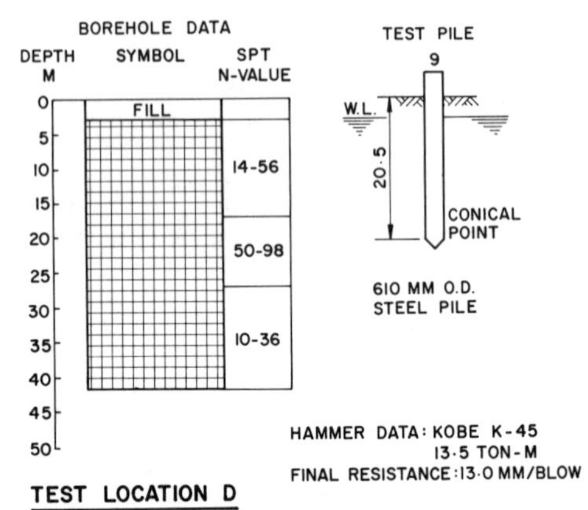

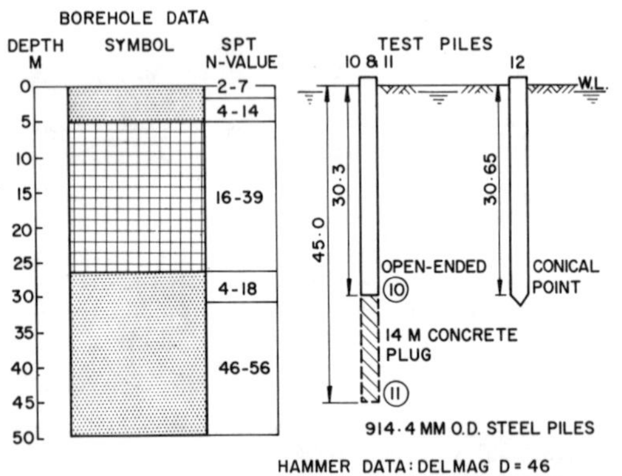

GROUND SURFACE

W.L. WATER LEVEL

CORAL

GRAY AND WHITE CORAL, WITH CAVITIES AND POCKETS OF SANDS AND SILTS.

MARINE SEDIMENTS

GRAY CARBONATE CORAL SANDS, SILTY SANDS, WITH SHELLS AND CORAL FRAGMENTS.

LIGHT GREEN TO GRAY CARBONATE SILTS AND SANDY SILTS WITH SHELLS.

ALLUVIAL DEPOSITS

GRAY-BROWN SLIGHTLY SILTY SAND.

Fig. 1 Subsurface and Pile Test Data

coral are present with SPT N values of 50 blows per foot and more.

The marine sediments covering the upper 5 to 8 meters of the sea bottom or overlaying the coral foundations have SPT N values of less than 10. The density of these carbonate soils increased with depth having SPT N values ranging from 14 to 50 blows per foot.

In the layers of brown alluvial soils, intermixed with carbonate sediments, the SPT N values were between 50 to 100 blows per foot.

4 PILE TESTING

Figure 1 gives details of the steel pipe piles installed for the testing programmes. The piles were driven, open or closed ended, to the penetration depths shown. The type and size of hammers used during installation are also shown together with the final sets, at which driving of the test piles was terminated.

The testing programmes comprised of maintained load (ML) tests in accordance with ASTM method D1143. The total test loads applied varied from 1½ to twice the maximum design loads.

At three test locations A, B and E, piles were extended and driven further down after failure occurred before the total test load was achieved.

Upon completion of the load test, the inside soil plug was removed in test pile 10 and a 14 m concrete plug installed. After a setting period of 7 days the pile was redriven to a penetration depth of 45.0 metres (test pile 11).

The steel pile at test location B was artificially plugged at 5.0 metres above the tip.

Conical points were used at other locations for closed ended piles.

5 TEST RESULTS

The results of the pile load tests are shown in Table 1. The ultimate compression capacities were derived from the load test results using various interpretation methods recommended by Fellenius (1975).

The ultimate tension capacities were easily identified as upward movement inincreased significantly at failure. The plastic deformations under the total test loads applied on test pile number 5 were négligible. Consequently, failure loads are larger than the test loads shown.

Table 1 summarizes also the ultimate point and shaft resistances calculated from the load test results.

Table 1. Summary of Pile Load Test Results

	Pile Load Test Failures		Ultimate Pile Bearing Capacities	
Test Pile No.	Compression (ton)	Tension (ton)	Point Load Q_p (ton)	Net shaft Resistance Q_s (ton)
1	667	315	390	277
2	800	300	542	258
3	235	-	195	40
4	590	260	354	236
5	>619	>300	>347	>272
6	180	-	85	95
7	236	-	204	32
8	260	-	170	90
9	300	-	233	67
10	338	-	187	151
11	673	-	437	236
12	610	-	457	153

Where no tension load tests were available, the net shaft resistance was calculated in accordance with following paragraph 6.2. The shaft resistances are net, that is maximum test load minus the weight of the pile.

The results of the tension test on test pile number 2 have not been included in the study. Failure occurred at 300 tons, which was 15 tons lower than that obtained for test pile number 1, having approximately 7.0 meters less penetration depth. The decrease is very likely due to movements of test pile number 1 during alternate compression/tension testing and redriving to the 44.35 m penetration depth. These movements occurring within a relative short period could have disrupted the contact between pile and coral material, resulting in lower total shaft resistance.

6 ANALYSIS OF RESULTS

6.1 Existing Theory

The static formula commonly used to determine the ultimate bearing capacity of a pile is shown in Figure 2. The ultimate bearing capacity consists of a point resistance Q_p and shaft resistance Q_s.

From the load test results approximate values will be derived for unit side friction f_s and unit end bearing q_p. The values are correlated with the SPT N values recorded along the pile shaft and below the pile tip.

6.2 Shaft Resistance

For the piles driven into cohesionless soils the unit skin friction f_s is calculated

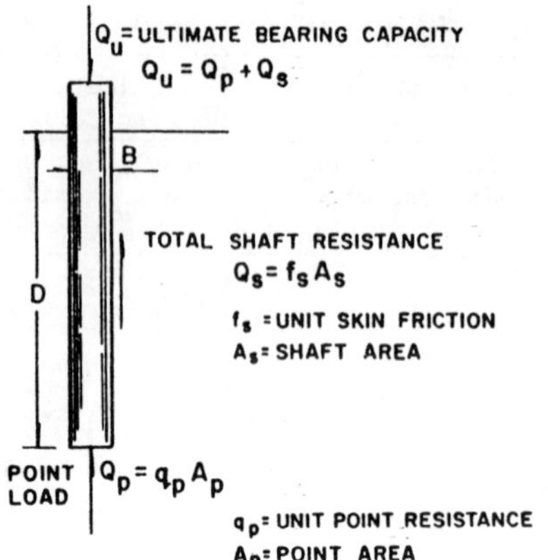

Fig. 2 Ultimate Bearing capacity.

with the following formula:

$$f_s = K\, p_o \tan \delta \leqslant f_1 \text{ where}$$

K = coefficient of lateral earth pressure
p_o = effective overburden pressure
δ = angle of soil friction on pile wall
f_1 = limiting value of unit skin friction.

However, the results of the tension tests show that the side friction for the driven steel piles is very low, and does not increase significantly with depth in accordance with above formula.

It is, therefore, more appropriate to assign a limiting value of unit skin friction, f_1, which remains constant after reaching a certain penetration depth D_c. McClelland (1974) and Datta et al (1980) show that a limiting value of 2.0 ton/m^2 has been used for medium dense calcareous sands.

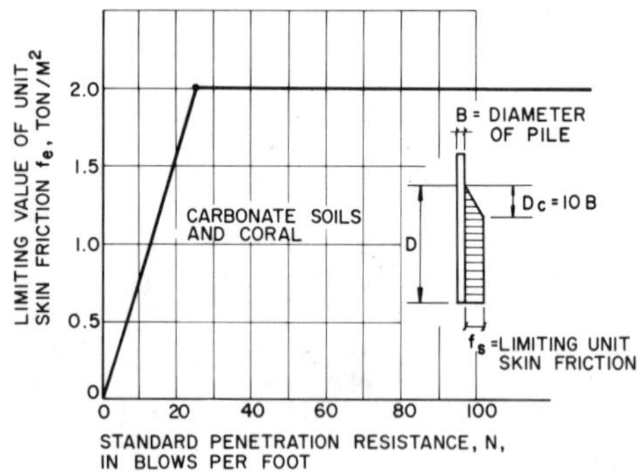

Fig. 3 Relationship between Limiting Unit Skin Friction and SPT N Values

The limiting value of 2.0 ton/m^2 as recommended by McClelland (1974) seems to correspond with the few tension load test results available. The average SPT N values along the embedded portion of the tension test piles varied from 27 to 40 blows per foot. Until further documentation becomes available it is recommended to use figure 3 for establishing shaft resistances both in compression and in tension.

The penetration depth at which the limiting skin friction values are reached should be taken as 10 times the diameter or larger.

6.3 Point Load

The unit point resistance q_p shown in figure 2 is calculated with the formula:

$$q_p = p_o\, N_q \leqslant q_1, \text{ where}$$

p_o = effective overburden pressure
N_q = bearing capacity factor
q_1 = limiting value of end-bearing

For the correlation of recommended bearing capacity factors the point loads Q_p shown in Table 1 were used.

Table 2 shows the results of the determination of the N_q values from above formula for the given ultimate point loads of the test piles.

Table 2. Analysis of Ultimate Base Resistance

Test Pile	D/B	Penetration Resistance N(blows)per foot) at pile tip	q_p (t/m^2)	p_o (t/m^2)	N_q
1	26.5	52	246	37.65	6.5
2	31.2	75	341	44.35	7.7
3	7.6	75	123	10.75	11.4
4	21.4	13	223	30.40	7.3
5	26.5	25	>219	37.65	>5.8
6	45.9	37	291	28.00	10.4
7	18.9	26	699	11.50	60.8
8	43.4	46	582	26.50	22.0
9	33.6	98	798	20.50	38.9
10	33.1	45	285	30.30	9.4
11	49.2	46	665	45.00	14.8
12	33.5	56	696	30.65	22.7

Figure 4 shows that a relationship could be established between the bearing capacity factors N_q and SPT N values. Two curves are shown, one for closed ended piles with penetration depths of at least 10 times the diameter, and a second for open-ended piles with a minimum penetration of 25 times the diameter.

To develop the end bearing values piles should be driven at least 4 to 6 pile diameters into the bearing stratum. Where weaker soils underly the bearing strata, a minimum of 3 pile diameters should be present below the pile tip to avoid punching through.

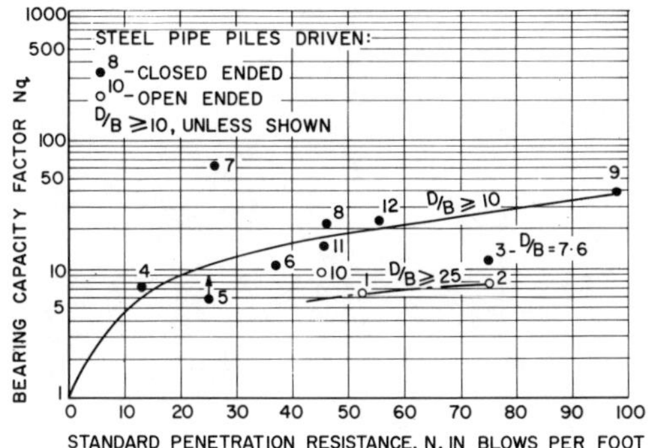

Fig. 4 Relationship between Bearing Capacity factors and SPT N Values

For the open-ended piles the support of the point has been taken over the full end area, that is by the soil plug. It may, therefore, be concluded that at a depth of 25 times the diameter the inside friction by the soil plug has not yet developed a resistance equal to the point load of a closed-ended pile.

The unit end bearing values q_p of Table 2 are derived from failure loads, and are to be used as limiting values for piles driven into the carbonate soils and coral with SPT N values as shown in figure 5. Again two curves are shown, one for closed-ended piles and the second for open-ended piles with reduced end-bearing capacities by the soil plugs.

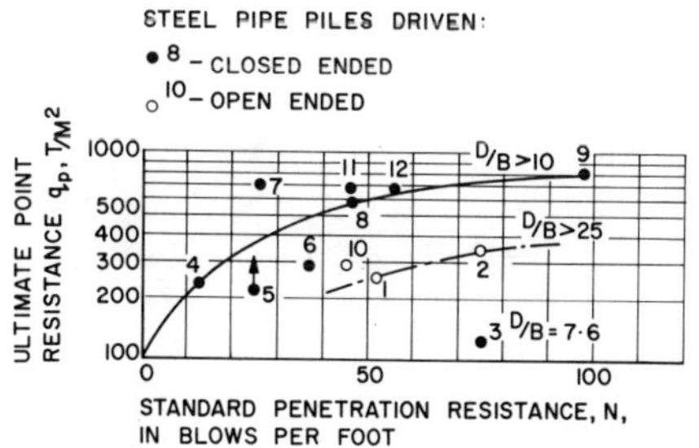

Fig. 5 Relationship between Limiting point resistance and SPT N Values

7 CONCLUSIONS

The standard penetration test (SPT) is being employed in most if not all subsurface investigations executed along the Red Sea coast of Saudi Arabia. A correlation with other types of penetration tests is therefore as yet not available.

Because of the heterogeneous subsoil conditions with the presence of coral formations, the SPT N values were not corrected for a particular overburden pressure.

The static method of analysis based on SPT N values recorded as presented in this paper, should be used to provide a preliminary indication of pile penetration depths to be expected for the ultimate pile capacities required. Pile load tests will still be necessary to verify the design.

Strain gauges or telltales should be placed along the pile shaft and near the pile tip in order to calculate the distribution of the shaft resistance and the total point load.

Carbonate content and degree of cementation are other important factors to be considered as suggested by Agarwal et al (1977).

8 REFERENCES

Agarwal, S.L., A.K. Malhotra and R. Banerjee 1977, Engineering Properties of Calcareous Soils affecting the Design of Deep Penetration Piles for Offshore Structures, 9th annual O.T.C., Houston, Texas.

Angemeer, J., E. Carlson and J.H. Klick 1973, Techniques and Results of Offshore Pile Load Testing in Calcareous Soils, 5th annual O.T.C., Houston, Texas.

Datta, M., S.K. Gulhatti and G.V. Rao 1980, An Appraisal of the Existing Practice of Determining the Axial Load Capacity of Deep Penetration Piles in Calcareous Sands, 12th annual O.T.C., Houston, Texas.

Fellenius, B.H. 1980, The Analysis of Results of Routine Pile Load Tests, Ground Engineering.

Hagenaar, J. and J. van den Berg 1981, Installation of Piles for Marine Structures in the Red Sea, X ICSMF, Stockholm, Sweden.

McClelland, B. 1974, Design of Deep Penetration Piles for Ocean Structures, Journal of the Geotechn. Eng. Div., ASCE, Vol. 100, No. GT7.

Stevenson, C.A. and C.D. Thompson 1978, Driven Pile Foundations in Coral and Coral Sand Formations, 7e Int. Havenkongres, K.V.I.V.

Proceedings of the Second European Symposium on Penetration Testing / Amsterdam / 24-27 May 1982

Drill rod energy as a basis for correlation of SPT data

J.R.HALL
Binary Instruments Inc., Wellesley, Massachusetts, USA

1 INTRODUCTION

Many factors play a role in the errors associated with the SPT (Standard Penetration Test) data. Research by Schmertmann and Palacios (1979), Kovacs, Salamone and Yokel (1981), and Shioi, Uto, Fuyuki and Iwasaki (1981) indicates that the most important factors are related to the hammer mechanism, hammer friction, fall height, impact efficiency, operator technique, rod size and rod length. The combined effect of all of these parameters may be accounted for by measuring the energy associated with the first compression pulse entering the drill rods. The above cited research also shows that for the same soil conditions the blow count, N, varies inversely with the compression energy delivered to the drill rods.

This paper describes an instrument that has recently been developed to automatically compute the energy associated with the first compression pulse from each hammer blow. Test results show that significant variations in compression energy may be expected for different drilling rigs even though they are operated according to standard procedures.

2 IMPACT BETWEEN A RIGID HAMMER AND AN ELASTIC ROD

2.1 Stress and Force Relationships

The impact between a rigid hammer and the end of a long rod of constant cross section produces an initial stress proportional to the impact velocity and impedance of the rod. The various quantities are related by the following formula:

$$\sigma_o = V_h \sqrt{E\rho} \qquad (1)$$

where σ_o = initial impact stress,
V_h = velocity of the hammer at instant of impact,
E = Young's modulus,
and ρ = mass density.

For the case of an infinitely long rod the force-time relationship will be

$$F_i(t) = A\sigma(t) = AV_h\sqrt{E\rho}\,\exp\left[-\frac{A\sqrt{E\rho}}{M_h}t\right] \qquad (2)$$

where $F_i(t)$ = force input to the rod as a function of time, t,
A = cross sectional area,
M_h = mass of hammer,
and exp(..)= base of natural logs raised to the power (..).

The energy associated with the impact force may be computed from the product of force and velocity of motion integrated over time. For one dimensional wave propagation the velocity may be computed from:

$$\frac{dx}{dt} = \frac{\sigma(t)}{\sqrt{E\rho}} \qquad (3)$$

2.2 Compression Pulse Duration-Computed

For a rod of finite length penetrating into soil the incident compression wave will be reflected from the bottom of the sampler as a tension wave. This wave returns to the top of the rod and pulls it away from the hammer. At this point in time the compression pulse is terminated. On this basis the compression pulse duration is given by

$$\Delta t = \frac{2\ell}{c} \qquad (4)$$

where ℓ = distance from point of impact to the bottom of the rods,

and c = compression wave velocity in the drill rods. ($c \doteq 5.1$ m/msec)

2.3 Compression Pulse Duration-Measured

Figure 1 shows recorded measurements by Shioi et al (1981) of the dynamic behavior between the SPT hammer and the top of the drill rod. The heavy solid line represents the motion of the top of the rods, the light solid lines represent computed motion of the top of the rods for various coefficients of reflection at the bottom of the sampler, and the heavy dashed line represents the motion of the hammer. In Fig. 1(c) the light dashed lines represent computed motion of the hammer for the case when a compression wave is reflected from the bottom of the sampler. This corresponds to the case when the sampler penetrates to rock and the reflected compression wave causes the hammer to bounce.

The results in Fig. 1(a) represent a relatively weak soil (N=6) at a depth of approximately 8m. After impact the tension wave separates the hammer from the rod at about 3.5 msec. The rod continues to penetrate into the soil even though the hammer has lost contact with the rod. Finally the hammer is again in contact with the rod but only after the penetration is nearly complete. The second hammer contact appears to contribute very little to rod penetration.

Figure 1(b) shows a condition of stiff soil (N=16) at a depth of approximately 40 m. Again the hammer loses contact with the drill rod because of the reflected tension wave. The second contact occurs at about 60 msec. Again the rod has essentially finished penetrating into the soil.

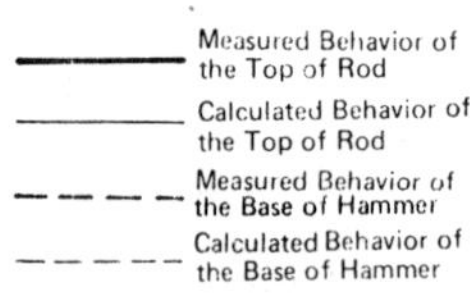

α: Coefficient of Reflection

$T = \frac{2l}{C}$, $C = \sqrt{\frac{E}{\rho}}$

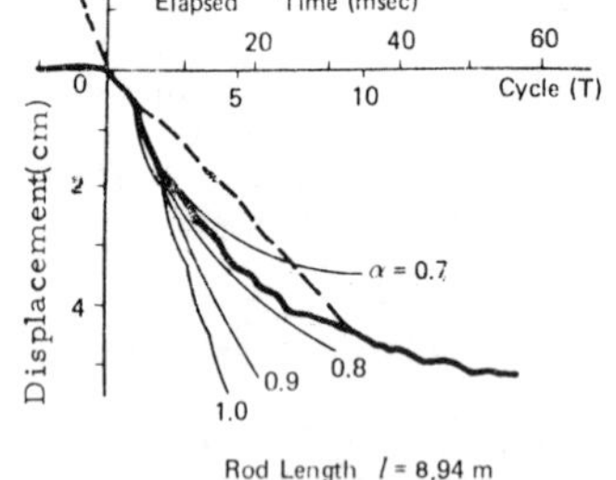

(a) Rod Length l = 8,94 m
Penetration S = 5.3 cm
Loose Silty Sand N = 6

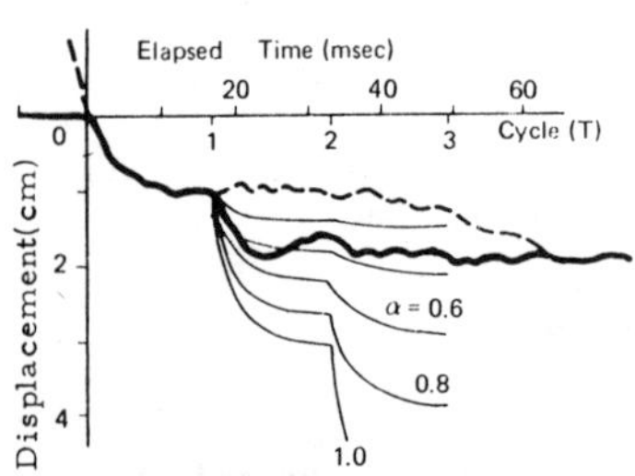

(b) Rod Length l = 41.94 m
Penetration S = 1.8 cm
Medium Sandy Loam N = 16

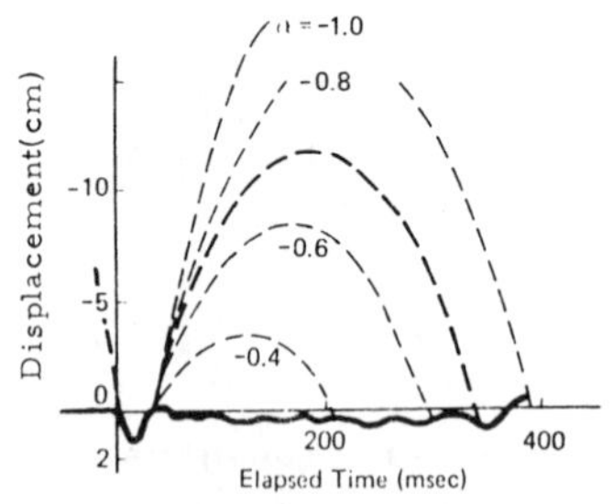

(c) Rod Length l = 48.93 m
Penetration S ≒ 0.3 cm
Very Dense Fine Sand N = 101

Fig. 1 Measured Hammer Impact and Rod Motion. (From Shioi, Uto, Fuyuki and Iwasaki (1981))

2.4 Energy Measurements

The above results indicate that the primary penetration energy transmitted to the drill rod is associated with the first compression pulse with a duration Δt controlled by the length of the rod from the point of impact to the bottom of the sampler. As a consequence, the input compression wave energy is independent of the soil type. Thus the compression wave energy formula may be derived from Eqs. 2, 3, and 4 to give:

$$E_i = \frac{1}{A\sqrt{E\rho}} \int_0^{\Delta t} F_i(t)^2 dt \qquad (5)$$

where E_i = compression energy input to the rod just below the point of impact.

In actual practice it is not possible to measure the force F_i directly at the point of impact. The force is generally measured with a load cell placed a minimum of 10 rod diameters below the point of impact. The first compression pulse energy at this point will be less than if the measurement were made at the point of impact. The difference may be computed with the following formula:

$$\frac{E_i}{E_\Delta} = \frac{1 - \exp\left[-4A\rho\ell/M_h\right]}{1 - \exp\left[-4A\rho\Delta\ell/M_h\right]} \qquad (6)$$

where E_Δ = measured energy at distance $\Delta\ell$ below the point of impact.

For rod lengths greater than 10 m the correction is generally less than three percent.

Another factor influencing the energy in the first compression pulse is the total rod length and its effect on the duration of the pulse. It is sometimes useful to be able to compare the energy delivered by various drill rigs and the infinite rod energy provides a convenient parameter for comparison. It may be computed directly from E_Δ using the following formula:

$$E_\infty = \frac{E_\Delta}{1 - \exp\left[-4A\rho(\ell - \Delta\ell)/M_h\right]} \qquad (7)$$

Calculations using the above formula show that for rod lengths greater than 12 m, E_∞ is approximately equal to E_Δ.

3 SPT ENERGY CALIBRATOR

Figure 2 shows a block diagram of the key elements comprising a system designed to measure the energy of the first compression pulse delivered by each hammer blow. The system consists of a strain gage load cell mounted in the drill rod string just below the point of hammer impact. The load cell is connected by a cable to an instrument that measures the dynamic force and performs the necessary computations for energy. The various elements of the instrument include an instrumentation amplifier, a force level detector, a squaring circuit, an integrator, control logic and a scaling circuit. The instrument reads out directly in per cent of the theoretical energy from a 140 pound hammer falling 30 inches. Internal calibration of the instrument is checked in the field through the use of a built in load cell calibration resistor and a crystal oscillator time base.

4 COMPARISON OF ENERGY DELIVERED BY VARIOUS DRILL RIGS

In a study by Kovacs et al (1981) the above described instrument was used for the measurement of energy delivered by fourteen different drill rigs with various hammers and operators. The values of E_∞ varied from 32 to 76 per cent. This indicates that the testing standards that presently exist are not sufficient to guarantee uniformity in the test results. The above results would represent a variation of more than two to one in the reported N-values for the same soil condition. The measurement of energy would eliminate the source of error associated with different drill rigs, hammers and operators by providing a means for correlation of the results.

5 CONCLUSIONS

The above described measurements of energy delivered by the first compression pulse from the SPT hammer indicate that improvements in SPT blow count correlation could be achieved by including energy measurements in the test results. This

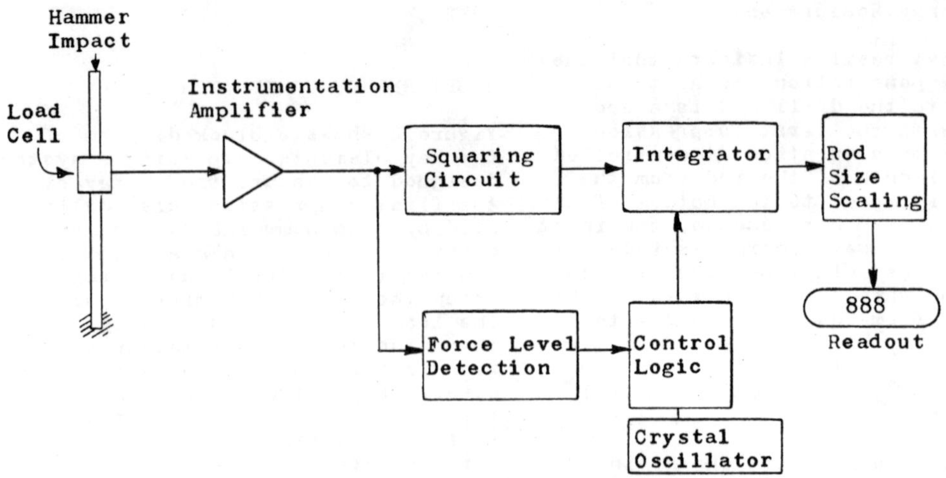

Fig. 2 Schematic showing key elements of an SPT Energy Calibration instrument.

additional parameter would remove the principal sources of error which are associated with variations in drill rigs, hammers, and operator techniques.

6 REFERENCES

Kovacs, W.D., Salomone, L.A. and Yokel, F.Y. 1981, Energy Measurment in the Standard Penetration Test, National Bureau of Standards Building Science Series 135.

Schmertmann, J.H. and Palacios, A. 1979, Energy Dynamics of SPT, ASCE Journal of the Geotechnical Engineering Division, August.

Shioi, Y., Uto, K., Fuyuki, M. and Iwasaki, T. 1981, Part III. Standard Penetration Test, Report by the Japanese Site Investigation Subcommittee on Present State and Future Trend of Penetration Testing in Japan, Prepared for Xth International Conference on Soil Mechanics, Stockholm, June.

Proceedings of the Second European Symposium on Penetration Testing / Amsterdam / 24-27 May 1982

Experience on a standard penetration test

HUANG SHI-MING
Design Institute of the Ministry of Textile Industries, Beijing, China

The standard penetration test can perform the same function as exploration as well as survey and test. The survey and test measure in situ is always to co-ordinate misering to take sample directly for evaluation of mechanics properties of stratum so that the exploration quality can be improved, exploration speed can be increased and the cost be reduced. In the early days of the 50's, the S.P.T. was employed to determine density of sand, at present time it has been widely used, has become one of the main measures adopted for power-sounding and has been brought into "Foundation design code for industrial and civil building (TJ-7-74)", "Engineering geologycal exploration code for industrial and civil building (TJ-21-77)", "Anti-earthquake design code for industrial and civil building (TJ-11-78)" etc., and also has been taken as the basis of the foundation design for medium and small projects.

During the more than recent ten years, many departments in China have been engaged in studying on this subject, they have not only achieved good experience on application of S.P.T. for the project, but also obtained satisfactory results of S.P.T. in theoretical research and automation of equipment.

1. Brief introduction to the equipment for S.P.T.:

   a) Height of hammer drop 76 cm
   b) 42mm outside dia. boring rod for the sounding rod
   c) Heavy hammer 63.5 kg

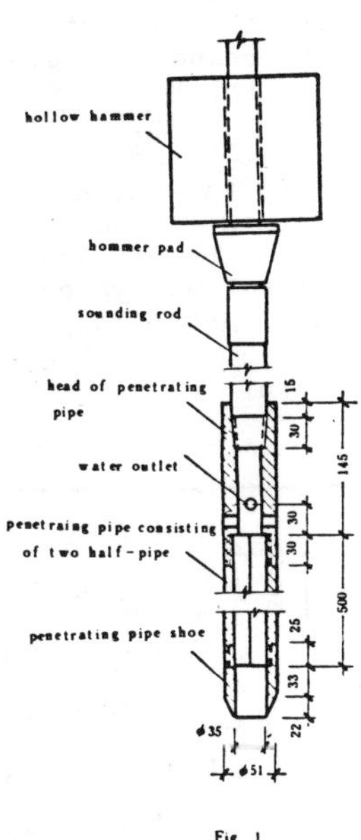

Fig 1

2. Penetrating mechanism of S.P.T.

It is generally considered that penetrating mechanism of penetrating pipe in stratum is similar to that of piling, the penetrating mainly emerges shearing and compacting physical mechanics process and a stress zone has been formed around the penetrating pipe, within the zone, soil or sand layer produces local shearing, so the resistance of stratum to the penetrating pipe is a comprehensive expression of shear strength of the stratum. The penetrating pipe is not continuously driving into the soil and inside bottom dia. of bore hole is greater than that of the penetrating pipe, so the condition around the pipe is likely to cause shearing damage.

3. The comparison between different operating methods

It is known that the operation method for S.P.T. is not yet standardized. The result that has been obtained with the same method in the same place is quite divergent, that is relative to driving style and the method for the controlling of hammer drop. The comparison test has been conducted by means of six operation methods in the area with uniform rock property and uniform structure, the result of the test has presented variation of soil property, but great difference to number of blows (refer to list 1).
When the casing is driven down, sand sometimes pours out, that means the casing strongly disturbs the natural state of the sand and this fact makes the penetrating index lower. If the hammer dropping is controlled by mechanical hoisting, it will make the penetrating index higher by 50-60% owing to the fact that inertial and friction resistance of the steel rope has consumed the energy for hammer dropping. From the curve 3 and 4 as well as 2 and 6 in Fig. 2 we can see the number of blows for casing knock-boring is smaller than that of slurry driving by 0.5 - 2 times under the condition of same controlling method of hammer dropping.

Therefore, there should be a regulation for the testing operating method of number of blows of S.P.T. if using above-mentioned operating method, that is, slurry driving and automatic-release shall be adopted for the controlling of the hammer dropping in S.P.T. For important projects, number of boring holes shall not be less than 5 at least and number of testing not less than 15 times for each layer of the soil.
Figure 2a shows S.P.T. which has been made respectively by three different departments with different operating methods, each line presents the average value of the testing for three bore holes, and the results are very divergent also.

List 1

soil situation			mud-knock-borring hammer drop to be controlledmechanically			casing-knock-borring hammer drop to be controlled manually			mud-rotary-borring hammer drop to be controlledmechanically			casing-knock-borring hammer drop to be controlledmechanically			mud-knock-borring hammer is freely falling			mud-rotary-borring hammer drop to be controlled manually			total average
depth (m)	thickness (m)	classification	max – min.	NO. of holes	average	max – min.	NO. of holes	average	max – min.	NO. of holes	average	max – min.	NO. of holes	average	max – min.	NO. of holes	average	max – min.	NO. of holes	average	
0~2.8	2.8	Sandy clay	10	1	10				8	1	8	10	1	10	3	1	3	8~6	3	7	7.7
2.81~3.83	1.05	clayey sand	10-8	4	9	14-3	3	7.5	14-6	5	9	8-5	3	6.5	7-4	4	6	7-3	3	5	7.1
3.86~14.60	10.75	silt	13-6	6	9	11-5	5	7.4	23-9	4	14.5	15-7	4	10.5	7-4	4	7.5	6	1	6	9.2
			53-25	6	34	20-15	5	16.5	49-38	5	44.5	25-14	4	19	21-14	4	17	31-24	3	28	26.5
14.61~26.60	12.0	fine sand	50-20	6	35	23-9	5	16.4	42-20	4	34	25-14	4	17.9	30-11	4	18	31-25	3	27	24.7
			94-57	6	79	82-35	6	61.8	100-47	5	67	37-21	3	27.8	65-36	4	44	86-61	3	76	59.3

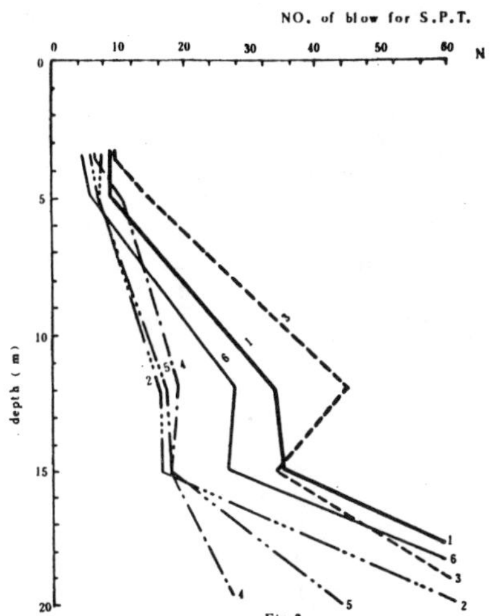

Fig 2

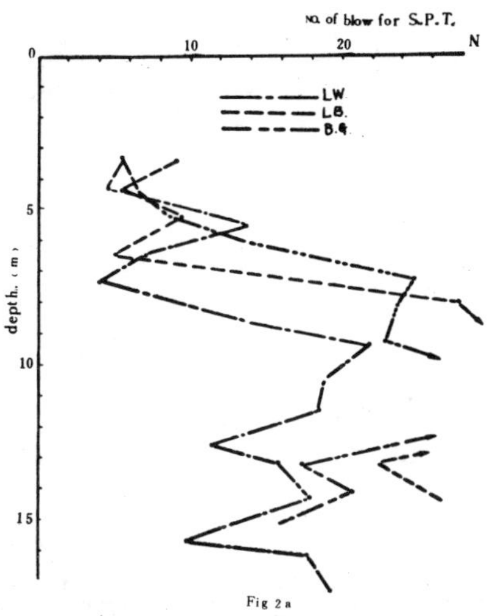

Fig 2a

There are more factors which give the effect to the result of S.P.T., the hammer dropping is the key point. The standard hammer used to be lifted through the rope by hand, by doing so, the labour intensity is not only great and the hammer dropping cannot be controlled accurately. Now we have employed mechanical power to lift the hammer, adopted automatic hammer dropping device to control the height of the hammer dropping and have achieved good result. Now we present several kinds of automatic hammer dropping devices which are used under normal condition in China (Fig. 3).

With comparison between the operating methods for automatic hammer dropping and manual hammer dropping. Blowing energy for automatic hammer dropping is greater than that of manual controlled hammer dropping, the relation between these two:

Nmanually=0.738+1.12 Nmechanically
correlation factor = 0.985
inspecting value 3.6 for regression factor under normal condition shall be calculated in accordance with following formula:

$$N_{mechanically} = N_{manually} \times (0.8 - 0.85)$$

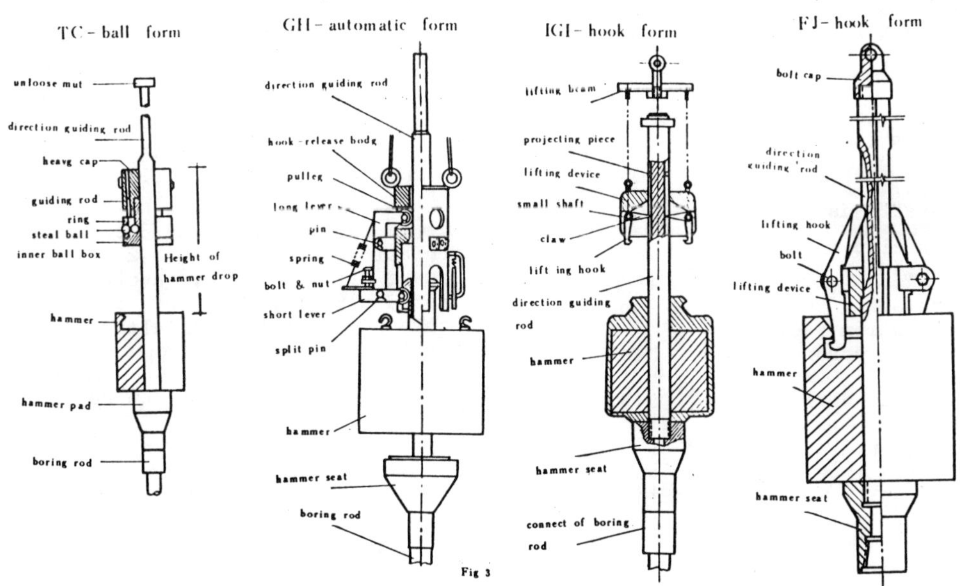

Fig 3

4. Application of S.P.T.
S.P.T. has been used in wide range of field in our country, we have already employed S.P.T. to solve following problems:

(1) To determine soil bearing capacity of foundation

Allowable soil bearing capacity (R) of sand, old cohesive soil (Q_{2-3}, $N > 15$) and common cohesive soil (Q_4, $N < 15$) is determined in line with number of blows N of S.P.T. and has already been included in "the Design code of the foundation for industrial and civil building" (TJ-7-74) (refer to list No. 2, 3).

allowable soil bearing capacity of sand [R] List 2

NO. of blow for S.P.T. N	10~15	15~30	30~50
allowable bearing copacity [R] (T/m²)	14~18	18~34	34~50

Allowable soil bearing capacity of old cohesive soil & common cohesive soil List 3

NO. of blow for S.P.T. N	3	5	7	9	11	13	15	17	19	21	23
allowable bearing capacity [R] (T/m²)	12	16	20	24	28	32	36	42	50	58	66

These data are obtained from the calculation according to the relation between proportional limit pressure P_o of 90 loading tests and number of blows N of 90 standard penetrating tests in nine typical districts through mutual co-operation:

common cohesive soil:

$P_o = 0.319 + 0.227\,N$

$(N < 18)$

correlation factor = 0.89

correlation factor error = $\pm$ 0.03

quadratic deviation = $\pm$ 0.368

old cohesive soil:

$P_o = 0.558\,N - 5.583$

correlation factor = 0.956

correlation factor error = $\pm$ 0.019

quadratic deviation = $\pm$ 0.37

above-mentioned data should be used in line with local construction practice, generally, it is economic and reasonable.

(2) To determine relative density of sand

The value N cannot only reveal the relative density of the sand, but also the value N is effected by self-gravity of the sand above testing depth, that is, number of blows is different for the test which is made in different depths because of different self-gravity around testing point in spite of same relative density of the sand.

The Railway Scientific Research Institute has made a test which is concerning the mutual relation between value N and relative density, the result of the test (Fig. 4) shows that this relation will be changed with the variation of effective cover pressure.

By making comparison on spot and statistics for 133 groups of tests, regression equation is obtained:

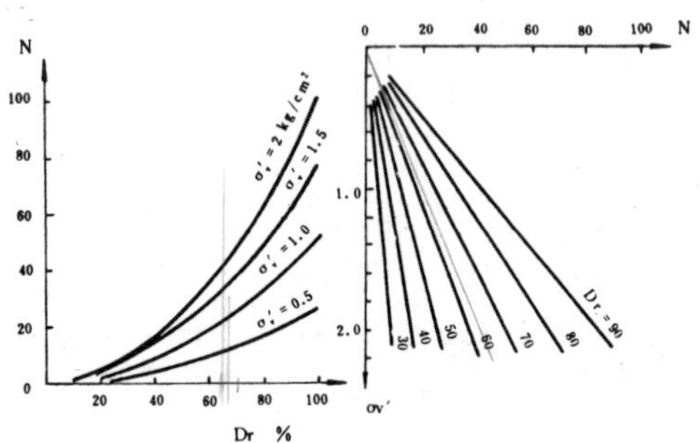

Fig 4

$$N = 5.22\,\sigma_v' \times Dr^2$$

in formula

σ_v' -- effective cover pressure, kg/cm^2

Dr -- relative density , %

N -- number of blows of S.P.T., number of blows/30cm

correlation factor = 0.892

standard error σ = 10.8 number of blows/30cm what is measured actually is coincide with the calculation, generally smaller than 1σ.

(3) Classification of the soils state :

List 4

State	liquid plasties	soft piastics	plastics	hard plastics	hard
specific penetrating resistance P_s	<4	4 - 10	10 - 30	30 - 50	>50
Liguidity index I_L	>1	1 - 0.75	0.75 - 0.50	0.5 - 0	<0
NO. of blow N machanically	<2	2 - 4	4 - 11	11 - 17	>17
unconfined compressive strength, qu kg/cm²	0.3	0.3-0.7	0.7-1.6		
cohesive force, c	0.14	0.14 - 0.35	0.35 - 1.20		

(4) Differentiation of foundation liquefaction

Number of blows for S.P.T. is one method for differentiation of the foundation liquefaction. Critical value N is stipulated in China's code (TJ-11-78) for differentiation whether saturated sand layer is liquefied in case of earthquake.

$$N' = \overline{N}'\left[1+0.125\ (ds-3)-0.05\ (dw-2)\right]$$

in formula

N'---- critical number of blows of S.P.T. for the sand liquefaction in case the depth where saturated sand being located is ds M and distance between ground surface to underground water level is dw M

$\overline{N}'$---- critical number of blows of the sand in case ds = 3M, dw = 2M

that is, N' = 6 in case of seismic intensity 7°

N' = 10 in case of seismic intensity 8°

N' = 16 in case of seismic intensity 9°

ds---- depth (M) where the saturated sand being located

dw---- the distance (M) from ground surface to the underground water level

The above-mentioned formula is stipulated in line with information concerning earthquake for 8 times and exploration analysis in China, also this formula is done in consideration of factors such as seismic intensity, density of sand as well as effective cover pressure. The experience demonstrates that this formula can be used for differentiation of the sand liquefaction within 15 M depth.

Besides, the penetrating resistance is different for the sand which has been obtained before and after the earthquake in Tangshan area. The average value of number of blows for the sand in the depth of 1-8M after earthquake is 2-8 less than that before earthquake, but on the contrary, the average value of number of blows for the sand in the depth below 10M after the earthquake, 11-23 more than that before the earthquake. This fact demonstrates that the saturated sand layer from ground surface to the depth of 8M after liquefaction in earthquake is more loose, but the sand below 10M is to be densified.

At present, some departments in China are now making careful study of further application and improving of technology of S.P.T., such as:

- -- Problem on number of blows effecting foundation in loose sand layer;
- -- Problem on softening of uneven distributed weathered rock stratification (eluvium) under water, inspected by means of S.P.T.;
- -- Problem on automatic equipment needed for S.P.T. and combination of automatic equipment and knock-boring misering, also problem on revision of length of the boring rod;
- -- Experience on application of special foundation soil and regional soil.

List 5

time \ $\overline{N}$ \ depth (M)	1.00~2.00	2.00~3.00	3.00~4.00	4.00~5.00	5.00~6.00	6.00~7.00	7.00~8.00	10.0~11.0	11.0~12.0	12.0~13.0	13.0~14.0	14.0~15.0	NO.of points
before earthquake	9.7	14.3	17.1	22.1	18.5	22.4	27.1	36.0	24.3	19.4	35.0	40.8	118
after earthquake	6.5	11.5	15.1	14.1	16.8	18.1	20.7	47.0	26.7	17.0	58.0	55.6	109
difference between NO. of blow	-3.2	-2.8	-2.0	-8.0	-1.7	-4.3	-6.4	+11.0	+2.4	-2.4	+23.0	+14.8	

We believe that the above-mentioned research work will make S.P.T. play a more important role if only developing advantage, overcoming disadvantage, and making testing methods are standardized.

REFERENCES

1. S.P.T. with respect to physical and mechanical properties of soils, Special Report, 1973.
2. Compilation of saturated sand liquefaction reference material, Special Report, 1977.

Proceedings of the Second European Symposium on Penetration Testing / Amsterdam / 24-27 May 1982

Correlation of N value with S-wave velocity and shear modulus

TSUNEO IMAI & KEIJI TONOUCHI
OYO Corp., Urawa Research Institute, Japan

1 INTRODUCTION

The velocity of S waves in the ground has come to be acknowledged as the most important geophysical property to be dealt with in seismic microzoning, earthquake engineering — which deals with the dynamic interrelationship between the ground and structures — as well as in the structural engineering fields of civil engineering and architecture. In 1967 the authors began research on the practical application of in situ S wave velocity measurement. There began the process of accumulating measurement data and checking out the reliability of measurement methods in a variety of grounds throughout Japan, with emphasis on the developed urban regions with their alluvial and diluvial deposits, but also including many other types of sites. As a part of this research, the "PS logging method", a kind of well shooting, was developed and applied for OYO's Borehole Pick, an inhole receiver that can be fixed to the borehole wall and detached at will at any depth. Measurement data from 400 boreholes in areas throughout Japan was accumulated.

In Japan, whenever site investigations are conducted, the standard penetration test is always carried out. The geotechnical index that is yielded by this test, N-value, is fundamental to evaluation of the dynamic properties of the ground and is indispensable in structural design as well.

In this paper, certain conclusions drawn on the basis of the large volume of data accumulated by the authors concerning the relationships between N value and S wave velocity and between N value and rigidity are given.

2 MEASUREMENT OF S WAVE VELOCITY

Heretofore, the refractive method was often used to measure velocity of elastic waves in the ground. However, this method requires that a large number of geophones be installed in the ground in a line to obtain arrival (travel) time of the wave. Using this data, 2 or 3 assumptions are made and velocity distribution analyzed. As such, its application is subject to the following limitations:

(1) Velocity values progressively increase from the surface downwards. Consequently, if there is tight, high velocity layer near the surface with a soft layer beneath, the lower layer will be undetectable.

(2) In multilayered ground, analysis becomes complicated. A large number of options are involved in the analysis, and as a result, error is great.

(3) The method requires that a measuring line along the surface 5 or 6 times the depth to which measurement is desired be set up. Often this is a practical impossibility in urban areas.

(4) The long measuring line required has an inherent disadvantage resulting from the limit to the energy of the S wave vibration source. The wave attenuates with distance, so that after a certain point it becomes impossible to ascertain phases, and therefore velocities of waves in deeper layers.

Unlike the refraction method, the logging method does not have these limitations. It can be used in any ground, and does not require an extended operations area. Velocity distribution of P and S waves is obtained by directly measuring arrival time values along the depth of the borehole. This method may be considered more effective than the refractive method. Figure 1 shows a variation on the PS logging method developed by the authors.

The inhole receiver consists of 1 vertical and 2 horizontal transducers, with a natural frequency of 28 Hz. If necessary, as optional equipment, a hydrophone to find changes in borehole water pressure or a compass to indicate the direction the horizontal transducers face as the probe rotates may be attached. One side of the probe is covered with a rubber tube. The rubber tube is inflated with water pumped from the surface. The probe may be in this way be fastened against or released from the borehole wall at any desired depth. The outer diameter of the probe is 48 mm, and is designed for use in boreholes of from 65 to 75 mm diameter.

Measurement is carried out at intervals of 1 or 2 meters. As shown in Figure 1. the plank hammering, weight dropping or blasting methods of wave generation may be used. The authors' experience has shown that for S wave velocity measurement, plank hammering is generally usable for depths of around 100 m, and has been used at depths of up to 350 m in quite areas. The uphole method, which uses blasting as the vibration source (Figure 1, right) may be employed for measurement at greater depths or sea bottom. Figure 1 shows examples of waveforms recorded by these methods.

3 VELOCITY OF P AND S WAVES IN THE GROUND

Using the above described method, measurement at approximately 250 sites, using 386 boreholes (total length, 16,610 m, with an average borehole length of 43 m) have been carried out. The sites are distributed throughout Japan. The data obtained can be considered to be representative of the ground in Japan's urban areas.

Figure 2 shows measurement data taken in alluvial, diluvial and some terriary ground, arranged according to type of wave measured (P or S), geological age and type of soil. This figure shows frequency of appearance of velocity values for each classification. The results show that there is a range of elastic wave velocity values, varying according to the type of geology or soil, and that this tendency is particularly marked for the velocity of P waves (Vp). Furthermore, there are no clear differences distinguishable for Vp between different soil types. In contrast to this, there is a tendency for S wave velocity values (Vs) to gather around a fixed value for each classification. Thus, it is to a certain extent possible to designate a range of values for different soil types for Vs, but this does not seem possible for Vp. To put it another way, there appears to be a correlation between Vs and soil type, but this is not the case for Vp. Rather, for Vp, other factors appear to be decisive.

This tendency is clearer when the distribution of Vp values are compared with those of Vs in each hole. There are many cases when a single layer as indicated by uniform Vp values amounts to several layers if one goes by Vs values, while no particular correspondence between Vp and Vs layers is discernable. Vp values show a clear distinction above and below the underground

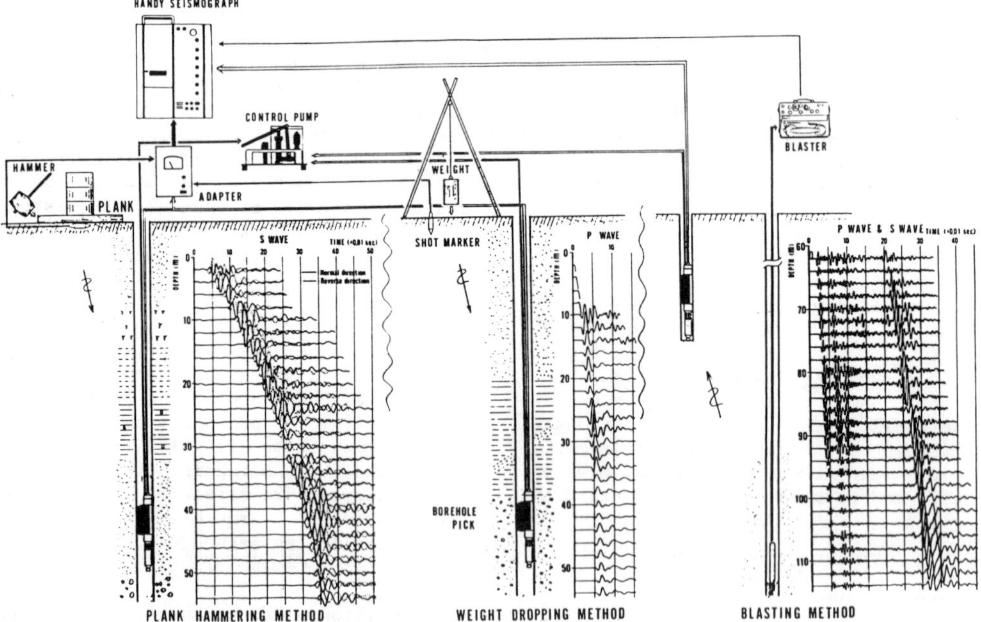

Fig. 1 Various methods of PS logging

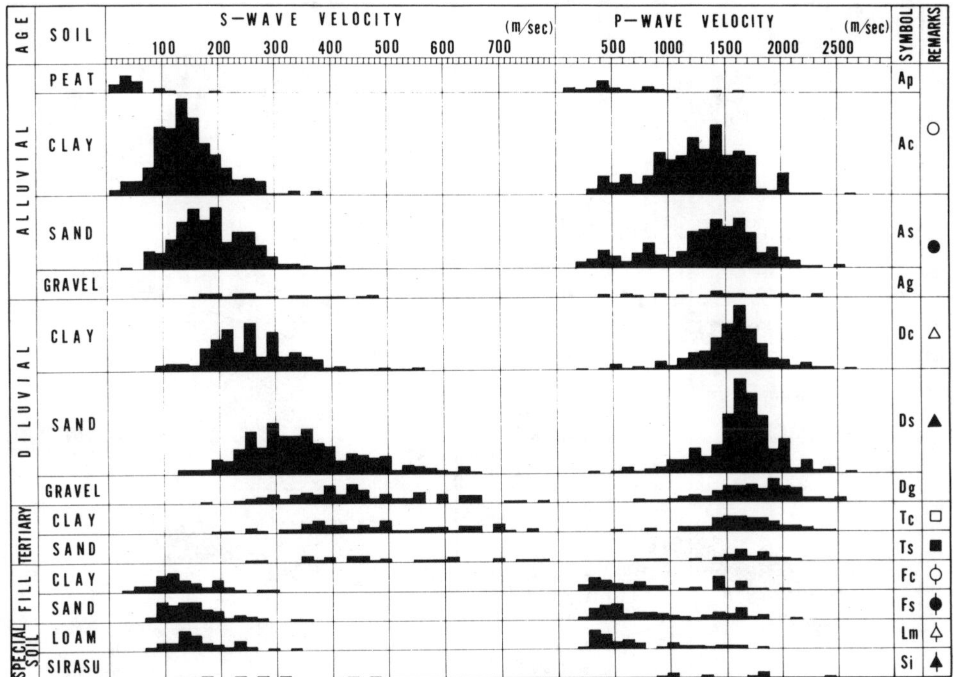

Fig. 2 Distributions of P- and S-wave velocities

water surface, with values below the waterline clearly within a uniform 1500 m/sec. In the case of Vs, there is a fairly good correspondence with soil mechanical data. This agrees with the explanation that Vp is determined by constituent grain and bulk modulus of pore water and that Vs is determined by structural elasticity under the influence of the size, form and firmness of particles.

From the above we may conclude that in using elastic waves to obtain dynamic information about soft ground, S waves, which directly tie in with dynamic properties of the ground are more important than P waves, which present a number of problems.

4 N VALUE AND S WAVE VELOCITY

Now, let us proceed to the relationship of S wave velocity values to N value as an index providing information on the dynamic properties of the ground. In the field of soil mechanics, dynamic ground properties may be determined by a variety of in situ and laboratory tests, which yield various index values that are recognized as being related to strength and deformation properties. Of these index values, N value, provided by the standard penetration test is commonly used, and is regarded as indispensable to structural design. Although the significance of N value from a theoretical point of view is unclear, as an empirical index of dynamic properties of the ground under uniform conditions, it is a highly practical tool.

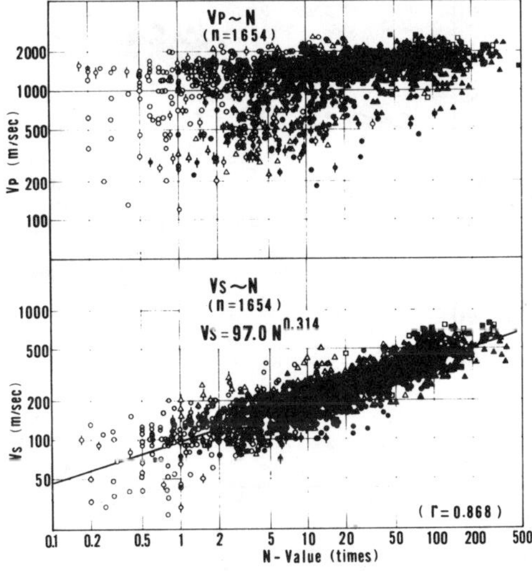

Fig. 3 N value and velocities (total data)

Figure 3 shows the relationship between N value and S wave velocity. At the top of this figure, a reference to P wave velocity is provided. The N values shown here are

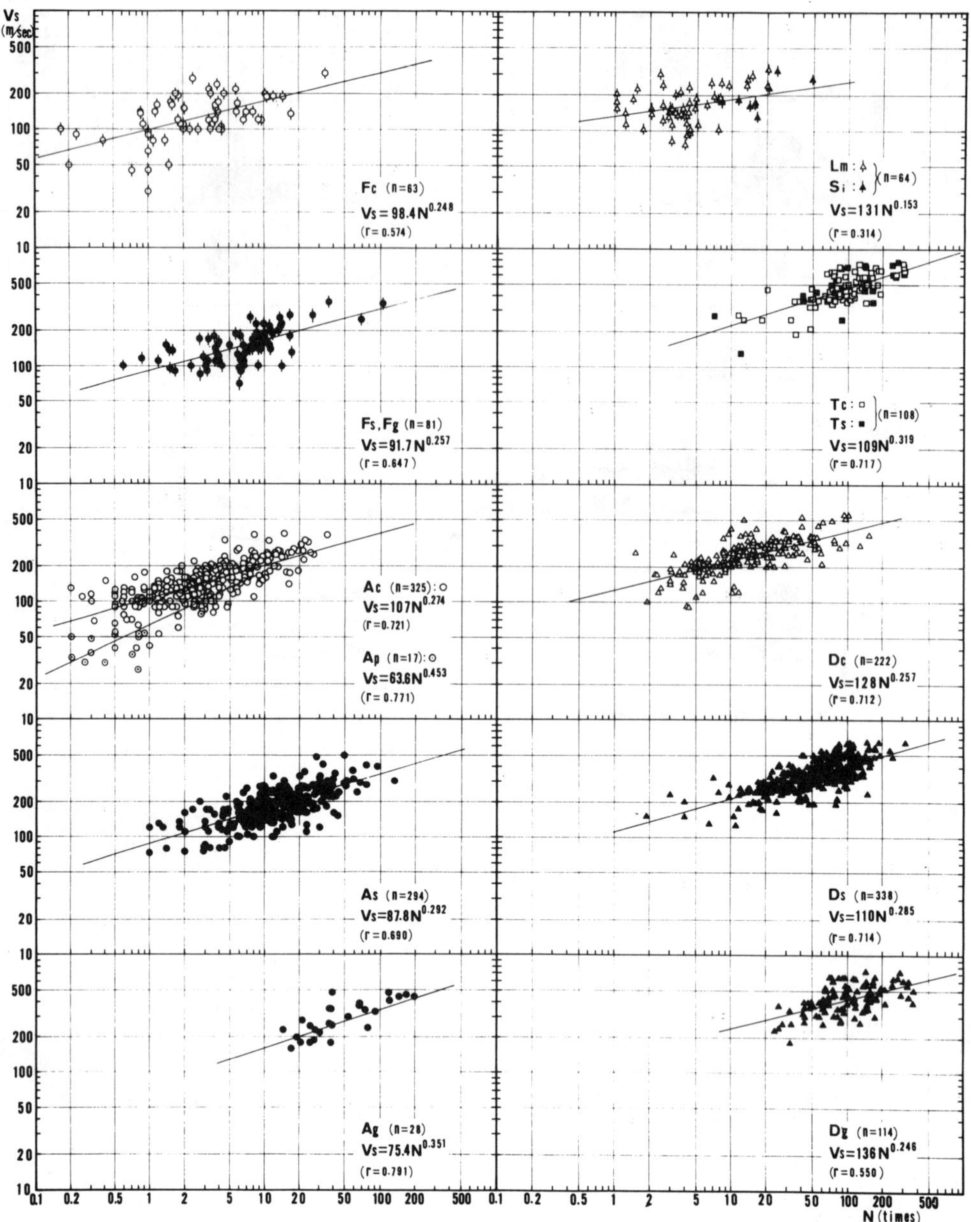

Fig. 4 N value and S wave velocity (soil type)

average values from a single velocity layer. N value above 50 are substitute for the number of blows required to achieve a penetration dapth of 30 cm from the actual amount of penetration achieved at 50 blows. The same was done for N values below 1. From the figure it can be seen that there is a direct relationship with S wave velocity and N value regardless of whether the ground is alluvial, diluvial or tertiary. However, there is no such relationship with P wave velocity apparent. This is in agreement with our previous findings.

In order to examine in more detail the relationship between Vs and N, data was arranged separately for geology and soil type

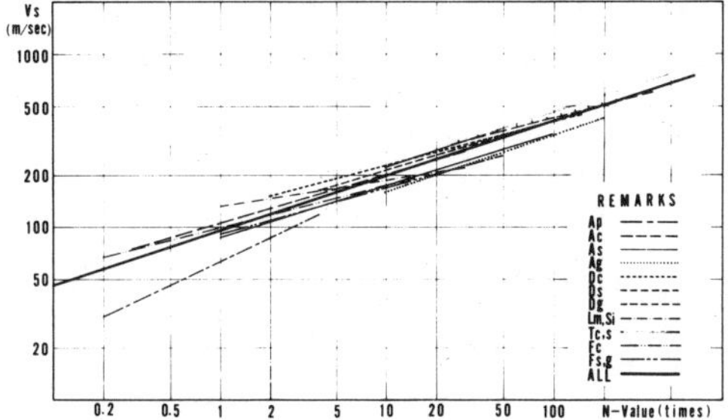

Fig. 5 Summarized relation of N to Vs

(Figure 4). We see from the figure that there are somewhat different tendencies according to geology. For the same N value, S wave velocity is greatest in Tertiary layers, less in diluvial layers and least in alluvial layers. When we make a distinction between clayey and sandy ground, we again find different trends; for the same N value, S wave velocities are higher in clayey than in sandy soil. Figure 5 summarizes the distribution ranges for each classification.

Disregarding the above distinctions between categories, the curve in the graph of Figure 3 represents the following experimental formula, derived from all 1654 bits of data available:

$$Vs = 97.0\ N^{0.314} \qquad (1)$$

5 N VALUE AND SHEARE MODULUS

Knowing S wave velocities and density (ρ) of the ground, the following formula from elastic wave theory allows us to determine shear modulus (Gd):

$$Gd = \frac{1}{g} \cdot \rho \cdot Vs^2 \qquad (2)$$

where, g; acceleration of gravity

Using in Formula (3) density values obtained either by density logging or by laboratory testing of samples and S wave velocity values mentioned above, Gd was determined. Figures 6 and 7 show the N--Gd relationship arranged according to the same classifications as for S wave velocity. Figure 6 shows the overall relationship, while Figure 7 arranges the data according to geology and soil type. Comparing these N--Gd relationships to the N--Vs relationship, there is a bit more of a tendency to scatter with the former. However, as Formula (3) also clearly shows, from the fact that Gd is proportional to the square of Vs, this is to be expected.

As we did with the relationship N--Vs, if we disregard differences for various categories, the formula for the data as a whole is:

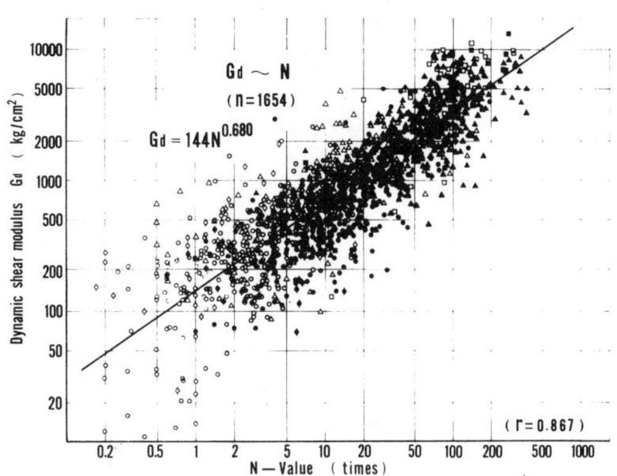

Fig. 6 N value and shear modulus (total data)

$$Gd = 144\ N^{0.680} \qquad (3)$$

6 AFTERWORD

Site investigation methods that have been used in urban areas where the ground is soft have almost exclusively relied on techniques from soil mechanics, rarely adopting physical investigation methods. However, in recent years there has been a rapid increase in the development of physical investigation methods in civil engineering and architectural engineering as well as the extension of the applications of these techniques from rock ground to softer type ground. In particular, as the field of earthquake engineering has developed, the important of elastic wave velocity, including S wave velocity has come to be recognized. This type of velocity information is closely related to the dynamic and mechanical properties of the ground. Already, elastic wave velocity data has become indispensable to the evaluation of the dynamic properties of the ground. It goes without saying that many problems with the analysis technique in this paper remain, and a great deal of work in the accumulation of data and theoretical and experimental analysis remain to be done.

In any case, for the purpose of attaining an overall grasp of the mechanical characteristics of the ground ranging from static to dynamic response, and to meet the requirements of structural design, in addition to the various testing methods adopted in the past, the type of data described in this paper on elastic wave velocity is of great help. The application of this kind of data may be considered to constitute an important contribution to the study of the mechanics of soils.

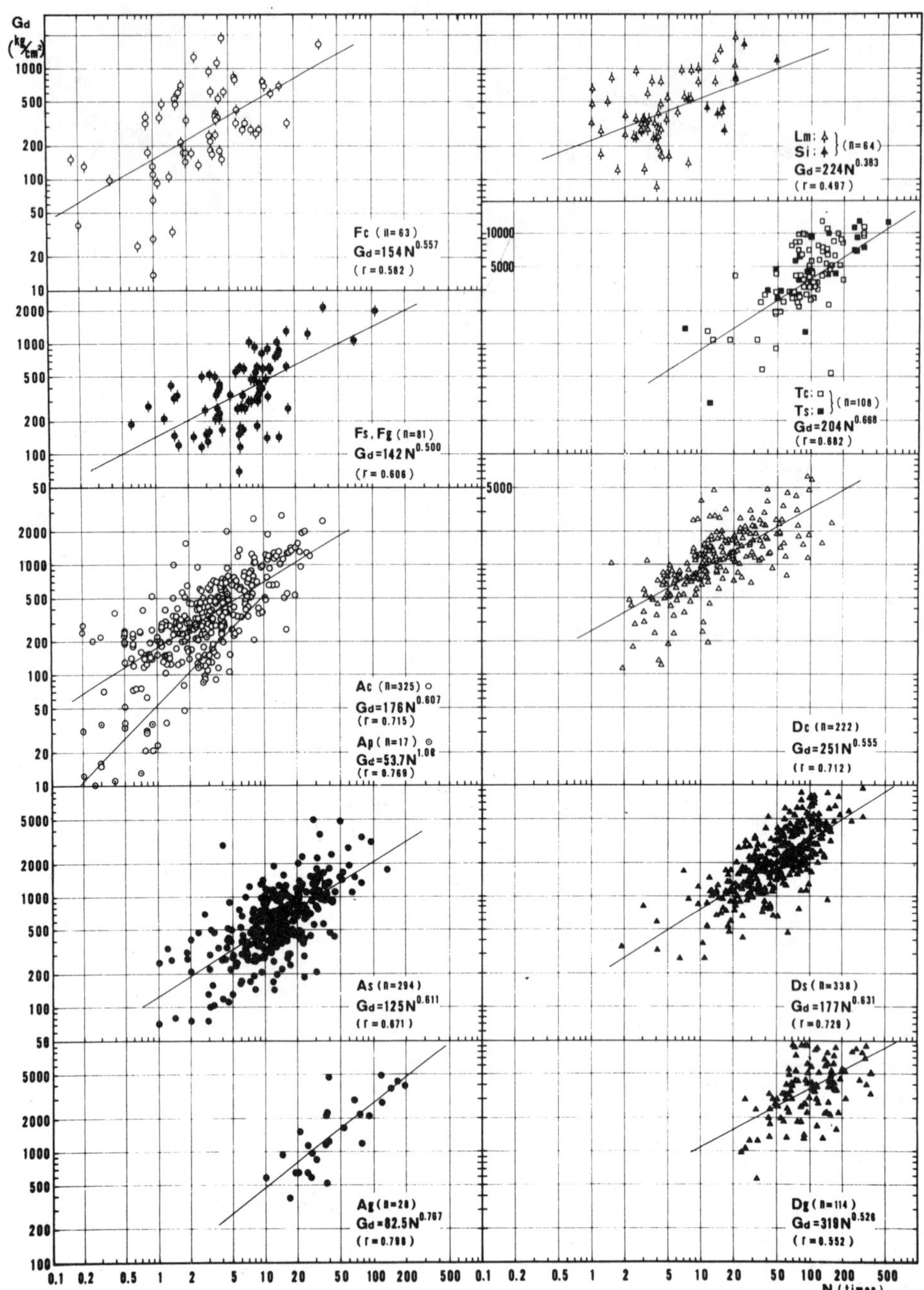

Fig. 7 N value and shear modulus (soil type)

7 REFERENCES

Imai, T. and M. Yoshimura 1972, Elastic wave velocity and physical properties of ground (in Japanese), BUTSURI-TANKO (Geophysical Exploration), vol. 25, No. 6.

Imai, T., H. Fumoto and K. Yokota 1975, The relation of mechanical properties of soils to P- and S-wave velocities in Japan (in Japanese) Proc. of 4th JEES.

Imai, T. 1977, P- and S-wave velocities of the ground in Japan, Proc. of IXth ICSMFE.

Imai, T., H. Fumoto and K. Tonouchi 1978, The relation of S-wave velocity and soil characteristics alluvial sand (in Japanese), Proc. of Vth JEES.

Proceedings of the Second European Symposium on Penetration Testing / Amsterdam / 24-27 May 1982

Relationships between N value and dynamic soil properties

TSUNEO IMAI & KOICHIRO YOKOTA
OYO Corp., Urawa Research Institute, Japan

1 INTRODUCTION

It is important for predicting ground behavior during earthquake to know what we call dynamic soil properties such as cyclic undrained shear strength and dynamic deformation characteristics. In order to grasp these dynamic soil properties a variety of direct or indirect measurement means have been so far utilized. Leading means to be enumerated among them for in-situ test are standard penetration, ram sounding, pressuremeter, seismic exploration and velocity log; ultrasonic velocity measurement, impact test, resonant column test and cyclic loading test for laboratory test.

Standard penetration test has features that system construction is simple and that sample can be recovered. Although it does not intend to directly seek for dynamic soil properties, it has achieved tremendous spread in U.S.A. and Japan for the purpose of grasping a rough idea about mechanical properties of ground and preparing geolobical section map. Blow count N value in standard penetration test is, needless to mention, considered to be measurement value obtained from result of a kind of dynamic penetration test and therefore presumed to have strong relation with dynamic soil properties measured by other means.

This paper is to report some relation the authors studied between N value and 3 dynamic soil properties--cyclic undrained shear strength obtained from cyclic triaxial test, shear modulus G obtained from in-situ PS log (Imai, 1977), resonant column test, cyclic triaxial test, cyclic torsional test, etc. and damping ratio h obtained from cyclic torsional test. In short, this paper is to discuss how aforementioned three dynamicparameters are expressed according to soil kind and N value.

2 TESTING METHODS AND PROCEDURES

2.1 Cyclic triaxial test

The apparatus used in the test was axial direction exciting, hydraulic servo type system to which sinusoidal and optional input waves can be loaded. The system allows frequency from 0.1 through 5Hz for cyclic loading. Fig. 1 schematically illustrates the cell used in the test.

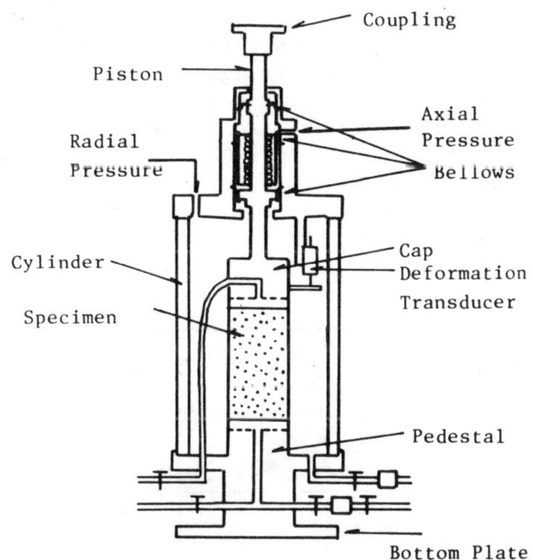

Fig. 1 Schematic view of the cell

The contact portion of the piston and the cell consists of three pieces bellows form seals. Axial pressure and lateral pressure can be independently controlled. Load cell is attached above and outside the cell. With this testing apparatus cyclic undrained shear strength of sandy soil, soil shear modulus G, soil damping ratio h are measured. Test specimen is 100mm high with diameter 50mm recovered all without disturbance in sampling. Cohesive sample was recovered by Thinwall sampler and sandy sample by sand sampler (Ohya, Ogura, Yokota, 1980). Immediately after the recovery sandy sample is frozen at site and transported to laboratory for storage in freezer.

The test sample prepared in given dimensions is set on the triaxial cell, circulated with water for saturation and loaded with back pressure. Then, it is isotropically consolidated under effective confining pressure equal to the effective overburden pressure. Yet, in case of frozen sandy soil the sample is once thawed under approx. 0.2kg/cm^2 vacuum, saturated and consolidated. After completion of the consolidation the test specimen is given liquefaction test or dynamic deformation characteristics test to obtain G and h under various strain levels.

The liquefaction test is to obtain dynamic strength ratio by knowing change in number of cycles required for liquefaction after changing stress ratio (cyclic stress divided by two-fold effective confining pressure) for three or four test specimens. Fig. 2 indicates examples of relation between stress ratio and number of cycles required for liquefaction, together with definition of dynamic strength ratio. In this paper dynamic strength ratio is defined to be such stress ratio that its double amplitude of axial strain may reach 5% after 20 load cycles as shown in the figure. (expressed as $R_{\ell 20}$). Frequently of cyclic loading in this test is 0.5Hz.

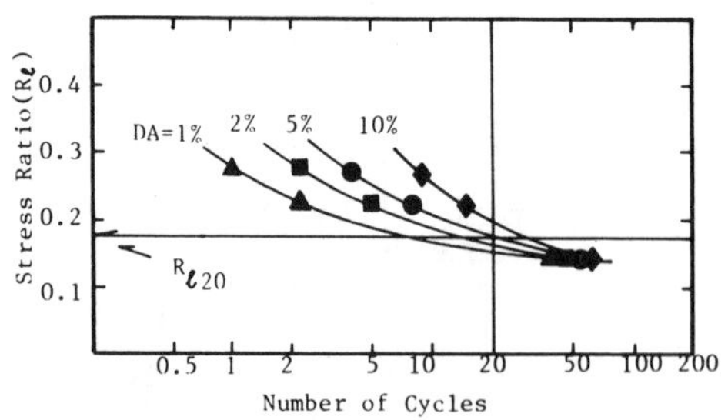

Fig. 2 Definition of $R_{\ell 20}$

Dynamic deformation characteristics test is to obtain G and h from the relation between load and deformation during cyclic loading. Test is conducted in small strain area around 10^{-4} in the beginning and continued in gradually greater strain area. Loading is performed by load control method with use of 0.25Hz sinusoidal wave and given cyclically ten times for same load level. Resultant record is load, deformation and pore water pressure recorded on electro-magnetic oscillograph and synthetic diagram of load and displacement on XY recorder, that is, hysteresis loop. For analysis 10th wave record is used. Sandy soil specimen was drained and cohesive soil specimen was undrained while being tested. In case of cohesive soil, however, specimen was drained whenever the test was completed at certain strain level so as not to produce excess pore water pressure. The results of cyclic triaxial test are shown in Fig. 3 with other test results. In the figure h is also defined.

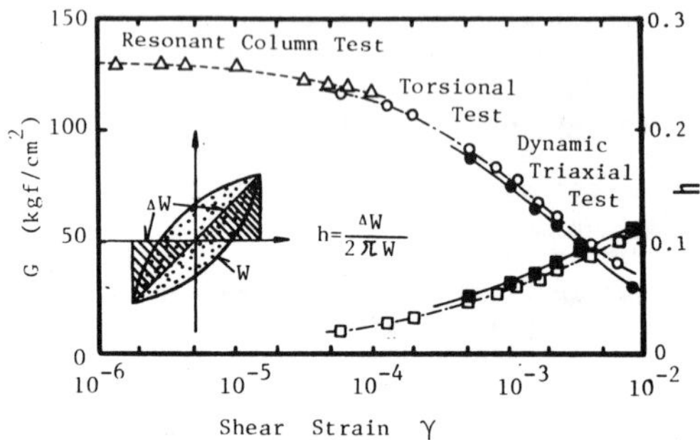

Fig. 3 Example of G, h∿γ relations

2.2 Resonant column test

The apparatus used in the test is designed to give torsional vibration at the top of the test specimen by supplying alternative current to the coil between permanent magnet. Basically, setting, consolidation, etc. of the specimen do not differ from those of cyclic triaxial test. For the specimen consolidated and completely prepared for the test excitation was first given with smallest possible input. With such input rotational velocity V of the top of the specimen in resonant state and reasonant frequency f of whole movable portion including the specimen are measured. Both V and f are digitally displayed. Determination of resonant state is done by watching the oscilloscope for when input force and response velocity waves correspond to a straight line lissajous curve.

Fig. 4 schematically illustrates the cell used in the test. The piston and the cell are basically jointed with only bearing and the friction between them is extremely small. V is measured with velocity sensor around excitation portion outside the cell. After a set of V and f has been measured in such manner similar measurement is performed with a bit increased input. The measurement is repeated until shear strain reaches 10^{-4} or so.

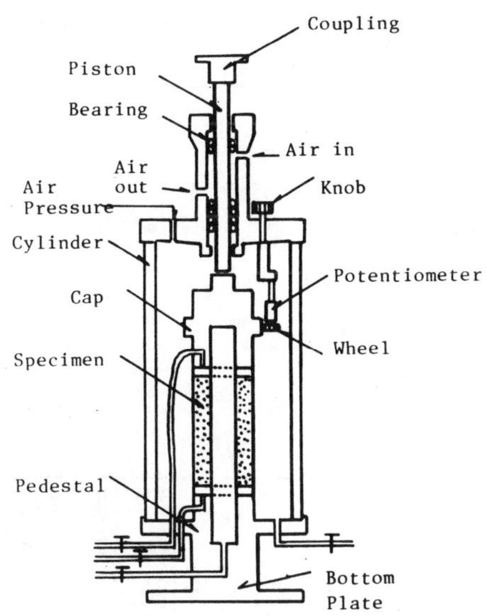

Fig. 4 Schematic view of the cell

From sets of V and f thus obtained, dimensions of the specimen and system's constants G and γ of the specimen can be calculated. (Drnevich, Hardin and Shippy, 1977) In Fig. 3 one example of the results is shown as provided in cyclic triaxial test.

2.3 Cyclic torsional test

The apparatus used in the test is hydraulic servo system where back and forth motion being caused by hydraulic pressure is converted to torsional force by means of a rack and a pinion. Difference from cyclic triaxial test is only whether loading mechanism is axial or torsional. Applicable wave form, frequency, etc. are almost same. As long as this paper is concerned the system was used only for the purpose of seeking for dynamic deformation characteristics.

The cell installed in the system is almost identical to that of reconant column test, (see Fig. 4) and the specimen used in resonant test is usually set as it is at the system for torsional test. Specimen set-up, consolidation and testing procedure are almost same with those of cyclic triaxial test. Items to be measured are torsional force T, torsional angle θ at the top of the specimen, and if necessary, pore water pressure.

The measuring systems, too has no difference from cyclic triaxial test. G is obtained from the torsional force and the dimensions of the specimen and h is obtained from hysteresis loop of T and θ. The test results were shown in Fig. 3 likewise in cyclic triaxial and resonant column test.

3 MATERIALS USED IN THE TEST

Fig. 5 gives distribution of sampling locations where the test specimens were recovered. Majority of the specimens were obtained from alluvial plain area close to seashore. Fig. 6 indicates histograms of N value, effective overburden pressure σ_v', and fine content, FC of the specimens used in the test. Since the sample used for liquefaction test is only alluvial sandy soil, distribution of the histogram for the test is biased, while it is fairly spread for other testing items.

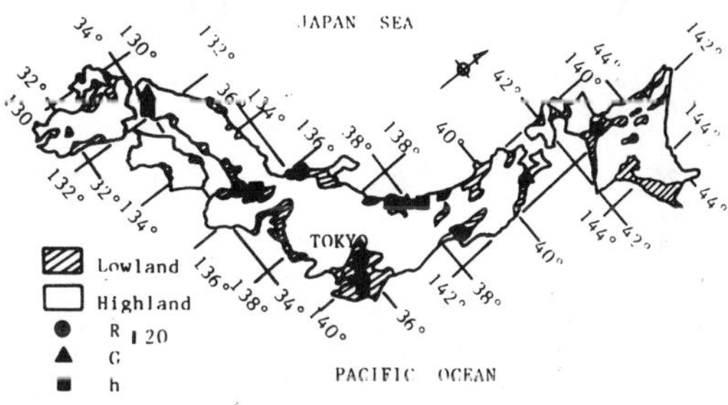

Fig. 5 Distribution of the investigation site

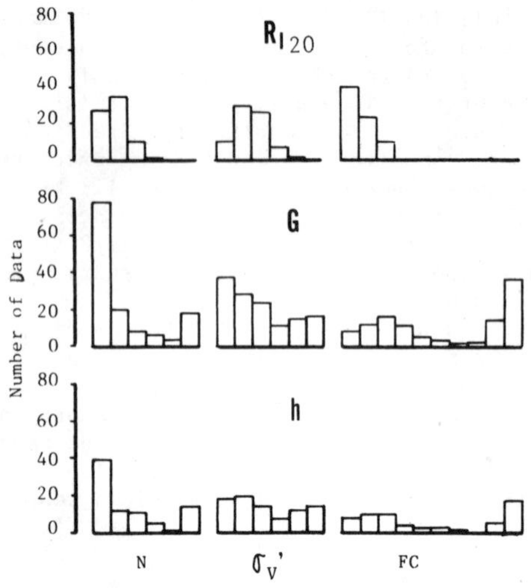

Fig. 6 Distribution of the sample

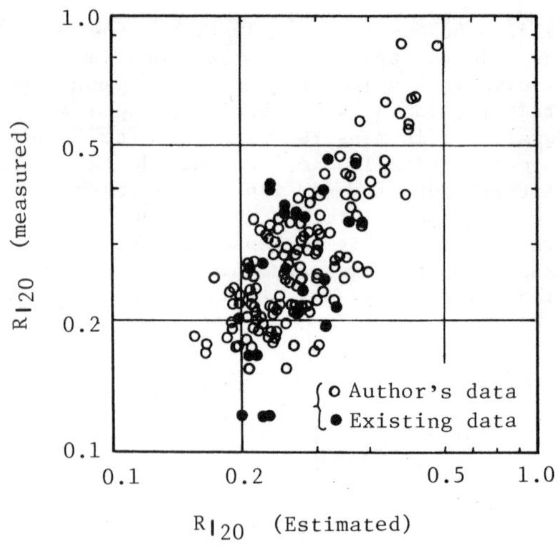

Fig. 7 Comparison $R_{\ell 20}$-measured with $R_{\ell 20}$-estimated

4 RESULTS AND DISCUSSIONS

4.1 Relationships between cyclic undrained strength and N value

The author have suggested one experimental formula on relation between cyclic undrained shear strength and N value. (Yokota, 1980) In this formula dynamic strength ratio, $R_{\ell 20}$ is defined as function composed of N value, effective overburden pressure σ_v', and fine content FC.

$$R_{\ell 20} = 0.164(1.012)^{N}(FC)^{0.0868}(\sigma_v')^{-0.441} \quad (1)$$

In determining the functional equation multiple regression analysis method was employed. Fig. 7 compares dynamic strength ratio computed with formula (1) $R_{\ell 20}$ (estimated) with actual dynamic strength ratio $R_{\ell 20}$ (measured). Yet, the data does not contain this test data. The two ratios have a fair agreement. Fig. 8 compares the results obtained from this liquefaction test with those computed with formula (1). Vertical axis in the figure represents $R_{\ell 20}$ multiplied by effective overburden pressure σ_v' (=effective confining pressure). It was shown as τ_{d20}.

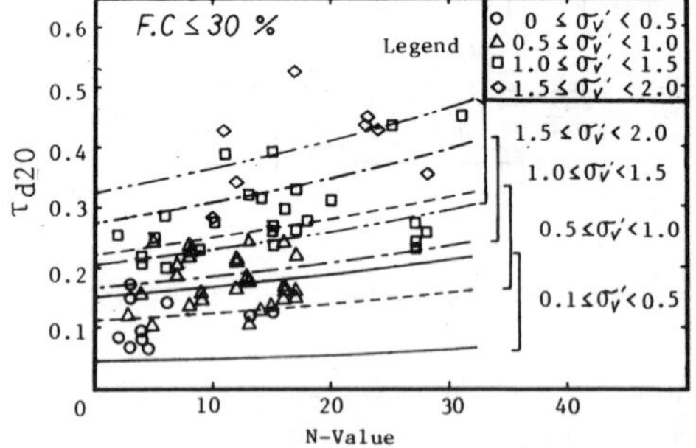

Fig. 8 Relationship between N-value and τ_{d20}

The relation between N value and τ_{d20} to be obtained by multiplying equation (1) by σ_v' is displayed in curve expression with use of σ_v' and FC as parameters. Then, the experimental data is plotted on it. Looking at the resulting figure these test results, though some dispersion is seen, apparently show good correspondence. However, it should be noted that the relation contains some effect caused by specimen compaction during sampling or adverse effect caused by specimen's relaxation. Considering the relation from view of the experimental formula we can say cyclic undrained strength changes, corresponding to N value in form of $(1.012)^N$.

4.2 Relationships between G and N value

Fig. 9 shows relation between N value and $G(=G_{RC})$ under condition that $\gamma=10^{-6}$ obtained from resonant column test and that between N value and G $(=G_{PS})$ obtained from PS log. G obtained from PS log corresponds to location where resonant column test was carried out. In Fig. G_{PS}-N average G_{PS}-N relation was computed from numerous PS logs conducted at various places in Japan. (Imai and Tonouchi, 1982)

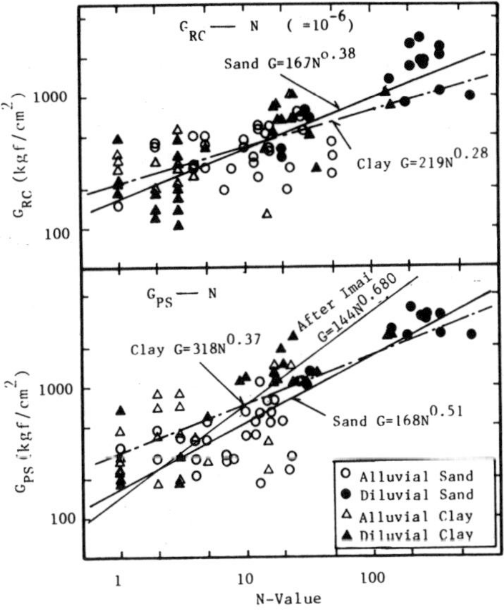

Fig. 9 Relationship between N-value and G

Compared with the average relation G in current data is rather small in the part where N value is large. In comparison between G_{PS}-N relation and G_{RC}-N relation G_{RC} is fairly smaller than G_{PS} in value of G in the portion where N value is large. This trend corresponds to the fact that $G_{PS} \geqq G_{RC}$ in diluvial soil of which N value is larger than that of alluvial soil. Possible reason is that sampling is harder in diluvial soil than in alluvial soil and easier to cause disturbance. Since Ko value is large in diluvial soil, higher confining pressure may be considered for loading to correspond to confining pressure of alluvial soil. In two figures, speaking of G - N relation itself, it has a good correlation especially in sandy soil. The straight line in the figure indicates related formula determined by least squares method only for sandy soil. In the computation N value less than 1 were omitted and not plotted in the figure.

4.3 Relationship between h and N value

In Fig. 10 relation between N value and maximum value of h, h_{max} is shown. As shown in Fig. 3 h increases as strain increases. In the test value of h_{max} was not necessarily seeked for, so that expediently here h value when $\gamma=10^{-2}$ is defined as h_{max}. In case when measurement was not conducted up to around $\gamma=10^{-2}$ the value was extrapolated. In such manner there seems no significant contradiction. The figure indicates value of h_{max} plotted per N value in each soil and confining pressure. The figure tells h_{max} tends to decrease as N value increases and that cohesive soil is smaller in h_{max} than sandy soil no matter what range N value may lie in. Its effect on confining pressure can not be clearly grasped from this figure alone.

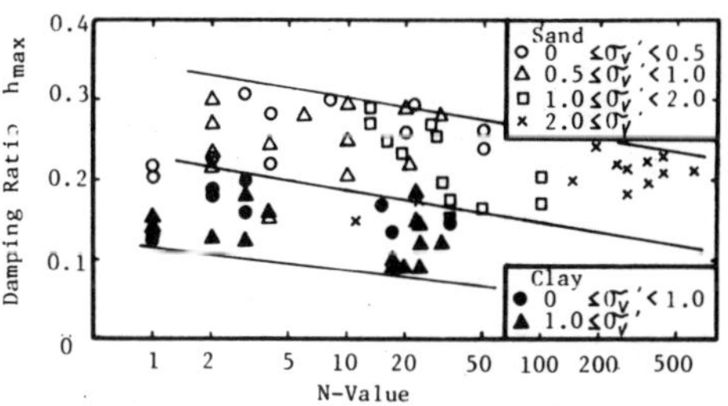

Fig. 10 Relationship between N-value and h

5 CONCLUSION

Cyclic undrained shear strength can be expressed in following equation with use of N value, σ_v', FC for sandy soil. Especially, N value is expressed as a function--$(0.012)^N$.

$$R_{\ell 20} = 0.164(1.012)^N(FC)^{0.0568}(\sigma_v')^{-0.441} \text{ or}$$

$$\tau_{d20} = 0.164(1.012)^N(FC)^{0.0568}(\sigma_v')^{0.559}$$

However, the aforementioned cyclic undrained shear strength is the strength to be determined by cyclic triaxial test and it should be noted that the strength does not immediately form cyclic undrained shear stress of the ground.

Speaking of G in low strain amplitude ($\gamma=10^{-6}$) both G_{PS} and G_{RC} can be approximated by estimate with N value for sandy soil. There will be no substantial difference between G_{PS} and G_{RC} for alluvial soil, but for diluvial soil of which N value is large the difference widens, in other words, $G_{PS} \geqq G_{RC}$.

In general, value of h_{max} decreases as N value increases, but its dispersion is so wide that it is hard to estimate h_{max} by knowing only N value. In wide range of N value sandy soil has larger damping than clayey soil does.

6 REFERENCES

Drnevich, V.P., B.O. Hardin and D.I. Shippy, 1977, Modulus and Damping of Soils by the Resonant Method. ASTM STP 654 91-125

Fukuoka, M.K. Ishihara and Y. Yoshida, 1978, Liquefaction Analysis of Sand Deposits, Special Research on Natural Calamities, Ministry of Education 41-66 (in Japanese)

Imai. T, 1977, P-and S-wave Velocities of the Ground in Japan. Proceedings of 10th ICSMFE, VOL 2, 257-260

Imai. T. and K. Tonouchi, 1982, Correlation of N-value with S-wave Velocity and Shear Modulus, Proceedings of ESOPT II

Ishizawa M, S. Nakagawa and I. Kurohara, 1977. Cyclic Undrained Triaxial Testing Using Undisturbed Sand Samples with Fine Content, Proceeding 12th Annual Meeting Japanese Soc. Soil Mech. Found. Eng. 397-400 (in Japanese)

Ohashi. A, T. Iwasaki, F. Tatsuoka and K. Miyata 1976, Technical Memorandum of the Public Works Research Institute. A Seismic Soil Exploration of Akebonobashi and Shintatumibashi of Tokyo Wangan Road (in Japanese)

Ohya. S, K. Ogura and K. Yokota, 1980. A New Sand Sampler-Twist Sampler and Its Application to Evaluation of Dynamic Soil Properties, Proceedings of 6th Southeast Asian Conference on Soil Mechanics 69-84

Yokota. K. 1980, Evaluation of Liquefaction Strength of Sandy Soils, Proceedings of 7th WCEE, VOL 3, 121-124

Proceedings of the Second European Symposium on Penetration Testing / Amsterdam / 24-27 May 1982

Prediction of liquefaction potential in Japan using N-values of SPT

TOSHIO IWASAKI & KEN-ICHI TOKIDA
Ministry of Construction, Ibaraki-ken, Japan

SUSUMU YASUDA
Kisojiban Consultants Co., Tokyo, Japan

FUMIO TATSUOKA
University of Tokyo, Japan

1 INTRODUCTION

Subsoils making up alluvial planes in Japan are found mainly in two layers: a loose-sandy layer in shallow depth and a soft-clayey layer underneath. Standard penetration test, SPT, has been prefered over other penetration tests for investigating these alluvial deposits. Because SPT can be conducted in both types of layers, and the SPT can obtain soil samples from these layers. For the above two reasons, the SPT is widely used in Japan.

The SPT was first introduced from the United States to Japan around 1951, and became a standard in 1961 as the first standardized penetration test. SPT has been used extensively since then, and the majority of design manuals compiled in Japan stipulate the use of SPT for site investigations.

In 1964, Niigata earthquake inflicted huge damage to buildings and bridges by liquefying loose-sandy soils. This was the first time the phenomenon of liquefaction became widely recognized. Koizumi(1966) and Kishida (1966) studied SPT-values in Niigata, and subsequently published critical SPT N-values was used to which differential the soils that had liquefied and that had not liquefied in Niigata. Though the nubmers of critical N-values have been slightly modified since then, they have been used widely in predicting liquefaction potentials.

Seed (1966) also studied the liquefaction occurred during the Niigata earthquake and suggested undrained cyclic shear strength of clean sands be evaluated by conducting cyclic triaxial compression tests. Based on the results of the cyclic triaxial tests, he and his colleagues proposed a simple method for estimating liquefaction resistance of sandy soils (Seed and Idriss 1971). The method used relative density, derived from SPT, as a prime soil parameter in determining liquefaction resistance.

In Japan, a great many numbers of cyclic triaxial tests have been conducted on undisturbed samples, and based on these test results, the authors have proposed a new method with which liquefaction resistance can be estimated directly from the N-values, instead of estimating in-situ relative density from N-values as Seed did in 1971 (Iwasaki et al., 1978). This method and its modifications have been used for the seismic stability assessment and design of highway bridges, tailings dams etc.

Most of the structures in Japan need to to be built on liquefaction-resistant ground as Japan is one of the most seismically active countries. And liquefaction resistance of ground has been frequently determined from the N-values.

In the following pages, this paper will give detailed explanation of liquefaction prediction method currently being used in Japan, and its future directions.

2 STANDARD PENETRATION TEST IN JAPAN

Standard penetration test has been standardized in Japan under the code name of JIS A 1219-1961. The SPT N-value is defined as the number of blow counts necessary for split-spoon sampler to penetration 30 cm by impart of a free falling hammer weighing 63.5 kg, falling 75 cm. The split-spoon sampler is stipulated to have dimensions as shown in Fig. 1(a); drilling rods are stipulated to have diameter of 40.5 mm or 42 mm as specified in JIS M 1409-Drill Rods for Core Drills. Knocking head and hammer widely used are shown in Fig.1(b) and (c). Though no special method for lifting the hammer and dropping is specified a rope-around-a-cathead method and a trip monkey method are used. In the former method, 13 to 16 mm in diameter rope is usually used.

The past experience shows that SPT N-values tested by different persons present

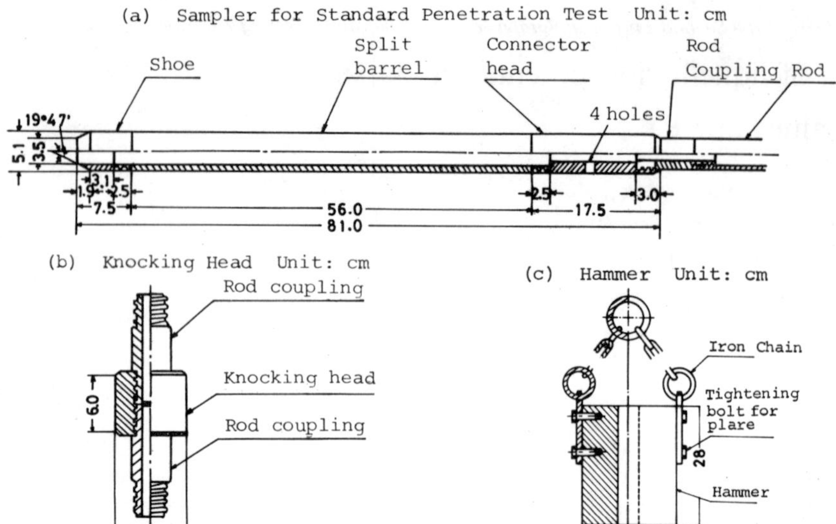

Fig. 1 Apparatus for standard penetration test in Japan

no major scatters in Japan.

3 JAPANESE LIQUEFACTION PREDICTION METHODS USING N-VALUES

Four methods are used in Japan to predict liquefaction potentials:

(1) Liquefaction potential is evaluated roughly based on topographical and geological information.

(2) Liquefaction potential is evaluated from N-values, grain-size distribution curves and location of groundwater table.

(3) Liquefaction potential is evaluated more accurately by conducting cyclic shear tests on undisturbed samples, and comparing the test results with the results of dynamic response analyses.

(4) Liquefaction potential is evaluated by conducting in-situ cyclic or blasting tests or shaking table tests.

The first method is used primary for pointing out liquefaction susceptible area for safety of river dykes and etc, or judging whether more detailed evaluation by the second to the fourth methods are needed or not.

The second method is used most commonly. When the judgedment by the second method is inadequate or site under study has special importance, the third method is applied.

The fourth method is useful for studying the effect of soil-improvement on liquefaction resistance and simulating behaviour of structures built on liquefiable ground. The application of the fourth method is rare due to its high cost.

N-value is used in the second method as a deciding parameter for evaluating liquefaction suseptability of ground. In the process of predicting liquefaction potential by N-value, there are two distinctily different procedures:

(A) Measured N-values are compared with critical N-values. When the measured N-value is less than the critical N-value, the ground is considered likely to liquefy.

(B) Liquefaction resistance of soil, R, is estimated using a formula discussed in the end of this section and the calculated value is compared with earthquake induced load, L. Liquefaction is likely if the liquefaction resistance factor, $F_L = R/L$, is less than 1.0.

The idea of the critical N-value was first introduced by Koizumi and Kishida soon after the Niigata earthquake of 1964. Koizumi (1966) compared the N-values measured before and after the earthquake at about 20 sites, and found out that the loose sands, whose N-values were smaller than some limiting number, later defined as critical N-value, densified and their N-values increased by the earthquake motion, and vise versa for dense sand. He thus concluded that the critical N-value is the limiting number differentiating ground from liquefying and not liquefing. Kishida (1966) studied the relationships between N-values and damage to buildings, and clari-

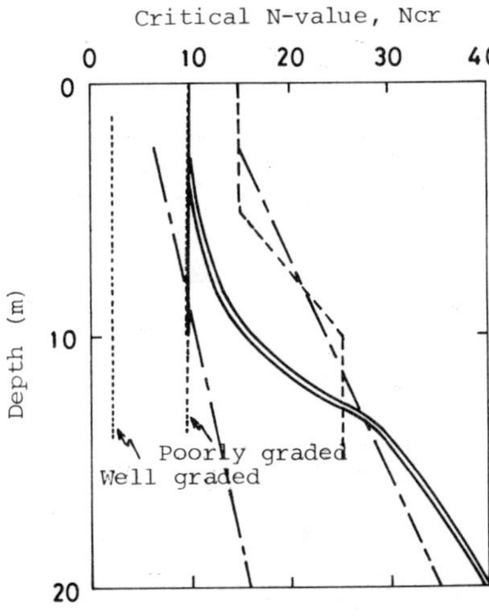

Legends

══ : Proposed by Koizumi (1966)
---- : Proposed by Kishida (1966)
········ : Harbor facilities code when acceleration is 160 gals
—— : Highway bridges (1972) and railway facilies codes (1974)
—·— : Buildings code (1974)

Fig. 2 Critical N-values proposed by Koizumi and Kishida and adopted to three regulations

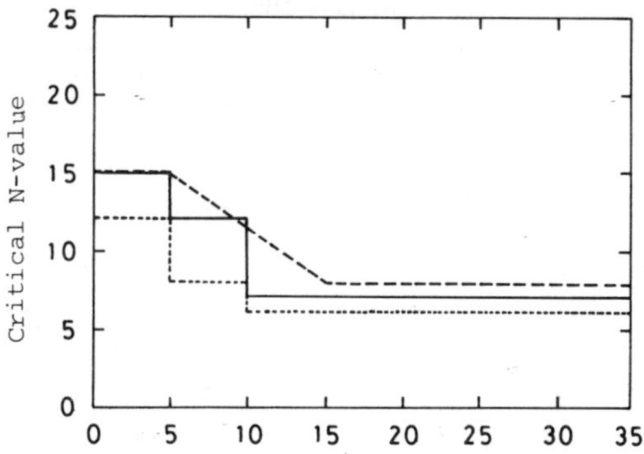

Percent of fines less than 74μ (%)

Legends

—— : Codes for oil tanks, near the edge of tanks (1978)
········ : Codes for oil tanks, under tanks (1978)
----- : Codes for LNG tanks (1979)
The critical N-values are constant over the depth.

Fig. 3 Critical N-values of two regulations for oil tanks and LNG tanks

fied that the buildings, whose ground had N-values smaller than the critical N-value, shown in Fig. 2 by the broken line, suffered liquefaction induced damage. Ohsaki (1970) verified the validity of the critical N-value using his experience from the Tokachi-oki earthquake of 1968.

Based on these studies, the methods for predicting liquefaction using the critical N-value were accepted into design codes of harbor facilities, highway bridges, buildings and railway facilities in the years from 1971 to 1974. The critical N-values specified in each code are shown in Fig. 2. It is seen in this figure that the critical N-values are different among the codes.

The critical N-values shown in Fig. 2 are, with exception for the harbor facilities code used independently of grain-size distributions. However, recently, it has been pointed out that though the N-values decrease with higher contents of fines, the cyclic resistances increase. Therefore, it is more rational to vary the critical N-values in accordance with the grain size. This idea was adopted in the design codes of oil tanks and LNG tanks in 1978 and 1979 respectively; the critical N-values for both codes varis with fines content as shown in Fig. 3.

Seed (1966) first proposed the liquefaction potential be evaluated by comparing cyclic resistance of soils with earthquake induced dynamic loads. In 1971 he and his colleagues proposed a simple method for estimating liquefaction resistance from N-values (Seed and Idriss, 1971). This method used N-values to estimate relative densities of soils; and the relative densities were in turn used to assess liquefaction resistance. In Japan, a large number of undrained cyclic triaxial tests have been conducted on undisturbed samples obtained from alluvial and reclaimed sands started from about 1975. By summarizing these test results, the authors derived a formula that unabled estimating of liquefacion resistance directly from the N-value (Tatsuoka et al., 1978), as follows:

$$R = 0.0882\sqrt{\frac{N}{\sigma_v' + 0.7}} + 0.225\log_{10}\frac{0.35}{D_{50}}, \quad \text{for } 0.04\text{ mm} \leqq D_{50} \leqq 0.6\text{mm} \qquad (1)$$

$$R = 0.0882\sqrt{\frac{N}{\sigma_v' + 0.7}} - 0.05, \quad \text{for } 0.6\text{ mm} \leqq D_{50} \leqq 1.5\text{ mm} \qquad (2)$$

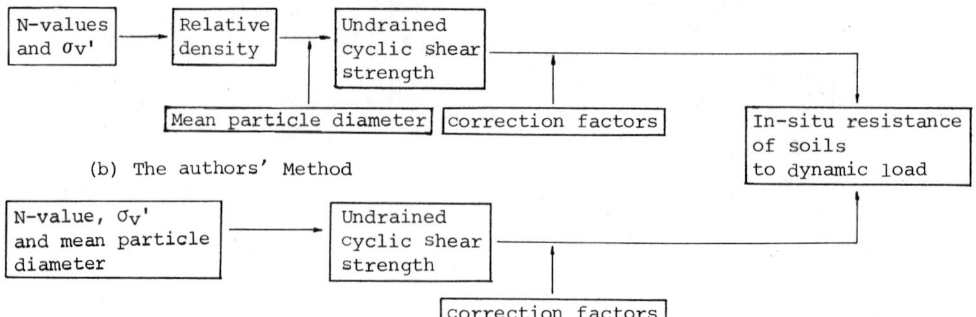

Fig. 4 Comparison of estimating processes by two methods

where R: resistance of a soil to dynamic load
σ_v': effective overburden pressure (kgf/cm^2)
D_{50}: mean particle diameter (mm)

Fig. 4 shows the differences in estimating processes proposed by Seed and Idriss and by the authors.

Eqs. (1) and (2) were adopted in the design codes of highway bridges, and tailings dams, with minor modifications in 1980. In these regulations, the second terms of Eqs. (1) and (2) are modified as shown in Fig. 5. Tailings materials in comparison with alluvial and reclaimed soils, have lesser resistance in the range of silt size.

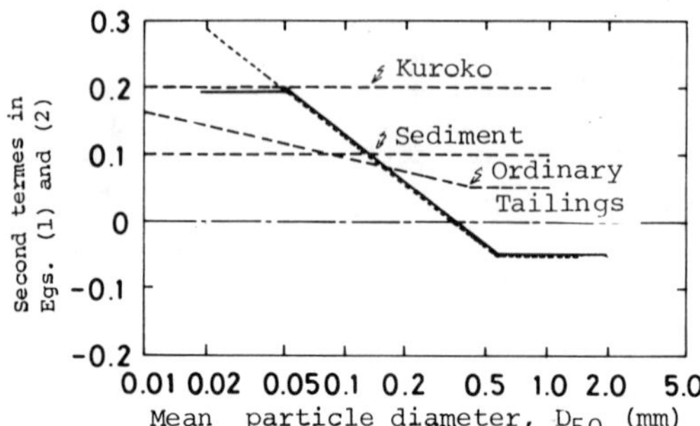

Legends

------- : Proposed by the Authors (1978)
——— : Now Code for Highways Bridges (1980)
---- : Code for Tailings Dams (1980)

Fig. 5 Second termes in Eqs. (1) and (2) proposed by tne authors and adopted to two regulations

4 AN EXAMPLE OF LIQUEFACTION PREDICTION USING AUTHORS' METHOD

As an example of predicting liquefaction potential using N-values, a microzonation of an area conducted by the authors will be presented. In Japan, a big earthquake is expected to occure in Shizuoka Prefecture in near future. As an forcast for likely damage to be inflicted on by this earthquake, the liquefaction potentials of preselected areas in the prefecture were studied by the authors (Watanabe, Iwasaki et al. 1980). First, 4000 boring data containing SPT N-values were collected and filed in a computer. The resistances to dynamic load at all depths down to 20 m were calculated using Eqs. (1) and (2) at all sites. The mean particle diameters were estimated from soil types following the general values shown in Table 1, because grain size distribution curves were not available at these sites. Liquefaction resistance factors, F_L, were calculated by dividing the resistances R by dynamic load L as shown in Fig. 6. Index of liquefaction potential, I_L, a weighted factor of safty over the depth of 20 m, was calculated at each site to express the severity of liquefaction at several different levels of assumed surface accelerations. And, finely, the minimum surface

Table 1 Mean particle diameters estimated for soil types (Watanabe, Iwasaki et al.,1980)

Soil type	Mean particle diameter (mm)
Silt	0.02
Sandy silt	0.04
Silty sand	0.10
Fine sand	0.20
Medium sand	0.25
Coarse sand	0.30
Gravel	0.60

accelerations which make I_L just exceed 15 were selected and named as critical acceleration required to cause liquefaction. Fig. 7 shows the critical accelerations averaged in each 1 km mesh in Shizuoka Prefecture.

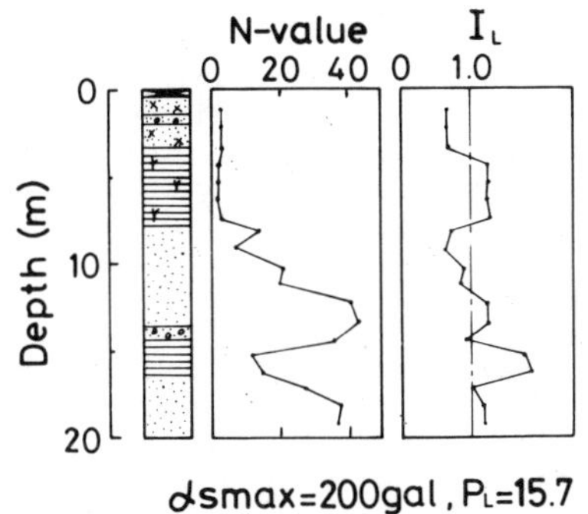

Fig.6 An example of calculated F_L and I_L (Watanabe, Iwasaki et al.,1980)

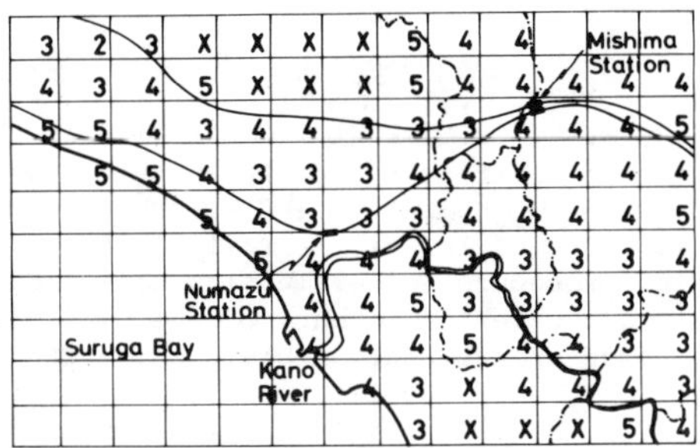

Note: Numbers in the map show the critical accelerations in 100 gals

Fig.7 The distribution map of the critical accelerations to induce liquefaction (Watanabe,Iwasaki et al.,1980)

5 LIMITATIONS IN THE PREDICTION METHODS USING N-VALUES

Through the experiences gained in predicting liquefaction potential using N-values, the authors consider that the predicting methods as mentioned above have following limitations:

(1) The liquefaction potential of gravelly ground can not be predicted by the above methods. For example, several coastal areas in Shizuoka prefecture are composed of gravelly ground, Eq. (2) could not be applied for large grain size particles.

(2) In dense sand or silty sand, the estimated resistances varies among predicting methods. Tokimatsu and Yoshimi (1981) compared the liquefactions resistance calculated by the method proposed by the authors

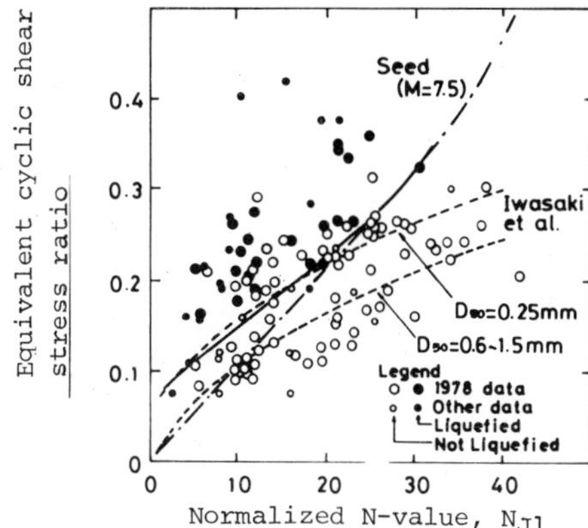

Legend
—— : Lower bound established by Tokimatsu and Yoshimi (1981)

Fig. 8 Comparison of the relationships between the N-values and the resistances proposed by the authors and Seed (Tokimatsu and Yoshimi, 1981)

and that calculated by method that proposed recently by Seed (1979) as shown in Fig. 8. In this figure, solid circles show the data where liquefaction were observed and open circles show the data where liquefaction were not observed. From this figure, Tokimatsu and Yoshimi made following comments that the authors' method tend to underestimate the resistance for large N-values, whereas the Seed's method tended to underestimate the resistance for small N-values.

The authors agree with Tokimatsu and Yoshimi. Eqs. (1) and (2) are derived originally based on the assumption that the undrained cyclic shear strength ratio increases linealy with relative density, however, according to the recent data obtained by cyclic triaxial apparatus, the strength increases rapidly with the relative density for dense sand as shown in Fig. 9. Therefore Eqs.(1) and (2) for dense sand needed to be modified.

Generally speaking, a soil strata having small N-values is likely to contain some fine particles and possesses some odded resistance to liquefaction. Seed's method does not take into consideration such effects of fine particles, therefore, it has the tendency to underestimate the cyclic resistance of soils containing fines.

(3) Liquefaction potentials of recent fills, densified by methods like sand compaction, pile driving, or placement of fills are difficult to evaluated accurately, because the N-values, coefficient of earth pressures at rest, and other factors change before and after the improvement work.

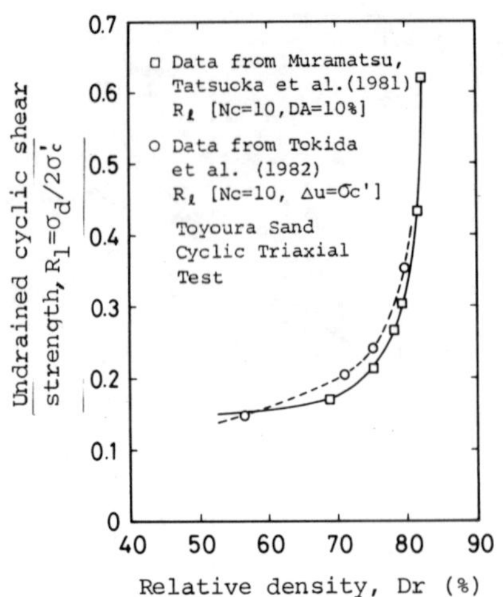

Fig.9 Relationships between undrained cyclic shear strength and relative density

6 CONCLUSING REMARKS

In Japan, several liquefaction prediction methods using SPT N-values are widely used, and they are specificalley stipulated into most of the current design codes for most structures with minor differences among the codes. This high degree of acceptance of N-value method rests on two facts. One: the standard penetration test has been the most popularly used sounding test over the last 30 years. Two: the liquefaction study of the 1964 Niigata earthquake investigated the relationship between liquefaction and SPT N-values. 15 years after the standarization of the SPT, the Dutch Double-tube Cone Penetration Test and the Swedish Weight Sounding Test were standardized in 1976. And, a few methods for predicting liquefaction using the CPT are proposed (Ishihara 1980, Fujinami et al., 1979). However, the majority of the prediction will continue to be conducted by N-values, with likely modifications for gravelly ground and dense sandy ground.

ACKNOWLEDGEMENT

The authors appreciate Dr. Kenji Mori of Kiso-jiban Consultnats Co., Ltd. for his kind review of the draft of this paper.

REFERENCES

Fujinami, T., Ohtsuka, M. & Yasuda, S. 1979, Liquefaction safe design of underground cabel, Proc. of the 6th Asian Regional Conf. on S.M.F.E. Vol. 1: 27-30

Ishihara K. 1980, Sate-of-the-art of seismic design and in-situ and laboratory dynamic test, Chishitsu to Chaosa, 3: 2-4 (in Japanese).

Iwasaki, T., Tatusoka, F., Tokida, K. & Yasuda, S. 1978, A practical method for assessing soil liquefaction potential based on case studies at various sites in Japan, Proc. of the 2nd Int. Conf. on Microzonation, Vol. 2: 885-896.

Kishida, H. 1966, Damage to reinforced concrete buildings in Niigata city with special reference to foundation engineering, Soils and Foundaions, Vol.6, No.1: 71-88.

Koizumi, Y. 1966, Changes in density of sand subsoil caused by the Niigata earthquake, Soils and Foundations, Vol.6, No.2: 38-44.

Muramatsu, M., Tatsuoka, F., Sasake, T. & Seki, S. 1981, Liquefaction of dense sand by triaxial and torsional simple shear tests, Proc. of the 16th Japan National Conf. on S.M.F.E.: 601-604(in Japanese).

Ohsaki, Y, 1970, Effects of sand compaction on liquefaction during the Tokachioki earthquake, Sois and Foundations Vol.10, No.2: 112-128.

Seed, H. B. 1966, Soil liquefaction in the Niigata earthquake, 2nd Japan Earthquake Engineering Symposium : 97-104.

Seed, H. B. & Idriss, I. M. 1971, A simplified procedure for evaluating soil liquefaction potential, Jour. of the Soil Mechanics and Foundation Division, ASCE, Vol. 97, SM9: 249-274.

Seed. H. B. 1979, Soil liquefaction and cyclic mobility evaluation for level ground during earthquakes, Jour. of the Geotechnical Engineering Division, ASCE, Vol.105, GT2: 201-255.

Tatsuoka, F., Iwasaki, T., Tokida, K., Yasuda, S., Hirose, M., Imai, T. & Kono, M. 1978, A method for estimating undrained cyclic strength of sandy soils using standard penetration resistances, Soils and Foundations, Vol.18, No.3: 43-58.

Tokida, K., Kondo, M. & Takamatsu, S. 1982, The relationship between undrained cyclic shear strength and relative density of sand, Proc. of the 9th Kanto Reg. Conf. on Civil Eng. in Japan : in press (in Japanese)

Tokimatsu, K. & Yoshimi, Y. 1981, Field correlation of soil liquefaction with SPT and grain size, Int. Conf. on Recent Advances in Geotechnical Earthquake Eng. and Soil Dynamics, Vol.1: 203-208.

Watanabe, S., Iwasaki, T., Tokida, K., Tatsuoka, F., Yasuda, S. & Sato, H. 1981, Prediction of liquefaction potential in Shizuoka prefecture using simplified procedure, Proc. of the 15th National Conf. on S.M.F.E. in Japan :1345-1348 (in Japanese).

Proceedings of the Second European Symposium on Penetration Testing / Amsterdam / 24-27 May 1982

Correlation between SPT values and electrical resistivities of subsurface strata

J.M.KATE
Indian Institute of Technology, Delhi, India

1 INTRODUCTION

Field engineers are well aware of Geophysical methods for aiding subsurface explorations. Electrical resistivity is one such non-destructive technique which is now being increasingly used due to its simplicity, low cost and quick output. So far, the use of this technique by geotechnical engineers is limited only to interprete subsurface profile and locate ground water table.

The ultimate aim of geotechnical engineer is to know the in-situ strength of every stratum contained in subsurface with a view to design suitable foundation. SPT is one of the test, which furnishes information regarding the resistances of various subsurface strata to penetration. No doubt, this test provides first hand information but besides being costly is laborious and time consuming.

In the light of this, in the present field investigation, an attempt has been made to establish a definite relationship, if exists between SPT blows and absolute resistivities of subsurface strata. The investigation comprised of conducting both the resistivity soundings as well as SPT for the same subsurface condition. The absolute resistivity of each of the stratum contained in subsurface profile was interpreted from sounding data. The values of absolute resistivities were plotted against the corresponding SPT blows both recorded as well as adjusted, to obtain the correlation.

2 LITERATURE REVIEW

2.1 Electrical resistivity

The resistivity test makes use of electric current passed through the ground to produce a potential drop proportional to the resistances offered by various strata; from which the apparent resistivities of these strata is calculated. The depth of current penetration into the ground depends upon the configuration of current and potential electrodes. By varying the electrode spacings the resistivities at different depths can be investigated and the process is known as resistivity sounding.

In Wenner's circuit array, 4 electrodes are equally spaced along a straight line passing through and symmetrical about the point of exploration. The apparent resistivity (ρ_a) in this arrangement is obtained by the following equation:

$$\rho_a = 2\pi \cdot a \cdot R \quad (1)$$

wherein a: spacing between consecutive electrodes
R: resistance

The depth upto which the current penetrates into the ground is equal to a.

The resistivity soundings provide the values of apparent resistivity. However, the application of resistivity data to relevent field problems such as identification of strata, locations of ground water table etc. has to be done only on the basis of absolute resistivities (ρ). In layered system apparent resistivity obtained at any depth is a function of absolute resistivities and thicknesses of various strata through which the current has passed. The absolute resistivities of various strata can be obtained by interpreting the apparent resistivity values by employing anyone or combination of the following methods. These methods are analytical (Onodera and Seibe 1970), numerical (Backus and Gilbert 1967), empirical (Moore 1945) and graphical methods such as standard curve (compagnie General de Geophysique 1963) matching, resistivity contouring, inverse slope (Sankar Narayan et al 1967), direct slope (Baig 1980) etc.

Each one of these methods has certain merits and drawbacks too, and any single method may not provide satisfactory solution for all types and orders of subsurface conditions. Certain field studies on electrical resistivity survey with ground truths have shown that Mean log resistivity technique (Kate 1981, Kate et al 1981) in combination with cumulative resistivity (ρ_c) plot (Moore 1961) can provide reliable results of absolute resistivities of each of the stratum contained in the subsurface profile.

2.2 Standard penetration test

Standard penetration test (SPT) conducted by means of the split spoon having specified dimensions, provides data regarding the resistances offered by subsurface strata to the penetration, in terms of number of blows. The number of blows by 65 kg weight falling freely through 75 cm height, required per 30 cm penetration of standard split spoon are recorded during the test. These recorded blows are then corrected for overburden (Gibbs and Holtz 1957) and dilatancy (Terzaghi and Peck 1948) effects wherever needed. Curves correlating these blows with relative densities, bearing capacities etc. for different types of soils are available in various publications.

3 FIELD INVESTIGATION PROGRAMME

3.1 Resistivity soundings

Electrical resistivity soundings were carried out at selected locations by using resistivitymeter and Wenner's circuit array arrangement. Electrode spacings were increased by a regular interval of 1.5 m and the observations were taken upto the current penetration depth of around 21 metres. At each location point observations were also taken by changing electrode line through 90° and the average values have been reported. The apparent resistivities at different depths were computed from the recorded values of resistances by using Equation 1.

3.2 Standard penetration test

At all these locations wherein resistivity soundings have been carried out,10 cm diameter bore holes were drilled and simultaneously standard penetration tests were conducted. These SPT were performed as per Indian Standard: 2131-1963 and at a regular interval of 1.5 metre. The SPT blows recorded (N_r) were corrected for overburden and dilatancy corrections as needed and the corrected values have been designated by adjusted SPT blows (N_a) in the text.

During boring process, soil samples from different depths were continuously procurred which were then tested in the laboratory for certain necessary tests.

4 RESULTS AND DISCUSSIONS

4.1 Electrical resistivity sounding

Figure 1 shows the variation of apparent resistivity (ρ_a) and Moore's (1961) cumulative resistivity (ρ_c) with depth for three different locations. The cumulative resistivity variation with depth has been used to determine the extent and number of strata contained in subsurface at each location.

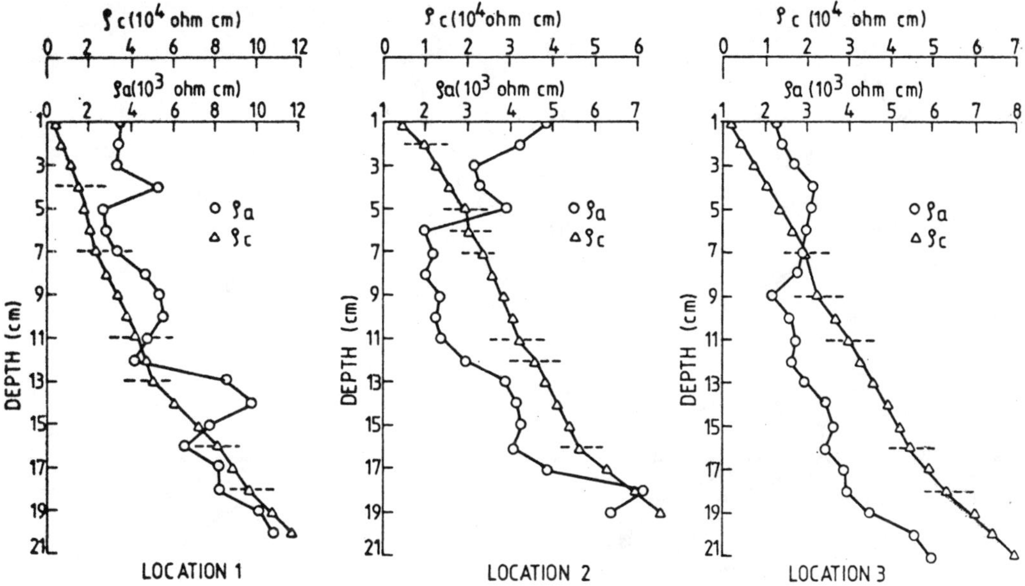

Fig.1 Variation of apparent resistivity(ρ_a) and cumulative resistivity(ρ_c) with depth at various locations

Table 1. Interpretation of resistivity data

Stratum extent (m)	thickness (m)	Mean ρ_a (10^3 ohm. cm)	ρ (10^3 ohm cm)
Location 1			
1 to 4	3	3.392	3.392
4 to 7	3	2.644	2.061
7 to 11	4	5.325	15.22
11 to 13	2	4.085	1.085
13 to 16	3	9.517	28.03
16 to 18	2	8.215	2.749
18 to 20	2	10.087	54.58
Location 2			
1 to 2	1	4.602	4.602
2 to 5	3	3.576	3.287
5 to 6	1	2.920	1.298
6 to 7	1	2.132	0.443
7 to 11	4	2.377	2.798
11 to 12	1	2.685	9.08
12 to 16	4	4.213	14.54
16 to 19	3	6.302	47.23
Location 3			
1 to 7	6	3.156	3.156
7 to 9	2	2.759	1.845
9 to 11	2	2.503	1.698
11 to 16	5	3.194	5.2
16 to 18	2	3.951	19.47
18 to 21	3	5.404	31.87

It can be seen from Fig.1 that there are 7,8 and 6 different strata at location 1,2 and 3 respectively, within the depths investigated. Table 1 shows the thicknesses of these strata thus obtained. Mean values of apparent resistivity for each of the stratum have been worked out and are reported in Table 1.

The absolute resistivity (ρ) of each of the stratum has been computed from the known thickness and apparent resistivity by using Mean log resistivity technique (Kate 1981) and the values are presented in Table 1. These values are ranging from 1.085 x 10^3 to 54.58x 10^3, 0.443 x 10^3 to 47.23 x 10^3 and 1.698 x 10^3 to 31.87 x 10^3 ohm cm at location 1,2 and 3 respectively. The lowest value of ρ gives an indication of existance of ground water table within that stratum. The ρ values of various strata at all these locations exhibit a general mixed sequence of strata i.e. soft strata underlained by relatively hard strata and vice versa.

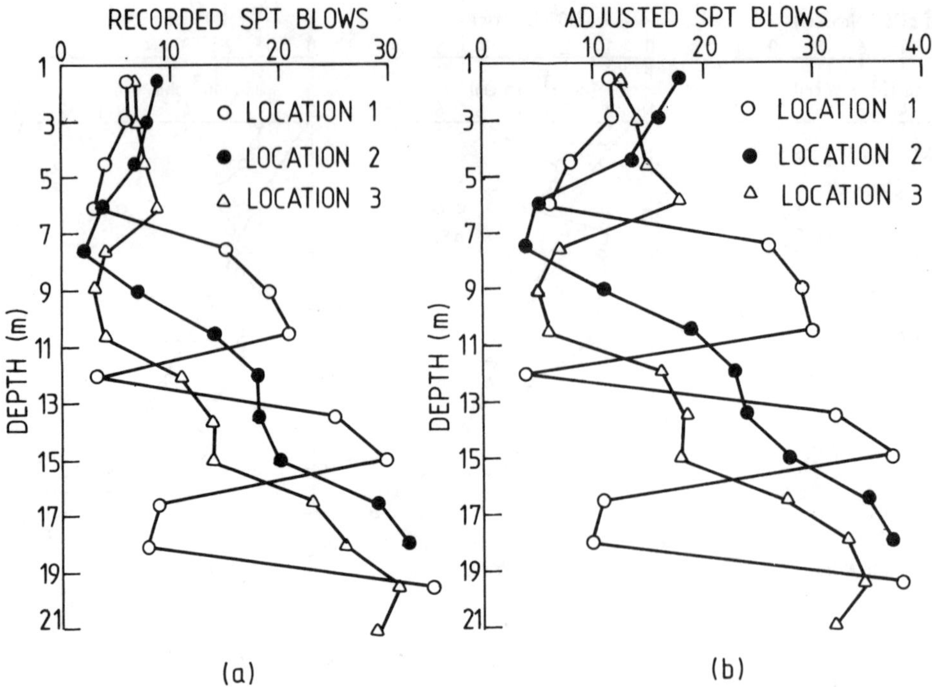

Fig.2 Recorded SPT blows (N_r) and adjusted SPT blows (N_a) versus depth curves at different locations

4.2 SPT and subsurface profiles

The variation between recorded SPT blows (N_r) with depth at various locations is shown in Figure 2(a) and that of adjusted SPT blows (N_a) with depth in Figure 2(b). The subsurface profiles at various locations shown in Figure 3, have been prepared from the laboratory classification tests conducted on samples collected from the bore holes during the tests. These profiles in general show the presence of layers of clay, silty clay, clayey silt, silt, sandy silt, silty sand, sand and gravel. The depths of ground water table recorded are 11.4, 6.4 and 9.6 metres at location 1,2 and 3 respectively.

It is interesting to note that the number of strata at each location, interpreted from cumulative resistivity plot are compatible with the bore hole profile. At location 2 and 3 the presence of ground water table in clay may be seperating it into two resistivity demarcations in terms of clay and saturated clay thus giving an additional stratum. The dotted lines on the subsurface profiles shown in Figure 3 show the thicknesses of strata obtained from cumulative resistivity plot. It can be seen that these thicknesses are more or less comparable with those from bore hole log. It may also be noted that the accuracy of thickness interpretation depends upon the electrode spacings; smaller the electrode spacings higher is the accuracy.

4.3 Correlation

SPT blows both recorded as well as adjusted, for various strata studied here are plotted against corresponding absolute resistivities as shown by semilog plot in Fig.4. It can be seen that both recorded as well as adjusted SPT counts exhibit bilinear relationship with absolute resistivity. Absolute

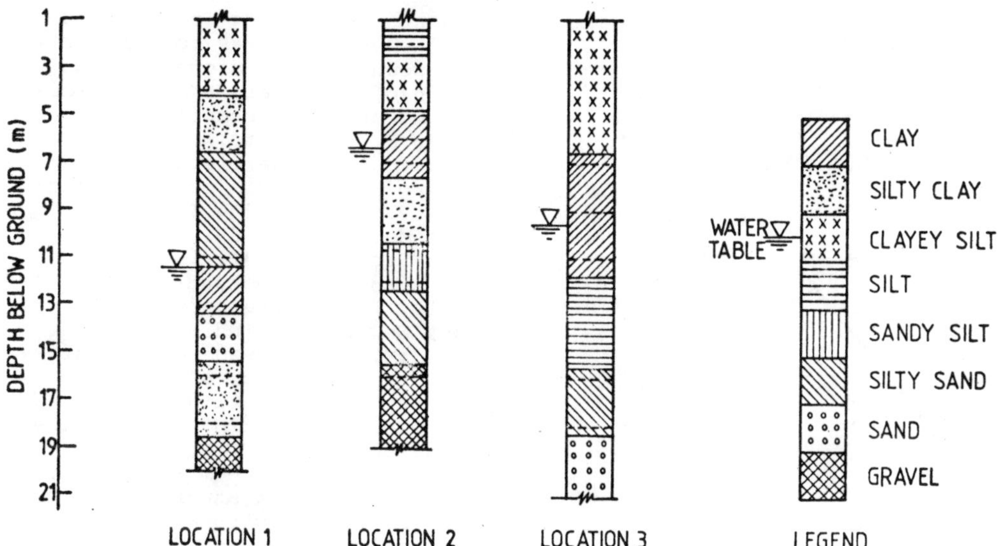

Fig.3 Subsurface profiles from bore hole data

resistivity increases with SPT blows much rapidly in soft (clayey) strata than comparatively harder strata. The following equations expressing the empirical relationship between absolute resistivity and SPT blows have been obtained from Figure 4.

For soft soils or clays:

$$\text{Log}\,\rho = 2.301 + 0.241\ N_r \qquad (2)$$

$$\text{Log}\,\rho = 2.301 + 0.1505\ N_a \qquad (3)$$

For medium to hard or coarse grained soils:

$$\text{Log}\,\rho = 3.1139 + 0.05\ N_r \qquad (4)$$

$$\text{Log}\,\rho = 2.9031 + 0.04762\ N_a \qquad (5)$$

It may be mentioned here that these relationships are approximate and may be valid within the resistivity ranges studied here.

5 CONCLUSIONS

On the basis of the studies conducted here the following conclusions have been summarised.

Semilog plot exhibits bilinear relationship between SPT blows, both recorded as well as adjusted; and absolute resistivities. This relationship distinguishes soft strata from that of comparatively harder strata in the form of two straight lines with different slopes.

The correlation may prove to be useful in similar subsurface conditions for interpreting SPT blows just by conducting electrical resistivity soundings.

6 ACKNOWLEDGEMENTS

The author expresses his sincere thanks to Geophysicists and Soil engineers of Centre of Studies in Resources Engineering, I.I.T.Bombay for valuable help and discussions. The author is very much thankful to the faculty and staff of Soil

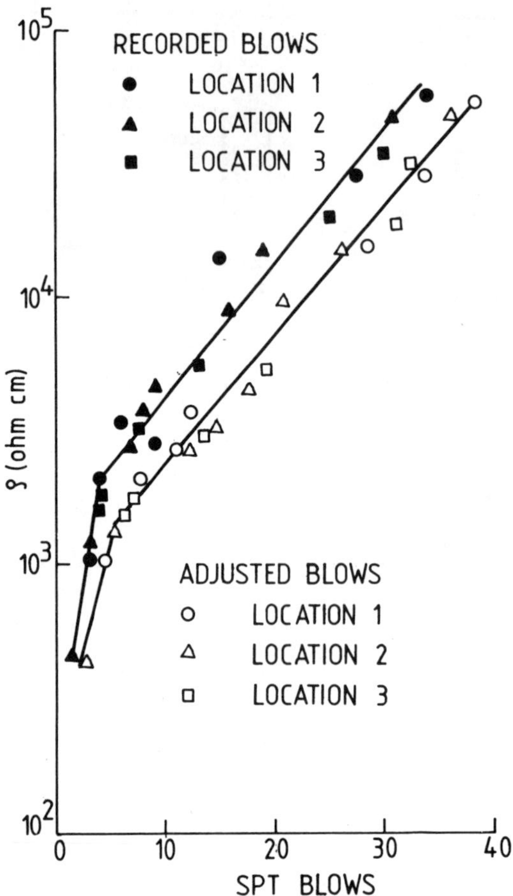

Fig.4 Relationship between absolute resistivity and SPT blows

and Rock Mechanics Division and Highway Engineering Section, Civil engg. deptt. of I.I.T. Delhi.

7 REFERENCES

Backus,G.E. and Gilbert,J.T.1967, Numerical applications of a formalism for geophysical inverse problems. Geophysics Journal, Royal Astro. Soc.13:247-276.

Baig,M.Y.A. 1980, Direct slope technique of determining absolute resistivity. Journal of Civil Engg. Div., Institution of Engineers (India) 61: 55-60.

Compagnie General de Geophysique 1963, Master curves for electrical soundings. European Association of Exploration Geophysicist, The Hague, The Netherlands.

Gibbs,H.J. and Holtz,W.G. 1957, Research on determining the density of sands by spoon penetration test. Proc. 4th Int. Conf. on Soil Mech. 1: 35-39.

Indian Standard 2131. 1963, Method for standard penetration test for soils. Indian Standards Institution, New Delhi, India.

Kate, J.M. 1981, Mean log resistivity technique for absolute resistivity determination. Submitted for publication, Indian Geotechnical Journal.

Kate, J.M., Govind Ram and Tokhi, V.K. 1981, Field studies on electrical resistivity with different circuit arrays. GEOMECH-81, Hyderabad, India.

Moore, R.W. 1945, An Empirical method of interpretation of earth resistivity measurements. AIME, Technical publications, 1743.

Moore, R.W. 1961, Geophysics efficient in exploring the subsurface. Journal of Soil Mechanics and Foundation Division, Proc. ASCE. 87, SM-3: 69-100.

Onodera and Seibe 1970, An analytic interpretation of apparent resistivity sounding curve for a multiple layered earth. Memoirs of the faculty of Engineering, Kyushu Univ. 29,2 : 107-120.

Sankar Narayan, P.V. and Ramanujachary, K.R. 1967, Short note-An inverse slope method of determining absolute resistivities. Geophysics, 32, 6 .

Terzaghi, K. and Peck, R.B. 1948, Soil Mechanics in engineering practice. New York, John Wiley & Sons.

Proceedings of the Second European Symposium on Penetration Testing / Amsterdam / 24-27 May 1982

SPT – CPT correlations

J.KRUIZINGA
Delft Soil Mechanics Laboratory, Netherlands

1. INTRODUCTION

In those cases where only SPT-results are available, engineers, who are more familiair with CPT interpretations, will translate the SPT N-values into CPT q_c-values. Therefor use can be made from correlation data published by several authors. A selection has been compiled in table 1. It appears that the scatter in sand is rather large $q_c = 3 - 10$ N.

The results of the site investigation at Marsden Point, New Zealand (North Island) offered the opportunity to check the above collected q_c - N correlation and in addition to find out if a f_s - N correlation is more realistic (f_s = CPT local friction).

2. SITE INVESTIGATION

In 1980 and 1981 hundreds of Dutch Cone Penetration Tests (CPT's) have been carried out mechanically, according to the ASTM standard D 3441 - 75 T. Besides cone resistance q_c at most of the locations local friction f_s was also measured. Based on the CPT-results 25 locations were selected for boring and sampling purposes. During the process of boring (wash boring type with rotary drill rigs and bentonite drilling mud) the Standard Penetration Test (SPT) was performed at intervals of approx. 1.5 m in accordance with BS 1377 : 1975. Alltogether 338 SPT N-values (blows/300 mm) could be compared with the CPT - - characteristics+

Table 1. q_c - N correlations

Soil type	q_c /N	Nr of comp. tests	Author
A	8-10		Schmertmann (1970)
A	18		Meigh-Nixon (1961)
B	5-6		Schmertmann (1970)
C	8		Meigh-Nixon (1961) and
C			Y. Lacroix (1971)
C	10	122	De Alencar Velloso (1959)
C	4		Meyerhof (1956)
D	3-4		Schmertmann (1970)
D	6	104	De Alencar Velloso (1959)
E	3-5	131	De Alencar Velloso (1959)
F	2	120	De Alencar Velloso (1959)
F	2		Schmertmann (1970)
F	4-5		Franki
G	3-5	202	De Alencar Velloso (1959)
G	2-3		Franki

q_c in kgf/$_{cm^2}$

N in blows/300 mm

A = sandy gravels and gravels
B = coarse sands and sands with little gravel
C = sand
D = clean, fine to medium sands and slightly silty sands
E = sandy silt
F = sandy clay, silty sand, cohesive silt-sand mixtures
G = clay, silty clay, clayey silt

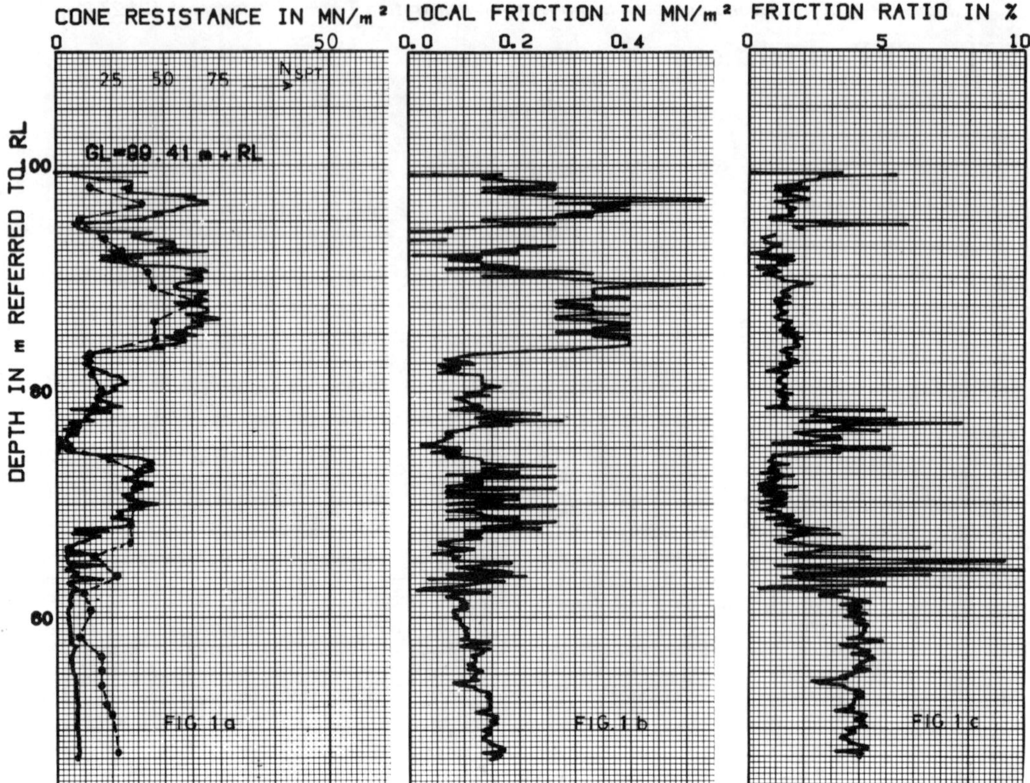

2 SOIL PROFILE

Very general the soil profile can be described as follows:

I 0 - 6 m medium fine sand, in a medium dense state

II 6 - 10 (15) m medium fine sand, in a dense state

III 10 (15) - 20 (25) m fine, silty sand with soft interbedded clayey silt layers

IV 20 (25) - 50 (55) m silty clay and clayey silt layers

The water table is about 1 - 2 m below grade level.

A typical CPT-result of the site has been presented in fig. 1 a,b,c, the corresponding field N-values are indicated as well.

The friction ratio follows the rules of Begemann:

Sandy toplayers (I, II):

$f_s : q_c$ is 1.5% taken on an average

Deep clayey silt layers (III):

$f_s : q_c$ is about 4% ($f_s = 1/25\ q_c$)

3 NUMBER OF COMPARATIVE TESTS AND TYPE OF SOIL

In the sandy toplayers 257 tests could be compared of which 22 have not been taken into account.Test levels at the transition of layers as well as tests with suspected SPT-values, were abandoned. Fig. 2a shows the type of sand involved.

In the silty and clayey silt (or silty clay) layers (fig. 2b, 2c) 81 tests have been carried out of which 16 were not suitable for comparison, in particular the very low N-values (N<5) in the silt layers. The abandoned values are corresponding with relatively too low q_c-values as well as too high q_c-values.

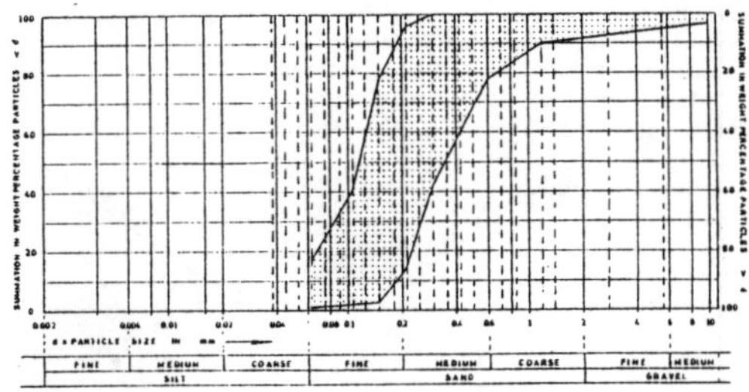

Fig. 2a (fine sand)

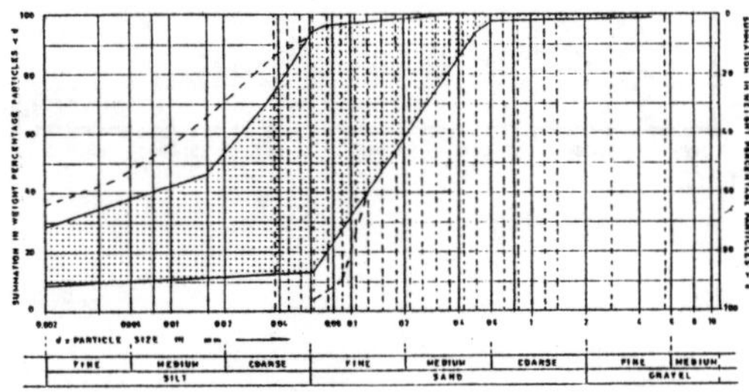

Fig. 2b (clayey silt)

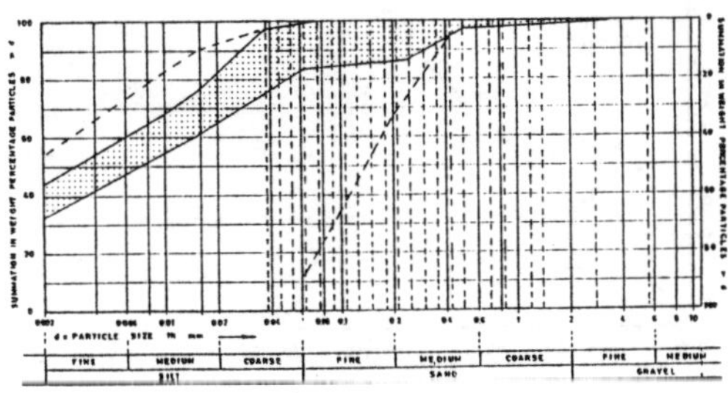

Fig. 2c (silty clay)

4 CORRELATIONS BY MEANS OF REGRESSION ANALYSIS

The selected field readings (N>5) were introduced into a computer programme REGRES to determine the best fitting curve. The degree of fitting is expressed by the standard deviation S and the index of determination ρ. If $\rho^2 = 0$, no regression exists; $\rho^2 = 1$ expresses a perfect fitting with no stochastic component.
The following functions were investigated.

Table 2. Results of regression analysis

Function y =	Sand (q_c and f_s in MN/m²)	
$a_o + a_1x$	$q_c=3.27+0.369N$	$\rho^2=0.52$
	$f_s=0.0668+0.005N$	$\rho^2=0.47$
$a_o + a_1x + a_2x^2$	$q_c=0.5+0.59N-0.0034N^2$	$\rho^2=0.54$
$a_1e^{a_2x}$	$q_c=7.248e^{0.02N}$	$\rho^2=0.45$
$a_1x^{a_2}$	$q_c=1.22N^{0.73}$	$\rho^2=0.53$
$\frac{x}{a_1 + a_2x}$	$q_c=\frac{N}{1.46+0.0177N}$	$\rho^2=0.54$
Function y =	**Silt and Clay**	
$a_o + a_1x$	$q_c=0.85+0.1N$	$\rho^2=0.61$
	$f_s=0.047+0.0033N$	$\rho^2=0.60$

Only in case of sand and q_c-correlations trancedental functions have been introduced.

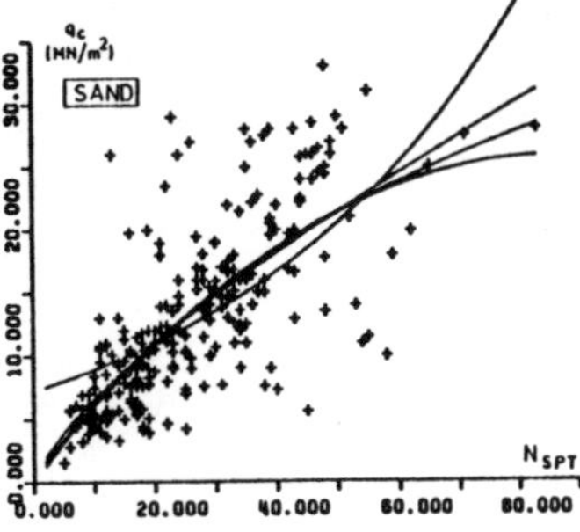

5 CONCLUSION

It can be concluded that the correlation q_c - N in sand is rather poor and is not much better in clayey silt or silty clay. It can also be concluded that the f_s - N correlation is not better than the q_c - N correlation.

If it is assumed that for $q_c = 0$ also N = 0, then the following simply polynomial function of the first order has been derived:

Sand	$q_c = 0.45N$	(fig. 3a)
	$f_s = 0.0068N$	(fig. 3b)
Silty clay	$q_c = 0.13N$	(fig. 4a)
Clayey silt	$f_s = 0.0048N$	(fig. 4b)

q_c and f_s in MN/m2
N in blows/300 mm

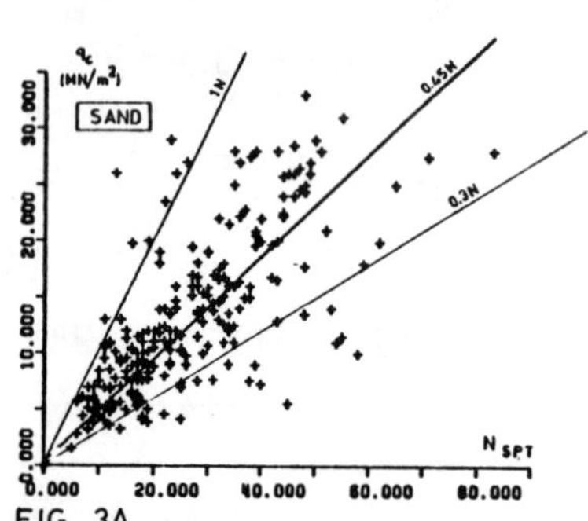

FIG. 3A

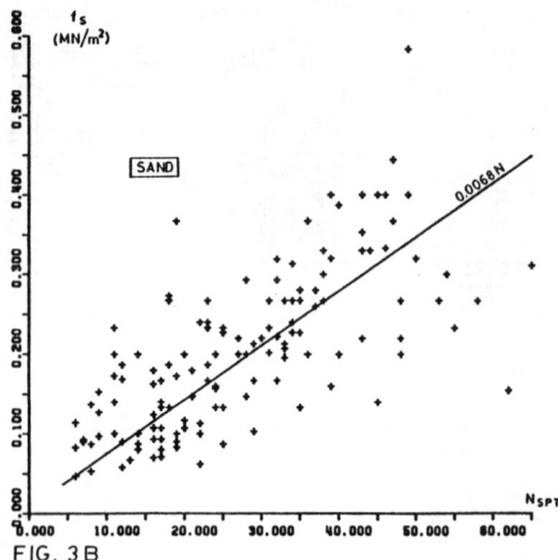

FIG. 3B

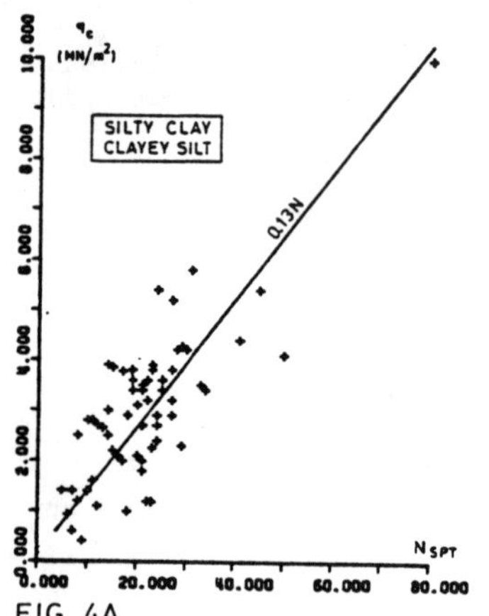

FIG. 4A

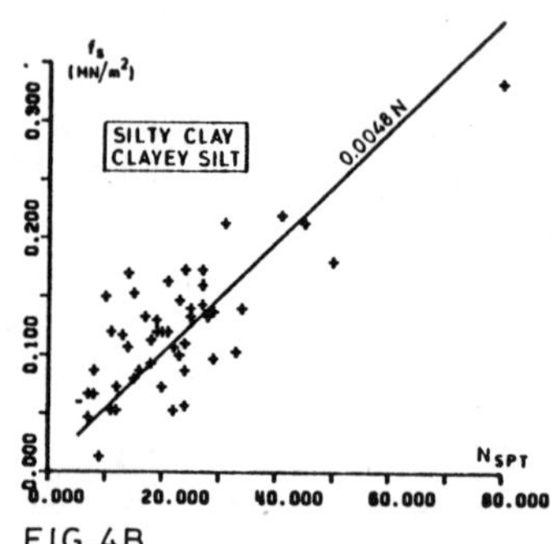

FIG. 4B

Proceedings of the Second European Symposium on Penetration Testing / Amsterdam / 24-27 May 1982

Direct and indirect determination of alluvial sand deformability

E.MARANHA DAS NEVES
Laboratório Nacional de Engenharia Civil, Lisbon, Portugal

1 INTRODUCTION

Of course the great importance of "in situ" geotechnical characterization of more or less thick formations of non cohesive soils is due to the difficulty there is in sampling this type of soils. The use of SPT and CPT test results (by far the most common tests) in foundation design is done by empiric methods so that it is important to assess the correlation on which they are based. In the case described advantage is taken of the execution of the Coimbra gated dam foundations to compare the deformabilities deduced from the results of the two kinds of penetration tests with those measured by plate load tests carried out at various depths. "In situ" unit weights were also measured as well as a grain size characterization. Assessment of sand deformability according to initial compactness was also attempted in laboratory.

2 SITE AND TEST CONDITIONS

The tests were only possible by virtue of the structure of the piers of the foundation of the dam pillars. As these are structural elements of an indirect foundation through an alluvial sandy medium they were materialized by diaphragm walls delimiting four cells in each one of them (Figure 1). After the construction of the walls the alluvial material confined by the diaphragm walls was removed and the cells, now only with water, were filled with concrete. The CPT and SPT tests were carried out at the upstream cell of pier 10 before dredging of the sand. Then, a manual excavation was made and plate load tests, sand unit weight determinations, and integral sampling (class 4) were carried out at different depths. Precautions were taken for the excavation, which was accompanied by pumping in order to keep the phreatic level deeper than the excavation level. Technical details can be found elsewhere (LNEC, 1981).

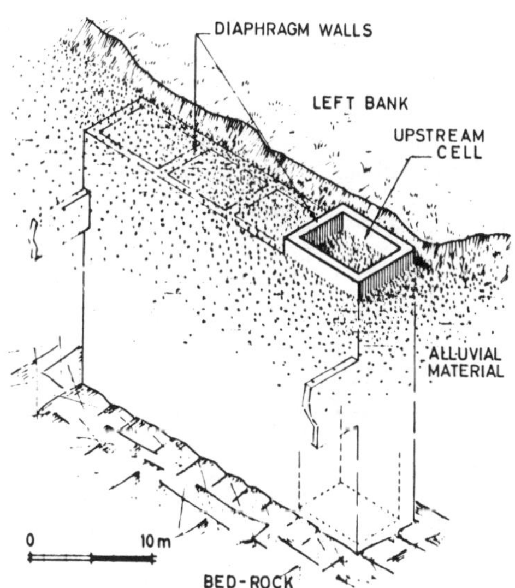

Fig. 1 Structure of no. 10 pier in Coimbra dam before dredging and concreting the cells. Cell tested

3 TEST RESULTS

The results of the tests are described hereafter and an interpretation of them is given in paragraph 4.

3.1 Indirect determinations. CPT and SPT tests

These tests were made before starting the excavation, with the sand of the cell still at level + 16.50. Two CPT tests (I and II) and one SPT

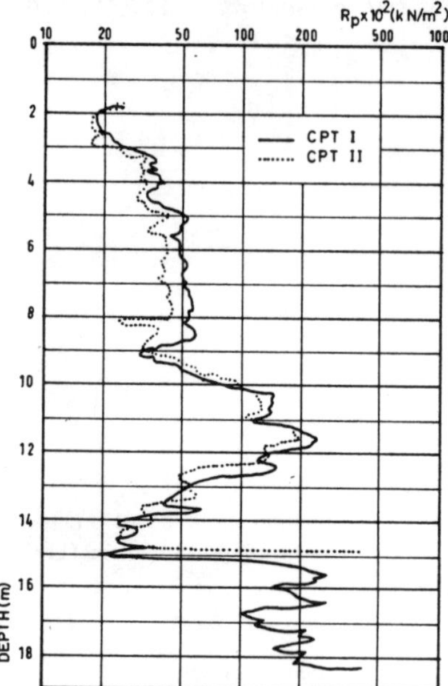

Fig. 2 Results of CPT tests

test were made. Results are given in Figures 2 and 3 respectively. The correlation between the

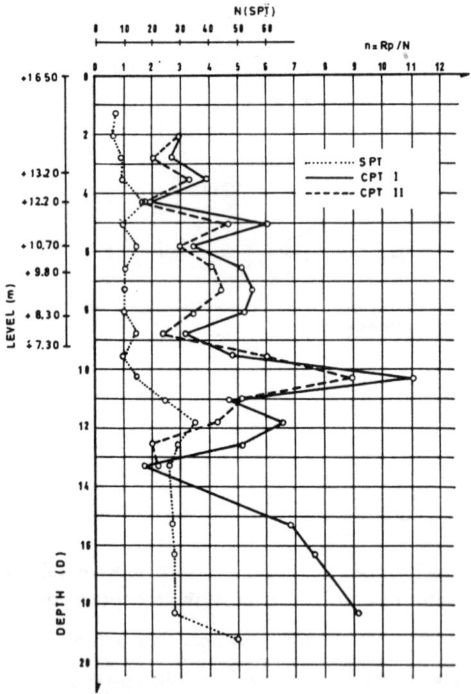

Fig. 3 Results of the SPT test and evolution of n with D

SPT (N) and the CPT (R_p) values at the same depth is expressed graphically in Figure 3.

3.2 Direct determinations. "In situ" and laboratory tests

Load tests with rigid, 0.55m diameter plates were carried out at different depths, as well as compactness determinations and integral sampling. Figure 4 shows the results of the plate load tests

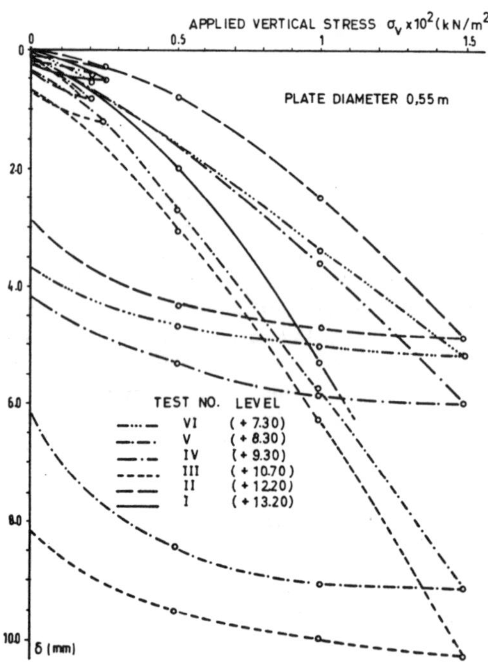

Fig. 4 Plate load tests at different depths

in the form of graphs (σ_v, δ). Figure 5 shows the grain size curves of the sandy materials of the integral samples. One-dimensional compression tests were carried out on these materials. The description of this type of tests (with no shear wall effects) can be found elsewhere (LNEC, 1977). Tests for three different initial dry unit weights (γ_i) were made for each sample. Figure 6 gives the results of the deformability tests on one of the samples.

As to the "in situ" unit weight determinations, the magnitude of the errors involved in the determinations did not make it possible to obtain quantitative information for correlation with the results of the penetration tests. Qualitatively, a well known beahaviour was found: denser sands have greater resistance to penetration.

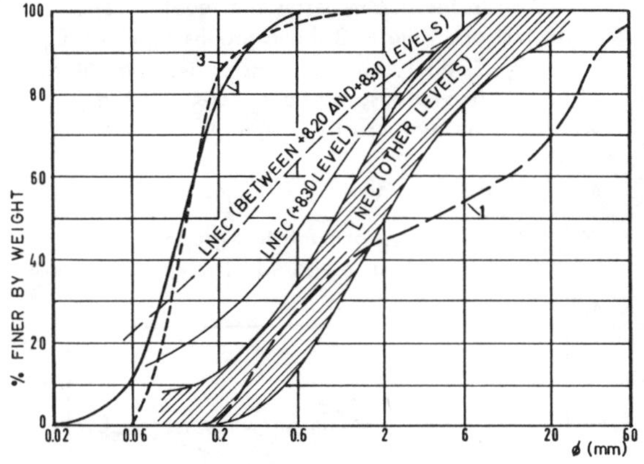

3 - - - - RODIN (FINE AND MEDIUM) NO.2 SITE.

1 ——— MEIGH E NIXON (VERY DENSE BROWN SILTY MEDIUM SAND) MOTHER WELL, LANARKSHIRE, UK.

1 — — — MEIGH E NIXON (MEDIUM GRAVEL WITH CORSE SAND) TABLE 2. DINSLAKEN, GERMANY.

(REF. IN THE TEXT)

Fig. 5 Grain size curves

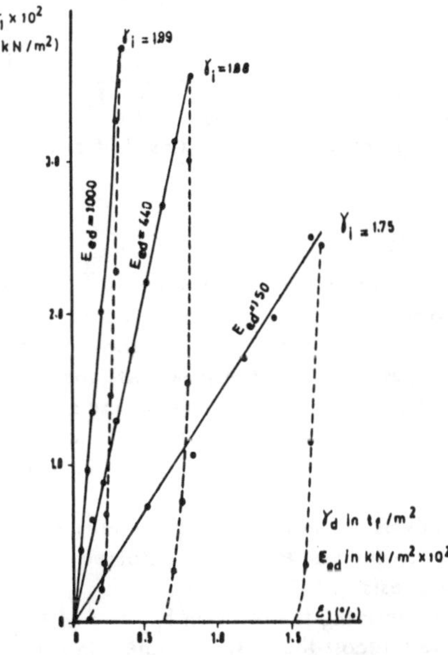

Fig. 6 One-dimensional compression test at three different initial unit weights (sand from + 13.2 level)

4 INTERPRETATION OF THE TEST RESULTS

This interpretation will take into account the possibility of correlating the values obtained by different experimental methods.

4.1 Correlation between CPT and SPT results

The matter has been much studied owing to the wide use made of these penetration tests (e.g. Folque, 1976). Figure 7 shows a generally accepted correlation, based on a large number of experimental results (Rodin et al, 1974). The grain size curves corresponding to the results obtained by two authors (Meigh and Nixon, 1961; Rodin, 1961) are presented in Figure 5. It gives an idea of how the representative diameter given in Figure 7 was established (the data on tests 1 to 7 may be found in Rodin et al., op. cit.). The same criterion was followed in the tests now carried out, the value used for n being the average of the results of the two CPT tests concerning the depth in question (Figure 2). The grain size characteristics are those presented in Figure 5.

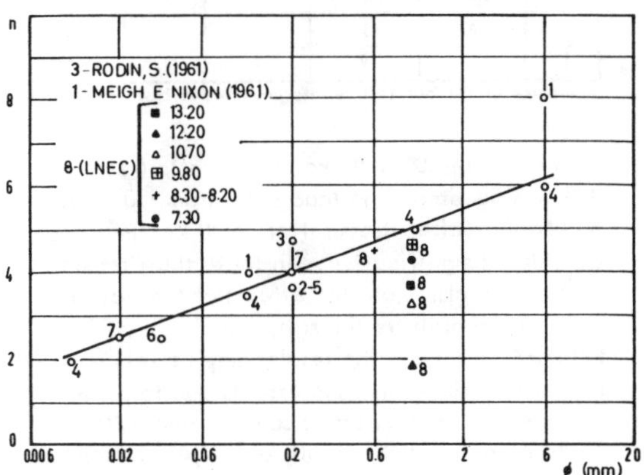

Fig. 7 n = f (Ø) plot for cohesionless soils

As can be seen the results of the tests carried out at greater depths are closer to the correlation, as opposed to the more superficial ones, particularly those at level + 12.20. By assuming that the grain size differences are not significative from a practical point of view, it is possible to analyse 35 correlations down to a depth of 18.3 m which gives a mean value of n equal to 4.5.

4.2 Correlation between the values of R_p and compressibility

Of the best known relations between R_p and the deformability E ($E = \alpha R_p$) we may quote those used by De Beer (1965) and Schemertmann (1970) in the forecasting of the settlements of structures with α equal to 1.5 and 2 respectively. E has the meaning of a modulus of deformability for a situation of uniaxial (E_{ed}) deformation. Based on the plate tests and on the theory of elasticity, a deformability modulus E_p may be deduced, by assuming vertical load and elastic, isotropic and semi-indefinite medium (Terzaghi, 1943). Table 1 shows values of E_p for different

Table 1. E_p values according different increments of σ_v and the corresponding SPT and CPT results.

PLATE LOAD TEST. NO.	$E_p \times 10^2$ kN/m² (ν = 0.3) $\sigma_v \times 10^2$ kN/m² 0-0.2(0.25*)	0-0.5	0-1.5 (1.0*)	N(SPT)	$R_p \times 10^2$ kN/m² a) I/II
I	200*	140	80*	9	35/31
II	350*	290	130	9	38/30
III	90*	100	70	14	49/42
IV	200	140	110	10	51/40
V	100	90	70	11	54/35
VI	150	150	130	12	50/53

a) Average of the four test results under each test level.

increments of σ_v. A Poisson ratio (ν) of 0.3 was adopted. The load step from 0 to 50 kN/m² (after stress relief of 20 kN/m²) was considered significative. In fact, the first step is meant to eliminate the effects of the adjustment of the plate to the soil; the thirth (0-1.5 k/N/m²), although the results may be less scattered, may already include effects due to plastifications (the yield stress of sand is about 60 kN/m²). Table 1 also gives the results of the penetration tests

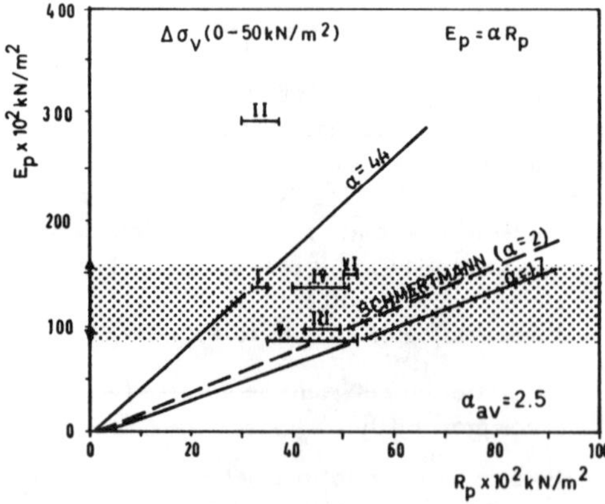

Fig. 8 Correlation between E_p from plate load tests and R_p from CPT tests

measured in the 0.6 m below the level of each plate test and Figure 8 shows graphs in which $E_p = \alpha R_p$ is analysed. By considering the value of test II anomalous, we have $\alpha_{av} = 2.5$, with extreme values of 1.7 and 4.4.

4.3 One-dimensional compression tests

With the values E_{ed} of each test graph $E_{ed} = f(\gamma_i)$ (Figure 9) was prepared. The maximum ver-

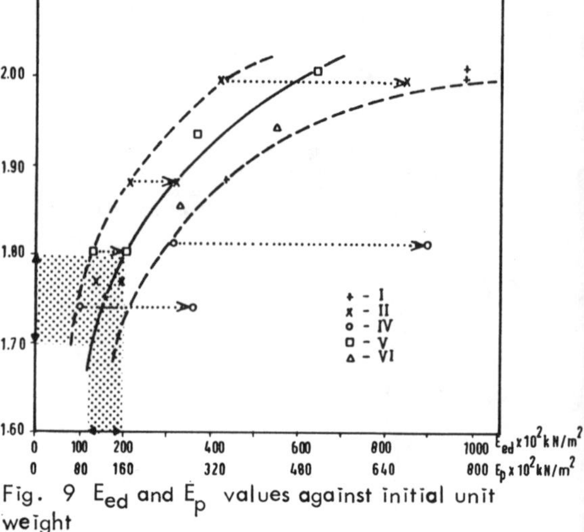

Fig. 9 E_{ed} and E_p values against initial unit weight

tical stress applied was of the order of 400 k N/m² and in some cases non-linear behaviour was observed. For this reason Figure 9 shows sometimes 2 values of E_{ed} for the same test, the E_{ed} values corresponding to the range of the smaller stresses of σ_v (which are quite similar to the σ_v of the plate tests) being considered more significative.

This analysis was meant to supply additional information on the values of α obtained in the plate load tests.

By characterizing the relation $E_{ed} = f(\gamma_i)$ we may try to forecast the most probable "in situ" deformation, taking into account the "in situ" compactness.

It was found that the range of the smaller values γ_i is the most interesting one considering the "in situ" compactness of those soils. For ν = 0.3 (resorting to the known Jaky relation with φ' = 35° deduced from the penetration tests) we get E (Young modulus) $\simeq 0.75\ E_{ed}$. The modulus E_p should show a value between E_{ed} and E. According to Barata (1973) $E_p \simeq 0.8\ E_{ed}$ so that in the present case E_p would have a value nearer to E (ν = 0.3).

Considering that the sand is loose to medium, ("in situ" measurements and indications of the penetration tests) the γ_i will be from 1.7 to 1.8 tf/m^3. In accordance with Figure 9 to these values correspond values of E_p comprised between 9000 and 16 000 kN/m^2 which agrees fairly with the values of E_p of Figure 8.

If in correlation $E = \alpha R_p$, E has the significance of a Young modulus then the value α will take on a slightly different value α'. Thus $\alpha' = \alpha \, E/E_p \simeq \alpha \, 0,93 = 2,35$.

5 CONCLUSIONS

From what has been said it can be concluded that:

1 Correlation n = 4 (Meyerhof, 1956) agrees closely with the average of the results obtained but this is not the case with the influence of the grain size as regards the work by Rodin et al. (1974, op. cit.) (greater n values for greater particle diameters). In fact only the soils located at greater depths behaved according to that proposal.

2 The experimental determinations of α may be done by comparing the values of R_p with the values of E deduced from "in situ" tests or from the settlements shown by structures founded at the place of the penetration tests. In the case described, "in situ' as well as laboratory tests were carried out.

For the saturated sands studied (medium to coarse with some gravel) a mean value of α equal to 2.5 was obtained.

6 ACKNOWLEDGEMENTS

The author wishes to thank Direcção Geral dos Recursos e Aproveitamentos Hidráulicos for having made this study possible, as well as Mr. Tavares de Castro, Resident-engineer at Coimbra dam for his helpful cooperation.

7 REFERENCES

Barata, F.E. (1970), Correlações importantes entre alguns módulos de deformações do terreno. IV COBRAMSEF, Rio de Janeiro, Vol. I, Tomo I: 100 - 116.

De Beer, E. (1965), Bearing capacity and settlement of shallow foundations on sand. Proc. Symp. Bearing Capacity and Settle.of Found., Duke University: 15 - 33.

Folque, J. (1976) Características mecânicas de solos deduzidas de ensaios de penetração. Geotecnia no. 17: 73 - 83.

LNEC (1977), Enrocamentos. Actualização de conhecimentos, estudos experimentais e aplicações em barragens e vias de comunicação. Lisboa.

LNEC (1981), Determinação da deformabilidade "in situ" de material aluvionar arenoso (açude--ponte de Coimbra). Lisboa.

Meigh, A.C., I.K. Nixon (1961). Comparison of in situ tests for granular soils. Proc. of the Fifth ICSMFE, Paris, Vol 1: 499 -507.

Meyerhof, G.G. (1956). Penetration tests and bearing capacity of cohesionless soils. Journal of Soil Mech. and Found. Div., ASCE, vol. 81, SM 1.

Rodin, S. (1961). Experiences with penetrometers, with particular reference to the standart penetration test. Proc. of the Fifth ICSMFE, Paris, vol. 1: 517 - 521.

Rodin, S., B.O. Corbett, D.E. Sherwood, S. Thornburn (1974). Penetration testing in United Kingdoom. Proc. of the Eurp. Symp. on Penetration Testing. Stockholm, vol. 1: 139 -146.

Schmertmann, J.H. (1970). Static cone to compute static settlement over sand. Journal of Soil Mech. and Found. Div., ASCE, vol. 96, SM 3: 1011 - 1043.

Terzaghi, K. (1943). Theoretical soil mechanics. John Wiley and Sons, Inc., New York.

Proceedings of the Second European Symposium on Penetration Testing / Amsterdam / 24-27 May 1982

Comparison of the results from static and dynamic penetration tests, in situ plate tests and laboratory compressibility tests

A.MARCU & M.POPESCU
Civil Engineering Institute, Soil Mechanics & Foundation Engineering Department, Bucharest, Romania

T.ABRAMESCU & C.BALACCIU
Institute of Studies & Projects for Land Reclamation, Bucharest, Romania

1 DETERMINATION OF FOUNDATION SOIL COMPRESSIBILITY CHARACTERISTICS

During the recent years, in the central part of Bucharest, a number of tall buildings provided with deep basements have been designed or were already built. Their direct founding on general raft foundation, at depths between 7 m and 16 m, put forward the necessity of performing studies on the entire thickness of the layers of recent age (Upper Pleistocene), consisting of interbeded sandy and clayey layers, under which - at a depth of about 40 m - a marly clay layer, less compressible, is present. Fig. 1 shows a typical profile for this area.

Due to the presence of sandy layers under the groundwater level, out of which undisturbed samples are difficult to be taken, in order to determine the compressibility characteristics, there have been used mostly standard penetration tests (SPT), cone static penetration tests (CPT), as well as plate bearing tests in large diameter boreholes. The cone static penetration tests were carried out using penetrometers of the "Dutch" type, while the standard ones were performed using the device lowered into the borehole and driven from outside by means of a cable. As regards the standard penetration tests carried out under the groundwater level, the corrections regarding the slowing down effects of the hammer falling into the water, were taken into account.

Within the area under study, a great number of cone and standard penetration tests were performed, allowing a statistically processing of data with taking into consideration stratification uniformity. Fig.1 presents the envelope diagrams and the variability diagram for the average value of the cone resistance q_c, as well as the variability diagram for the average value of the blowing number, N_{med}, resulting from standard penetration tests.

Static penetration tests could not be continued beneath the + 60 elevation, because of the great depth (more than 20 m below the surface) and of the high resistance on the cone in the sandy-gravel layer.

Alongside the penetration tests, plate tests were also carried out, in two boreholes, at every two meters depth, for determining the compressibility of the layers below the foundation level of the building under design. The plate diameter was 28.2 cm. Undisturbed samples were also extracted from the clay layer ; their compressibility was determined in the laboratory by means of oedometer and triaxial compression tests.

In Romania, the method for determining soil compressibility by plate tests directly into the borehole or into open pits has conducted to good results in estimating building settlements. The settlements of direct foundations (isolated foundations, continuous foundations or rafts) calculated on the basis of the elastic semispace model or of the finite thickness elastic layer, using the values of the linear deformation modulus, determined by plate tests, proved to be close to the measured ones (Marcu, 1977).

Plate tests were carried out according to the scheme showed in fig.2 ; after plate loading up to a unitary pressure, equal with the geological pressure, σ_g, the loading with 0.05 ... 0.1 MPa pressure increments has been started, maintaining a constant pressure during the interval necessary for the settlement, s, to get stabilized.

The E deformation modulus was determined in the proportionality domain of the relationship between pressure and settlement, using Schleicher's formula regarding the settlement of a rigid plate, resting upon an elastic semispace.

The E values, determined by plate tests at different depths are showed in fig.1 ; a variation quite similar to that of N and q_c values is to be noted. The comparison of the E and q_c range of values allows checking of certain correlations, statistically established. Thus, as regards clayey

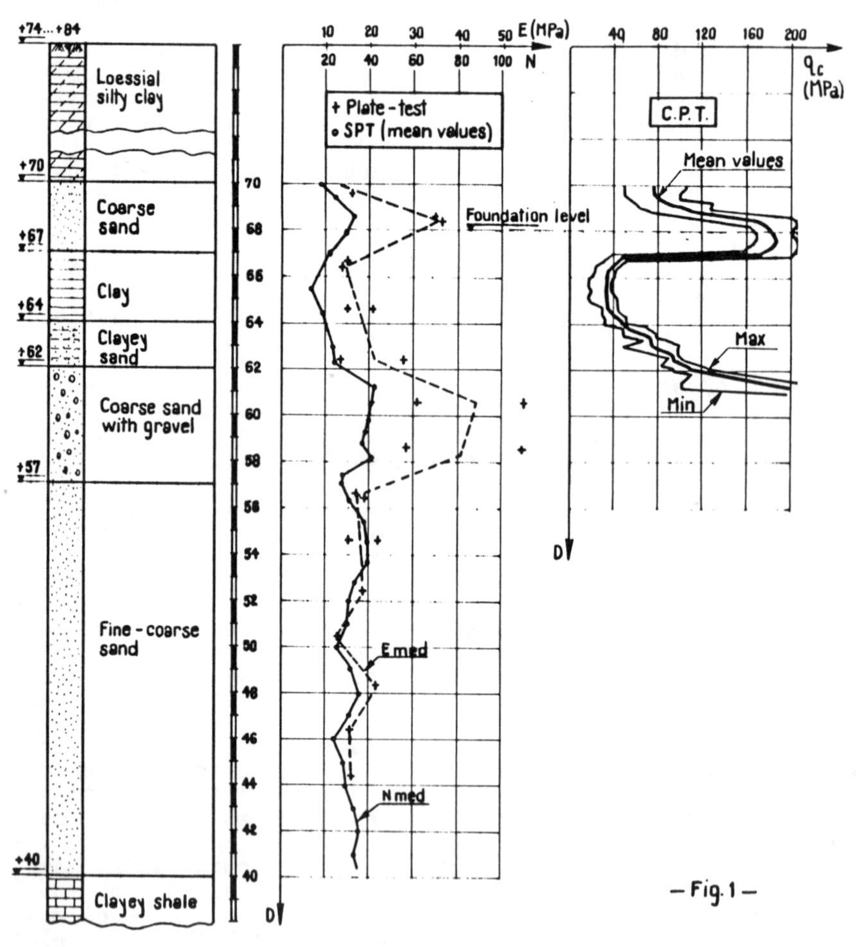

– Fig. 1 –

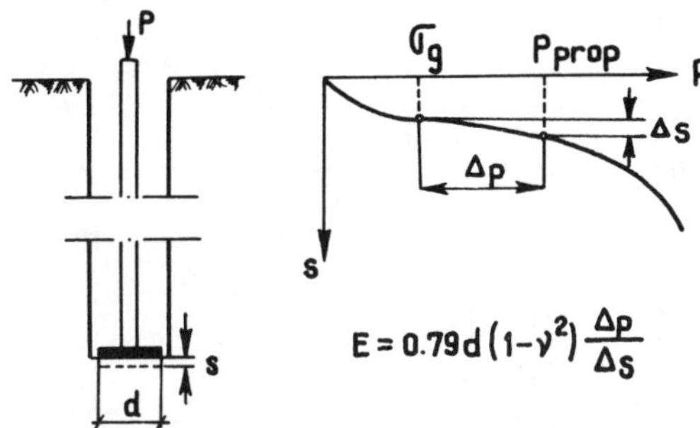

Fig. 2

soils of quaternary age, a statistic relationship was previously established (Marcu, 1977) on the basis of the results provided by 25 plate tests, carried out in parallel with static penetrations, on different locations in the southern part of Romania (The Danube Plain). Taking into account the influence of the consistency index, I_C, the geological pressure σ_g (in MPa), at the testing level and the soil porosity, n, (%) the statistic correlation (for a reliability level $\alpha = 0.95$) was obtained :

$$E = 3.8\ q_c + 2.84\ I_C - 12.7\ \sigma_g$$

$$- 0.553\ n + 25.9 \pm 0.425$$

The following statistic correlation coefficients have resulted :

- the bulk correlation coefficient : $R = 0.979$
- the partial correlation coefficients: $r\ (E, q_c) = 0.937$; $r(E, I_C) = 0.140$; $r(E, \sigma_g) = 0.294$; $r(E,n) = 0.801$

As the partial coefficients of correlation show the reduced influence of the I_C and σ_g values, the following simplified correlation may be accepted :

$$E = 3.8\ q_c - 0.55\ n + 26 \qquad (1)$$

The average value of the cone resistance q_c at the testing level, deduced from the static penetration tests, was represented in fig.3 in accordance with each E value determined by means of the plate tests in the clay layer. The dependance resulted from the relation (1) was represented in the same graph for an average porosity n = = 43 % :

$$E = 3.8\ q_c + 2.5 \qquad (1')$$

A good grouping of the experimental points around the straight line defined by the relation (1') is to be noted.

For comparing, the linear deformation modulus was also determined in the laboratory by testing undisturbed samples taken from the clay layer in oedometer and in the triaxial compression device.

The modulus of linear deformation was determined in oedometer within the range of pressures between the geological pressure, σ_g, at the depth from which the sample was taken, and a pressure equaling

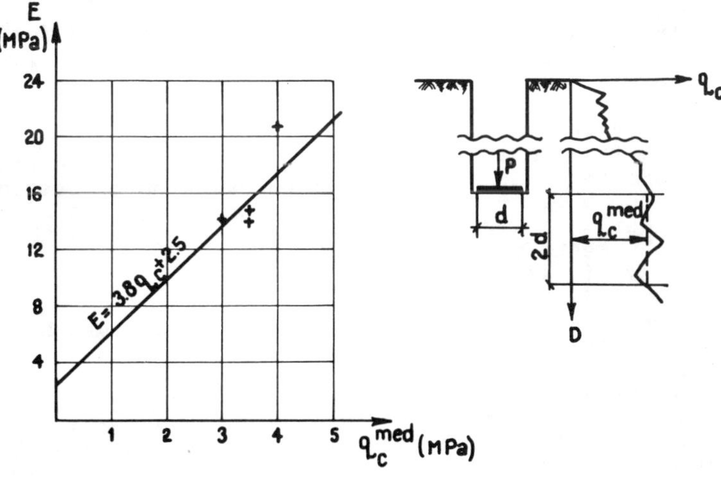

Fig. 3

σ_g + 0.2 MPa, because the increase of the vertical stress at the level of the clay layer, due to the raft foundation load will be of approximately 0.2 MPa.

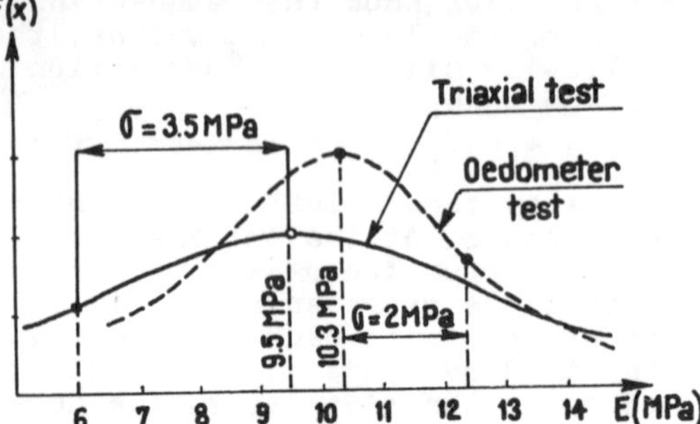

Fig.4

In the triaxial compression device the saturated clay samples were initially consolidated under the principal stresses $\sigma_1 = \sigma_g$ and $\sigma_3 = K_o \cdot \sigma_g$ (K_o being at rest pressure coefficient), following an increase of the efforts in 4 ... 6 pressure increments : $\Delta\sigma_1$ = = 0.05 MPa and $\Delta\sigma_3 = K_o \cdot \Delta\sigma_1$. During each pressure increment the vertical and volumetric strain were registered until their stabilization. In the range of the above mentioned pressure values, a linear stress-strain relationship has resulted, thus permitting determination the Poisson coefficient, ν, and of the linear deformation modulus, E (Caquot and Kerisel, 1966).

The diagrams of the density distribution for the E values resulted from the statistical processing of the data obtained by means of oedometer and triaxial compression tests are presented in fig.4. It must be noted the fact that the average values are in close agreement for both methods of determination, but are lower than the E values determined by means of the plate bearing tests (values which vary between 14 and 21 MPa). The effect of disturbing natural structure during soil sampling and performing the tests in the laboratory, is thus emphasized.

For the sand layers, the correlation between the experimental E values, determined by plate tests and the q_c values, is shown in fig.5. In the same figure there

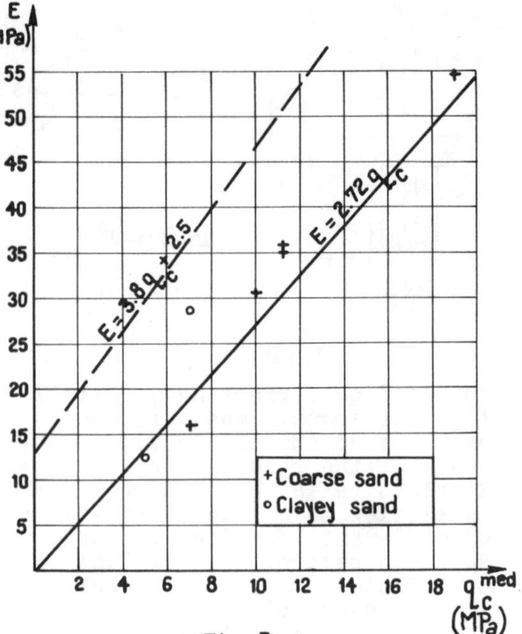

Fig. 5

are also represented some correlations mentioned in the technical literature.

The correlation recommended by Vesic (1970)

$$E = 2(1 + I_D^2)q_c \qquad (2)$$

becomes for an average relative density I_D = 0.6 (determined on the basis of the SPT data) :

$$E = 2.72\ q_c \qquad (2')$$

The correlation obtained by Trofimenkov and Vorobkov (1974) for saturated alluvial sands is :

$$E = 3.4\ q_c + 13 \qquad (3)$$

It is noted that for the intercepted sandy formations, including the layer of clayey sand, the experimental data are grouped satisfactorily around the straight line of correlation corresponding to equation (2'), remaining lower than the values shown by the equation (3).

2 ESTIMATED AND MEASURED SETTLEMENTS

The data obtained by the plate bearing tests in boreholes, correlated with the results of the static penetration tests, allowed the determination of the average values of the linear deformation modulus, E, for every stratum, as well as the domain of variation of these values for the construction site. On the basis of these values the average settlement and the probable unequal (differentiated) settlement were computed.

The high building for which the above mentioned study was achieved will be founded at + 68 elevation on a general raft foundation with a square-like shape with the side of 60 m (having trimmed corners and a central cavity).

The adopted computation model was that of the elastic layer of finite thickness (Egorov, 1961) resting at + 40 elevation on the marly clay layer considered as nondeformable. Admitting that the settlement due to the weight of the underground part of the structure will take place during the execution phase, the settlement due to the superstructure which brings an increment of pressure of 0.08 MPa on the raft foundation was calculated.

An average settlement s = 4 cm has resulted, of which 1 cm corresponds to the clay layer located close beneath the foundation level.

The estimation of the settlement directly from the data of the penetration tests was also performed on the basis of the empirical relationships recommended in the technical literature (Szechy,Varga, 1978). As the static penetration tests were not executed on the whole depth of the compressible layers, the calculation was made using only the data of the standard penetration tests.

In accordance with the relationship recommended by Meyerhoff (1965) after changes necessary for the transition to the international system of units (SI), the settlement, of the foundation is :

$$\bar{s} = \frac{2.84\ p}{N} = 9.88\ \text{mm}$$

where : p = 80 kPa
N = 25 - the average number of SPT blows over the deforming zone.

It was noticed that $\bar{s} << s$, showing that the above mentioned relationship cannot be applied in case of the raft foundations of large dimensions.

Beneath the foundation of the structure under discussion, deep settlement marks are to be placed, to measure the effective settlement of each layer and determine the real values of the linear deformation modulus.

Presently we have the measured settlements of a 21 stores hotel located in the neighbourhood of the investigated zone and founded on general raft foundation, on the

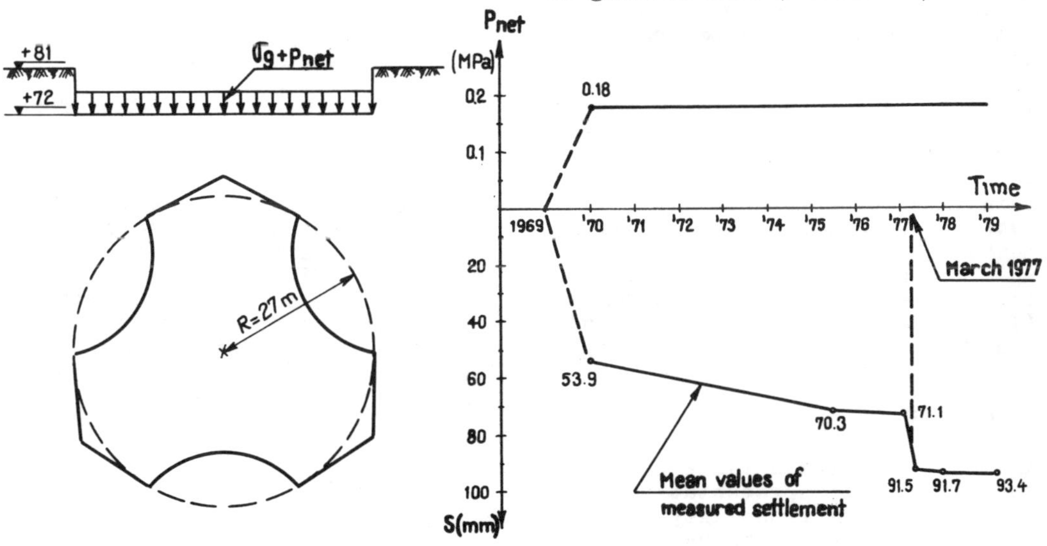

— Fig.6 —

same complex of layers.

The variation in time of the foundation average settlements is shown in fig.6, the measurement being commenced after the execution of the underground part of the structure, so that after a pressure approximately equal to the geological pressure existing before the execution of the excavation was transmitted (the settlements were recorded by "PROIECT BUCURESTI" Institute).

The settlement calculated for this raft foundation loaded with the distributed pressure p = 0.18 MPa, by the method of the elastic layer of finite thickness, considering the values $E_{average}$ established by the plate bearing tests on the location of the new structure (fig.1), was of about 7.5 cm, that confirms the validity of these values of the linear deformation modulus.

It is also noted the supplementary settlement of about 2 cm suffered by the existing structure as a consequence of the earthquake of March 4, 1977. At the above mentioned date, the S-Eastern part of Europe was shaken by an earthquake with the epicenter at the distance of about 150 km from Bucharest. The magnitude of the earthquake in the focus was of 7.2 on the Richter's scale, the intensity in the Bucharest area of about 8.5 degrees of the Mercalli's scale and the period of the seismic movement of about 1 second. The amount additional settlement (less than 0.1 % of the total thickness of the sandy layers), shows a slight influence of the seismic effect on the sand density. This is also correlated with the lack of evident phenomena of liquefaction of the saturated sand and is explained by the relatively high values of the relative density (I_D = 0.5 ... 0.8) which were deduced from the SPT tests.

3 CONCLUSIONS

The "in situ" tests allow the determination of representative values for the deformability characteristics of soils, values which are confirmed by the effectively measured settlements of the structures.

The correlations (1) and (2) between the values of cone resistance, q_c and the values of the linear deformation modulus, E, determined by plate bearing tests, are in good agreement on different locations consisting of alluvial layers of quaternary age, normally consolidated.

The empirical formula of estimating the settlements, directly from the data of the penetration tests, can be applied only for formations and foundations similar to those for which they were initially obtained, the generalization of their application for all types of direct foundations being unjustified by the practical results.

4 REFERENCES

Caquot, A. & Kérisel, J. 1966, Traité de mécanique des sols, Paris, Gauthier - Villass.

Egorov, K.E. 1961, Stress and strain distribution in an elastic layer of finite thickness, Sbornik, N. I.I. Osnovanii - Mekhanika Gruntov N 43, Moskow, Stroiizdat (in Russian).

Marcu, A. 1977, A comparison between laboratory and field values of cohesive soil compressibility characteristics, Proc. of the 5th Danube Conf.on Soil Mech.and Found.Eng., Bratislava : 195 - 203.

Meyerhoff, G.G. 1965, Shallow foundations, Journ. Soil Mech.Found. Div., ASCE, vol.91, SM2.

Széchy, K. & Varga, L. 1978, Foundation engineering. Soil Exploration and spread foundations, Akadémiai Kiadó, Budapest.

Trofimenkov, I. & Vorobkov, L. 1974, Field methods for investigating geotechnical properties of soils, Moskow, Stroiizdat (in Russian).

Vesic, A,S. (1970), Tests on instrumented piles, Ogeeche River Site, Proc.of the ASCE, vol.96, SM 2, March.

Proceedings of the Second European Symposium on Penetration Testing / Amsterdam / 24-27 May 1982

Effects of rod diameter in the standard penetration test

KAZUAKI MATSUMOTO
Port & Harbor Research Institute, Ministry of Transportation, Yokosuka, Japan

MIKIO MATSUBARA
OYO Corp., Urawa Research Institute, Japan

1 INTRODUCTION

Of the many sounding methods in use in Japan today, the Standard Penetration test is the most commonly used. This method was originated by H.A. Mohr about 1927. The first samplers used in the test were somewhat smaller than that which is used today, the present standard size having originated by an improved design by the Raymond Concrete Pile Co.

In a work written by Terzaghi and Peck in 1948 comparative experimental results showing relationships between N values and density of sand, and between N value and allowable bearing capacity of sand and clay ground were presented, leading to the general acceptance of the Standard Penetration Test as an in-situ testing method. Although this method came into use for site investigations in Japan from around 1951, any number of variations on the procedure were used due to the fact that the Terzaghi and Peck reference only presented a general outline of the test method, with no detailed specifications. Accordingly, the Japan Society of Soil Mechanics and Foundation Engineering's Sounding Committee applied itself to the drawing up of a proposal to standardize the test, resulting in the Japan Industrial Standards (JIS) specifications that are in use today.

In 1965 a Subcommittee for the Standardization of Penetration Testing was formed by the European International Society of Soil Mechanics and Foundation Engineering. The subcommittee's activities continued until it had prepared a proposal for the standardization of both static and dynamic sounding, the Standard Penetration Test and the Swedish Sounding. These standards were presented at the International Conference on Soil Mechanics and Foundation Engineering held in Tokyo in 1977. Of these, the specifications for the Standard Penetration Test do differ somewhat from the JIS standards, although they are similar. Specifically, the European standards call for a rod diameter of 43.7 mm when testing is to a depth of 15 m or less, while for greater depths either steadies are to be attached every 3 meters or a 54.0 mm diameter rod is to be used. In contrast to this, the JIS specifications call for a rod diameter of 40.5 mm or 42 mm, amounting to a major discrepancy.

In order to investigate the effects on results obtained in the Standard Penetration Test, if any, by using rods of differing diameters, the authors conducted comparative experiments in the field using JIS standard rod diameter 40.5 mm, and rods of 50 and 60 mm diameters. This paper gives the results of those experiments and discusses what rod diameter is desirable in the Standard Penetration Test.

2 EXPERIMENTAL OBJECTIVES AND PROCEDURES

2.1 Comparison of Standard Penetration Test results obtained using rods of different diameters

A number of boreholes were drilled at 5 meter intervals in the same type of ground. In each of these boreholes, the Standard Penetration Test was carried out using rods of different diameters and the results were compared. Table 1 shows the specifications of the rods used. The test site was ground reclaimed from the sea, with an elevation of -2 meters. The first 25 meters of this ground was soft clay. From 25 to 37 meters was a hard diluvial layer, below which was weathered rhyolite. The comparative experiments were carried out between 37 and 45 meters. The boreholes were all 86 mm in diameter.

Table 1. Rod specifications

Nominal Diameter	Outer Diameter	Inner Diameter	Weight	Coefficient of Elasticity
mm	mm	mm	kg/m	kgf/cm^2
40.5	40.5	31.8	4.34	$2.18x10^6$
50	50.2	37.9	7.28	$1.97x10^6$
60	60.2	48.0	9.14	$1.81x10^6$

A 40.5 mm diameter rod, 50 cm in length was connected to the sampler used in the experiments. To this were connected other rods, variously of the same diameter and 50 and 60 mm in diameter. The hammer guide rod had a diameter of 40.5 mm. Every effort was made to remove abstractions from the free fall of the hammer. Number of blows and amount of corresponding penetration figures were automatically recorded.

2.2 Measurement of energy transmitted by the rod

In the Standard Penetration Test, the potential energy of the hammer is transmitted to the knocking head, which is in turn transmitted by the rod to the sampler, resulting in the penetration of the sampler into the ground. The energy transmitted by the rod was measured to see if it differed according to rod diameter. In these experiments, 2 of the boreholes used in the Standard Penetration Test result comparison experiments were used in the reclaimed ground noted above.

Method of measurement

The same method used by Uto et al. (1973 a, 1973 b) in their investigation of the mechanism of penetration of the rod in the Standard Penetration Test was employed. In this method, the strain in the rod by the hammer blow is measured as a waveform by a strain gauge attached to the rod. From these strain waveforms, the energy transmitted by the rod is known. Details of this method are given by Uto et al. (1973 a, 1973 b, 1975). Half of the energy transmitted by the rod is strain energy and half is energy of movement. In the experiments conducted by the authors, only strain energy was measured, and this figure was doubled to obtain the energy transmitted by the rod.

Rods and strain gauges used in the experiments

Table 1 lists the 3 types of rods used. The foil strain gauges used had a resistance of 500Ω, gauge factor was 2.1 and gauge length was 2 mm. Strain gauges were attached to four places around the red to find out the effects of bending (Fig. 1). In order to avoid electrical noise, aluminum foil was used to completely shield the strain gauges and all lead wire was shielded. In addition, epoxy resin and a rubber costing was applied to the gauges to protect them from borehole water and the shocks of the hammer blows.

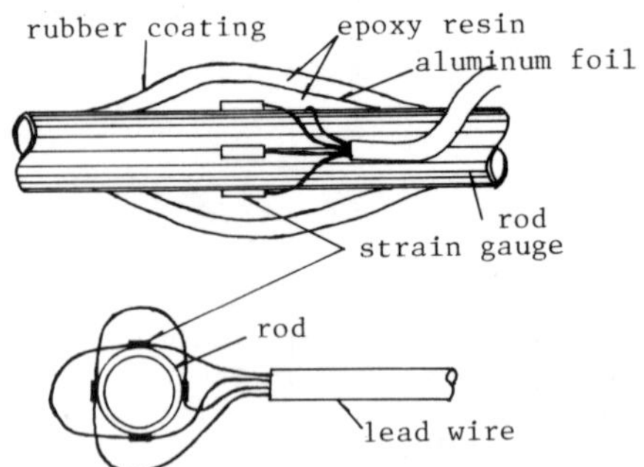

Fig. 1 Disposition of strain gauge

Measurement system

Figure 2 shows the measurement system employed. Data obtained was directly fed into a computer. Because the strain waveforms being measured are produced by direct impact, it was necessary for the measuring system to be able to cover a range from direct current to high frequencies. The frequency range of the amplifier is from DC to 50 KHz. The data recorder is from DC to 10 KHz. All measurement data was recorded while being monitored using a storage oscilloscope.

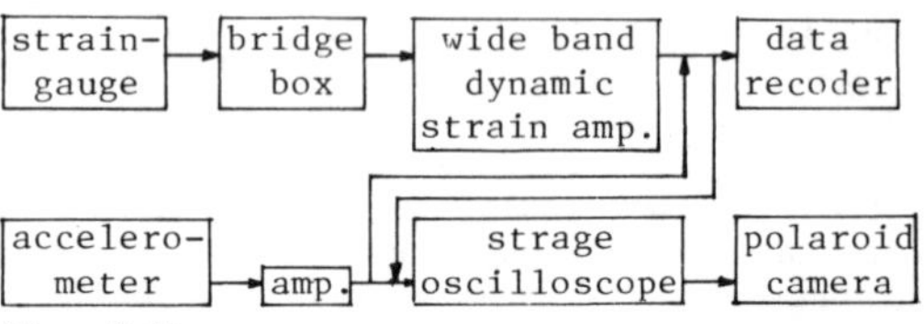

Fig. 2 Measurement system

Field measurement

In order to minimize the effects of reflection of strain from the end of the rod and to measure the amount of energy transmitted as accurately as possible, it is desirable to use as deep a borehole as possible. Accordingly, the experiments were carried out in boreholes of 46 and 56 meter depths. Figure 3 shows rod lengths and location of strain gauges. Standard procedure was followed in carrying out the Standard

Penetration Test. However, it should be noted that the following two points represent minor variations on the standard procedure:

(1) The rod was not passed through the boring machine spindle in order to avoid friction, which would disturb the measured strain waveform.

(2) Because reflection, etc. from the strain waveform of the hammer guide rod has an effect, the guide rod was made of wood and rubber packing was used in connecting to the knocking head.

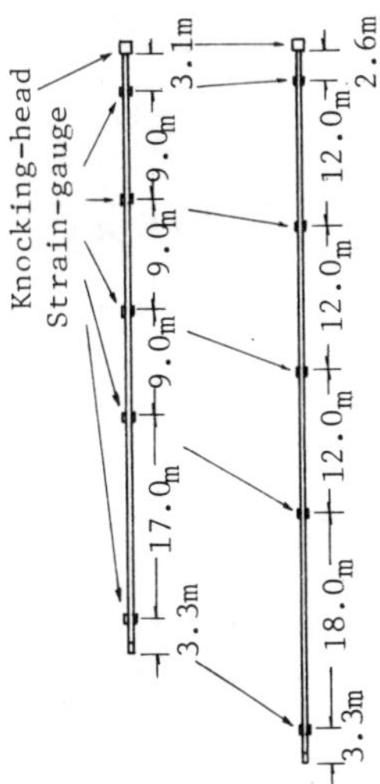

Fig. 3 Location of strain gauge

Analysis of strain waveforms

The following formulas were used to determin the strain energy transmitted by the rod from the measured strain waveforms. As noted above, the value yielded by these formulas is doubled to obtain total energy transmitted by the rod.

$$\sigma = E\varepsilon \tag{1}$$

$$U_p = Act\sigma^2/2E \tag{2}$$

where, σ: stress produced in the rod
ε: measured strain
U_p: strain energy transmitted by the rod
E: elasticity of the rod
A: area of the rod
c: velocity of longitudinal wave transmitted by the rod
t: duration of σ stress

An HP-5451c Fourier analyzer was used in analysis of data. Analog measurement data from the data recorder was subjected to A/D transformation at a sampling time of 50 μs and processed within the Fourier analyser.

3 EXPERIMENTAL RESULTS AND THEIR SIGNIFICANCE

3.1 Results of standard penetration test measurements conducted with differing rod diameters

Due to the fact that the measurements were carried out in weathered rhyolite, the N values (Standard Penetration Test results) obtained were all 50 or above. Results were compared on the basis of amount of penetration after 50 blows These results are shown in Figure 4. It can be seen from this figure that there is considerable scatter of penetration distances for each depth. Although somewhat smaller penetration distances can be noted for the 40.5 mm rod, it must also be realized that such factors as lack of homogenity within the layer, unevenness of weathering, etc. Thus, no clear difference due to differences in rod diameter can be claimed to be apparent.

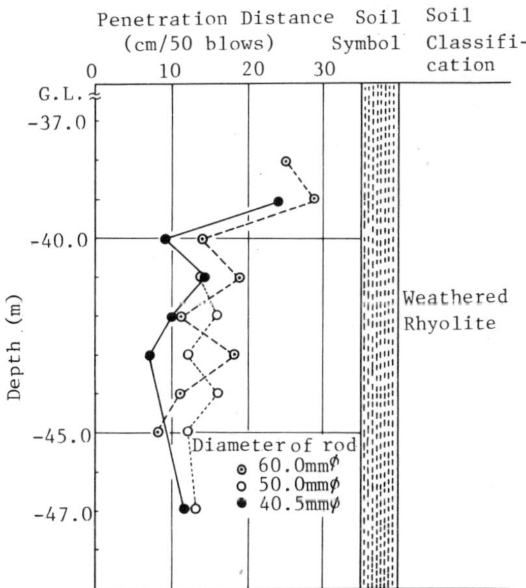

Fig. 4 Comparison of penetration distance for different rod diameters

Koreeda et al. (1980, 1981) conducted the Standard Penetration Test using two diameters of rod, 40.5 mm and 50 mm in boreholes of 30 m and 47 m depths in two different ground types. They found from the results of these tests that there is no statistical difference in results yielded by the different rods either for sand or clay ground or in ground more or less than 15 meters in depth. For all practical purpose, it was concluded, there are no differences at all. This agrees with the results obtained by the authors.

3.2 Energy transmitted by the rod

Using the strain waveforms measured by the strain gauges attached to the 5 gives an example of measured strain waveforms. As mentioned, the total energy transmitted by the rod is twice that of strain energy. Figure 6 (a) shows the relationship between the total energy transmitted by the rod from the points at which strain energy was measured and distance from the knocking head. In the analysis, data from the strain gauge closest to the end of the rod was not used because of the great influence due to reflection. Figure 6 (b) shows the relationship between potential energy of the hammer (4.763 kgf.cm) and all measured energy and distance from the knowcking head. It can be clearly seen from these figures that for all rod diameters 90% of the potential energy of the hammer is transmitted by the rod. In other words, there is no significant difference according to rod diameter.

In order to find the amount of attenuation present in the rod, the energy measured by the uppermost strain gauge was designated 1 and Figure 6 (c), which compares this energy with that measured by the other strain

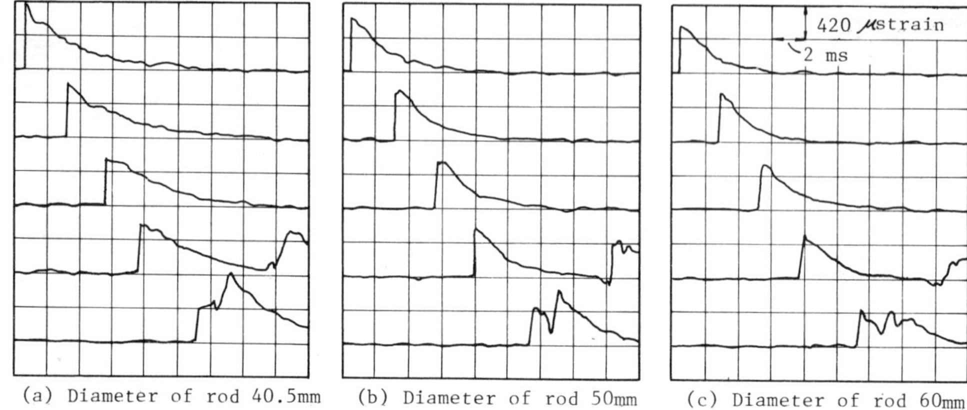

Fig. 5 Example of measured strain waveform

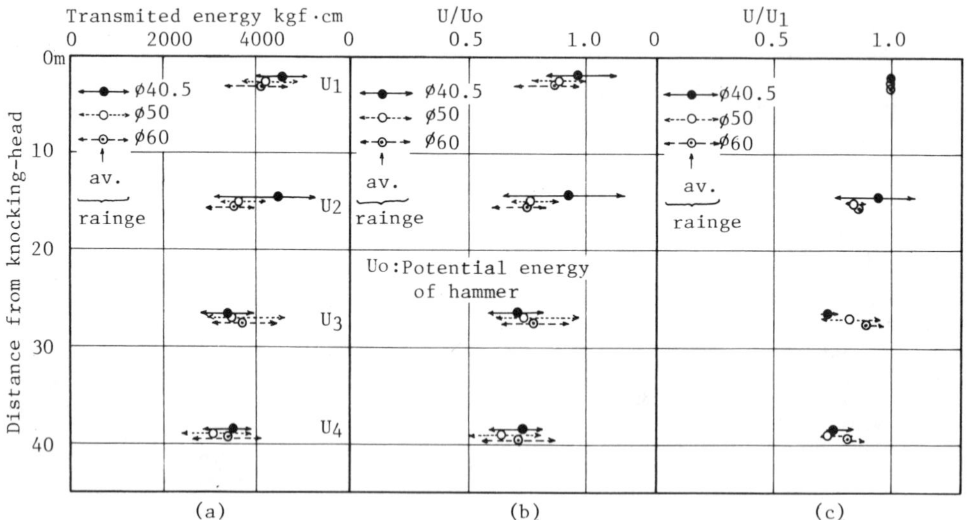

Fig. 6 Energy transmitted by rod

gauges, was produced. No difference in amount of attenuation due to differences in rod diameter can be identified from this figure. These results agree with those obtained by Uto et al. (1973 b) in which a 40.5 mm rod was placed horizontally and blows equivalent to those in the Standard Penetration Test were administered, producing the results shown in Figure 7.

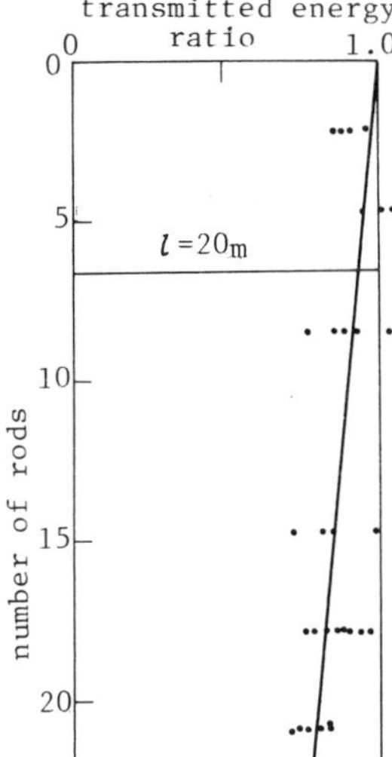

Fig. 7 Changes in transmitted energy ratio for various rod length (from Uto et al. 1973)

It has been noted above that the amount of energy produced by a blow to the rod does not differ according to the diameter of the rod. However, Figure 5 shows that the measured strain waveforms do differ. The smaller the diameter of the rod the longer the time that strain continues and vice-versa. Figure 8 shows the results obtained by Uto et al. (1973 a) in which various boring rods were administered blows equivalent to that of the Standard Penetration Test and strain waveforms from the rod obtained. These results also agree with those of the authors. Thus, as pointed out by Uto, et al. (1973 a), even if amount of energy applied is uniform, failure of the ground at the end of the rod can be expected to differ as a result of differences in change with time of stress produced by different rod diameters. Consequently, it can easily be considered that from this, differences in rod diameter could hava na effect on Standard Penetration Test results. However, there are many unclear points at present regarding the mechanism by which the ground is disturbed. Thus, it cannot be judged how much influence is exerted by stress waveforms.

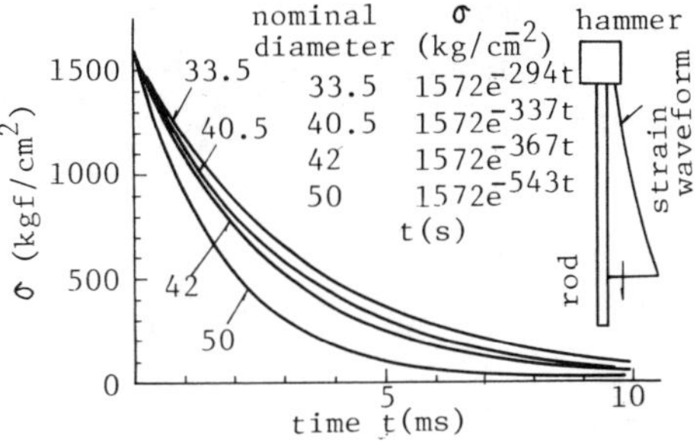

Fig. 8 Effect of rod diameter (from Uto et al. 1973 a)

4 CONCLUSIONS

From the above, the following conclusions may be drawn:

(1) Regardless of differences in rod diameter, the total amount of energy transmitted by the rod from the hammer in the Standard Penetration Test is the same.

(2) Regardless of differences in rod diameter, there is no difference in attenuation of energy in the Standard Penetration Test.

(3) From conclusions (1) and (2), it may be further concluded that there is no reason that rods of different diameters should be used for different depths. Neither is there any reason why steadies must be attached for greater depths.

(4) Although the amount of energy transmitted by the rod is the same for all diameters, stress waveforms, i.e., change with time of stress, differs. From this, it may be surmised that failure of the ground at the end of the rod will differ according to rod diameter or that the results of the Standard Penetration Test would be affected. Consequently, a variety of rod diameters should not be used in the Standard Penetration Test.

(5) There is no significant difference to the Standard Penetration Test in changing rod diameter within the range of 40 to 60 mm.

(6) On the basis of conclusions (1) through (5) it is our judgement that the rod diameter of 40.5 mm that has been used to the present to accumulate the great volume of Standard Penetration Test data that now exists should continue to be used.

It was noted that the use of a rod or greater thickness caused on extreme drop in the efficiency of the conduct of the Standard Penetration Test. The use of a too-thick rod cancels out the ease of the test, which is its original advantage.

5 ACKNOWLEDGEMENT

We received valuable guidance and advice throughout the experiments from Prof. Uto and Mr. Fuyuki of Tokai University. For this we are deeply grateful. In addition, operations in the experiments were shared by the Port and Harbor Research Institute Ministry of Transportation personnel as well as those OYO Corporation's Urawa Research Institute and Okayama Office. To these individuals we also express our thanks.

6 REFERENCES

Koreeda, T., Yoshihashi, T. and Muromachi, T. (1980) Comparing SPT N-value with JIS standard rod and ather rod, Proc. Sounding Symp., 1980 JSSMFE: 53-58 (in Japanese)

Koreeda, T., Yoshihashi, T. and Muromachi, T. (1981) Effect of drill rod stiffness on N-value, Proc. 16th Japan National Conference on SMFE: 77-80 (in Japanese)

Matsumoto, K. and Matsubara, M. (1980) Effect of rod diameter on N-value, Proc. Sounding Symp. 1980 JSSMFE: 59-64 (in Japanese)

Uto, K., Fuyuki, M., Kondo, H. and Morihara, M. (1973a) Experimental study on Test (1), Proc. of the Faculty of Engineering, Tokai Univ. 1973-1: 73-94 (in Japanese)

Uto, K., Fuyuki, M., Kondo, H. and Morihara M. (1973b) Experimental study on the mechanism of dynamical penetration of rod in the Standard Penetration Test (2), Proc. of the Faculty of Engineering, Tokai Univ. 1973-2: 55-65 (in Japanese)

Uto, K., Fuyuki, M., Kondo, H. and Morihara, M. (1975) Fundamental studies on the mechanism of dynamical penetration of rod from a viewpoint of the wave theory, Proc. of the Faculty of Engineering, Tokai Univ. Vol. II: 9-30

Proceedings of the Second European Symposium on Penetration Testing / Amsterdam / 24-27 May 1982

Validity of existing procedures for the interpretation of SPT and CPT results

T.K.NATARAJAN & D.S.TOLIA
Central Road Research Institute, New Delhi, India

1 INTRODUCTION

It is widely recognised that SPT and CPT values are reflections of both relative density and geostatic stress. Earlier, Terzaghi and Peck(1948) did not mention the effect of overburden pressure on SPT and CPT values. Gibbs and Holtz(1957) were among the first to give the well-known charts, Fig. 1, showing the effect of overburden pressure on SPT values at different relative densities on sands. Subsequently, Schultze and Menzenbach (1961), Schultze and Melzer(1965), Meyerhof (1965), Bazaraa(1967), Peck and Bazaraa (1969) and others had also confirmed the effect of overburden pressure on the SPT value. Mohan, Aggarwal & Tolia(1971) confirmed the chart given by Gibbs and Holtz(1957) after carrying out controlled laboratory tests on sands.

Effect of overburdon pressure on static cone penetration test(SCPT) resistance were discussed by Chaplin(1963), Schultze and Melzer(1965), Dahlberg(1974) and Tolia (1976). A chart, Fig. 2, was given by Tolia(1976 & 1978), showing the overburden pressure effect on static CPT resistance at various relative densities of sands. Similar effect of overburden pressure on dynamic cone resistence values (62.5mm Ø and 60° apex) were found by Mohan, Aggarwal and Tolia(1971), Fig. 3. The chart shows the effect of overburden pressure on dynamic cone penetration resistance (Nc) value at different relative densitites in sandy soils.

It is a well-known fact that the soil pressure charts given by Terzaghi & Peck (1948) and Peck, Hanson and Thornburn(1974), Fig. 4, underestimate the allowable soil pressure. Accordingly, several authors like Coffman(1960), Teng(1962), Thornburn (1963), Alpan(1964), Bazaraa(1967),

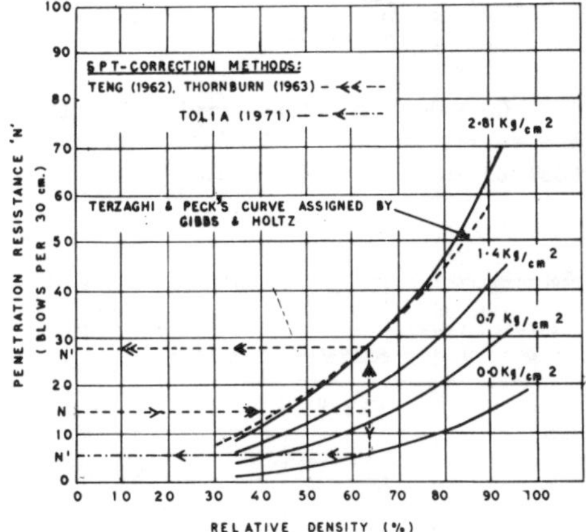

Fig. 1 Correlation between relative density and SPT at different overburden pressures in sands(After Gibbs and Holtz 1957)

Peck & Bazaraa(1969), Alam Singh(1975) and others suggested modifications to the effect that the measured N value should be first corrected to reflect the influence of effective overburden pressure. The method of correction proposed initially always involved an increase upon the observed N value, Fig. 1, which already represented an exaggerated value on account of the influence of overburden pressure.

Peck, Hanson & Thornburn(1974) again gave another modification i.e. SPT-correction chart assuming that the N value at a depth corresponding to an effective overburden pressure of 1 kg/cm^2 as the standard. This was presumably not based on any scientific reasoning. Why should not the standardization be done at zero kg/cm^2 i.e. at the

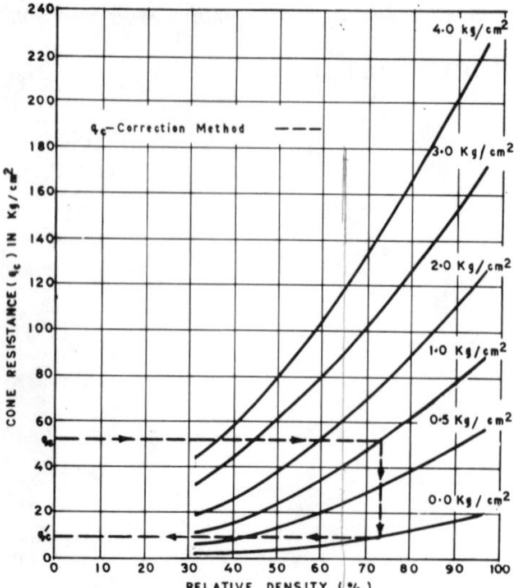

Fig. 2 Correlation between relative density and cone resistance at different overburden pressures in sands(After Tolia 1978)

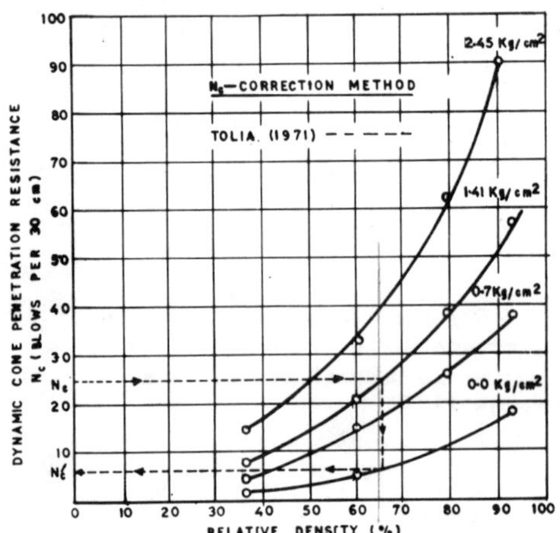

Fig. 3 Correlation between relative density and dynamic cone resistance at different overburden pressures in sands (After Mohan, Aggarwal and Tolia 1971)

ground surface from where the N value starts increasing linearly with depth. Why should we keep on suggesting modifications after modifications only to keep the particular chart valid. de Mello(1971) had stated that most authors mentioned above, followed the line of suggesting "corrections to the

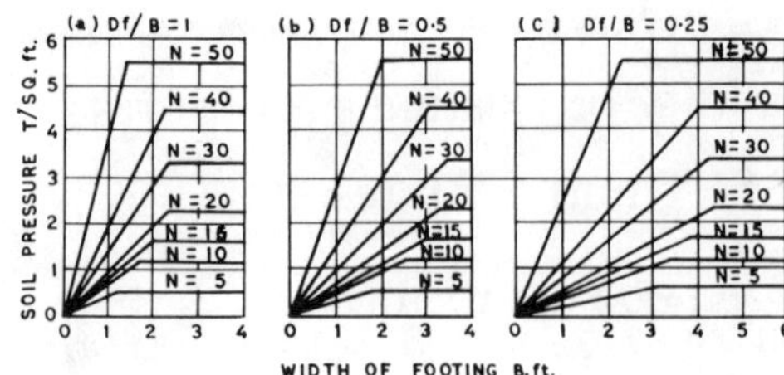

Fig. 4 Design chart for proportioning shallow footings on sand (After Peck et. al. 1974).

correlations", which were really not correlations, but were more truly 'amendments'.

2 VALIDITY OF CHARTS GIVEN BY TERZAGHI AND PECK

Terzaghi & Peck(1948) and Peck, Hanson & Thornburn(1962, 1974) correlated the surface plate bearing pressure values at the footing depth whereas the bearing capacity found by plate bearing test may be constant throughout the depth for a homogeneous stratum as illustrated in Fig. 5. From Gibbs & Holtz(1957) chart, Fig. 1, corresponding to 90% relative density, we find a linear increase in N value with depth whereas for the same relative density, a constant bearing pressure will be obtained with depth for this homogeneous stratum. Hence a constant bearing pressure cannot be directly correlated with the variable N value at different depth.

Further reservations may also arise, as

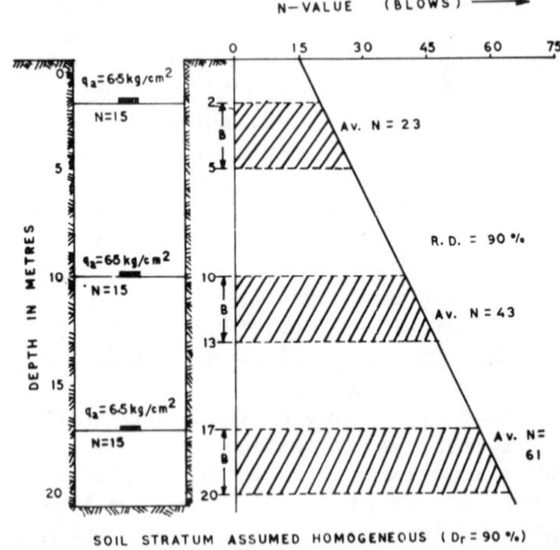

Fig. 5 Validity of Terzaghi & Peck's Chart (An illustrative example)

Bazaraa(1967) reported that the Terzaghi & Peck(1948) correlation was based on limited plate loading test data and some of the N values were obtained using a smaller diameter spoon than the one used in the SPT test. These facts clearly invalidate both the soil pressure charts mentioned above for future use. On the basis of Terzaghi & Peck(1948) chart, Meyerhof(1965) gave three empirical equations for predicting allowable bearing pressures for spread footings and rafts. Due to the above said fact, these equations too are found to underestimate the allowable bearing pressures. Tolia (1971) and Natarajan & Tolia(1972) had analysed the validity of the above charts in detail, with different illustrations.

As a plate bearing test is normally conducted with the requirement that there should not be any surcharge around the plate, any correlation of the allowable pressure so determined with the measured N value at that depth is only possible when the measured N value at that depth is free from the influence of surcharge effects. Therefore based upon the plate bearing test, N value observed on the surface at that level and using Terzaghi's settlement expression, Mohan, Aggarwal & Tolia(1971) proposed a new penetration allowable pressure chart which was further extended by Tolia(1978) for static cone penetration test(CPT) values, Fig. 6.

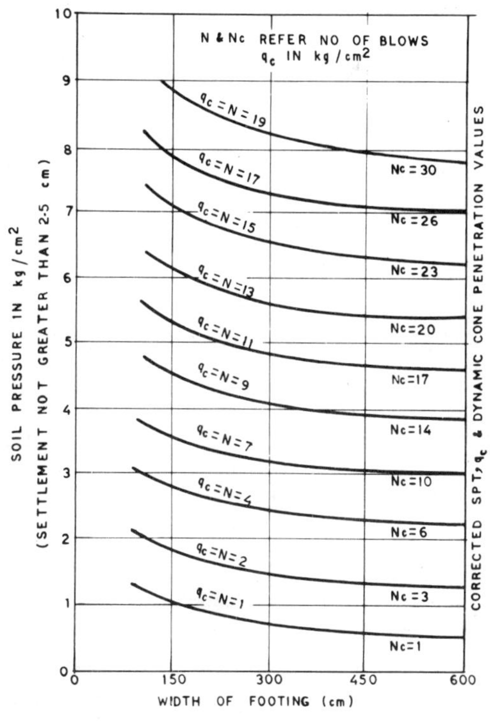

Fig. 6 New penetration-allowable pressure chart for footings on sands (After Mohan, Aggarwal & Tolia 1971)

3 VALIDITY OF EXISTING SPT-CORRECTION METHOD

Gibbs & Holtz(1957) did not describe or recommend any SPT-correction method but only produced chart, Fig. 1, showing the effect of overburden pressure on N value. Several authors mentioned in Article 1 above used this chart, Fig. 1, for SPT-correction method. The method of correction is always based on increasing the measured N values by drawing a line vertically upwards from the intersection of the point representing the measured N value and overburden pressure at that depth. Fig. 1, to intersect the Terzaghi & Peck's dotted curve assigned by Gibbs & Holtz(1957). The point of intersection so obtained is then projected horizontally to give the corrected N value. Peck & Bazaraa (1969) reported that this dotted curve was never published by Terzaghi & Peck and the numerical values were given by Gibbs & Holtz for a limiting range of N values, However, all the methods suggested involve increasing the measured N value which is already reflecting an increase due to the effect of overburden pressure at the particular depth. Therefore, there is no scientific reasoning behind these methods which, at best, may be called'best-fit' methods.

Peck, Hanson & Thornburn(1974) gave yet another 'best-fit' SPT correction chart, Fig. 7, by assuming N value as standard at an effective overburden pressure of 1 kg/cm^2. It is generally found that even if the whole stratum is assumed to be homogeneous, the SPT value is apt to increase linearly with depth from the ground surface due to the effect of overburden pressure. Hence, the N value at the surface of the ground ought to be taken as standard, which is free of surcharge effect. At greater depths, the N value will change with the change of relative density and will increase with depth.

Hence, based on the standard N value at the ground surface, Tolia(1971) had given SPT correction method by adapting the observed N value for zero overburden pressure i.e. by removing the effect of overburden on measured N value at that particular depth by using the Gibbs & Holtz(1957) chart, Fig. 1. In this method, a line is drawn vertically downwards from the intersection of the point representing the observed N value and the overburden pressure curve at that depth, Fig. 1, so as to intersect the zero overburden pressure curve. The point of intersection so obtained is then projected horizontally to give

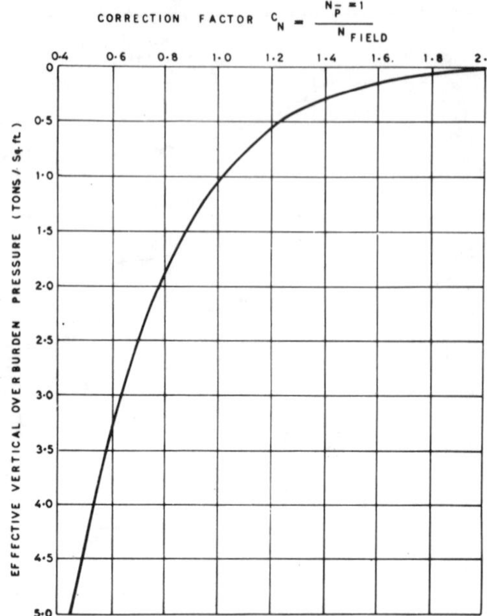

Fig. 7 Chart for correction of N values in sand for influence of overburden pressure. (After Peck, Hanson & Thornburn 1974)

the corrected N value. On the basis of this method of correction, an empirical equation was given below, which can be used for SPT correction for higher overburden pressures also.

$$N' = N(\frac{0.7}{P + 0.7}) \quad \text{--- (1)}$$

where 'N'' is the corrected N value, 'N' is the in-situ observed N value at an effective overburden of 'P' in kg/cm^2.

Similarly, the in-situ observed static cone penetration test(SCPT) and dynamic cone penetration test(DCPT) resistance values are corrected by the similar procedure as mentioned above using Fig. 2 and Fig. 3 respectively.

4 INTERPRETATION OF OBSERVED SPT AND CPT RESULTS

Based on the foregoing discussions, it can be said that the strict use of the charts given by Terzaghi & Peck(1948) and Peck, Hanson & Thornburn(1974) is erroneous and even misleading tending to give an incorrect estimate of the allowable bearing pressure and settlement. Therefore, the soil pressure chart, Fig. 6 given by Mohan, Aggarwal & Tolia(1971) can safely be used to estimate the allowable soil pressure using corrected N value, qc value and Nc value. The new correction method mentioned in the above para adapts all the observed N value, qc value and Nc value to zero overburden pressure. Validity of this new method had been confirmed by Natarajan & Tolia(1972) by comparing the actual plate bearing value with the values found by corrected SPT value conducted at the same level. They also confirmed by an analysis of plate bearing tests, conducted above and below the ground water-table that the effect of submergence is already reflected in the measured SPT and CPT values.

5 CORRELATIONS OF SPT, CPT AND E VALUES

From the work of Dunn(1974), it is indicated that the ratio qc/N for one particular type of soil should increase with increasing depth since overburden pressure may effect qc and N values by different amount being static and dynamic nature of testing procedure. This is confirmed by Tolia(1978) comparing the results of overburden pressure effect on N, qc and Nc values from Fig. 1 to 3. The ratio of qc/N is 1 at the zero overburden pressure and increases to 4 when the overburden pressure is 2.8 kg/cm^2. This clearly shows that one single correlation for one particular type of soil can not be recommended with increasing depth.

Because of the above said facts, various authors Meyerhof(1956), Webb(1969), Schmertmann(1970), Sanglerat(1972), Simons(1972), Lacroix & Horn(1973) and Alperstein & Leifer (1976) have given different conflicting results. Similar reasoning may also be given for the ratio qc/cu, which ranges from 15 to 30 found by the various investigators. Hence, it is also not possible to present one single relationship between E and qc or N values. Natarajan & Tolia(1979) have proposed new correlations between E and qc or N values which are regarded as more realistic, since the correlations have been achieved by comparing the E values with corresponding qc and N values at comparable overburden pressures and at comparable relative densities.

6 CONCLUSIONS

The study indicates that the present recommendations for the interpretation of the measured N, q_c and Nc values at different depths are erroneous and even misleading. The measured N, qc and Nc values at particular depths should first be corrected for zero overburden pressure by the method given by Tolia (1971), then used in Fig. 6 to predict the allowable soil pressure based on settlement criterion at the relevant depth. Though the parameters E, N, qc and Nc are influenced by the overburden pressure, the overburden effect is found to be different at different depths in each case, corresponding to varying degrees of relative density. Hence correlation between these parameters is possible only when one parameter is compared with the other corresponding parameter at comparable overburden pressures and at comparable relative densities.

7 ACKNOWLEDGEMENT

The paper is published with the kind permission of the Director, Central Road Research Institute, New Delhi, India.

8 REFERENCES

Alpan, L. 1964, Estimating settlements of foundations on sands, Civil Engg. and Public Works Rev., Vol. 59: 1415-1418.

Alam Singh 1975, Soil Engineering in theory and practice, 2nd Edition, Vol.1: Asia Publishing House, Bombay & London.

Alperstein, R. & S.A.Leifer 1976, Site investigation with static cone penetrometer, J.Geotechnical Engg. Divn., ASCE, Vol. 102, No. GT5: 539-555.

Bazaraa, A.R.S.S. 1967, Use of the standard penetration test for estimating settlements of shallow foundations of sand, Ph. D. Thesis, University of Illinois, Urbana.

Chaplin, T.K. 1963, The compressibility of granular soils, with some applications to foundation engineering, European Conf. on Soil Mech. & Found. Engg., Wiesbaden, Vol. 1: 215-219.

Coffman, B.A. 1960, Estimating the relative density of sands, Civil Enggineering, A.S. C.E., Vol. 30, No. 10:78-79.

Dahlberg, R. 1974, The effect of the overburden pressure on the penetration resistance in a preloaded natural fine sand deposit, Proc. European Symposium of Penetration Testing, Stockholm, Vol. 2:89-91.

Dunn, C.S. 1974, Settlement of a large raft foundation on sand, B.G.S. Conf. on Settlement of Structures, Cambridge, 14-21.

Gibbs, H.J. & W.G. Holt$_z$ 1957, Research on determining the density of sands by spoon penetration testing, Proc. 4th Int. Cong. on Soil Mech & Found. Engg., London Vol. 1 : 35-39.

Lacroix, Y. & H.M.Horn 1973, Direct determination and indirect evaluation of relative density and its use on earthwork construction projects, Evaluation of Relative Density and its Role in Geotechnical Projects Involving Cohesionless Soils, ASTM-STP 523,

de Mello, V.F.B. 1971, The standard penetration test, State-of-the-Art-Papers, Proc. 4th Panamerican Conf. on Soil Mech. & Found. Engg., San Juan, Puerto Riço, Vol. 1 : 1-86.

Meyerhof, G.G. 1956, Penetration tests and bearing capacity of cohesionless soils, Jour. Soil Mech. & Found. Div., A.S.C.E., Vol. 82, SM1: 866/1-19.

Meyerhof, G.G. 1965, Shallow foundation, Jour. Soil Mech. & Found. Div., A.S.C.E., Vol. 91, SM2: 21-31.

Mohan, D., V.S. Aggarwal & D.S. Tolia 1971, Bearing capacity from dynamic cone penetration tests, Indian Geotechnical Journal Vol. 1, No. 2: 133-142.

Natarajan, T.K. & D.S. Tolia 1972, Interpretation of standard penetration test results, Proc. 3rd Southest Asian Conf. on Soil Engineering, Hong Kong, Vol. 1: 53-57.

Natarajan, T.K. & D.S. Tolia 1979, Modulus of elasticity of sandy soils by sounding methods, Proc. 6th Asian Regional Conf., Singapore, Vol. 1: 51-54.

Peck, R.B. & A.R.S.S.Bazaraa 1969, Discussion, Jour. Soil Mech. & Found. Div., A.S.C.E., Vol. 95, SM3: 905-909.

Peck, R.B., W.E. Hanson & T.H.Thornburn 1974, Foundation Engineering, 2nd Edition, Wiley & Sons, New York, 115.

Sanglerat, G. 1972, The penetrometer and soil exploration, Elsevier Publ. Co., Amsterdam, 464.

Schmertmann, J.H. 1970, Static cone to compute settlement over sand, Jour. Soil Mech. & Found. Div., A.S.C.E. Vol. 96 No. SM3: 1011-1043.

Schultze, E. & E.Menzenbach 1961, Standard Penetration test and compressibility of soil, Proc. 5th Int. Conf. on Soil Mech. & Found. Engg., Paris, Vol. 1: 527-532.

Schultze, E. & K.J.Melzer 1965, The determination of the density and the modulus of compressibility of non-cohesive soils by soundings, Proc. 6th Int. Conf. on Soil Mech. & Found. Engg., Montreal, Vol. 1: 354-358.

Simons, N.E. 1972, Prediction of settlement of structures of granular soils, Ground Engineering, Vol. 5, No. 1: 24-26.

Teng, W.C. 1962, Foundation design, Prantice Hall, Inc., New York.

Terzaghi, K. & R.B. Peck 1948, Soil mechanics in engineering practice, Wiley & Sons, New York, Also 2nd Edition, 1967.

Tolia, D.S. 1971, A critical review of Terzaghi's penetration-allowable pressure chart, St. Jour. Institution of Engineers (India), No. 4: 69-71.

Tolia, D.S. 1976, The interpretation of static cone penetration test with particular reference to the influence of over-

burden pressure, M. Sc. Dissertation, submitted to the University of New Castle upon Tyne, England.
Tolia, D.S. 1977, Interpretation of dynamic cone penetration tests with particular reference to Terzaghi & Peck's chart, Ground Engineering, England, Vol. 10, No. 7: 37-41.
Tolia, D.S. 1978, Interpretation of static cone penetration tests, Indian Geotechnical Journal, Vol. 8, No. 3: 152_168.
Thornburn, S. 1963, Tentative correction chart for the standard penetration test in non-cohesive soils, Civil Engg. and Public Works Review, Vol. 58, No. 683: 752-753.
Webb, D.L. 1969, Settlement of structures on deep alluvial sandy sediments in Durban, South Africa, B.G.S. Conf. on In-situ Investigations in Soils and Rocks, London, 173-179.

Proceedings of the Second European Symposium on Penetration Testing / Amsterdam / 24-27 May 1982

Mechanics base of standard penetration test values and its application to bearing capacity prediction

Y.NISHIDA, K.YOKOYAMA, H.SEKIGUCHI & T.MATSUMOTO
Department of Civil Engineering, Kanazawa University, Japan

1 INTRODUCTION

The standard penetration test (SPT) has very widely been used in the site investigation practice in Japan. Of the related reasons the following are to be noted: the simplicity of testing procedure; the ability of sampling; the adaptability to a wide range of subsoil conditions; and in particular the accumulation of plenty of case records which correlate bearing capacities of soils under consideration with their blow counts from SPTs.

In contrast, there are very few studies which throw a light on fundamentals of the mechanics of SPT. This tendency may be attributed to the complexities which result from the fact that SPT is a kind of soil-structure interaction test involving repeated impact loading with interrupted partially-drained soil behavior. Despite such apparent complexities, the authors are of the opinion that SPT has a potential as a rheology test from which in-situ stress-strain and strength parameters of soil are able to be back-analyzed.

In what follows, as a first step toward analyzing SPT as a boundary value problem of mechanics, a theoretical study will be made by assuming quasi-static soil behavior and by introducing a new version of cavity expansion theory in relation to the penetration mechanism of SPT sampler.

2 THEORETICAL BASIS OF SPT N-VALUE

When SPT sampler is driven into a deposit of dense sand, the lower part of the sampler tends to be plugged with a mass of heavily compacted soil as shown schematically in Fig. 1 (Nishigaki, 1980; Miki et al., 1980). In that situation the mechanism of penetration of SPT sampler is judged to be the same as that of a driven pile with a closed end. Therefore, keeping in mind the application to end-bearing capacity prediction for pile foundations, we will restrict our following discussion to the situation illustrated in Fig. 1.

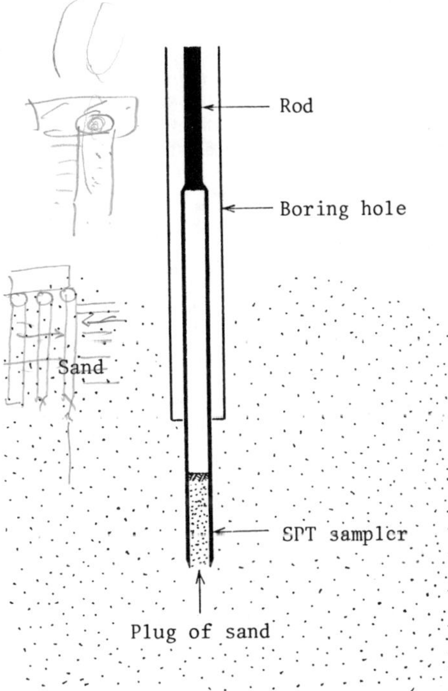

Fig. 1 Schematic expression of a SPT sampler driven into sand

2.1 Conservation of energy

Figure 2 shows the displacement-time pattern of the sampler tip during a given blow, together with the time-dependent change in the end-bearing resistance.

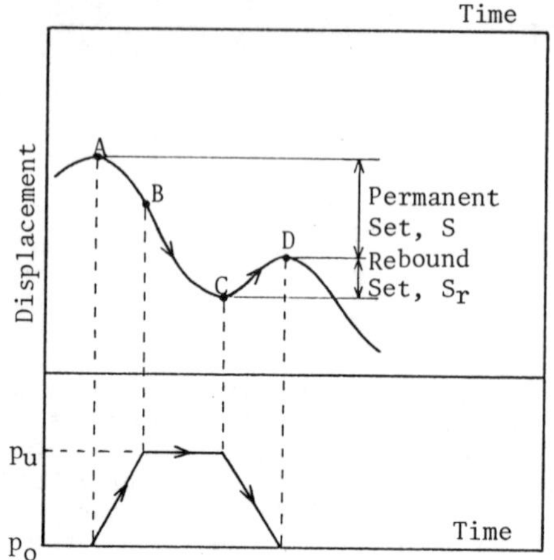

Fig. 2 Displacement-time pattern of a SPT sampler tip

Since the energy per blow delivered to the sampler is consumed as the energy which is required for the penetration and subsequent rebound of the sampler, it follows that

$$e_f\, W_h\, H \left(\frac{W_h + e^2\ W_p}{W_h + W_p} \right) = \pi a^2 p_u\, S + E_r \qquad (1)$$

where e_f is the efficiency of driving, W_h is the weight of hammer, H is the height of free fall of hammer, W_p is the sum of the weights of rod and sampler, e is the coefficient of restitution, a is the radius of sampler, p_u is the ultimate end-bearing pressure, S is the permanent set, and E_r is the rebound energy.

In the subsequent sections, we will show how the parameters p_u and E_r are related to soil properties.

2.2 Prediction of p_u based on cavity expansion theory

Figure 3 shows characteristic regions of soil which are assumed here to exist around the penetrating SPT sampler. Region 1 simulates a highly compacted soil immediately below the sampler tip and is idealized here as a hemispherical cavity subjected to an internal pressure, p_u. Region 2 is a thick-walled hemispherical region whose soil

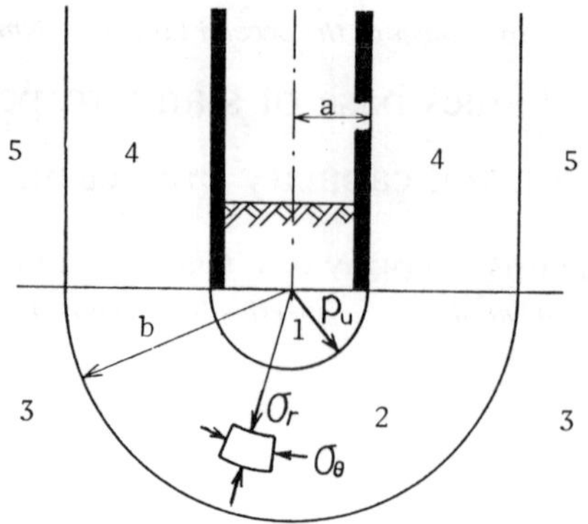

Fig. 3 Model of hemispherical expansion a cavity

elements are in states of failure by the action of pressure, p_u. The failure criterion used here is the following Mohr-Coulomb law:

$$\sigma_\theta - \sigma_r = (\sigma_r + \sigma_\theta) \sin \phi \qquad (2)$$

where σ_r is the effective radial stress, σ_θ is the effective hoop stress and ϕ is the angle of internal friction in terms of effective stress. Region 3 is a semi-infinite region whose soil elements remain in elastic states. It is obvious that soil elements in regions 2 and 3 should satisfy the following equilibrium equation:

$$\frac{d\, \sigma_r}{d\, r} + \frac{2\ (\sigma_r - \sigma_\theta)}{r} = 0 \qquad (3)$$

Regions 4 and 5 are plastic and elastic regions, respectively, which simulate the lateral deformation pattern of soil adjacent to the side surface of sampler. It may be appropriate here to mention that in the subsequent analysis, regions 4 and 5 do not play an explicit role.

Use of Eqs. (2) and (3) together with elastic constitutive law for regions 2 and 3 allows the ultimate end-bearing pressure p_u to be expressed as follows (Vesic, 1972; Yamaguchi, 1973):

$$p_u = \frac{3\, p_o (1 + \sin \phi)}{3 - \sin \phi} \left(\frac{b}{a}\right)^{\frac{4 \sin \phi}{1 + \sin \phi}} \qquad (4)$$

where p_o is the initial mean effective stress in the soil and b is the radius of elastic-plastic boundary to be determined subsequently.

Let us now consider the volume change behavior of soil around the penetrating

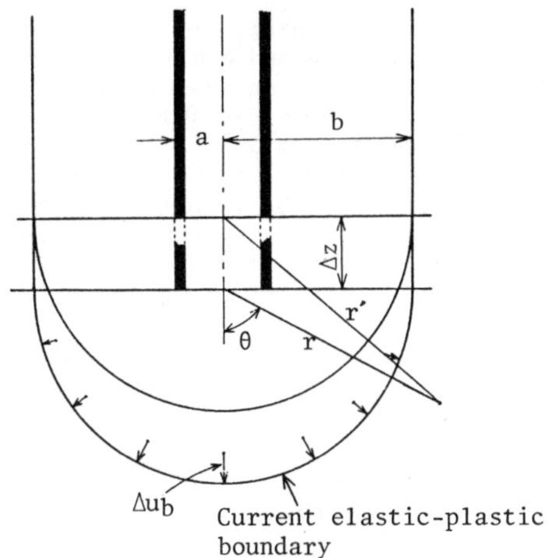

Fig. 4 Movements of soil particles during penetration of a SPT sampler

sampler. When the sampler tip penetrates further by the amount Δz from a depth z, the associated volume of penetration is obviously equal to $\pi a^2 \Delta z$ (see Fig. 4). Also, the volume of penetration should be compensated by the outward movements $\Delta u_b(\theta)$ of soil particles forming the current elastic-plastic boundary, because the volume changes of soil elements in the hemispherical plastic region are negligible after they have reached the failure or critical states. Thus, it follows that

$$\pi a^2 \Delta z = 2 \pi b^2 \int_0^{\pi/2} \Delta u_b(\theta)\, d\theta \qquad (5)$$

where a is the radius of sampler and b is the radius of the current elastic-plastic boundary which is associated with the sampler locating at the depth $z+\Delta z$. Then, the afore-mentioned cavity expansion theory with consideration of such translation of hemispherical plastic region as shown in Fig. 4, leads to

$$\Delta u_b(\theta) = \frac{3 p_o \sin \phi}{G(3 - \sin \phi)} \Delta z \cdot \cos \theta \qquad (6)$$

where θ is the angle defined on Fig. 4. For details of deriving Eq. (6), reference should be made to Nishida et al. (1980).

Substituting Eq. (6) into Eq. (5), we get

$$\frac{b}{a} = \sqrt{\frac{3 \sin \phi'}{6(1+\sin \phi)} \frac{G}{p_o}} \qquad (7)$$

Thus, the ultimate end-bearing pressure P_u is completely expressed by Eqs. (4) and (7).

2.3 Prediction of E_r based on cavity expansion theory

The rebound energy E_r can be regarded as equal to the work which is required to displace the elastic-plastic boundary until the radial stress σ_r acting on it increases to σ_b (see Fig. 5).

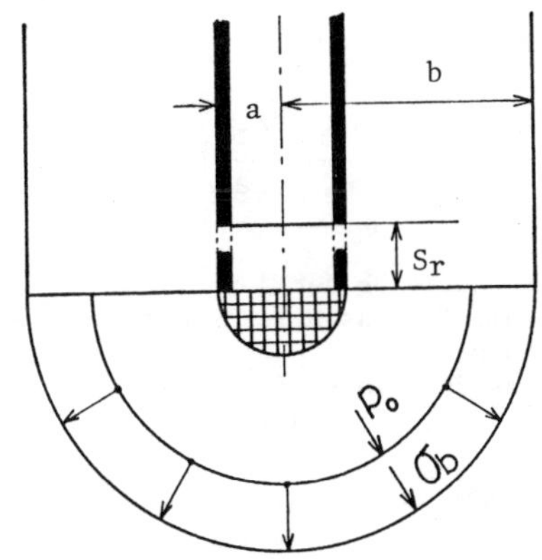

Fig. 5 Elastic displacement of elastic-plastic boundary

Then, the application of the cavity expansion theory to this situation illustrated in Fig. 5 leads to

$$E_r = 2\pi b^2 [(p_o+\Delta p)\Delta u_{r,1} + (p_o+2\Delta p)\Delta u_{r,2} + \cdots + (p_o+n\Delta p)\Delta u_{r,n}] \qquad (8)$$

where n is the number of subdivision of stress increase, Δp is equal to $(\sigma_b - p_o)/n$ and $\Delta u_{r,n}$ is the displacement increment which generates while σ_r changes from $p_o + (n-1)\Delta p$ to $p_o + n\Delta p$.

Since the mean effective stress remains at p_o in the elastic region, it follows that

$$\Delta u_{r,n} = b\, \Delta p / G \qquad (9)$$

Substituting Eq. (9) into Eq. (8), putting $n \to \infty$ and then using Eq. (7), we obtain

$$Er = \frac{p_0 (3 + \sin \phi) \pi a^3}{3(3 - \sin \phi)} \left(\frac{b}{a}\right) \qquad (10)$$

2.4 Theoretical relationship between N-value and soil parameters

The theoretical results obtained in the preceding sections can be summarized as follows.

$$p_u = \frac{e_f' W_h H}{\pi a^2 (S+C/2)} \tag{11}$$

where

$$e_f' = e_f (W_h + e^2 W_p)/(W_h + W_p) \tag{12}$$

and where

$$C = \frac{(3 + \sin\phi)\, a}{9\,(1 + \sin\phi)} \left(\frac{b}{a}\right)^{\frac{1 - 3\sin\phi}{1 + \sin\phi}} \tag{13}$$

Here, it should be noted that Eq. (11) is of the same form as Hiley's dynamic formula while the parameters p_u and C have been derived as functions of soil parameters [see Eqs. (4), (7) and (13)].

Since the SPT N-value is defined as the number of blows per S=30 cm, Eq. (11) can be rewritten in the form:

$$N = \frac{60}{2(e_f' W_h H/\pi a^2 p_u) - C} \tag{14}$$

3 POSSIBLE VALUES OF PARAMETERS b/a AND C

From a practical viewpoint, it is of interest to investigate possible range of the value of parameter b/a and that of parameter C.

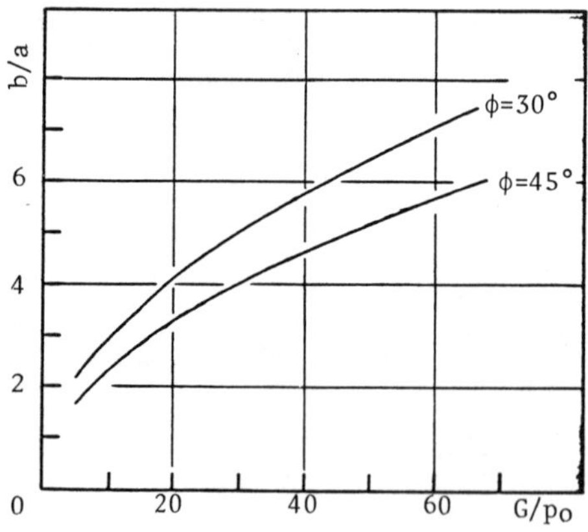

Fig. 6 Relationship among b/a, G/p_o and ϕ

Fig. 6 shows relationship among the parameters b/a, G/p_o and ϕ which are predicted from Eq. (7). An analysis of available experimental data has shown that the value of ϕ ranges from 30° for loose sand to 45° for dense sand, and that the value of G/p_o ranges from 10 for loose sand to 70 for dense sand (Nishida, 1975). As a result, the possible value of b/a is estimated to range from 2.9 for loose sand to 6.2 for dense sand.

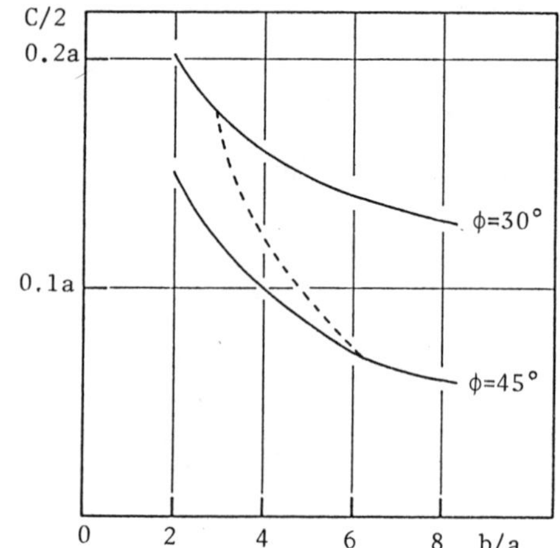

Fig. 7 Relationship among C/2, b/a and ϕ

Fig. 7 shows relationship among the parameters C, b/a and ϕ which are predicted based on Eq. (13). Since the possible range of b/a has just been estimated to range from 2.9 for loose sand and to 6.2 for dense sand, it is seen from Fig. 7 that the parameter C/2 ranges from 0.18a for loose sand to 0.07a for dense sand.

4 DISCUSSION OF END-BEARING CAPACITY OF DRIVEN PILE

When SPT sampler is plugged with a highly compacted soil, it can be regarded as being a driven pile with a diameter of 5 cm. Eqs. (4) and (7) show that the ultimate end-bearing pressure p_u does not depend on the diameter of the penetrating body under consideration. Therefore, Eq. (11) is understood to be used for predicting the end-bearing capacity p_u of a given pile having any diameter in the following two ways: one is based on the use of W_h, H, e_f, a, S and C in practice, the other is based on the use of SPT N-value through the following relation

$$p_u = \frac{243 e'_f N}{30 + 0.25N} \qquad (\mathrm{kgf/cm^2}) \qquad (15)$$

where the following relation are used; W_h= 63.5 kgf, H=75 cm and a=2.5 cm.

Fig. 8 shows comparisons between end-bearing capacity of piles obtained from load tests carried out by PCB Committee of Japan (1964) and by Vesic (1964), together with computed curves based on Eq. (15). In the load tests by PCB Committee, the closed-end piles having a diameter of 200 mm were driven into either medium dense sand or dense sand. Vesic (1964) carried out the load tests in a very thick deposit of naturally-moist medium to fine sands by using the closed-end steel pipe piles having a diameter of 100 mm.

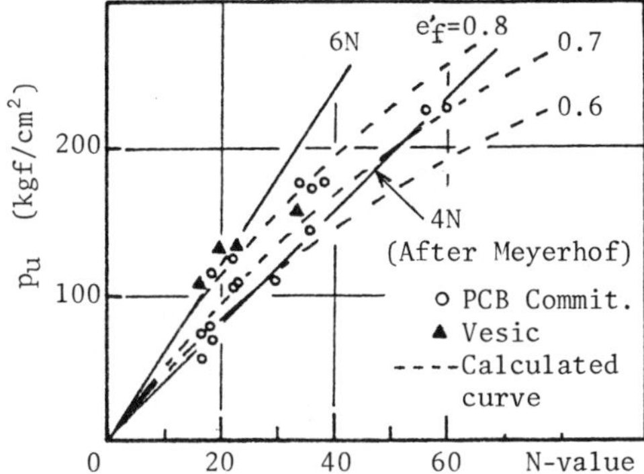

Fig. 8 p_u versus N-value relation

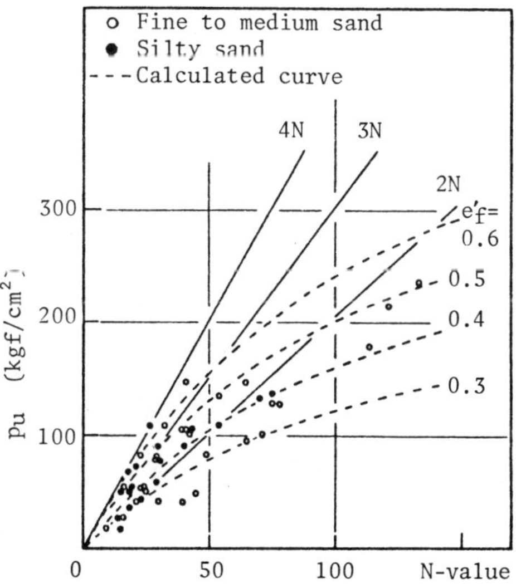

Fig. 9 p_u versus N-value relation

Fig. 9 compares measured values of bearing capacity obtained from load tests carried out by Komatsuda et al. (1974), together with computed curves based on Eq. (15). It is clearly seen from Figs. 8 and 9 that the end-bearing capacity p_u does not increase linearly with increasing N-value but tends to be concave downward with increasing N-value as predicted by the present theory.

5 CONCLUSIONS

A new version of the cavity expansion theory has been developed in this work to clarify the penetration mechanism of SPT sampler driven into a homogeneous deposit of sand. Based on the analysis of collected case records of pile loading tests, it has been shown that p_u does not increase linearly with increasing N-value but tends to be concave downward as predicted by Eq. (15).

6 REFERENCES

Komatsuda, S. et al. 1974, On the loading tests at the base of a bored hole and the end-bearing capacity of bored piles, Proc. 9th Japanese Conf. Soil Engrg.: 537-540.

Meyerhof, G.G. 1956, Penetration tests and bearing capacity of cohesionless soils, Proc. ASCE, Vol. 82, SM1: 1-19.

Miki, K. et al. 1980, The study of the large size penetration test, Proc. of Sounding Symposium, the Research Sub-Committee on Site Investigation, Japanese Society of Soil Mechanics and Foundation Engineering : 223-228.

Nishida, Y. et al. 1974, An experimental study on bearing capacity factor for deep foundations in sands, Memoirs of the Faculty of Technology, Kanazawa University, Vol. 8, No. 2: 15-25.

Nishida, Y. 1975, Calculation of the range of compaction caused by piles, Tsuchi-To-Kiso (JSSMFE), Vol. 13, No. 8: 33-75.

Nishida, Y. et al. 1980, An estimation of the bearing capacity of the pile base in sands and calys, Memoirs of the Faculty of Technology, Kanazawa University, Vol. 13, No. 2: 129-139.

Nishigaki, Y. 1980, Consideration of N-value obtained on sandy soil ground, Proc. of Sounding Symposium, the Research Sub-Committee on Site Investigation, Japanese Society of Soil Mechanics and Foundation Engineering: 109-114.

PCB Committee 1964, Experimental studies of bearing capacity of piles in sand, p. 1-201.

Vesic, A.S. 1964, Investigations of bearing capacity of piles in sand, North American Conf. on Deep Foundation (Duke Univ. SML3).
Vesic, A.S. 1972, Expansion of cavities in infinite soil mass, Proc. ASCE, Vol. 98, SM 3: 265-291.
Yamaguchi, H. 1973, An elastic-plastic analysis of a spherical expansion of a cavity in infinite soil mass and its application, Technical Report, No. 15: 1-11, Dept. of Civil Engrg., Tokyo Institute of Tech..

7 APPENDIX

The assumption of hemispherical cavity expansion associated with the penetrating SPT sampler has stemmed from observations of the deformation pattern of sand around the penetrating, solid pile (Photo 1). For details of the materials used and testing procedure, see Nishida et al. (1974).

Photo 1

Proceedings of the Second European Symposium on Penetration Testing / Amsterdam / 24-27 May 1982

Relationships between N-value by SPT and LLT measurement results

SATORU OHYA
OYO Corp., Technical Center, Tokyo, Japan

TSUNEO IMAI & MIKIO MATSUBARA
OYO Corp., Urawa Research Institute, Japan

1 INTRODUCTION

The Standard Penetration Test is the most commonly used method among the many types of sounding methods practiced in Japan. It is carried out as a matter of standard practice whenever soil investigation borings are carried out. The LLT (lateral load tester) uses a rubber tube and what is generally called a pressuremeter, an inhale loading test apparatus. The LLT was developed in 1965 by Suyama et al. on the basis of a monocell type K value tester developed in 1959 by Fukuoka et al. The LLT is now the most commonly used measuring apparatus of its kind in Japan.

The Standard Penetration Test and the LLT test differ completely, both in principle and in quantities measured. Theoretically there is no basis for comparison of the results from these two tests. Nevertheless, if a correlation between the two could be demonstrated empirically, it would provide an extremely useful means for expanding the range of applicability to design of the Standard Penetration Test. This paper, based on a vast quantity of LLT measurement data that has been accumulated since the development of the apparatus, examines the relationship between N values obtained by the Standard Penetration Test and coefficient of ground resitution, coefficient of elasticity in pseudoelastic regions and yield stress as provided by the LLT.

2 LLT

As shown in Figure 1, the LLT consists of a gas tank, a pressure-volume meter, a probe and nylon tubing. Nitrogen has for pressurization goes from the tank through the regulator to the volume meter. This volume meter is actually a tank containing water that is connected to the probe's cell by the nylon tubing. Gas pressure forces water inside, expanding the cell. The volume of water sent to the cell can be read off as a lowering in the level of the stand pipe in the front part of the volume meter. Both the gas pressure conveyed from the gas tank to the volume meter and water pressure within the cell may be read off. The probe generally used is 90 cm long with a diameter of 80 mm, but probes of 60 and 70 mm are also available. The length of the cell is 60 cm, and it consists of an inner and outer rubber tube.

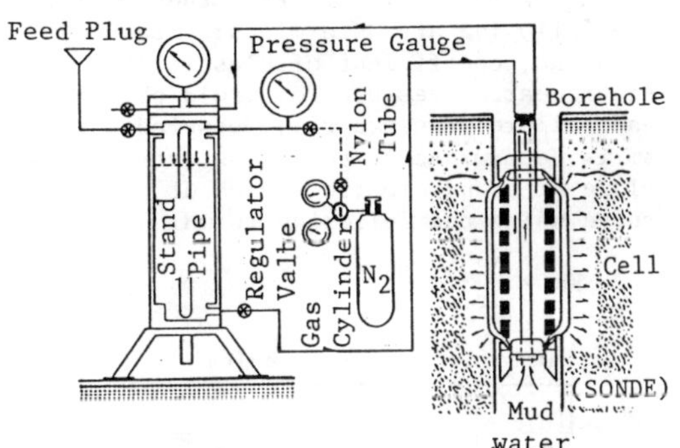

Fig. 1 LLT apparatus

In the LLT test, time and loading weight are controlled. Loading speed is uniform, loads increasing at 2 minute intervals. Deformation is measured at intervals of 15, 30, 60 and 120 seconds at each pressure stage.

Arrangement and analysis of data is carried out in the following way: First, effective pressure Pe is read from the pressuremeter. Pe is the water pressure within the cell, which is the pressure that is actually applied to the ground. Next, corrections are made for hydrostatic pressure, which is determined by the restitution of the rubber (a factor that depends on state of expansion), depth, borehole water level, etc. The objective is to find ground deformation or change in radius of the borehole in relation to various values of pressure Pe. Borehole radius is equivalent to the outer radius of the cell. The amount of expansion of the cell after each loading is found by reading the drop in water level in the stand pipe after 120 seconds. This value is taken as the deformation of the rubber tube, and the resulting outer radius of the tube is calculated. The ground is not an elastic body in that it deforms instantly in response to pressure. Rather, it is like a viscoelastic body in that its deformation usually proceeds with time. Speed of deformation changes with the load. By investigating this type of change, such dynamic properties of the ground as failure load can be found. The standard value, speed of deformation ΔH is determined by noting the progress of deformation at intervals from 30 to 120 seconds after loading at each stage.

Effective pressure Pe and borehole radius r, determined by the above process, are used to prepare the Pe--r curve. In addition, Pe and speed of deformation ΔH are used to prepare the Pe--ΔH curve. Examples of these are shown in Figure 2. From these curves the following are determined: Measured K value Km, coefficient of elasticity E, static earth pressure Po, yield pressure Py and failure pressure $P\ell$. Measured K value Km is determined as the almost straight line slope of pseudoelastic region of the Pe--r curve using the following formula:

$$Km = \frac{\Delta P}{\Delta r} \tag{1}$$

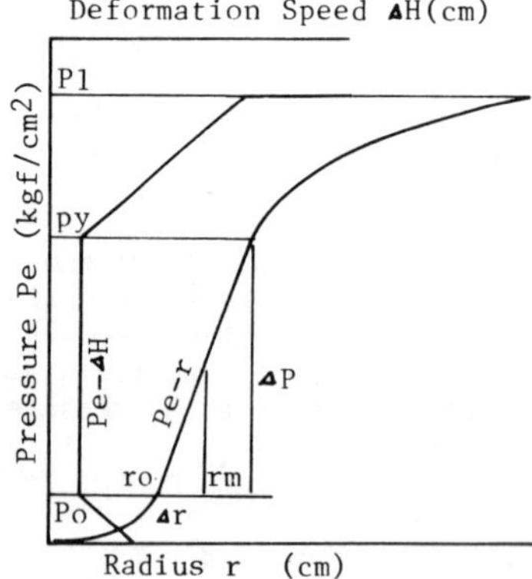

Fig. 2 Deformation process

Coefficient of elasticity E is determined using elastic theory. Strain is assumed to be 2-dimensional. In the following formula, rm is the median radius of the curve of calculated K values and ν is Poisson's ratio:

$$E = (1+\nu) \cdot rm \cdot Km \tag{2}$$

A detailed description of the LLT appears in a paper by Suyama, Ohya and Imai (1982).

3 MEASURED VALUE USED IN COMPARATIVE INVESTIGATION

The authors have accumulated several tens of thousands of bits of LLT measurement data and N values from the Standard Penetration Test from all over Japan. It would, however, not be practical to attempt a correlation of all this data. Thus, in this paper, we have limited ourselves to Standard Penetration Test and LLT data taken from alluvium and diluvium, the commonest type of ground for foundation design, in three representative urban negions throughout Japan, Tokyo, Nagoya and Osaka, as well as the Sakaide region, which includes a variety of types of sandy ground.

4 CORRELATION BETWEEN STANDARD PENETRATION TEST N VALUES AND LLT MEASUREMENTS

LLT measurement results provide various constants from which can be obtained deformation and strength characteristics of the ground. Of these, we attempted to find relationships between ground restitution coefficient K, coefficient of elasticity in pseudoelastic areas E and yield stress Py from the LLT test and N value from the Standard Penetration Test. K value is used in calculating lateral bearing capacity of piles. This capacity varies according to width and displacement of the pile. Also, K value as measured by the LLT (Km) differs according to the diameter of the probe used, even in the same ground. For these reasons, the value Km as such cannot be used as an index of ground deformation characteristics. In this regard, Imai (1969, 1970) proposed a special definition for K value with reference to ground characteristics only, specific K-value Ko, or that value of K when the ground is displaced 1 cm with a loading width of 1 cm. This is expressed in Formula (3). This formula takes into account probe diameter, amount of displacement and correction for the fact that the LLT applies a load over 360° while pioe loading is in one direction only.

$$Ko = \frac{\pi}{2} \sqrt[4]{2ro \cdot (rm-ro)^2} \cdot Km \quad (3)$$

where, ro: initial radius (see Fig. 2)
rm: intermediate radius, determined from measured K value Km

In this paper, we have used this comparative K value Ko as coefficient of ground restitution for comparison with N value.

In showing relationship between N value and LLT data, results are shown separately for clayey and sandy soil. Figures 3 and 4 show the relationship between N value and specific K value Ko; Figures 5 and 6, N value and coefficient of elasticity E; and Figures 7 and 8 show the relationship between N value and yield stress Py. When penetration of the sampler in the Standard Penetration Test was 30 cm or more the value for N at 30 cm was calculated and used.

In all of these graphs a certain amount of scatter is present. However, it is clear that there is an overall tendency to follow a proportional relationship. In all of the relationships, smaller values of N show more scatter, a trend that is especially marked in clayey soil. One possible explanation for this is that in soft or loose ground measurement accuracy in the Standard Penetration Test in low. Clayey alluvial soil in Japan is very soft and sensitive. Thus there is every reason to expect that there would be this much scatter the dynamic type Standard Penetration Test and the static type LLT test. Also, in all the relationships, given the amount of scatter there is, no special difference can be seen different regions or between alluvial and dilluvial ground.

Among the relationships between N value and specific K value Ko and N value and coefficient of elasticity E, directly proportional relationships can be seen both for clay and for sand. However, the proportional constant differs greatly between the two. For the same N value, specific K value Ko and coefficient of elasticity in clayey soil are 3 to 4 times as great as those for sandy soil. For the relationship between N value and yield pressure Py. almost no difference can be seen for clayey and sandy soil.

The above may be broadly summarized in the following formulas:

Clayey soil: $N=1/2\ Ko\ (Kgf/cm^3)$,
$N-1/15\ E\ (Kgf/cm^2)$,
$N=2$ to $3\ Py\ (Kgf/cm^2)$

Sandy soil: $N=1.5\ Ko\ (Kgf/cm^3)$,
$N=1/4\ E\ (Kgf/cm^2)$,
$N=2$ to $3\ Py\ (Kgf/cm^2)$

5 CONCLUSIONS

Although the principles and the quantities measured by the Standard Penetration Test and the LLT are completely different, there is a clear correlation between the results of the two tests, after allowance is made for the scatter that does appear. This relationship does not vary in accordance with region or whether the tests were conducted in alluvial or diluvial ground. However, there is a great difference in the relationships between N value and specific K value Ko and N value and coefficient of elasticity E for sandy soil and for clayey soil. For the relationship between N value and yield pressure Py, there is no significant difference between sandy soil and clayey soil.

The establishment of a relationship between results from the Standard Penetration Test and LLT measurements has a great significance in the expansion of applications of N value to design. For example, the way to using the relationship between N value and specific K value for rough estimating lateral bearing capacity of piles in pile design may be regarded as having been opened.

6 REFERENCES

Imai, T. (1969) Studies (3) of transverse K value of ground - K value used in design, Soil and Foundation, vol.17,11: 13-18 (in Japanese)

Imai, T. (1970) Studies (4) of transverse K value of ground - method of calculation of transverse behavior of pile based of measurement results by LLT, Soil and Foundation, vol.18,1:11-16 (in Japanese)

Suyama, K., Imai, T. and Ohya, S. (1982) Lateral Load Tester (LLT) - its method and accuracy, Proc. Symp. The Pressurement and its marine applications

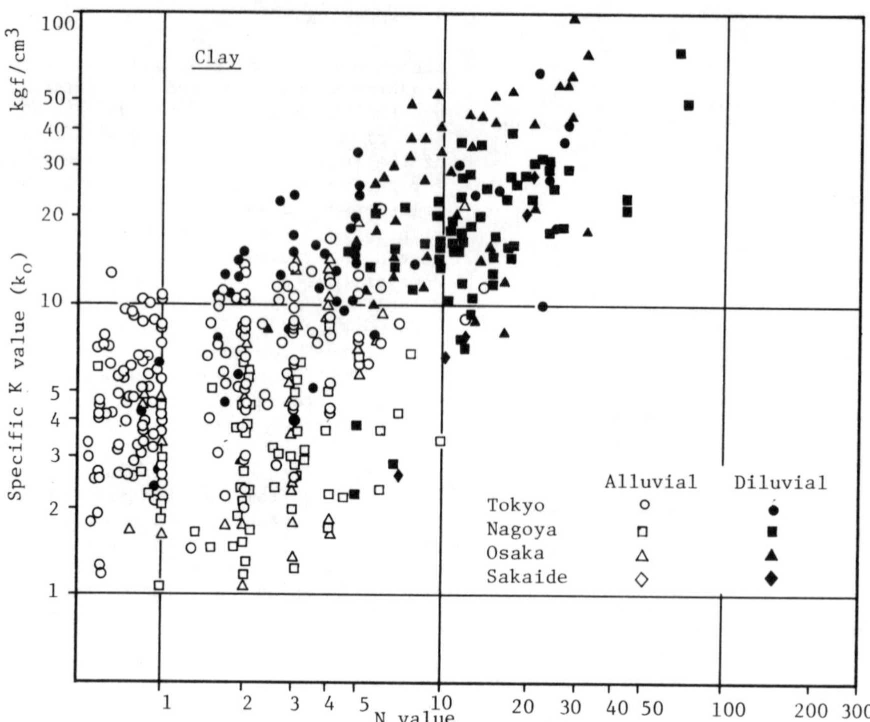

Fig. 3 Relationship between N value and specific K value (clay)

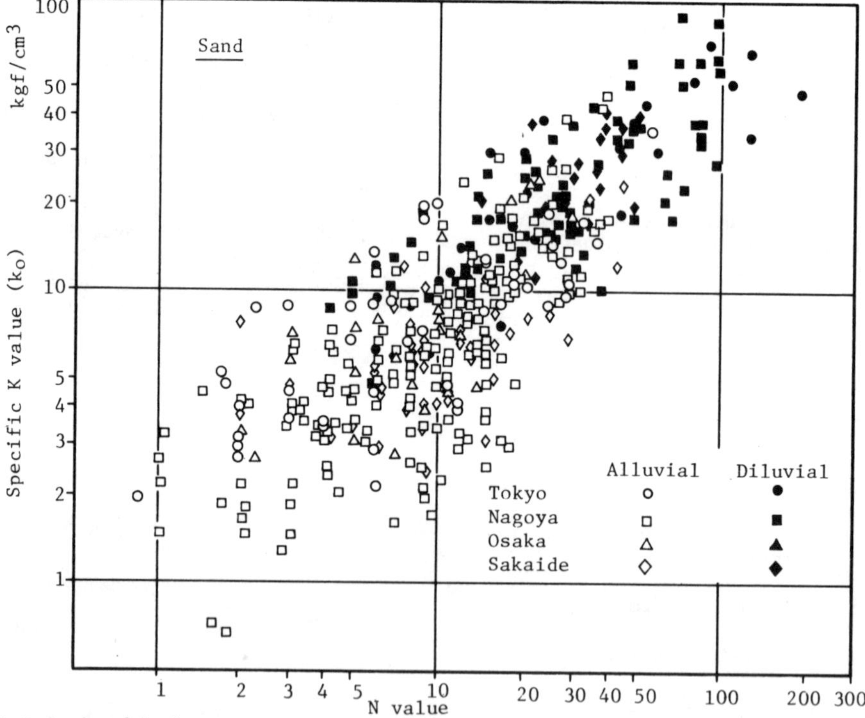

Fig. 4 Relationship between N value and specific K value (sand)

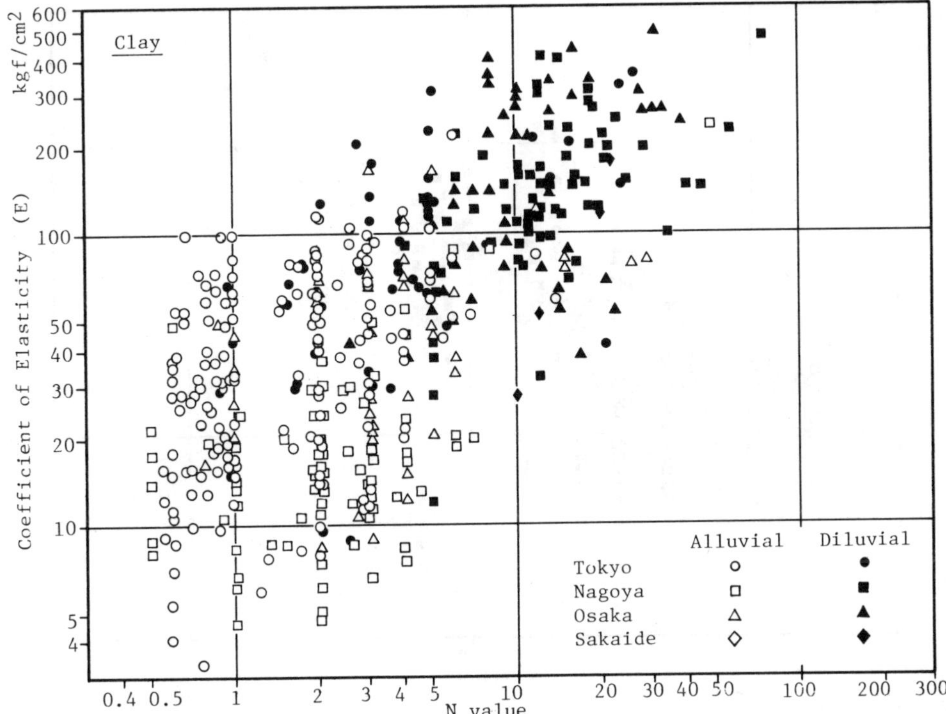

Fig. 5 Relationship between N value and coefficient of elasticity (clay)

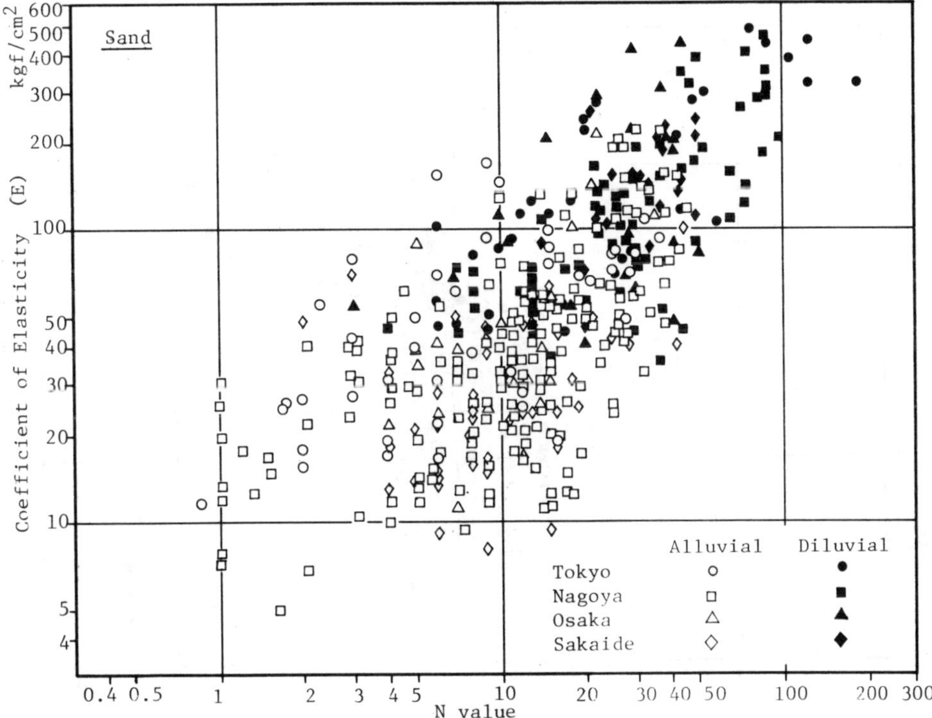

Fig. 6 Relationship between N value and coefficient of elasticity (sand)

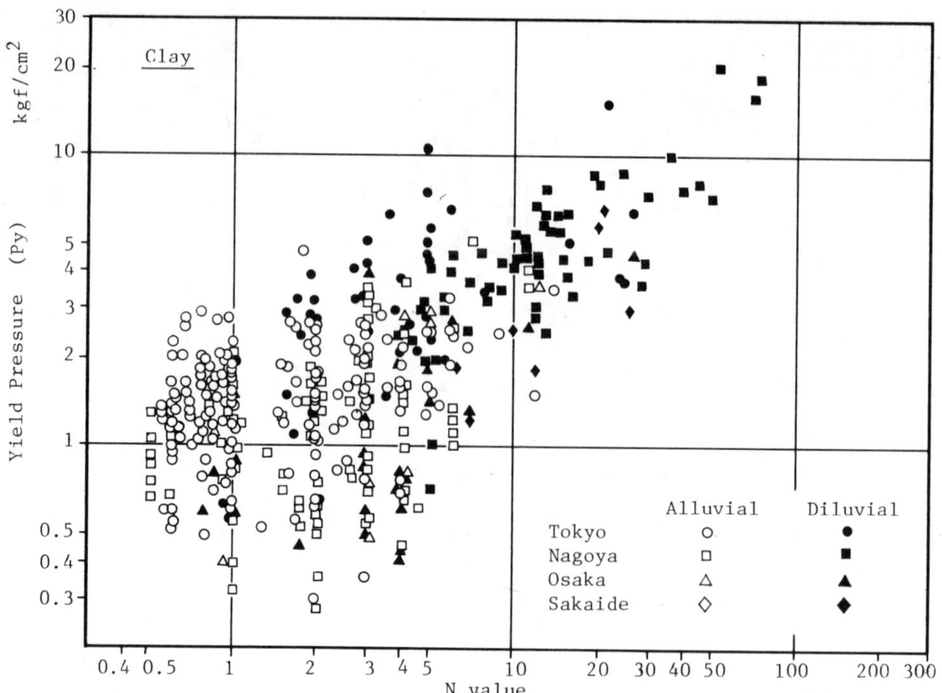

Fig. 7 Relationship between N value and yield pressure (clay)

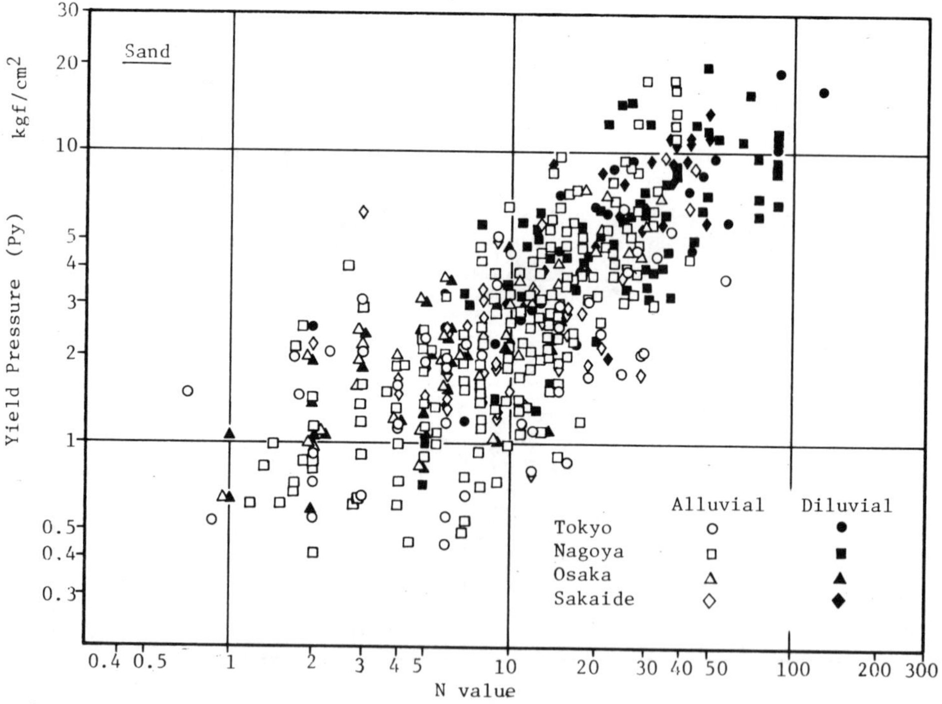

Fig. 8 Relationship between N value and yield pressure (sand)

Proceedings of the Second European Symposium on Penetration Testing / Amsterdam / 24-27 May 1982

Assessment of indepth densification of sandfill due to compaction in a reclamation area by cone penetration resistance

S.D.RAMASWAMY & K.Y.YONG
Department of Civil Engineering, National University of Singapore

1 INTRODUCTION

In order to assess the densification of sand fill by dynamic consolidation in a large reclamation area, static cone penetration tests (CPT) and standard penetration tests (SPT) were used extensively. In view of the requirement of a large number of such tests within a short period of time and since CPT could be carried out rapidly and economically, an attempt was made to arrive at a working relationship between the SPT and CPT values. The correlation obtained on the basis of observed test results, which in conjunction with the Gibbs and Holtz (1957) criteria, was found to be useful in evaluating the relative density attained at various depths. The advantages, reliability and the economy inherent in the use of CPT method for large scale applications are also discussed within this context.

2 PROPERTIES OF SAND FILL

The grain size distribution curves and other pertinent data for the sand fill are shown in Fig. 1. Uniformity coefficient shows a fairly large scatter from 6 to 50, indicating that the poorly graded sand fill belong to the medium to coarse sand region and is similar to the coarse sand used by Gibbs and Holtz (1957). At most SPT boreholes, the N values recorded indicate that the medium to coarse sand has an initial relative density of 60% to 70% above the existing water level and less than 60% below it.

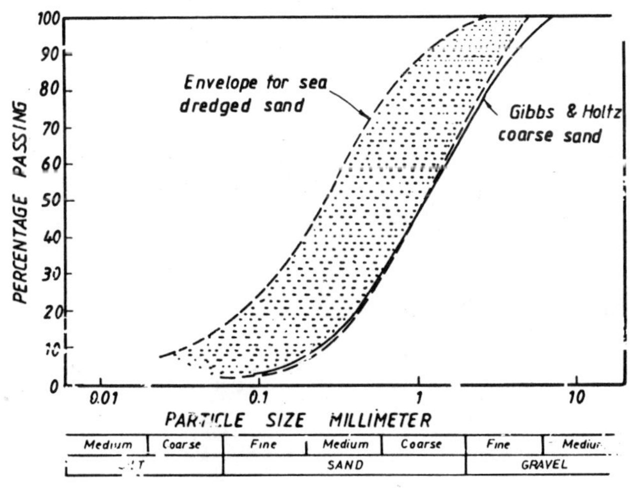

γ_{max} - 17.46 kN/m³

γ_{min} - 15.10 kN/m³

G_s - 2.65

Dominant mineral - Quartz

Shape - sub-round to sub-angular

Fig. 1 Characteristics of sand fill

3 ASSESSMENT OF INDEPTH DENSIFICATION OF SAND FILL

3.1 Methods for quality evaluation

Results of studies reviewed by Marcuson (1978) concluded that the SPT method is not sufficiently accurate to be recommended for final evaluation of the relative density of a site unless specific correlations are developed. Besides, few correlations are available for CPT with density. In fact, a correlation for a given deposit of sand with any of the sounding methods cannot be universal and hence the best approach would be to produce one's own correlation for the particular sand fill and sounding device. However, to facilitate the evaluation of the in-situ relative density, it was decided to use the quantitative correlation between the SPT value, N, and the relative density, D_r, which includes the influence of the effective overburden pressure as developed by Gibbs and Holtz (1957). A typical set of SPT data is plotted as shown in Fig. 2. Each point plotted is the average of at least 18 SPT readings at that depth after treatment. It should be noted that the low overburden pressure and the low penetration resistance values were obtained in the extrapolated portions of the Gibbs and Holtz relationship. A similar correlation of N and D_r developed by Bazarra (1967), but which gives curves more conservative than those developed by Gibbs and Holtz is also shown for comparison. It is interesting to note from Fig. 2 that most of the points plotted have relative densities greater than 75% based on the Gibbs and Holtz criteria and greater than 70% based on Bazarra correlation.

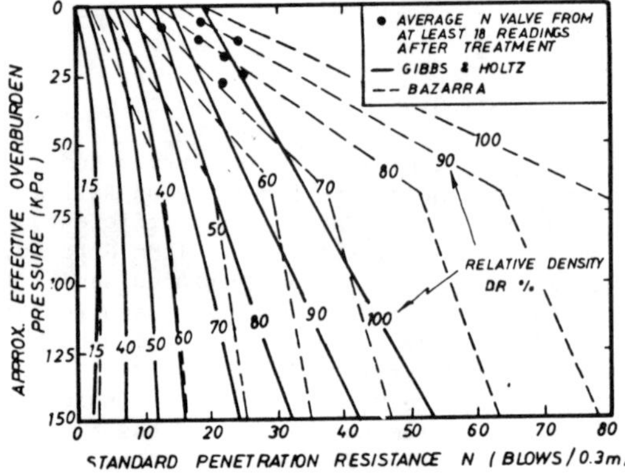

Fig. 2 Correlation between Relative density, effective overburden pressure and standard penetration resistance.

3.2 Test results

SPT and CPT results were obtained in order to assess the improvement achieved under the dynamic compaction. SPT were conducted before and after treatment, initially with water and later with bentonite slurry in the boreholes in accordance with ASTM Penetration Test and Split Barrel Sampling of Soils D1586-67. The CPT utilizing the 60° cone point with 1000 mm² base area, also known as the Dutch Cone, with friction

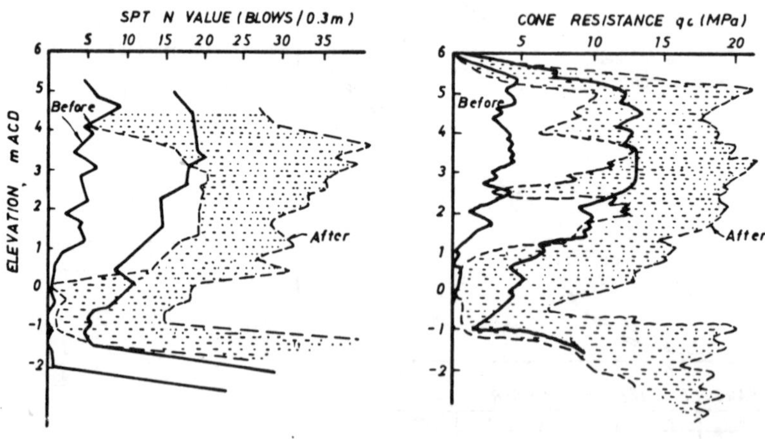

Fig. 3 Improvement in Density by SPT and CPT in reclamation area

sleeves, were also conducted virtually at the same location so that influence of change in material on the test results is minimised. The cone bearing values were measured by both hydraulic pressure gauges and electrical pressure transducers at the cone.

Typical results of SPT and CPT similar to those obtained by Choa et al (1979) before and after improvement, are shown in Fig. 3. Between 1.5 m and 3 m below the surface of the sand fill, average N value of 8 to 10 was observed before treatment which subsequently rises to an average value of 24 after treatment. Below the mean sea level, values of N as low as 3 to 4 were measured before treatment and after compaction, it increases to about 10. CPT bearing values, q_c, vary widely with the penetration depth indicating both variation in particle size and density. Values of q_c as high as 14 MPa and 20 MPa were observed at 2 m depths before and after treatment respectively.

3.3 Correlation of test results

To correlate the CPT and SPT values, average results of both the tests at corresponding locations and depths after the application of compaction are plotted as shown in Fig. 4. From the results plotted, a reasonable working relationship of $q_c = 0.6N$ can be assumed with $q_c = 0.4N$ and $q_c = 0.8N$ as the lower and upper bounds. The scatter in results may be attributed to a number of factors such as grain size and shape, amount of fines and the different testing procedures. Furthermore, residual lateral stresses developed during heavy tamping of fill may influence the results, in particular, the SPT values. In fact, Laicroix and Horn (1972) and Saito (1977) have shown that high N values are the result of the high lateral stresses developed during compaction, which can be greater than the overburden pressure.

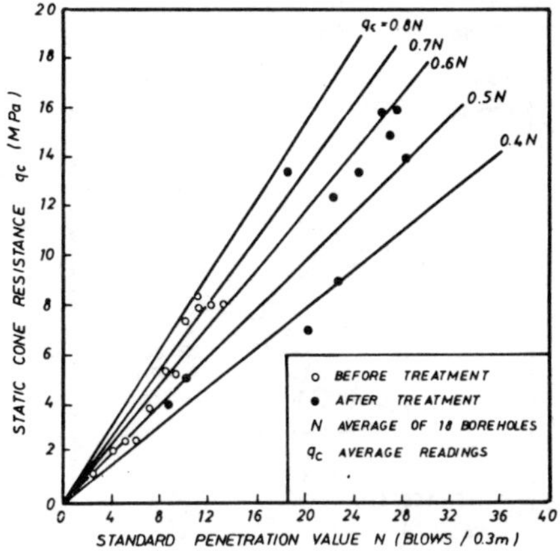

Fig. 4 Correlation between SPT-CPT results

Nevertheless, the above findings conform to the relationship obtained by a number of authors as shown in Table 1 in the form q_c = CN where q_c is the CPT bearing value in MPa, N is the SPT value in blows/0.3 m and C is the correlation factor.

A typical set of cone bearing results after treatment is shown in Fig. 5. All cone bearing curves plot within the envelopes shown. To compare, SPT data have been plotted using the scale factor $N = q_c/0.6$. Gibbs and Holtz relationships are also shown in order to readily see the variation in relative density indicated by the Dutch cone envelope. In general, the criteria for compaction discussed earlier were achieved with dynamic consolidation operations at the site by varying the number of passes between 2 and 5.

4 ADVANTAGES OF CPT METHOD TO EVALUATE DENSIFICATION

A number of limitations inherent with the SPT method were encountered during its usage. In most of the areas requiring treatment, the sand fill stratum was only 3 m to 4 m thick and it was therefore

Table 1: Correlation factors for q_c (MPa) and N (blows/0.3 m)

Reference	Soil Description	Correlation factor, C
1. Sanglerat (1972)	coarse sand	0.4 - 0.6
2. Lacroix and Horn (1972)	sand	0.4 - 0.6
3. Schmertmann (1975)	coarse and gravelly sand	0.5
4. Radhakrishna and Ramaswamy (1979)	medium to coarse sand	0.5

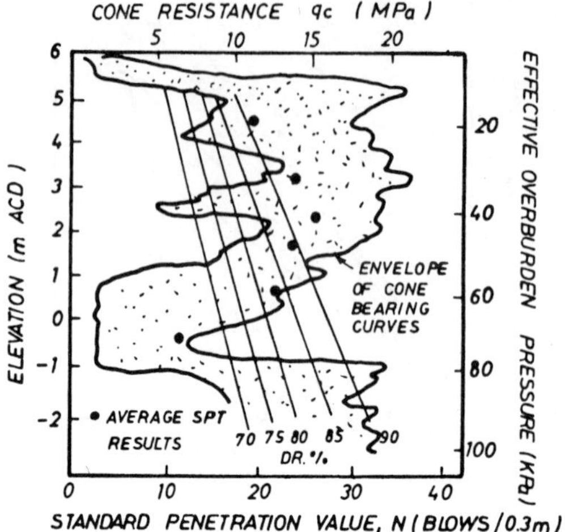

Fig. 5 Comparison of relative densities using standard penetration and Dutch cone sounding data

impractical to conduct such tests at intervals of depth less than 1.0 m. Furthermore, evaluation by Selig and Ladd (1972) on SPT methods indicate that N is influenced by factors such as particle bonding and angularity, vertical and lateral stress levels, and test procedure. Boring operations are likely to cause disturbance to the sand stratum and this could have an effect on the test results. Any loose coarse sand accumulated due to insufficient washing of borehole bottom should be recognised as yielding low N values. Because of these shortcomings, SPT method should only be used as an indicator for relative density and the Dutch cone was employed to supplement the SPT method and to monitor the progress of the dynamic consolidation operations.

Unlike the SPT method, CPT can be operated in a length of few centimeters. The SPT method would tend to average the results of the strata within the range of sampling length. On the other hand, CPT gave continuous q_c values at close intervals (0.2 m) and is also more sensitive in picking up wide variation in the density of the stratification.

It is possible to obtain a large number of q_c values rapidly to provide approximate qualitative indications of the degree of compaction. These measurements ensure that the uniformity of work can be evaluated rapidly during dynamic consolidation so as to ensure consistency of application of method specifications. On the average, five CPT profiles per equipment can be easily achieved compared to one SPT profile per equipment in one day. The cost saving from the reduced time taken to obtain a profile can be increased further when one considers the approximate expenditure in carrying out the SPT and CPT tests at the site, which was in the ratio of 2 : 1 per profile.

5 CONCLUSION

The CPT method is found to be a reliable, economical and less time-consuming approach in assessing the densification of the sand fill. In addition, the quasi-static test is very sensitive to variations in particle size, and the test procedure does not seriously alter the structure of loose sands.

It is suggested that to minimise the distance between the SPT and CPT soundings, the latter should be performed prior to sinking a borehole for the SPT method. Also, care should be taken to avoid disturbance at the bottom of borehole otherwise liquefaction in poorly graded loose fine sand will affect the correlation.

6 ACKNOWLEDGEMENTS

The authors are grateful to Messrs Technique Louis Menard S.A., Singapore, for providing the CPT and SPT data from various jobs on sandfill compaction by dynamic consolidation carried out in Singapore and for permission to use the data to prepare this paper.

7 REFERENCES

Bazaraa, A.R.S. 1967, Use of the standard penetration test for estimating settlements of shallow foundations on sand. PhD Thesis, University of Illinois, Urbana.

Choa, V., G.P.Karunaratne, S.D.Ramaswamy, A.Vijiaratnam and S.L.Lee 1979, Compaction of sand fill at Changi Airport. Proc. 6th Asian Regional Conference on Soil Mechanics and Foundation Engineering. 1:137-140.

Gibbs, H.J. and W.G.Holtz 1957, Research on determining the density of sands by Spoon Penetration Testing. Proc. 4th Int. Conf. on Soil Mech. and Fdn. Eng., London. 1:35-39.

Lacroix, Y and H.M.Horn 1972, Direct Determination and Indirect evaluation of relative density and its use on earthwork construction projects. ASTM STP 523:251-280.

Marcuson, W.F.III 1978, Determination of in-situ density of sands. Dynamic Geo-

technical Testing. ASTM STP 654:318-340.
Radhakrishna, R. and S.D.Ramaswamy 1979, In situ measurements for deep compaction control of dredged sand fill. Proc. Int. Symposium on In-situ Testing of Soils and Rocks and Performance of Structures. Roorkee, 316-319
Saito, A. 1977, Characteristics of Penetration Resistance of a reclaimed sandy deposit and their change through vibratory compaction. Soils and Foundations Vol. 17:4:30-43.
Sanglerat, G. 1972, The Penetrometer and Soil Exploration. Development in Geotechnical Engineering Vol. 1, Elsevier Publishing Co.
Selig, E.T. and R.S.Ladd 1972, Evaluation of Relative Density Measurements and applications. ASTM STP 523:487-504.
Schmertmann, J.H. 1975, Measurements of in-situ shear strength. Proc. Conf. on in-situ Measurement of Soil Properties, North Carolina. 57-138.

Proceedings of the Second European Symposium on Penetration Testing / Amsterdam / 24-27 May 1982

Pressuremeter correlations with standard penetration and cone penetration tests

S.D.RAMASWAMY
Department of Civil Engineering, National University of Singapore

I.U.DAULAH & Z.HASAN
Techniques Louis Menard S.A., Singapore

1 INTRODUCTION

The Pressuremeter test (PMT) is the most versatile in-situ testing device available to-date to the geotechnical engineers in determining the soil strength and deformation parameters. During the early period of development of the Pressuremeter, its usage were more or less confined in France. Recently Pressuremeter tests are receiving more and more appreciation outside France. However, most geotechnical engineers are still more adapted to other in-situ testing methods such as Standard Penetration Test (SPT) and Static Cone Penetration Test (CPT). Therefore a reasonable correlation between Pressuremeter test results with that of SPT and CPT will be of assistance to the users.

Attempts have been made in the past to establish correlation among these three methods by various authors, in different types of soils, with varying degree of success. In this paper, the authors have attempted to establish correlations in granular soils from three different sites and examined their findings with those obtained by others.

Three projects listed in this paper involved soil improvement by Dynamic Consolidation technique. Two of these projects were located in Singapore; one at Changi for the new International Airport and the second was in an island of Singapore (Pulau Ayer Merbau) for a Petrochemical Complex. The other project was located in Bangladesh for a fertiliser factory.

A large number of PMT, SPT and CPT were carried out to establish the pre-treatment and post-treatment densities of the granular deposits. The properties of the deposits in the three different sites are discussed more elaborately in the following sections.

2 SUBSOIL CONDITIONS

2.1 Changi International Airport, Singapore

The project is located on the eastern coast of the Singapore island on reclaimed land. The reclaimation was done with hydraulic sandfill dredged from the sea using cutter suction dredgers. The hydraulic sandfill thickness within the project area varied from 6m to 8m. Most of the material in the fill was sub- rounded, sub-angular quartz in the medium to coarse sand region according to M I T classification. The bottom part of the sand fill at the interface of the sea-bed can be classified as Sand-silt-clay loam.

2.2 Petrochemical Complex at Pulau Ayer Merbau, Singapore

The project was for a warehouse construction for the high density polyethelene plant for the Petrochemical complex. The site was reclaimed from the sea by hydraulically dredged sand sized particles. Thickness of the fill ranged from 4m to 5m. The hydraulic fill was predominatly Calcarius sand with only 5-10 percent of quartz sand grains. The mineralogical analysis by means of X-ray diffraction method indicated 90-95 percent clacide and aragonite contents present in the form of carbonate skeletal material from various organism, e.g. shells, echinoderms, foraminifera.

On visual examination, the sand particles appeared almost as light as sawdust.

2.3 Ashuganj Fertiliser Factory, Bangladesh

The site consisted of approximately 5m to 8m of recently placed hydraulic (dredged) fill of fine sand overlying natural alluvial soil which extended to great depths. The natural alluvial soil at the site were low energy, river deposit of adjacent MEGHNA River and its plain. This material were well sorted (poorly graded) silty fine sand and sandy and clayey silts, often lying in closely bedded, elongate lenses of sands and silts. The presence of at least trace amount of MICA was encountered throughout the soils including the hydraulic fill. In the study reported, the thickness of the described layer extended to a depth of 20m including 5m hydraulic fill.

3 FIELD TEST

In-situ field test, i.e. SPT, CPT and PMT were carried out systematically on all the three sites before, during and after application of Dynamic Consolidation. The results presented here shows only the before treatment and after treatment values in order to examine any change of correlation between loose and dense state of the granular deposit in the given site condition.

In choosing the different SPT, CPT and PMT data, which are presented here in the form of statistical means, great care was taken to select only those data that present uniform subsoil and boundary conditions.

Pressuremeter in all the three sites were carried out by driving the probe to test elevation, following standard procedure as recommended by Menard (1976). Tests were carried out at 1m interval.

CPT were carried out using standard static cone penetration equipment. Begeman type sleeve-friction cone having an apex angle of 60 degree and nominal diameter of 36mm and base area of 10cm square was used. Cone was pushed at a speed of 20mm per second and both cone resistance and jacket friction were measured on a Bourdon gauge.

SPT were carried out at a regular interval of 1m following ASTM specifications.

The results of all the three different types of in-situ tests for the three sites are presented in fig. 1, 2 and 3. Typical grain size distribution curves representing the soil profile is shown along with respective data.

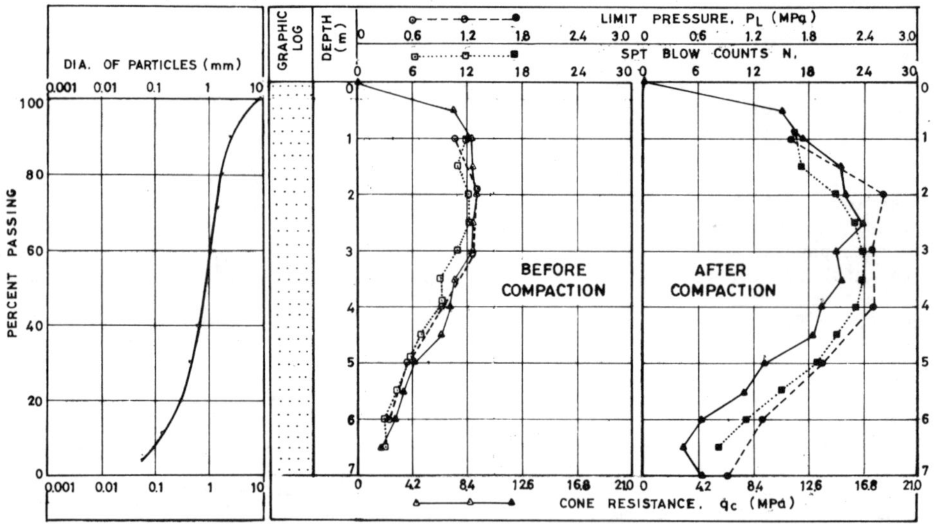

Fig. 1 Correlation of PMT, CPT and SPT - Changi International Airport, Singapore

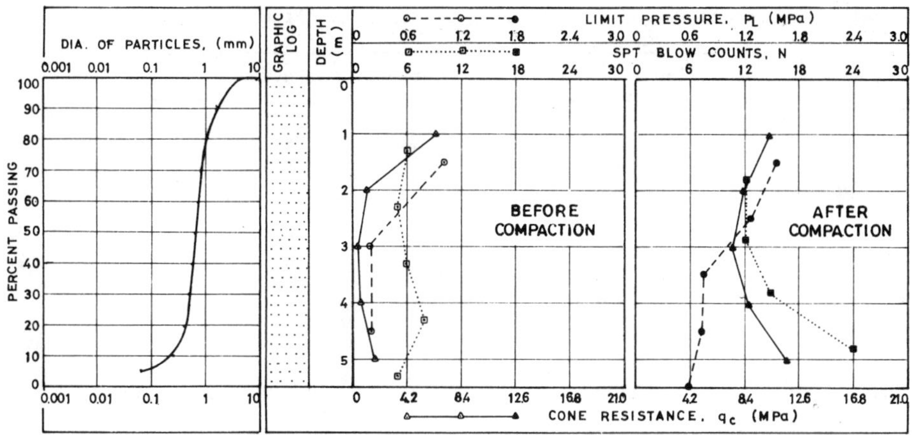

Fig. 2 Correlation of PMT, CPT and SPT - Petrochemical Complex at Pulau Ayer Merbau

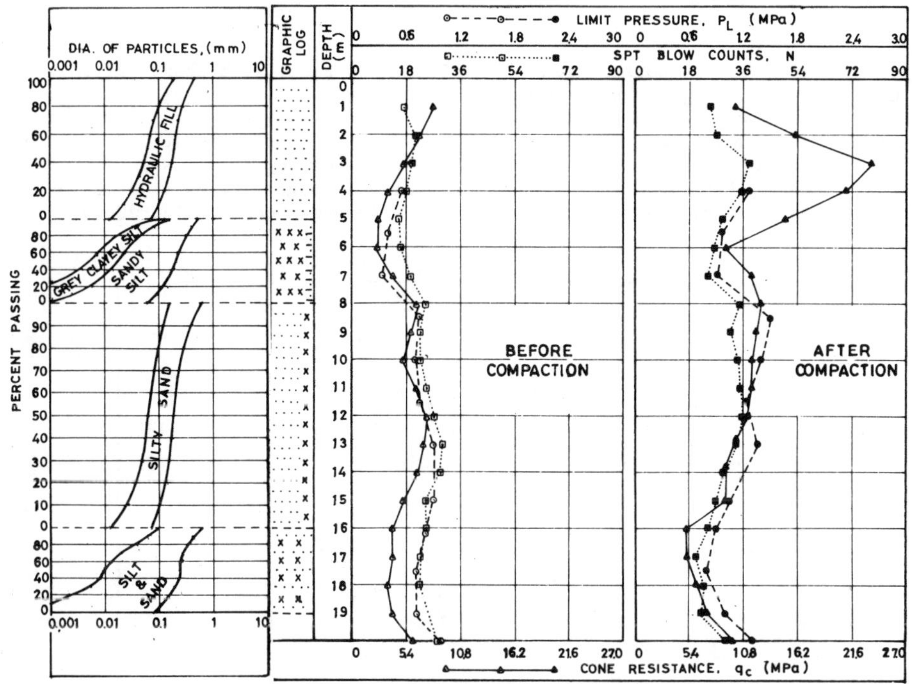

Fig. 3 Correlation of PMT, CPT and SPT - Ashuganj Fertiliser Factory, Bangladesh

4 DISCUSSION

The three sites presented here consisted of predominantly granular deposits. However, the properties of the grains varied considerably within the three sites. (As described earlier)

A close examination of figure 1, 2 and 3 reveals that same correlation factor cannot be applied for all the three soil categories examined.

Findings from the figures are tabulated in Table I.

Table I. Correlation factors for different kinds of tests at different sites.

Soil Category	Location	Correlation Factor		
		N/Pl	q_c/Pl	q_c/N
Coarse to Medium Sand:	Changi, Singapore			
Loose state		10	7	0.7
Dense state		10	7	0.7
Calcarius Sand:	Pualau Ayer Merbau, Singapore			
Loose state		35-45	2.5-7	0.1
Dense state		10-35	7 - 12	0.5-0.7
Silty Sand	Ashuganj, Bangladesh			
Loose state		30	6-9	0.2-0.3
Dense state		30	6-9	0.2-0.3

Note: Pl, q_c are expressed Mpa

It is interesting to note that correlation in case of Changi and Ashuganj did not show significant differences with change in the degree of compaction as would be normally expected from observation made by Cassan (1978). However, marked difference in correlation factor is observed as in case of Calcarius sand, with change of density of the sand.

In the following paragraph, the authors examine the various correlations observed on the three different sites in light of the available correlations reported by others. ((Van Wambecke (1975, 1962),

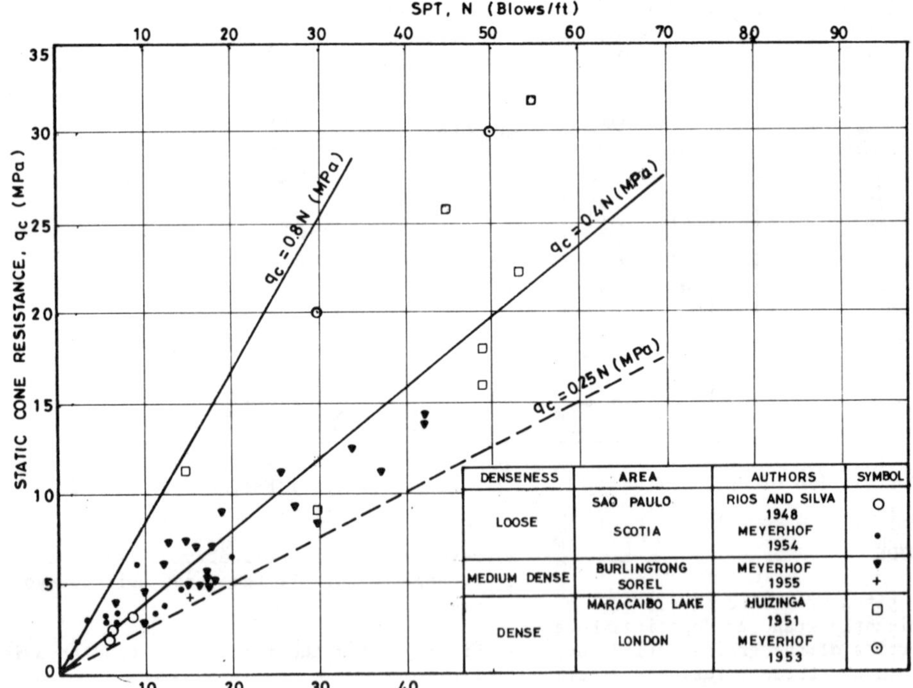

Fig. 4 Correlation factors between q_c/N proposed by various authors

Sanglerat G. (1965), Menard (1963), Cassan (1978)))

4.1 q_c versus N

Various authors have proposed different correlation for q_c and N ranging between $0.25N \leqslant q_c \leqslant 0.8N$ as represented in fig.4

Schertmann (1970) proposed a more comprehensive correlation between N and q_c for different soil types as presented in the table below.

Table II. Relationship between static point resistance q_c and N-values of SPT

Soil Type	q_c/N
Silts,sandy silts, and slighty cohesive silt-sand mixtures	0.2
Clean, fine to medium sands and slightly silty sands	0.3-0.4
Coarse sands and sands with little gravel	0.5-0.6
Sandy gravels and gravel	0.8-1.0

The authors' finding of $q_c = 0.7N$ for Changi sand and $0.2N \leqslant q_c \leqslant 0.3N$ for Ashuganj sand compares well with the proposed correlation in the above table.

No correlation is reported by others in Calcarius sand. The correlation observed by the authors is within the range shown in fig. 4, when the sand is in dense state. However, when the sand is in loose state, the scatter was too wide to propose any correlation factor.

4.2 q_c versus Pl

The correlation factor reported by different authors is within the range of $6\ Pl \leqslant q_c \leqslant 10\ Pl$ (Van Wambecke (1962), (1975)) for sand.

The correlation observed by the authors in the Changi sand is $q_c = 7\ Pl$. The authors expected a correlation for the silty sand to be somewhat different than that of pure sand. However, surprisingly the correlation, $6\ Pl \leqslant q_c \leqslant 9\ Pl$ obtained from Ashuganj is well within the range reported for pure sand.

It is also interesting to note that the change in the granulometry of the deposit (Ref. fig 3) did not have any appreciable effect on the correlation factor.

In case of Calcarius sand, a marked difference in the correlation factor was noticed between the loose and dense state of the sand. There is also a wide scatter of results. However, for the loose and dense state, following range of correlations can be proposed:

Loose sand - $2.5\ Pl \leqslant q_c \leqslant 7\ Pl$

Dense sand - $7.0\ Pl \leqslant q_c \leqslant 12\ Pl$

4.3 N Versus Pl

Fig. 5 shows the results obtained by some

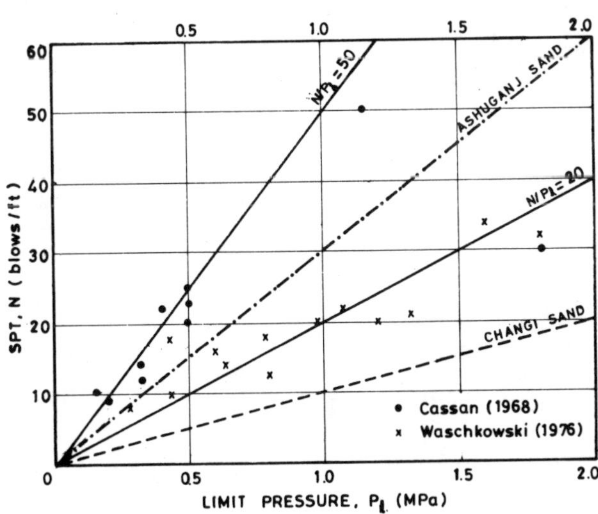

Fig. 5 Correlation of N versus Pl

authors (Cassan, 1968), Waschkowski (1976). Cassan (1968) carried out tests in Lencute and Dunkerque sands, and Waschkowski (1976) carried out tests in silty sand of Blois region. A large scatter is noted. Findings by the authors for Changi and Ashunganj sands have been plotted on this figure. It may be inferred from this figure that correlation for Ashuganj sand is well within the range as reported earlier. However, the correlation factor (pl = 0.1N) of Changi sand is found to be different.

In case of Calcarius sand, similar scatter in correlation factor was observed as reported for q_c versus Pl and qc versus N. Allowing for such scatter the following correlation is proposed:

$$10\ Pl \leqslant N \leqslant 45\ Pl$$

5 CONCLUSION

It is established from study of the SPT, CPT and PMT results of the three sites that meaningful correlations do exist among the three types of in-situ tests and the observed correlations are generally in good agreement with those reported by others.

However, the properties of soils,specially the granulometry and mineralogy have important bearing on the correlation factor as observed for the three different types of sands, i.e. in Changi, Pulau Ayer Merbau and Ashuganj.

Therefore it is imperative that the soil properties are carefully examined before adopting any particular correlation.Even for soils with indentical soil properties, the correlation may be different as observed for the Changi sand in case of N versus Pl. It is therefore recommended that the proposed correlation for any given type of soil should be counterchecked in the field before adopting the same for the full scope of investigation in the particular site.

6 REFERENCES

Cassan, M. (1968) Les essais in situ en mecanique des sols. Construction, no. 10 Octobre: 337 - 347

Cassan, M. (1978) Les essais in situ en mecanique des sols. Book published by Edition Eyrolls, 61 Boulevard Saint-Germain, 75005 Paris: 438 - 442

Huizinga, T.K. (1951) Application of results of deep penetration test to foundation piles Building Research Congr. London, 1 (3): 173 - 176

Menard, L. (1963) Calcul de la force portante des fondations sur la base des resultats des essais pressiometriques seconde partie: Resultats experimentaux et conclusions, Sols Soils no. 6: 9 - 291

Menard, L. (1976) Regles relatives a l' execution des essais pressiometriques, Sols Soils no. 27: 524

Meyerhof G.G. (1953) An investigation for the foundation of a bridge on dense sand Proc. Int. Conf. Soil Mechanic Foundation Eng. 3rd, Zurich 2 : 66 - 70

Meyerhof, G.G. (1954) Recent studies of foundation behaviour Eng. J. Can. 37: 123 - 127

Meyerhof G.G. (1955) The influence of the roughness of the base and ground water condition on the ultimate bearing capacity of foundations Geotechnique, 5 (3): 227 - 242

Rios, L. and Silva, F.P. (1948) Foundation in downtown San Paulo (Brazil) Proc. Int. Conf. Soil Mechanic Foundation Engrg 2nd, Rotterdam 4: 69 - 72

Sanglerat G. (1965) Le penetrometre et la reconnaissance des sols. Dunod.

Schertmann (1970) Static cone to compute static settlement over sand. Proc. ASCE J. Soil Mechanic Foundation Div: 3

Van Wambecke A. (1962) Methodes d'investigation des sols en place, etude d'une campagne d'essais comparatifs, Sols Soils no. 2.

Van Wambecke A. (1975) Les correlations entre caracteristiques geotechniques. Annales des travaux publics de Belgique no. 4.

Waschkoeski, E. (1976) Comparaisons entre les resultants des essais pressiometriques et le SPT - Rapport de Recherche du Laboratoire Regional des Ponts et Chaussees de Blois, FAER 1.05.23.5 Juin, (Unpublished)

Proceedings of the Second European Symposium on Penetration Testing / Amsterdam / 24-27 May 1982

Pitfalls of the SPT

GUY SANGLERAT
Ecole Centrale de Lyon, France

THIERRY-ROBERT A.SANGLERAT
Soils Investigations, Lyon, France

SUMMARY

Two adjacent sites were considered for the location of a proposed nuclear power plant in France. The geotechnical results, for both sites, were very similar, except for the reported values of S.P.T., which varied by a factor of three.

According to the interpretation of the N blow count, the liquefaction problems at the first site investigated, were nominal. They were considered serious at the other side. Additional S.P.T. were made in conjonction with static-dynamic penetration test. The results showed that erroneously low S.P.T. had been reported due to a considerable disturbance of the sandy soils below the water table.

1. INTRODUCTION

The site chosen for the construction of a nuclear power plant is located on a river bank. The preliminary site investigations included : cored samples, grain size analysis, standard penetration tests (S.P.T.), radiographic diagraphy, plate loading tests pressumeter tests, permeability tests and sismic tests over a depth of 50 meters in sandy soils below the water table.

Since the location of the power station has been slightly changed, a new site investigation was performed still down to a depth of 50 meters.

During those two site investigations, where similar testing equipment and team were in action, the N blow count values were found to be very different, while every layer seemed to have very close characteristics according to all other investigation s and laboratory tests.

In order to estimate correctly the liquefaction risks under the earthquakes, it was in fact necessary to find out which of the two sets of results was the good one.

The geological data were as follows :

- from 0 to 1 m. depth
  brown sandy loam
- from 1 to 2,5 m. depth
  silty sand and compact pebbles
  water table between 1 and 2 meters
- from 2.5 to 10.5 m. depth
  yellow sand and gravel
- from 10.5 to 20 m. depth
  sand and gravel
- from 20 to 34.5 m. depth
  fine yellow sand (lightly argillaceous)
- from 34.5 to 37.5 m. depth
  yellow sand
- from 37.5 to 50 m. depth
  grey sand (more or less argillaceous)

2. ANALYSIS OF S.T.P. RESULTS

The results of the two investigations on the site can be summarized as follows :

- First site investigation

$$44 < N < 98$$

with a few exceptions yielding 21 or 34
Average blow count (N) = 63

In some soundings, N, which was usually between 53 and 114, could rise up to even 121 or 220.

- Second site investigation

$13 < N < 46$

Exceptionnally sometimes reaching 60 or 80 at considerable depth.
Average blow counts (N) = 29

The N values from one investigation to the other therefore varied by a factor of three or more.

The results of the first investigation indicated no danger of liquefaction. On the other hand, the second investigation did indeed indicate the opposite. So, it was of the utmost importance, for the construction of this nuclear power plant, to understand from where this important difference was coming from.

One of the first proposed explanations was that the operators had been changed. Furthermore, the drive of the hammer has been found not to correspond to Terzaghi and Peck's recommandations (1967). Therefore, a 18% decrease of the blow count N was necessary for the results of the second investigation. But even with this correction factor, the values yielded of this investigation were far below the results of the first investigation.

Through seminars, lectures and various papers, the authors emphasized the pitfalls of S.P.T., some of which are : discontinuity and setting of equipment (Sanglerat 1979). The results are often unreliable as many very skilled and experienced contractors found out, despite their efforts.

In order to verify if the soils differed between the two locations of the investigations, a third investigation was suggested and performed, using the static and dynamic penetrometer (Sanglerat 1974) and additional S.P.T.

This third investigation was performed under the author's authority, under the following conditions :

- S.P.T. will not be performed in the gravel layer, since very high blow count is little relevant
- the hammer drop will be constant thanks to a frame which exactly adjusts the height to 76.5 cm

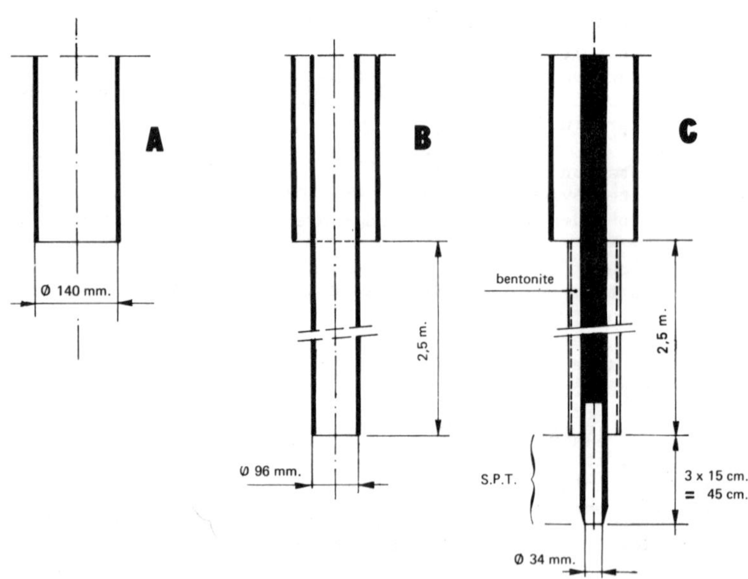

FIG. 1 Correct S.P.T. Performing

a) large diameter (Ø : 194 mm) sample tube in the near surface alluvium deposit

b) rotary sample boring, Ø = 140 mm, 3 m. long with betonite slurry (fig. 1A)

c) core boring with a Ø = 96 mm core taker (fig. 1B)
vibropercussion over 2.5 m. deep, when sampling followed by S.P.T. (fig. 1C) ; over 3 m. deep, if no S.P.T. performed, immediately after.

d) S.P.T. to be performed at three successive 15 cm depths with the first penetration record beeing overlocked as normal (fig. 1C). The soundings rods were driven with a 140 lb hammer drop of 76.3 cm.

Results yielded :
- Third site investigation

$$37 < N < 83$$

with lowest values of 18 and highest value of 118
Average blow count (N) = 57

These values are certainly similar to the ones obtained for the first investigation.

The static-dynamic penetration tests confirmed that the two previously tested sites had very similar soil conditions.

Therefore, the results from S.P.T. yielded by the second investigation were totally wrong.

3. CONCLUSIONS

The very low N values obtained during the second investigation came from an error in performing the tests.

In fact, in order to save time in the course of deep testing, the boring foreman decided to modify the testing procedure, without notifying to the supervisor.

Instead of performing the test after using a core taker of Ø = 96 mm, (fig.1C) the boring foreman reamed out with a core taker of Ø 140 mm (fig.2A), therefore causing stress release in the fine sand under the water table, thus giving N values from a remolded soil, considerably smaller than the undisturbed N probable values.

Fig. 1 and 2 present a sketch of the two processes used for the test :

- Fig. 1 shows the proper way of performing S.P.T. used for the first and third investigation

- Fig. 2 shows the improper setting of equipment used during the second investigation

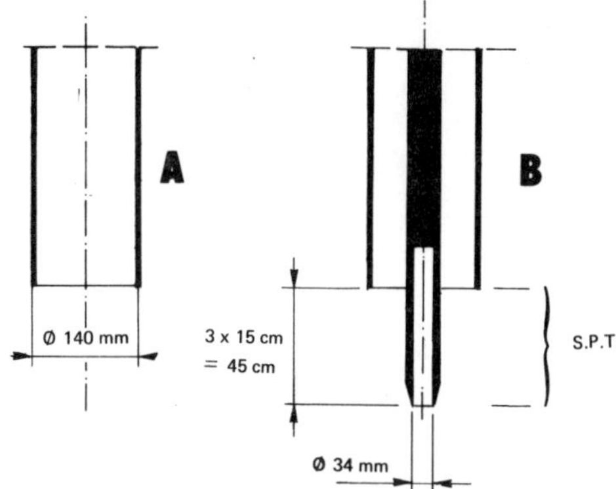

FIG. 2 Incorrect setting of core taker

These very serious errors in the tests procedure could have led to unnecessary and drastic decisions, if all investigations had been performed in the same way as the second investigation.

This demonstrates that site inspectors should strictly control and supervise the S.P.T. test procedure, in order to avoid all possible human errors ; otherwise, the results are bound to be meaningless.

4. REFERENCES

Sanglerat, G. 1974
Penetration testing in France
State of the art report
Esopt 1, Stockholm Vol. 1. pp 47-58

Sanglerat, G. 1979
The penetrometer and soil exploration
Elsevier Amsterdam, second edition 464 p.

Terzaghi K. and Peck R.B., 1967
Soil mechanics in engineering practice
Wiley, New York, N.Y. 2nd ed. 566 p. .

Proceedings of the Second European Symposium on Penetration Testing / Amsterdam / 24-27 May 1982

Penetration tests for determination of characteristics of flood dike materials

DŽ.SARAČ
Faculty of Civil Engineering, Sarajevo, Yugoslavia
M.POPOVIĆ
Institute for Geotechnics & Foundation Engineering, Sarajevo, Yugoslavia

1 INTRODUCTION

Reconstruction and heiqhtening of flood dikes along Sava river is planned due to modified hydraulic conditions and unreliable behavior. The length of the embankments under reconstruction is about 200 km, while their height amounts mainly 4-6 m. These embankments were constructed in a long period by local materials, varying in wide range from silts and CL clays to CH clays. Subsoil consists of similar materials, underlaid by sands and gravels. The embankments were mainly constructed 30 years ago, with subsequent partial repairs and heiqhtenings. Some parts of the embankments were constructed with nonadequate technology and without geotechnical control of material placement. For above reasons, the embankments have very heterogeneous composition and nonuniform quality and it was necessary to perform comprehensive soil investigations, as a basis for project of the reconstruction.

With regard to very long embankments and required extent of exploratory works, it was planned to use mainly penetration tests, being faster and more economical than boring and laboratory soil testing. On a limited number of characteristic cross - sections, situated on mutual distance of about 2-3 km, borings and laboratory testing of soil samples were carried out together with penetration testing of soil. 5-6 borings were performed for each cross-section, with standard penetration tests (SPT) and sampling at intervals of about 1.0 m. In addition, 3-4 cone penetration tests were carried out close to borings (Fig. 1). Dutch static cone penetrometer was used, with capacity of 100 kN. Classification and mechanical properties were tested on the soil samples.

It was planned to apply mainly penetration tests for the embankment parts between characteristic cross-sections. Therefore, dependances between parameters, obtained by penetration tests and other soils properties, were analysed.

Investigations of the embankments are in progress, and in this paper, the results are given for an embankment section, about 15 km long, where the works are finished.

2 SOIL CONDITIONS

Quarternary river deposit extends in the region of the embankment, to the depth of about 20 m. Tertiary overconsolidated clays appear under river deposit. The thickness of the tertiary sediments was not established by borings, but it is well known that it amounts over one hundred meters. Upper part of the river deposit to the depth of about 10 m consists of silty and sandy clays, with plasticity varying in very broad limits, so that CL, CI and CH clays are presented. Dense sands and gravels are situated below the clays to the tertiary subsoil. Immediately to the surface, the clays have high plasticity, which decreases with depth due to increase of sand fraction percentage. Groundwater level

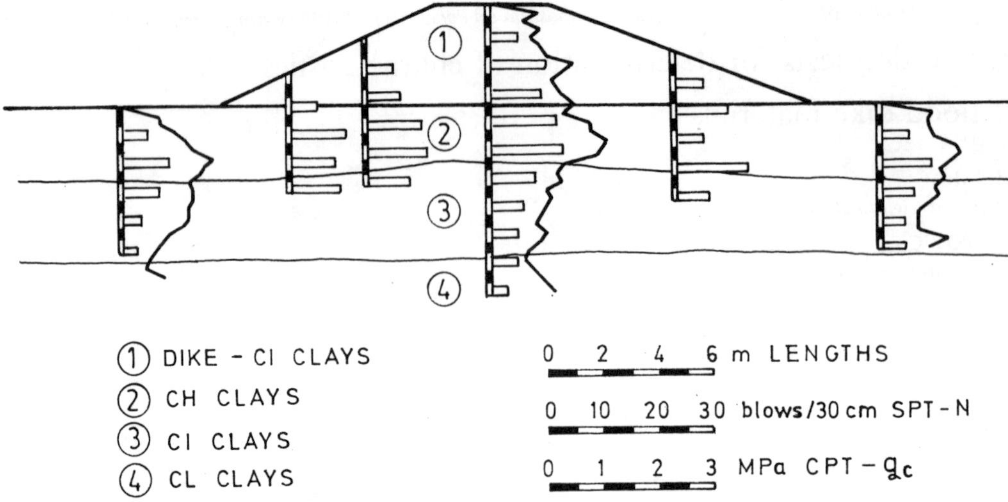

Fig. 1 Characteristic cross-section with exploratory works

varies, in dependence on the river level, so that in flood periods it comes to the soil surface. Exploratory works were performed during low water levels, with groundwater level at about 6-7 m from the soil surface.

Basic results of soil investigations are presented in Fig.2, for central borehole in the cross-section from Fig.1. It is obvious that relatively good agreement exists between the results, obtained with different testing methods.

Laboratory tests on the samples from subsoil and embankment have shown that the clay properties vary in very wide range. In addition to abovementioned variations in plasticity, rather high differences in water content and consistency are obtained, so that consistency index (I_c) varies predominantly between 0.10 and 1.10. On account of that, mechanical properties also vary significantly, what can be seen from Fig.3, where dependance is given between consistency index and unconfined strength (q_u) of clays.

By statistical analysis a linear correlation is obtained:

$$q_u = 28.7 + 356.80\ I_c$$

with correlation coefficient $r = 0.751$.

3 CORRELATIONS OF PENETRATION AND LABORATORY RESULTS

The application of penetration test results by introduction of correlations with soil properties was considered at the European Symposium on Penetration Testing, held in Stockholm, 1974. In summaries of group discussions, devoted to interpretation of results of static and dynamic penetration tests, De Beer (1974) and Schultze (1974) have given a review of the most frequently used correlations.

With regard to the considered problem, it was interesting to establish density, consistency and strength of materials in embankment and subsoil. For clay materials, it was not possible to obtain direct correlation between penetration results and material density. However, analysis of obtained results has shown that it is possible to establish correlation of consistency and strength with penetration test results. This is obtained for both static and standard dynamic penetration tests, with application of linear correlations.

In Fig. 4 dependance between cone resistance (q_c) and consistency index (I_c) is given. By statistical evaluation of the results, the correlation:

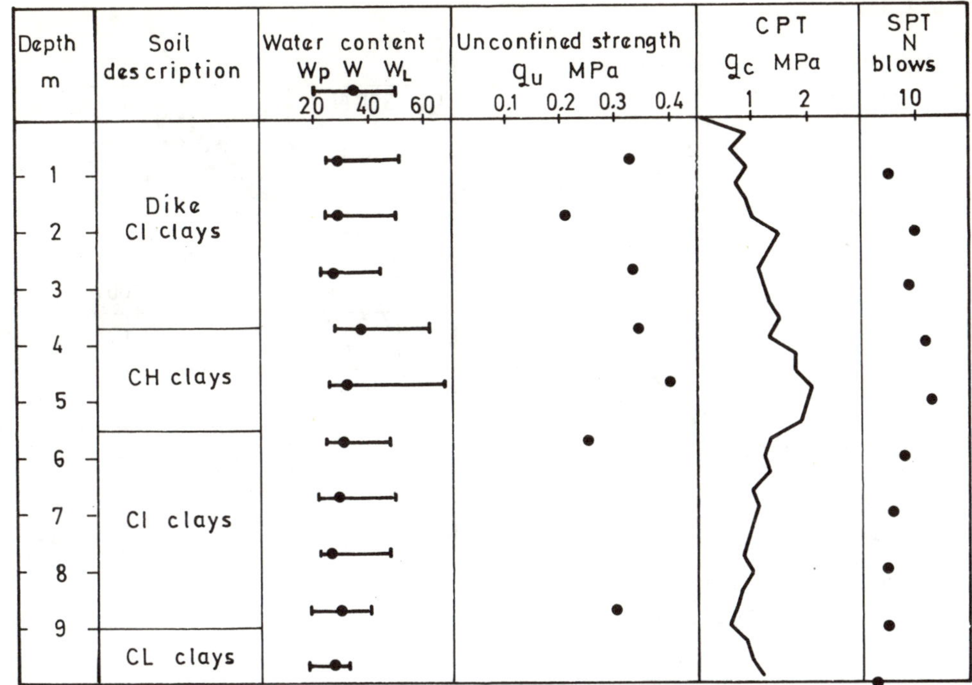

Fig. 2 Soil conditions on the characteristic cross-section

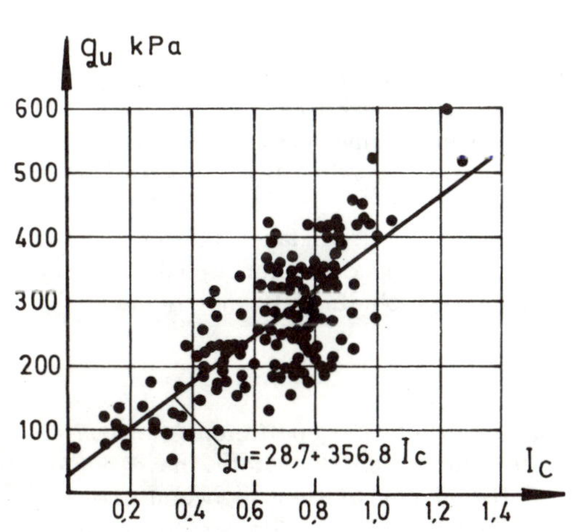

Fig. 3 Dependance q_u - I_c

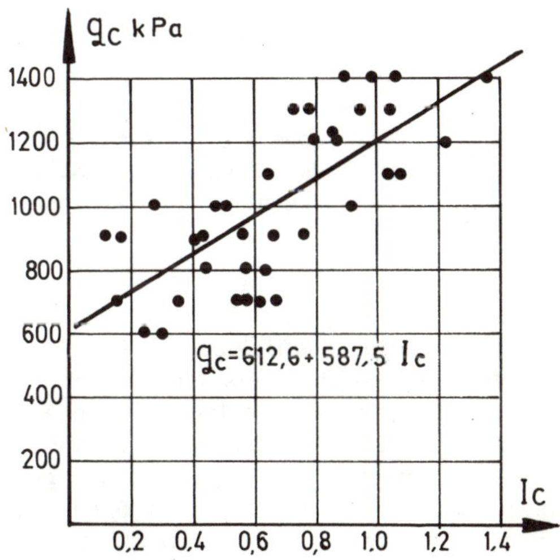

Fig. 4 Dependance q_c - I_c

$$q_c = 612.6 + 587.5\ I_c$$

is obtained, with correlation coefficient r = 0.729. For soft to medium clays, this correlation is in good agrement with values, given by Stefanoff and Bejkoff (1974), while for stiff clays somewhat lower values q_c are obtained.

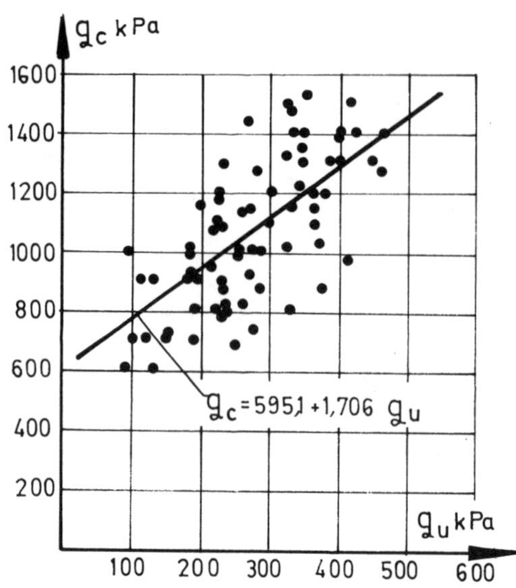

Fig. 5 Dependance $q_c - q_u$

Dependance of cone resistance and unconfined strength is shown in Fig. 5. A linear correlation is also obtained, in the form:

$$q_c = 525.1 + 1.706\ q_u$$

with correlation coefficient r = 0.693. It should be noted that this form of correlation corresponds to a limited range of unconfined strengths, from 100 to 450 kPa. However, ratio q_c/c_u (c_u = 0.5 q)varies in rather broad limits - from 5 to 15, what generally corresponds to previous investigations (De Mello, 1969; De Beer, 1974).

It was mentioned before that certain dependances were obtained also for results of standard dynamic penetration tests.

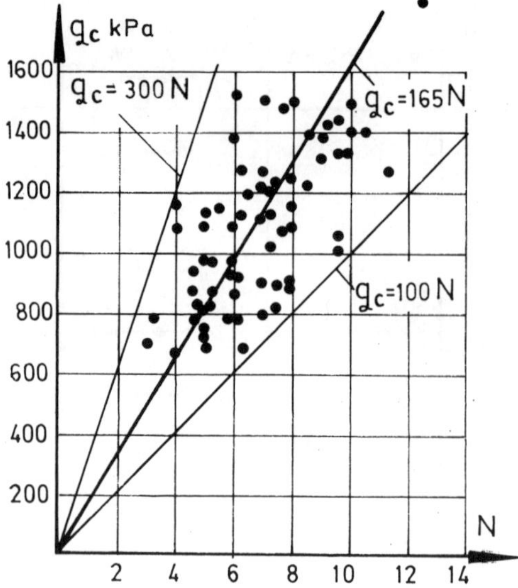

Fig. 6 Ratio between q_c and N

Direct comparison between static and dynamic penetration, presented in Fig. 6, has given ratio q_c/N between 100 i 300 (q_c in kPa), while average value of this ratio amounts 165. This value is somewhat lower than the value q_c/N = 200, which is often related to clay materials (Sanglerat, 1972), but the difference is not high. It should be noted that Schmertmann (1970) reports that for clays this ratio could reach as low as about 100. Moreover, some previous investigations have given q_c/N = 162, for similar materials (Sarač and Lalović 1979).

In Fig. 7 dependance between SPT blowcount N and consistency index is given, which was obtained in the form:

$$N = 2.894 + 6.953\ I_c$$

with correlation coefficient r = 0.765. The autors, unfortunately, have not found published correlations between these parameters, and the comparison is not possible.

Finally, Fig. 8 shows dependance N-q_u, for which linear correlation was also established, in the form:

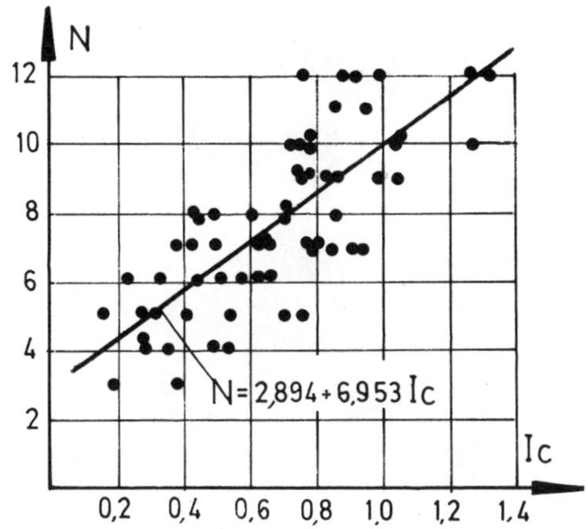

Fig. 7 Dependance N - I_c

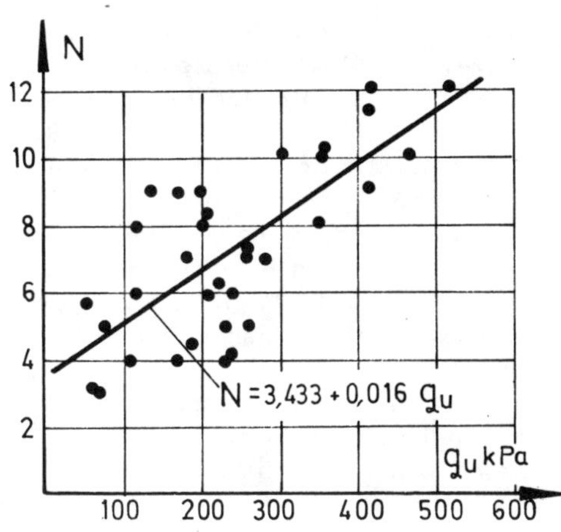

Fig. 8 Dependance N - q_u

$$N = 3.433 + 0.016\ q_u$$

with correlation coefficient $r = 0.730$. The obtained ratio between q_u and N differs considerably from the correlation, given by Terzaghi and Peck (1948). Namely, for given SPT blowcount N, unconfined strength according to the correlation in Fig. 8 is remarkably higher. It should be noted that for various clay deposits this correlation may vary considerably (See for ex. De Mello, 1969) and that presented correlation (Fig. 8) is in good agreement with correlation given by Desai et al. (1974). However, due to relatively small number of the tests, definite conclusions regarding to above correlation could not be given.

4 CONCLUSION

In spite of rather high scattering of the results, it is shown by statistical evaluation that there exist linear correlations of penetration test parameters with consistency and strength of soil. For involved number of tests, high significant correlation coefficients are obtained. This is the case for both static and standard dynamic penetration tests.

This analysis has enabled predominant application of penetration tests for determination of quality of constructed embankments. It was found that this method is faster and more economic than borings and laboratory testing of soil samples.

5 REFERENCES

De Beer, E.E. 1974, Interpretation of results of static penetration tests, Proc. of the ESOPT, Vol. 2:1, p. 62-72. Stockholm, National Swedish Building Research.

De Mello, V.F.B. 1969, Foundation of buildings in clay, Proc. Int. Conf. on Soil Mech. and Found. Eng., State of-the-art Volume, p. 49-136. Mexico.

Desai, M.D., G.R.S.Jain, Swami Saran and P.K.Jain 1974, Penetration testing in India, Proc. of the ESOPT, Vol.1, p.167-184. Stockholm, National Swedish Building Research.

Sanglerat, G.1972, The penetrometer and soil exploration. Amsterdam, Elsevier.

Sarač, Dž. and B.Lalović 1979, Pulling tests on precast concrete piles in silty clays, Proc. of the 6th Asian Regional Conf. on Soil Mech. and Found. Eng., Vol. 1, p. 349 - 352. Singapore.

Schmertmann, J.H.1970, Static cone to compute static settlement over sand,

Proc. ASCE, 96, SM3: 1011-1043.
Schultze, E. 1974, Interpretation of results of dynamic penetration tests, Proc. of the ESOPT, Vol. 2:1, p. 73-79. Stockholm, National Swedish Building Research.
Stefanoff, G. and M.Bejkoff 1974, Penetration testing in Bulgaria, Proc. of the ESOPT, Vol. 1, p.19-25. Stockholm, National Swedish Building Research.
Terzaghi, K. and R.B. Peck 1948, Soil mechanics in engineering practice. New York, John Wiley and Sons.

Proceedings of the Second European Symposium on Penetration Testing / Amsterdam / 24-27 May 1982

Prediction of engineering behaviour of sands from standard penetration tests

K.R.SAXENA & G.T.SRINIVASULU
A.P.Engineering Research Laboratories, Hyderabad, India

INTRODUCTION

The engineering characteristics of cohesionless soils, such as relative density (Rd) and the angle of shearing resistance (∅),are invariably estimated from the results of Standard Penetration Test(SPT). The functional relationships, SPT=f(Rd) and ∅=f(Rd) were first proposed by Terzaghi in 1948. The results of Gibbs and Holtz research, at USBR, in 1957, indicated the effect of one more parameter, the effective vertical stress(P) on SPT and established the functional relationship SPT=f(Rd,P). The general validity of such a functional relationship was endorsed by the research results of Schultze(1961),Bazaraa(1967) Gawad(1976) & Marcuson et al(1977). While there is a very close quantitative agrement between the results of Gibbs & Holtz and Schultze the situation is not so with the results of other investigators. In applying Gibbs & Holtz research results to the field SPT in borings, Vesic (1967) concluded that the Gibbs & Holtz correlation for SPT = f(Rd,P) underestimates the Rd values;whereas, Peck and Bazaraa(1969)concluded that the same relationship of Gibbs & Holtz overestimates the Rd values. La Rochelle(1970), Victor de Mello (1971) and Marcuson (1977) have concluded that the use of SPT= f(Rd,P) relationship may result in unacceptably great error in the estimation of Rd of sand deposits.

The purpose of this paper is to highlight the discrepancies that would result in the estimation of Rd and ∅ values based on the relationship SPT=f(Rd,P) and ∅ = f(Rd). This study does not however provide any alternative solution for this problem.

2 SPT=f(Rd,P) RELATIONS-COMPARED

The research of Gibbs&Holtz,Gawad , and Marcuson indicated that for any given Rd,the N values of SPT increases with increase in the vertical effective stress (P). The research results of Gibbs&Holtz are usually expressed by the relationship:

$$Rd = 21\sqrt{\frac{N}{P+0.7}} \quad \ldots \quad (1)$$

in which P is in Kg/cm^2.

Equation(1) is graphically presented in Fig.(1), in which the ordinate is N/P+0.7 and the abscissa is Rd. The experimental data of Schultze(1961) also perfectly fits with curve(1) shown in the figure, which is essentially drawn for the Gibbs& Holtz data.

Bazaraa(1967) assumed that the Terzaghi correlation for SPT =f(Rd) is valid for a P value of about 0.8 Kg/cm^2 and developed the following correlations to correct the SPT values for the depth effects:

$$N_c = \frac{4(N)_m}{1+2P} \quad \ldots \quad (2)$$

for $P \leq 1.5$ $Kips/ft^2$,

and

$$N_c = \frac{4(N)_m}{3.25+0.5P} \quad \ldots \quad (3)$$

for $P \geq 1.5$ $Kips/ft^2$.

where N_c & N_m correspond to corrected and measured N values and P the effective vertical stress at test elevation.

The relative densities estimated from the Terzaghi relationship for SPT = f(Rd), for the corrected SPT values as per equations (2) and (3) are plotted in figure(1)& curve (2) is obtained.

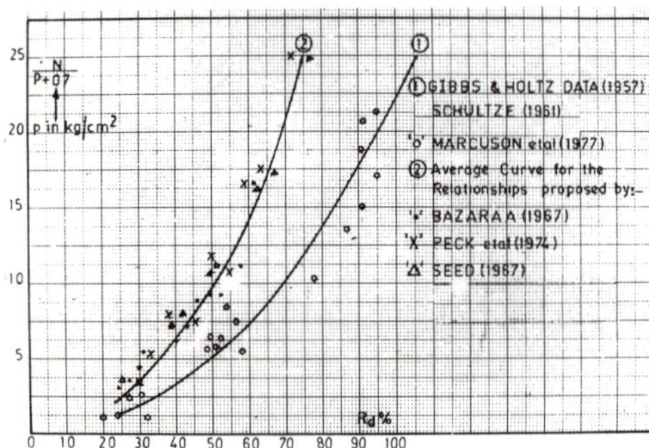

Fig.1: Comparison of SPT=f(Rd,P) relationships

Seed(1967) and Peck et.al (1974) suggested the following equations to correct the SPT values for the depth effects:

Seed:

$$N_c = N_m \left| 1 - 1.25 \log \frac{P}{P_1} \right| \quad \ldots (4)$$

in which P is in Kg/cm^2 and

P_1 is 1 Kg/cm^2.

Peck et. al:

$$N_c = N_m \left| 0.77 \log \frac{20}{P} \right| \quad \ldots (5)$$

in which P is in Kg/cm^2.

The Rd estimated from the Terzaghi relationship for SPT=f(Rd), for the corrected SPT values as per equations (4) & (5) are also plotted in Fig.(1) & these values are found to be very close to curve(2).

The data of Marcuson et.al (1977) is also plotted in Fig.(1) and the dispersion of the data around curve (1) may be seen from the figure. The data of Gawad(1976),(not shown in the figure), also showed similar dispersion as with Marcuson's data.

From Fig.(1)it may be found that the difference in the Rd estimated from curve (1) & (2) is of the order of 20% at 40% Rd range and about 30% in the 80% Rd range. The dispersion of data points of Marcuson (1977), (and also of Gawad not shown in the Fig.) suggest that the individual Rd values may differ by as much as 15% from the values estimated from curve (1).

While comparing the Rd values from curves(1)&(2) it is necessary to remember that curve (1) and the data points of Marcuson (and also of Gawad) around it are based on the laboratory experiments; whereas field observations are the basis for curve (2) and the data points around it. Thus curve (1) may be termed as laboratory curve & curve (2) as the field curve. It is apparent that the Rd estimated from curve (1) is consistently greater than that estimated from curve (2). Only Vesic(1967) has reported some scant data which shows that the Rd estimated from curve (1) are lower than those actually measured in the field. It appears, at first glance, that this case may be disregarded in view of the large data supporting curve (2). However, it appears that such a situation do occur with SPT at deeper depths.

3 Ø = f(Rd) RELATIONS - COMPARED

Victor de Mello (1971) has analysed the Gibbs & Holtz data and worked-out the following relationship for SPT = f(O/,P):

$$SPT = 4 + 0.015 \frac{2.4}{\tan\phi}\left[\tan^2\left(\frac{\pi}{4}+\frac{\phi}{2}\right)e^{\pi\tan\phi} - 1\right] + P\tan^2\left(\frac{\pi}{4}+\frac{\phi}{2}\right)e^{\pi\tan\phi} \pm 8.7 \quad \ldots (6)$$

Equation (6) is presented graphically in Fig.(2). Since the Gibbs & Holtz data for the functional relationship, SPT = f(Rd,P) gave a best fit when plotted between N/P+0.7 & Rd, equation(6) has also been plotted between N/P+0.7 & Ø. It is further felt that such representation of data would facilitate an easy comparison with other similar data.

Also shown in Fig.(2) are data of Meyerhof (1959) - summarised in the relationship Ø =28+0.15 Rd, Bjerrum (1961), Cornforth (1964), Bishop (1966), D.Appolonia et al(1968) and Alpan(1964). The data points of the present study are also shown in the figure. This data is collected from two sites, the Obra and Polavaram Projects, in India.

If Ø is considered as the more significant first order parameter and is used to find out Rd, the data presented in Fig.(2) shows that the Rd estimated from the relationship of different investigators will vary by about ± 20%. On the other hand if Rd is considered as the more significant first order parameter, and is used to find out Ø, the maximum difference in Ø values estimated will be of the order of about 5°, for any Rd. The line shown in Fig.(2) may be considered as the average relationship for Ø = f(Rd) function. This line is drawn passing through(Ø)min., which is assumed as about 29° (the angle of repose for clean, dry, medium sands); (Ø)max., assumed as 42° and the average points for the data, shown in the figure. For any Rd

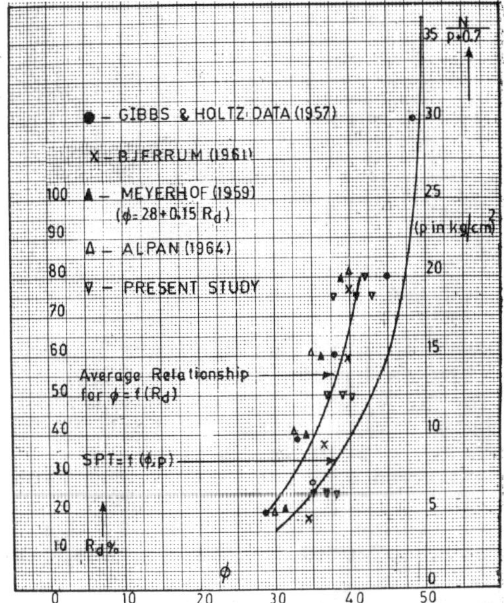

Fig.2: Comparison of SPT = f(Ø,P) and Ø = f(Rd) relationships

the value of Ø estimated from the line may differ ± 2° from Ø = f(Rd) relationships of different investigators. Similarly the Rd estimated from Ø values may differ by about +10% for Ø = 35° and by about -10% for Ø > 35°.

4 CASE STUDY

A number of SPT were conducted at a site proposed for the construction of an earth-cum-rockfill dam across the river Godavari, where the alluvium consists of medium sand upto a depth of about 30 m. The results of SPT conducted in about 20 borings along and in the vicinity of the proposed dam line are summarised in Fig.(3), which is a plot of depth vs N.

From Fig.(3) it may be observed that the N values did not increase appreciably with depth, a large number of values being in the range of 10 - 25 throughout the depth of the deposit.

For all the data points shown in Fig. (3), the Rd were computed based upon Gibbs & Holtz relationship,(as shown in Curve(1) of Fig. (1)and as well as from the average relationship of Bazaraa, Seed and Peck (as shown in Curve (2) of Fig. (1)).The vertical effective stress at each test elevation was computed assuming a saturated unitweight of 2.00 gm/cc for the sand strata. The vertical effective stress computed in this manner, at each test elevation, is slightly in error, because, for simplicity of presentation, a uniform level ground is assumed where-as the ground level for different boring is different, the maximum difference being of the order of 4m. This difference is considered to have insignificant effect on the estimated Rd. The computed Rd values were statistically analysed and the results of analysis are shown in Fig.(3). Curve(1) in Fig. (3) is the plot of the regression equation between Rd and depth, for which the Rd values were estimated from curve (1) of Fig.(1). Curve(2) is a plot of the regression equation between Rd and depth, for which the Rd were estimated from curve (2) of Fig.(1). The Rd were found to decrease with increase in depth. For obvious reasons such a trend is not acceptable.

For the same situation the Terzaghi relationship for SPT = f (Rd), would have indicated a slightly increasing Rd with increase in depth, for the relationship of Terzaghi does not take the depth effect into consideration.

The Rd were also determined by direct measurement of insitu unit weight of the sand and then determining the maximum and minimum unit-weights of the sand in the laboratory. The insitu density measurements were done by block sampling of the sand in five trial

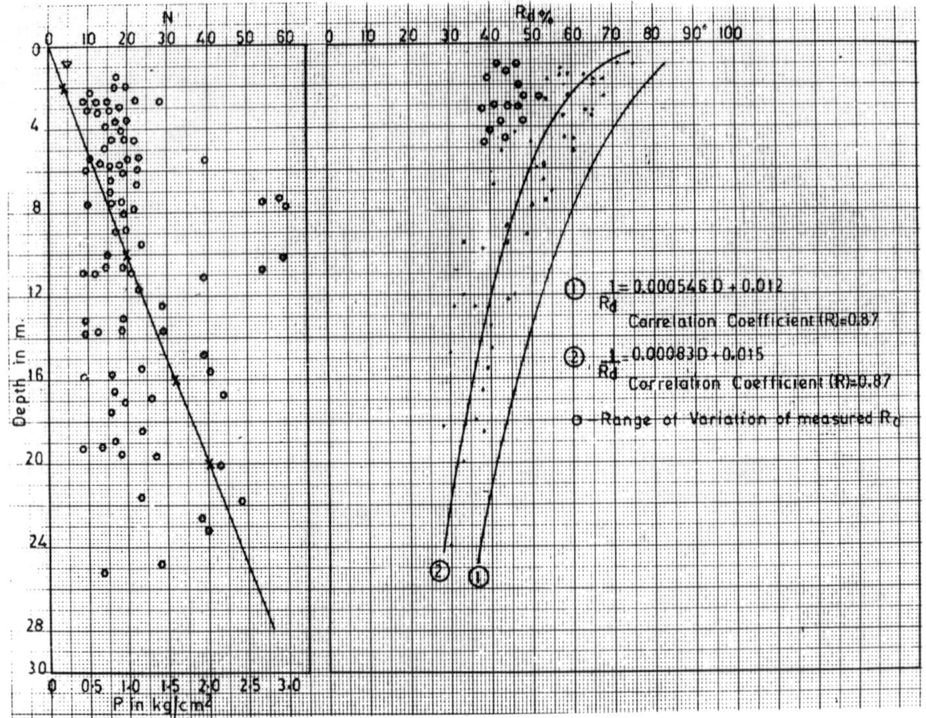

Fig.3: (a) N vs depth (b) Rd vs depth

pits excavated upto a maximum depth of about 5 m. Some selected data after discarding very low values of the order of 20% from these direct measurements is also shown in Fig.(3). It may be observed that the measured values of Rd are close to the values estimated from the Terzaghi's relationship.

The measured values of Rd are much smaller than those estimated either from Gibbs&Holtz or Bazaraa relationships. This difference between the measured and estimated Rd is obviously due to the depth corrections, to the measured N values proposed in the above methods.

Before further proceeding to examine whether or not the measured N values should be corrected for the depth effect, it is necessary to examine the relative correctness of measured and estimated Rd values. The estimated Rd values from Gibbs and Holtz relationship are in the range of 70 to 85% upto 5 m. depth. Such high Rd are improbable in alluvial sand deposits which may be considered as young and freshly laid deposits as they are periodically subject to flood flows and fluctuations in ground water table. Therefore there seems to be sufficient reason to believe that the estimated Rd from Gibbs and Holtz relationship are on the higher side and these high values may be the result of depth correction applied to the measured N values. The measured values of Rd seem to be the most probable values for such deposits. Further the measured values are closer to the values estimated from the Terzaghi relationship which do not consider the depth effects on N values. Therefore the question of correcting the N values for the depth effects needs further consideration.

4 IS DEPTH CORRECTION NECESSARY?

The authors feel that the stress at the bottom of a bore hole is not that due to the overburden above the test level. Even if it is assumed that the stress is equal to the horizontal stress $P_h = K_o \gamma H$, where K_o is the coefficient of earth pressure at rest, this stress also may not have any influence on the penetration of the spoon, for the sand can displace upword into the empty space in the casing, around the spoon, rather than trying to find its way into the surrounding sand mass. This is rightly applicable to loose sands, as found in alluvial deposits subjected to displacements by activity of rivers.

Hence the applicability of laboratory penetration test results to predict the field Rd on the basis of SPT conducted in borings needs further careful examination. In this connection it is worth remembering that the experiments of Gibbs & Holtz, Schultze, Gawad and Marcuson were not conducted for the holing condition in sand. Instead, the above researchers applied different stresses over the surface of the sand, in a tank, and conducted the SPT right over the surface of the sand. The results of such tests under 'no holing' condition do not obviously bear any resemblence to the field SPT in bore holes.

5 SUMMARY & CONCLUSIONS

The purpose of this paper is to highlight the discrepancies in the estimation of engineering behaviour of cohesionless soils based upon the SPT. The following conclusions are drawn from the foregoing study.

1. The estimation of Rd of alluvial sand deposits based upon the correlations of the form SPT=f(Rd,P), developed as a result of laboratory experiments, would result in unacceptably great errors. As a result of the use of these correlations, the Rd are overestimated at shallower depths & underestimated at deeper depths.

2. The estimation of the Rd via the functional relationships of the form SPT =f(Ø,P), and Ø=f(Rd), as shown in Fig.(2), gives less error than when the Rd is determined from the relationship of the form SPT=f(Rd,P), but even this error may be of considerable magnitude.

3. The main reason, if not the sole reason, for the discrepancies as mentioned in (1) and (2), appears to lie in the method of correcting the measured N values in the field for the depth effects. The overburden effect, which is found in the laboratory experiments conducted for 'no holing' condition, cannot be expected to prevail and influence the field SPT in borings. Even if it is still felt that the depth should have some effect on the penetration resistance, the effect is not certainly due to the vertical effective stress, γH, but may be due to the horizontal stress, $K_o \gamma H$. The correction to the measured N values in the field, proposed by Bazarra, Eq.(2)&(3); Seed, Eq.(4) and Peck, Eq.(5) appears to have taken this fact into consideration to certain extent (see curve (2) of Fig.(1)). The authors feel that even the effect of the horizontal stress needs re-examination.

6 REFERENCES

Alpan, I. 1964, Estimating settlements of foundations on sands, Civil Engineering and Public Works Rev., Vol. 59, p. 1415.

Applonia, D.J. et al. 1968, Settlement of Spread Footings on Sand, ASCE, Vol. 94, SM.3, p. 735.

Bazaraa, A.R.S.S. 1967, Use of the Standard Penetration Test for estimating settlement of Shallow foundations on sand, Ph.D.Thesis, University of Illinois, Urbana.

Bjerrum, L., S. Kringstad and O. Kummeneje, 1961, The Shear Strength of fine sand, 5th ICSMFE, Paris, Vol. I, p.29.

Gawad, T.E.A. 1976, Standard Penetration Resistance in Cohesionless Soils, Soils &Foundations, Vol. 16, December, No.4, p. 47.

Gibbs, H.J. & W.H.Holtz, 1957, Research on Determining the density of sands by spoon penetration testing, 4th ICSMFE, London, Vol.I, p. 35.

Marcuson, W.F. & W. A. Bieganousky 1976, Laboratory Standard Penetration Tests on fine Sands, Jr. of Geotechnical Engineering Divn. ASCE, Vol.103, No.GT 6, p. 565.

Marcuson, W.F. et al, 1977, SPT & Relative Density in Coarse Sands, ASCE, Jr. of Geotechnical Engineering Division Vol. 103, No. GT 11, p. 1295.

Meyerhof, G.G. 1965, Shallow foundtions, ASCE, SM.2, Vol. 91. March, p. 21.
Peck, R. B. 1963, General Report, Soil Properties, Field Investigations, 2nd Panam, CSMFE, Brazil, Vol. 2, p. 449.
Sanglerat, G. 1972,The Penetrometer and Soil Exploration, Elsevier Publishing Company, Amster dam, Netherlands.
Seed, H.B. 1976, Evaluation of **Soil** liquefaction effect on level ground during earthquakes, ASCE, Annual Convention and Exposition, Liquefaction Problems in Geotechnical Engineering, Philadelphia, PA, p. 1.
Schultze, E and Menzebach, E. 1961, Standard Penetration Test & compressibility of Soils, 5th ICSMFE Paris, Vol.I, p. 527.
Vesic, A.S. 1967, General Report on Shallow Foundations, deep foundations without Piles, Pile foundations, 3rd Panam. CSMFE, Caracas, Vol. III, p. 181.
Vesic, A.S. 1967,A study of bearing capacity of deep foundations, Final Report, Georgia Institute of Technology, Atlanta, Georgia.
Victor, F.B. de Mallo, 1971, State of the Art Report on the Standard Penetration Test,4th Panam,CSMFE, San Juan, Puerto Rico, Vol. I, p. 1.

Proceedings of the Second European Symposium on Penetration Testing / Amsterdam / 24-27 May 1982

Application of N-value to design of foundations in Japan

YUKITAKE SHIOI & JIRO FUKUI
Public Works Research Institute, Tsukuba Science City, Japan

1 PREFACE

When a structure is constructed on a certain area of ground, boring and sounding are generally done as an in situ investigation for evaluating the dynamic properties of that ground design of the foundation.

Today, in Japan, the most popularly used method of sounding is the standard penetration test (S.P.T.). This is because S.P.T. is a simple test. At the same time, it permits sampling and has a wide range of applications in the aspects of the kind of ground, depth, etc. Further, a number of ground constants can be easily estimated from the N-value measured by the S.P.T.

However, neither the mechanism of the S.P.T., accuracy of N-value measurements nor accuracy of estimates of the ground constants from the N-value have been discussed much.

In this report, the authors will introduce the present conditions on application of the N-value. Also, to be introduced are some of the results of investigations and research on the accuracy N-value measurements and of estimates of the ground constants in Japan. The safety factor in the calculation formula of the bearing capacity of pile in road bridge foundations will be examined.

2 PRESENT CONDITIONS OF APPLICATION OF N-VALUE IN JAPAN

Those sounding methods used extensively in Japan include the ones listed in Table 1. Of these, the S.P.T. is the first soil survey method specified in the Japanese Industrial Standards in 1961 as JIS Designation: A1219-1961 (J2SA 1219-1976) "Penetration Test for Obtaining N-value to Know the Relative Values of Soil Hardness and Coherence in Its Original Position." At about the same time, rules concerning the design and execution of the foundation structures using the N-value were shown in the guidelines and criteria of the various enterprises. Thus the S.P.T. as a soil survey method came to be used extensively.

Table 1 Kinds of sounding

Systems	Name
Tube type dynamic penetration	Standard penetration test (S.P.T.)
Cone type dynamic penetration	Dynamic cone penetration test
Static penetration	Portable penetration test
	Dutch cone
	Swedish sounding
Vane	Simple vane test
	Vane test
Pull-out	Iskymeter test

Further, from experience, some methods of evaluating the ground constants from the N-value obtainable by this method have been developed, and their accuracy of estimation has improved.

Figures 1 through 3 show excerpts of items concerning the S.P.T. from survey results on the methods of choosing the foundation type conducted by the Technical Research Council of the Ministry of Construction in 1975 for about 2800 foundations of road bridges, gates and weirs. Figure 1 is the result of the sounding survey. About 85 percent has sounding done in one way or another, and of them, about 2/3 used the

S.P.T. Figures 2 and 3 are the results of the survey of calculations of the bearing capacity of piles and that of the coefficient of ground reaction in caisson. In larger cases, the bearing capacity and coefficients of ground reaction are calculated in use of the N-value obtained from the S.P.T.

As we have seen, the S.P.T. is an indispensable method of survey for design of structures.

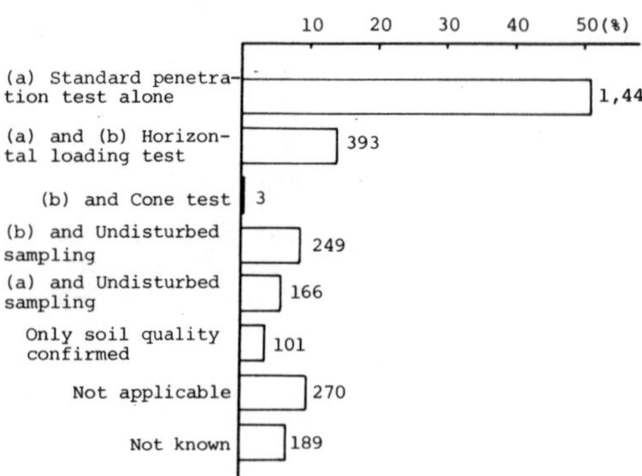

Fig. 1 Sounding

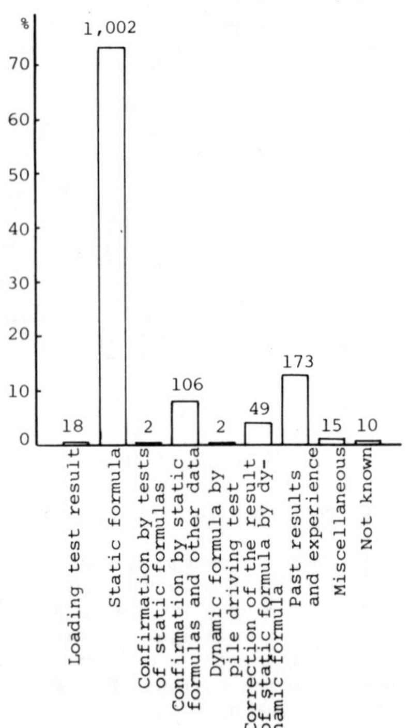

Fig. 2 Pile bearing capacity

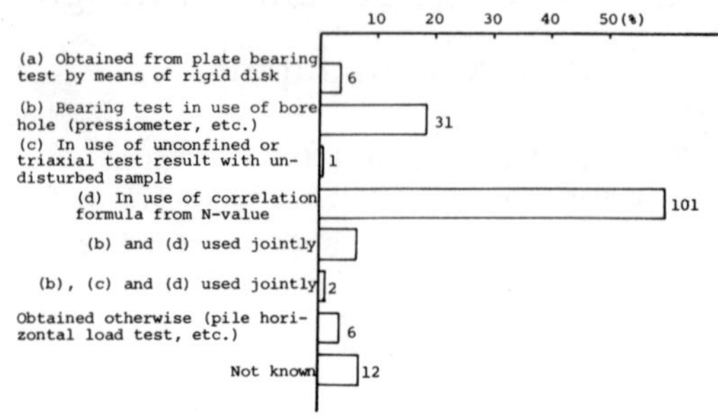

Fig. 3 Subgrade reaction coefficient of caisson

3 CALCULATION OF GROUND CONSTANTS IN USE OF N-VALUE

In the following are shown typical ground constants calculated by using the N-value and their calculation formulas presently employed in Japan.

a) Internal friction angle of soil

$\phi = \sqrt{15N} + 15$ (deg)
Road Bridge Specifications

$\phi = 0.3N + 27$ (deg)
Design Standards for Structures, Japanese National Railway

b) Cohesion of soil

$c = 0.6 \sim 1.0N (t/m^2)$
Road Bridge Specifications

$c = 0.6\ (t/m^2)$
Terzaghi & Peck

c) Modulus of deformation of the ground

$E = 28N\ (kg/cm^2)$
Road Bridge Specifications

$E = 25N\ (kg/cm^2)$

Design Standards for Structures, Japanese National Railway

d) Coefficient of ground reaction

$k = 0.15N\ (kg/cm^3)$
Technical Standards for Port and Harbor Facilities

$k = 0.1 \sim 1.0N\ (kg/cm^3)$
Structural Design Criteria, Building Standards

In addition to the above, the N-value is also used for determining the unconfined compression strength of clay, the relative density of sand and ground fluidization at

6 STUDY ON THE ACCURACY AND SAFETY FACTORS OF THE FORMULAS FOR CALCULATING THE BEARING CAPACITY OF PILES USING THE N-VALUE

In paragraph 4, several formulas for calculating the bearing capacity of piles using the N-value were introduced. Here, their accuracy will be examined from load test data. Figures 7 through 9 show histograms of the ratio on the value from measurement to the value calculated from the N-value of the ultimate bearing capacities of driven, cast-in-place and bored piles.

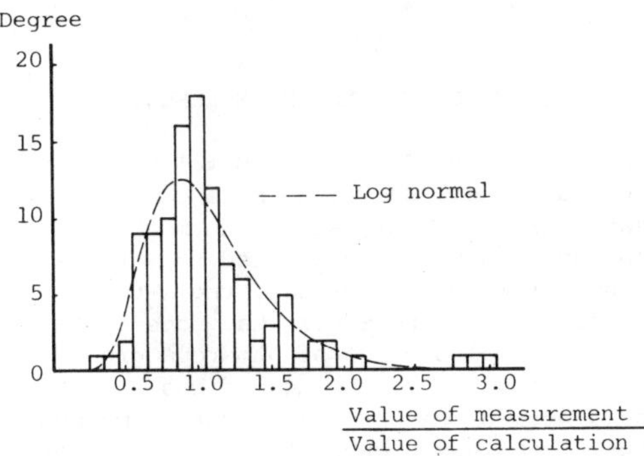

Fig. 7 Accuracy of the values of calculation of driven piles

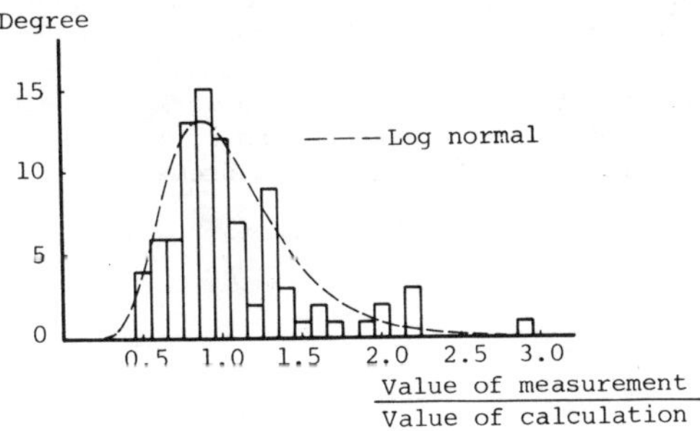

Fig. 8 Accuracy of the values of calculation of cast-in-place piles

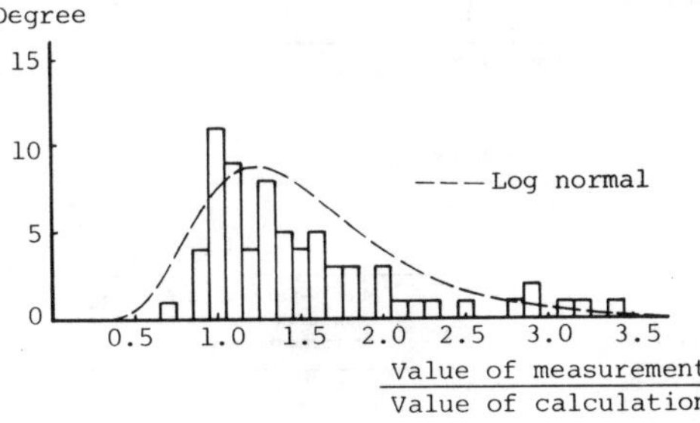

Fig. 9 Accuracy of the values of calculation of bored piles

Distribution of this ratio (values of measurement/values of calculation) is nearly logarithmico-normal for all types of piles. The coefficient of variation representing the accuracy of the formula is the smallest with the driven pile and largest with the bored pile. The driven pile is a method used since a long ago and much work has been carried out on it so that it seems the accuracy of its formula has been improved. The bored pile is a relatively new method and has not been worked on so much as the driven pile. Further, it has a greatly scattered bearing capacity due to the method of work, resulting in deteriorated accuracy. But, if much work is carried out on it in the future, the workability will be improved and so the accuracy of estimation.

In the case of the road bridge foundations in Japan, design is done with a safety factor of 3 used for the bearing capacity. The safety factor should normally be determined in consideration of the load, variation of the resistance of the structure, analytical error, error in work and uncalculable factors. Here, the relation between the safety of ultimate bearing capacity of the pile according to the calculation formula and the variation of the N-value was formulated upon the reliability theorem using the above data on the accuracy of N-value measurements and load tests, and the safety factor of the calculation formulas of the bearing capacity specified in the Road Bridge Specification were examined.

When the N-value is formulated as a governing probability variable for the actual ultimate bearing capacity of the pile, it is expressed as below.

$$R = P \cdot (N/N') \cdot R_n$$
$$P = R_e/R_n$$

where R: Real ultimate bearing capacity;
R_e: Ultimate bearing capacity by load test;
R_n: Ultimate bearing capacity by calculation formula;
N: N-value of actual pile; and
N': N-value of test pile.

From the above formulas, the mean value R_m of R and the coefficient of variation V_R are obtainable as below.

$$R_m = P_m \cdot R_n$$
$$V_R = \sqrt{V_p{}^2 + V_N{}^2 + V_{N'}{}^2}$$

According to the reliablity theorem, the resistance (or bearing capacity) R is expressed as a function of the load action Q_i (for example, live load Q_L and dead load (Q_D) as below.

the time of an earthquake.

4 CALCULATION FORMULA OF BEARING CAPACITY USING N-VALUE

The N-value is used not only for calculation of ground constants but also for calculation of the bearing capacity of the foundation.

1) Pile foundation

Driven pile:

$$Ru = qdAp + (1/5\bar{N}_S L_S + \bar{N}_C L_C)\phi$$

Cast-in-place pile:

$$Ru = 300Ap + (1/2\bar{N}_S L_S + \bar{N}_C L_C)\phi \quad \text{(Sandy ground)}$$

$$Ru = 3quAp + (1/2\bar{N}_S L_S + \bar{N}_C L_C)\phi \quad \text{(Clayey ground)}$$

Bored pile:

$$Ru = 10NAp + (1/10\bar{N}_S L_S + 1/2\bar{N}_C L_C)\phi \quad \text{(Sand layer)}$$

$$Ru = 15NAp + (1/10\bar{N}_S L_S + 1/2\bar{N}_C L_C)\phi \quad \text{(Gravel layer)}$$

where Ru: Ultimate bearing capacity (t);
Ap: Area of pile end (m^2);
qd: Ultimate bearing capacity per unit area at the end of the pile (t/m^2) (Fig. 4);
$\bar{N}_S$: Mean N-value of sand layer;
$\bar{N}_C$: Mean N-value of cohesive soil layer;
L_S: Thickness of sand layer (m);
L_C: Thickness of cohesive soil layer (m);
ϕ: Circumference of pile (m); and
qu: Unconfined compression strength (t/m^2)

2) Spread foundation

Where the load inclination is small and most of the structure is not high, the allowable bearing capacity may be estimated from Table 2.

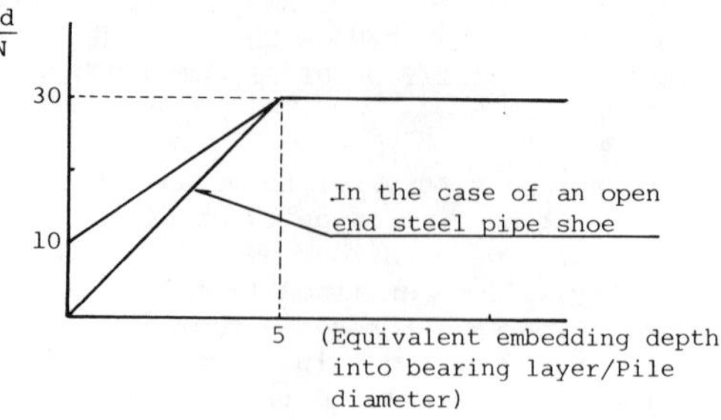

Fig. 4 Calculation diagram of ultimate bearing capacity per unit area qd of pile end ground

5 ACCURACY OF N-VALUE MEASUREMENTS

Initially, the S.P.T. was developed as a method of soil survey to be applied to sandy ground. But, since the method is so easy, it has been applied to various kinds of ground. Further, the N-value obtained from the S.P.T. has come to be used for calculation of numerous ground constants. Thus, the method has been regarded as a universal soil survey method. But, there has been little discussion of fundamental problems such as the mechanism of the standard penetration test and the accuracy of N-value measurements. Recently, however, reflecting the situation or movement for safer and more reasonable design of structures, some research work has been carried out on the mechanism of the S.P.T., the accuracy of N-value measurements, etc.

Table 2 Allowable bearing capacity per unit area for estimation

Kind of foundation grounds		Normal (t/m^2)	During earthquake (t/m^2)	Standard values: N value	Standard values: Unconfined compression strength (kg/cm^2)	Remarks
Bedrock	Uniform hard rock with few cracks	100	150	-	100 or higher	
	Hard rock with many cracks	60	90	-	100 or higher	
	Soft rock, mud stone	30	45	-	10 or higher	
Gravel layer	Compact and solid	60	90	-	-	
	Not compact and solid	30	45	-	-	
Sandy ground	Dense	30	45	30 ∿ 50	-	**When N-value of S.P.T. is less than 15, not suitable as foundation ground.**
	Medium	20	30	15 ∿ 30	-	
Cohesive soil ground	Very hard	20	30	15 ∿ 30	2.0 ∿ 4.0	
	Hard	10	15	8 ∿ 15	1.0 ∿ 2.0	
	Medium	5	7.5	4 ∿ 8	0.5 ∿ 1.0	

For example, Y.Nishigaki proposed that the hammer drop method would involve great variation of the loss of drop energy whether it was made by the cone pulley method or the Tombi method with the dropping made at a height as high as 50-75cm so that the automatic drop method of less variation should be used.

Also, J.Takenaka stated that the main cause for poor accuracy of N-value measurements is the energy loss in the drop of the hammer. By using an automatic drop apparatus, it would be possible to raise the accuracy of N-value measurements to about ±14 percent.

On the other hand, the Technical Committee of the Kanto Geological Surveyors Association conducted comparative tests of the N-value measurements using four kinds of impact methods, that is, 1) Tombi method, 2) cone pulley method, 3) cone pulley method with stopper and 4) automatic drop method at two points of alluvial and diluvium grounds. Figures 5 and 6 show the N-value measurements by these tests. As the result of these tests, they reached the following conclusion.

The items shown in Table 3 are considered as factors affecting the N-value. But, as far as the test results are concerned, no appreciable difference is noted in the N-value from one method to another. The reasons given are consideration of such factors as 1) minute differences in the instruments and excavation techniques used, 2) lack of uniformity of the ground sites and 3) are human factor.

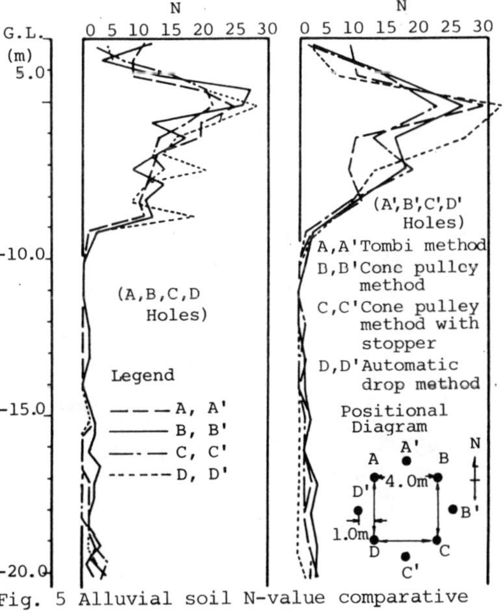

Fig. 5 Alluvial soil N-value comparative graph

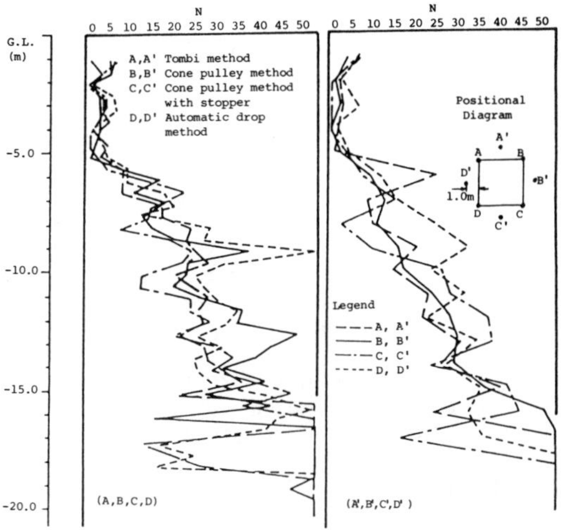

Fig. 6 Diluvium soil N-value comparative graph

Table 3 Factors affecting N-value

Factors	Items
Mechanical factors	• Parts differences {Rope, pulley rod, sampler shoe, Sampler drain port, Monkey weight} • Differences in machines and excavation implements {Pump, engine, core tube}
Ground factors	• Cohesive soil • Sand, gravel • Groundwater level
Human factors	• Partnership {Age, experience, skill} • Psychological and physical elements
Environmental factors	• Temporary works • Season, weather • Conditions in the vicinity of work site

$$R_m e^{-\alpha\beta V_R} = R_n \cdot P_m e^{-\alpha\beta V_R}$$

$$\fallingdotseq e^{\alpha\beta V_E} \sum_i (1 + \alpha\beta V_Q) Q_{im}$$

$$= \gamma_o(\gamma_D Q_D + \gamma_L Q_L)$$

(in the case of live or dead load only)

where $\alpha(=0.55)$: Separation factor
$\beta(=3)$: Safety index;
V_E : Coefficient of variation of error in analysis; and
V_Q : Coefficient of variation of load action Q_i.

On the other hand, according to the allowable stress design method, the resistance and load are expressed as below.

$$1/\nu \cdot R_n = (Q_D + Q_L)$$

where ν: Safety factor

Accordingly, ν may be expressed as

$$\nu = \frac{\gamma_o(\gamma_D Q_D + \gamma_L Q_L)}{m \cdot e^{-\alpha\beta V_R}(Q_D + Q_L)}$$

The relation between ν and Q_L/Q_D calculated in use of the values of P_m and V_p obtained from the load test data with the coefficient of variation of N-value assumed to be 0.15, is shown in Figure 10. For γ_o, γ_D and γ_L, 1.0, 1.1 and 1.4 were used respectively (LRFD Criteria, U.S.A.).

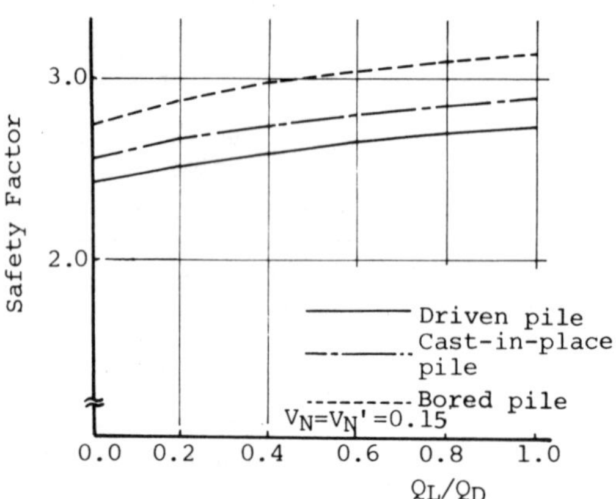

Fig. 10 Safety factor versus Q_L/Q_D

In the foundation structures, the value of Q_L/Q_D is small so that a safety factor of 3 used in the Road Bridge Specification is a reasonable value.

7 CONCLUSION

From this paper the following are summarized.

1) S.P.T. is so popular as to be indispensable for the survey of ground condition and measured data of N-value have been stocked for various type of grounds.

2) The correlations between N-value and other design coefficients of foundation have been studied and accuracy of estimated design coefficients is becoming higher.

3) The safety factor of bearing power of pile calculated with N-value is evaluated in the relation of variation of N-value and it is found that the values of safety factor specified in some specifications seem to be fairly reasonable.

8 REFERENCES

Japanese National Railway 1974, Structure Design Criteria.

Japan Building Society 1974, Building Foundation Structure Design Standards.

Japan Ports and Harbors Association 1979, Technical Criteria for Port Facilities.

Japan Road Association 1980, Road Bridge Specification IV "Substructures".

Nishigaki, Y. 1974, Mechanism of S.P.T. (Part 1), Jour. of Soil Mechanics and Foundation Engineering.

Nishigaki, Y. 1974, Mechanism of S.P.T. (Part 2), ibid.

Public Works Research Institute 1977, Survey of the Methods of Choosing the Type of Structure Foundation, Report of the 31st Technical Meeting, Ministry of Construction.

Shioi, Y. and Sugimura, Y. 1980, Present Condition and Prospects of the Use of S.P.T. Results, Sounding Symposium, In Situ Survey Research Committee, Society of Soil Mechanics and Foundation Engineering.

Takenaka, J. 1974, N-value and Soil Characteristics (Part 1), ibid.

Takenaka, J. 1974, N-value and Soil Characteristics (Part 2), ibid.

Technical Committee of the Kanto Geological Surveyors Association 1976, Comparative Experiments of the Impact Methods in S.P.T., Jour. of Soil Mechanics and Foundation Engineering.

Uto, K. 1980, Review of the Present Conditions of S.P.T., Sounding Symposium, In Situ Survey Research Committee.

Proceedings of the Second European Symposium on Penetration Testing / Amsterdam / 24-27 May 1982

Wave propagation effects induced by standard penetration tests

J.H.TRONCOSO
Universidad Católica de Chile, Santiago

1 INTRODUCTION

Any dynamic disturbance created inside a soil mass generates stress waves that propagate, as dilatational P-waves and shear S-waves, from the point of disturbance through the soil. As the waves travel outward from the source, attenuation of stresses occur due to hysteretic damping, proper of the material, and to simple radiation or geometric damping. Performance of Standard Penetration Tests (SPT) generates such wave propagation effects, and, therefore, an SPT may be used to measure shear wave velocities and possibly other dynamic soil properties provided that adequate receiving stations are installed to record passage of the waves.

2 EXPERIMENTAL STUDY

Information related to the characteristics of wave propagation effects associated with performance of SPT was obtained as part of a program devoted to development of in-situ impulse tests for determination of shear moduli of soils as a function of strain (Troncoso, 1975). In one of the stages of that program, Standard Penetration tests were performed inside a compacted sand fill which had previously been instrumented with particle velocity transducers. These transducers, located at different distances from the vertical axis of SPT, in a horizontal plane of the compacted fill, at elevation - 1.52 meter below ground surface, recorded the motions of the soil as the split spoon sampler was driven in several tests, starting at the ground surface and proceeding downward. Standard equipment and procedures were used in the performance of the tests, as shown in Figure 1. Results of these tests which are considered interesting for understanding of wave propagation phenomena, associated with performance of SPT, are presented below.

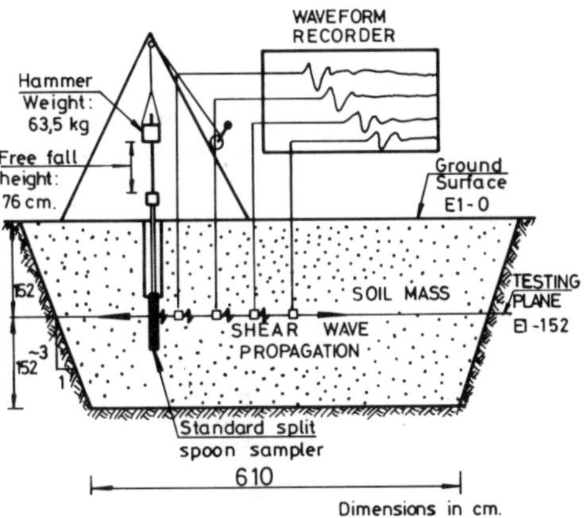

Fig. 1 Testing set-up in compacted soil fill

3 SOIL PROPERTIES

The test fill was constructed with silty fine to medium sand which had 50 percent effective diameter equal to 0.26 mm and 16 percent silt. Specific gravity of these soils was 2.59, average water content was 11 percent and the shape of the particles was subangular. Shear wave velocities, determined in cyclic triaxial tests, varied between 122 meter per second for 10^{-2} percent shear strain to 106 meter per second for 10^{-1} percent shear strain, in specimens compacted to a dry unit weight of 1.7 kg/dm^3, and under effective confining pressure of 0.43 kg/cm^2. Damping ratios between 5 and 10 percent were measured in same cyclic triaxial tests for same strain rate.

The standard penetration resistance, measured in the compacted fill, closely above and below the main testing plane, was 9 blows per foot from elevation - 1.34 meter to elevation - 1.64 meter and it was 12 blows per foot from elevation - 1.82 meter to elevation - 2.12 meter, below ground surface. Average degree of compaction of the fill was 86.3 percent of maximum Modified Proctor density.

4 TEST SET-UP AND EQUIPMENTS

The experimental program was performed in the compacted sand fill, constructed inside a pit with a set-up as shown in Figure 1. The pit was excavated with the shape of a truncated cone of 9 meter diameter at the ground surface, 6 meter diameter at the bottom, and 3 meter in depth. The backfill operations were performed by filling in 15 cm loose lifts. Each layerwas manually spread and levelled and compaction was accomplished by six passes of two vibratory plates routed in concentric and overlapping circles.

The main plane of testing was defined at mid-depth of the fill, 1.5 m. below the ground surface. Receiving stations were installed in the soil mass, as the fill was built up, in locations as shown in Figure 2. Most receiving stations consisted in one vertical transducer, Mark Products type L-10 AK, cylindrical in shape, 3 cm in diameter and 3 cm in length. A few horizontal velocity transducers and accelerometers were also installed to provide supplementary information.

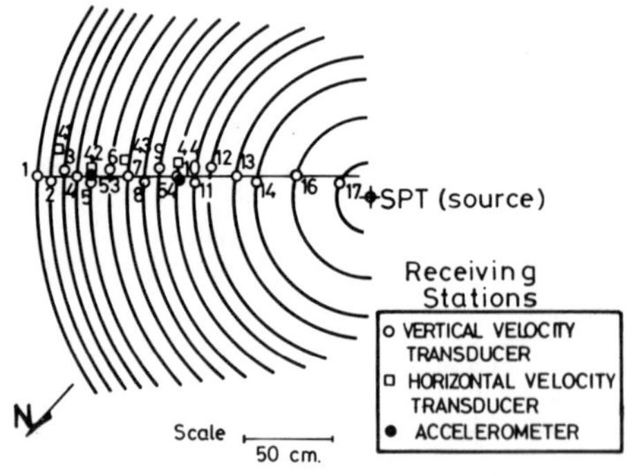

Fig. 2 Shear wave propagation induced by standard penetration test. Receiving stations at elevation - 1.5 m for tests in compacted soil fill.

Once the fill was finished several series of shear wave propagation tests were performed using a new impulse anchor source under development; result of those tests have been published elsewhere (SW-AA 1975; Troncoso 1975). At the end of the program the standard penetration tests were performed. The position of the axis of the SPT relative to each receiving station is shown in Figure 2. The responses of the receiving stations to every blow of the SPT were recorded as velocity time historiesby a 4 channel pulse recorder, (Biomation Model 1015 Wave form Recorder), capable to store the pulses, in digital form, with a precision of one word every 0.01 millisecond. Processing of these data permitted to obtain particle velocity-time histories, shear wave velocities and other data pertaining to the wave propagation phenomena, inside the soil mass.

5 TEST RESULTS

Four particle velocity pulses could be recorded each time the split spoon sampler was driven by each standard impact. Each impact could be considered as an impulse test, in which the sampler was the wave generating source. This source changed in elevation, with reference to the plane of the receiving sensors, from one test to another. Table 1 lists vertical particle velocities, measured at different radial distances, r, from the vertical axis of the split spoon sampler, for tests performed at different elevations. The depth, z, is the distance between the tip of the sampler and the plane of the sensors, after the impact.

Analyses of the results in Table 1 have to take into consideration that dilatational wave components of motion are included in some of the vertical particle velocity pulses, because the position of the center of the generating source is generally above or below the plane where the recording stations are installed, and that penetration resistance to each impact on the sampler may determine the magnitude of generated strains.

As indicated in Table 1, maximum particle velocity, measured at 19.6 cm from the center of the sampler, was 3.33 cm/seg.This velocity was reached when the tip of the sampler was driven 2.5 cm in one blow to a final depth of 7.6 cm below the plane of the sensors. It is also observed that particle velocities smaller than 0.2 cm/seg were measured, in general, for points beyond

1.2 meter above or below the tip. Shear wave velocities, generated by penetration of the sampler from 7.6 to 8.9 cm below the plane of the sensors, varied between 129 meter per second, measured between radial distances of 19.6 and 77.5 cm, 149 meter per second between 19.6 and 124 cm, and 154 meter per second between 19.6 and 197 cm.

It is clear from these results that shear wave velocity increases and that particle velocity decreases as the wave travels outward from the generating source.

Table 1. Vertical particle velocities, v_1 (cm/seg), measured in standard penetration tests in a compacted sand fill

Depth Z (cm)	Radial distance, r, (cm)					
	19.6	77.5	89.9	102.8	124	196.6
110.5	0.10	0.08				
82.6	0.94	0.41				
58.4	1.04	0.46			0.36	0.13
11.4	2.03	0.25				
- 7.6	3.33	0.94				
- 8.9	3.07	0.94			0.64	0.64
- 12.7		1.22	0.96	0.81	0.60	
- 61.0	1.80	0.81			0.79	0.76
-152.4	0.23	0.18			0.12	0.08

Shear strains, γ, and shear modulus, G, may be estimated from particle velocity and wave velocity, by the equations:

$$\gamma = \frac{v_1}{V_s} \qquad (1)$$

$$G = \rho V_s^2 \qquad (2)$$

Where V_s : shear wave velocity

v_1 : particle velocity

ρ : mass density

Therefore, shear strains generated by SPT ranged between 4.1×10^{-4} and 2.6×10^{-2} percent and corresponding shear moduli ranged between 384 and 270 kg/cm^2, for the compacted sands of the tests.

The decrease of particle velocity is a consequence of geometrical damping and hysteretic damping. Assuming that geometrical damping is inversely proportional to the radial distance, r, the logarithmic decrement of material damping, δ, may be estimated by the following equation (Stokoe et al 1979):

$$\delta = \frac{V_s}{f} \; \frac{\ln \dfrac{r_1 v_{11}}{r_2 v_{12}}}{(r_2 - r_1)} \qquad (3)$$

where f is the frecuency of the wave and v_1 is maximum particle velocity at a given radial distance. This equation is valid only in a range of radial distances from the center of the sampler where shear wave pulses have been recorded with-out superposition of dilational wave pulses. Application of equation 3 to data obtained for sampler penetration ending in z =-12.7 cm, gives logarithmic decrements of 0.35 to 0.44, which correspond to approximate damping ratios of 5 to 7 percent for shear strains ranging between 3.8×10^{-3} to 0.8×10^{-2} percent. These results confirm that hysteretic damping increases as shear strain increases.

It is therefore observed that soil dynamic properties computed on the basis of wave propagation effects generated by SPT compare fairly well with same properties measured in the laboratory for these compacted silty sands.

6 FINAL REMARKS

Shear strains, induced by performance of Standard Penetration tests, in a compacted silty sand deposit, have been found to vary between 2.6×10^{-2} and 4.0×10^{-4} percent at points located between 0.2 and 2 meter away from the center of the split spoon sampler. Shear disturbances generated by SPT travelled as shear waves with velocities in the order of 129 to 154 meter per second in this range of strains.

These results are in good agreement with same soil properties determined in laboratory and other in-situ tests. They indicate, therefore, that SPT may be used as a source of waves for determination of wave velocities and soil moduli. To obtain large strains in these determinations, it is necesary to instal receiving stations very close to the split spoon sampler.

Damping characteristics of the soils may be estimated also by substracting the effects of geometrical damping from total attenuation.

Because shear wave velocity, shear modulus and damping ratio are strain dependent properties, reports of wave velocity tests should always be supplemented with corresponding strain values to allow comparisons and eventual correlations to be established with other soil properties.

The good agreement between laboratory and in-situ measurements of soil dynamic properties could be established in this case because the soils tested were compacted soils of similar structure and stress history insitu and in the laboratory.

7 ACNOWLEDGEMENTS

The experimental work analyzed in this paper was performed in Seattle, USA, under contract between the U.S. Nuclear Regulatory Commission and the Joint Venture of Shannon & Wilson Inc. and Agbabian Associates to develop an In-Situ Impulse Test for Determination of Shear Modulus of Soils. Special acknowledgement is given to Mr. Fred R. Brown, who together with the author performed the analyzed tests, and to Mr. Stanley D. Wilson and Dr. Raymond P. Miller, from Shannon & Wilson Inc., who guided and directed the In-Situ Impulse Test project.

8 REFERENCES

Troncoso, J.H., 1975, In Situ Impulse Test for Determination of Soil Shear Modulus as a Function of Strain, Ph.D. Tesis, University of Illinois at Urbana-Champaign.

SW-AA, 1975, In-Situ Impulse Test. An Experimental and Analytical Evaluation of Data Interpretation Procedures, A Joint Venture of Shannon & Wilson, Inc. and Agbabian Associates.

Stokoe, K.H., R.J. Hoar, and R.F. Ballard Jr. 1979, Cross-Hole and Down-Hole Vs by Mechanical Impulse, Discussion, ASCE Journal of Geotechnical Engineering Division, Vol. 105, N°GT 10.

Proceedings of the Second European Symposium on Penetration Testing / Amsterdam / 24-27 May 1982

Comparison between N-value and pressuremeter parameters

H.TSUCHIYA & Y.TOYOOKA
Kiso-Jiban Consultants Co. Ltd., Tokyo, Japan

1 INTRODUCTION

K. Terzaghi and R.B. Peck proposed relationships between N values (blow-count /0.3m) of standard penetration tests and other soil characteristics such as unconfined compression strength, relative density of sand and consistency, as well as other factors such as bearing capacity of footings.

These have proven particularly useful and many similar relationships have been subsequently proposed.

In Japan, the standard penetration test was introduced around 1950 and due to the practical and useful nature of the test as well as the advantage it enjoys over other methods regarding observation of soil samples, it has become very popular in a short period of time and is now included in most test borings.

As the N value method has become so popular in Japan, the relationship between this value and the parameters of other tests have been established for practical purposes. Among these is the ability to check the reliability of the results of various tests .

As for pressuremeter testing, R. Yoshinaka proposed a relationship between the N values of the standard penetration test and the deformation modulus of the pressuremeter test. Subsequently, a number of design standards have adopted a similar relationships to predict deformation of foundations.

While N values are related to the strength of soils against transient vertical loads, the deformation modulus, E_p, of the pressuremeter test is related to deformation under static horizontal loads. As the natures of these values are inherently different it might be considered extremely difficult to correlate them with one another or that the N values might be better correlated to the strength parameter than to the deformation modulus.

This paper discusses the relationship between the N value of the standard penetration test and the deformation modulus, E_p, and yielding pressure, P_f , of the pressuremeter test based on more than 500 data sets obtained at the same location for various types of soils.

2 TEST DATA ANALYZED

The data sets analyzed are the results of tests in which both standard penetration and pressuremeter tests were carried out in the same bore hole at similar depths.

The types of soils tested were determined based on the boring log. Grain-size distribution tests were also carried out for half of the data sets and these results were also taken into consideration. The number of the data sets analyzed for each soil type is shown in Table 1. Data sets for special soils of limited distribution are omitted here. Also omitted are data sets in which N values are less than unity.

The standard penetration tests and grainsize distribution tests were carried out in accordance with procedures specified in the Japanese Industrial Standard (JIS). The standards for these two tests are nearly identical to those currently in use in most countries with the exception of the diameter of the boring rods. As no specifications are provided in the standard for penetration depths less than 30cm, in the present investigation, when the depth after 50 blows was less than this figure, the blow count was estimated from the test results assuming that the penetration depth was proportional to the blow counts.

Procedures of the pressuremeter test are

also similar to those currently in use elsewhere. In the present study the deformation modulus, E_p, was obtained from the inclination of the linear portion of the pressure-volume relationship, and the yeilding pressure, P_f, corresponded to the level at which the linear portion of the inclination of the pressure-volume relationship showed an abrupt change.

Table 1. Data sets analyzed

Soil type		Number of data sets	Remarks (In case grain-size distribution test results exist)
Alluvial	Clayey soils	130	Fine content ≧ 50%
	Snady soils	175	Fine content < 49%
	Gravelly soils	50	Content of coarse material (D ≧ 2mm) > 30%
	Organic soils	20	Water content > 150%
Diluvial	Clayey soils	71	Fine content ≧ 50%
	Sandy soils	52	Fine content < 49%
	Gravelly soils	50	Content of coarse material (D ≧ 2mm) > 30%
Teritory	Mudstone	38	
Total	——	586	——

3 PROCEDURES OF ANALYSIS

It was supposed that the relationship between N values and the values of E_p or P_f for the data sets previously mentioned could be approximated by the formula:

$$E_p \text{ or } P_f = a.N^b \quad ..... (1)$$

where a and b are constants and take different values for E_p and P_f. This formula is the same as that adopted by R. Yoshinaka to approximate the N - E_p relationship. The constants in question are determined by first plotting the E_p or P_f values of a data group against corresponding N values on a graph in which the horizontal axis corresponds to Log(N) and the vertical axis to Log(E_p or P_f). Next, a line of least error is obtained from this graph which a pproximates the relationship between E_p or P_f and the N values, that is, the sum of the square of the vertical distances between the line and the plotted points. Constant a is the value of E_p or P_f on this line corresponding to the N value of unity while constant b is the value of the inclination of this line. Both constants a and b were obtained for all data sets as well as for each individual soil type.

4 RESULTS

The relationship between N and E_p or P_f for all data sets is shown in Fig. 1. accompanied by a line which approximates the same.

As will be noted in this figure, the value of b for E_p is different from that for P_f. While the value of b for E_p is almost equal to unity, indicating that it is almost proportional to the N value, the value of b for P_f is smaller, indicating that it does not increase as much as E_p does when the N value increases. This variation reflects characteristic differences between E_p and P_f.

It will be further noted from Fig. 1 that E_p and P_f scatter to a similar degree when plotted against the N value, although the relationship between N and P_f was expected to show better correlation than that of E_p.

Similar relationships for each individual soil type are shown in Figs. 2(a)~(h). It seems from these figures that the degree of approximation of formula (1) for all data sets and for individual soil types does not differ greatly and no specific advantage is found in separating soil types in determining constants a and b.

Judging from Figs. 1 ~ 2, the values based on formula (1) scatter in a fairly wide range being 1/3-3 of the value actually obtained in the pressuremeter tests. This scattering must be taken into consideration when applied to practical situations.

All constants obtained regarding Figs. 1 ~ 2 are summarized in Table 2 in which the values for the N - E_p relationship given by R. Yoshinaka are also shown. The constants obtained for all data sets are almost the same as those devided by R. Yoshinaka although the plotted relationships are scattered. Both a and b for each individual soil type are different from each other.

Fig. 1. N values - pressuremeter parameter relationships (all deta sets)

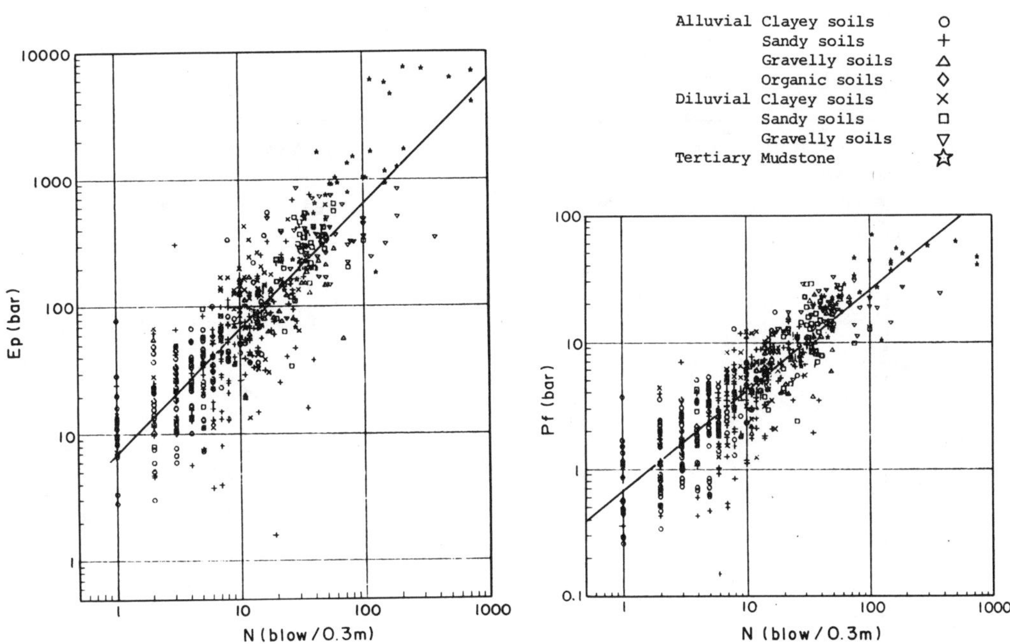

Fig. 2.(a) N values-pressuremeter parameter relationships (alluvial clayey soils)

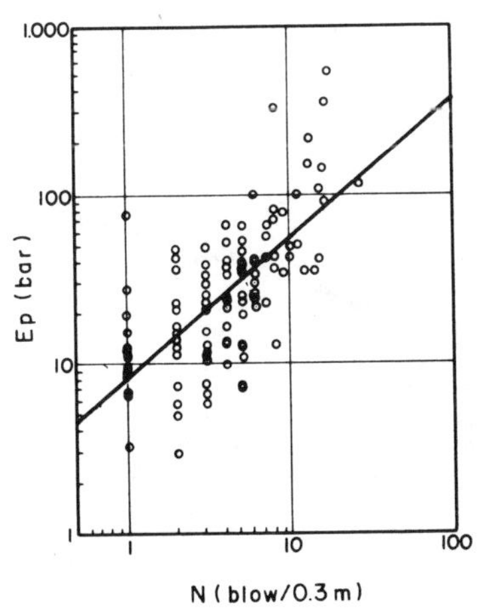

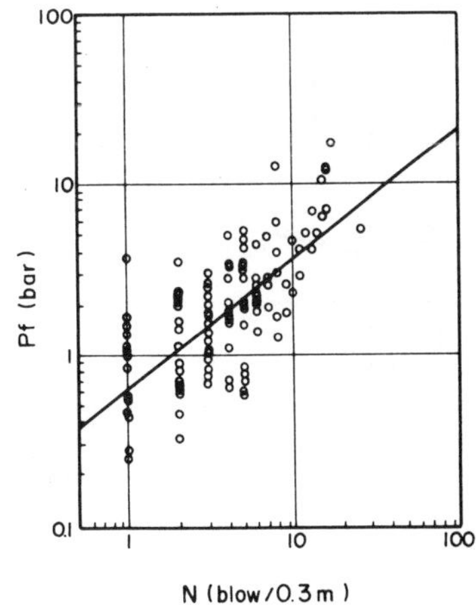

Fig. 2.(b) N values-pressuremeter parameter relationships (alluvial sandy soils)

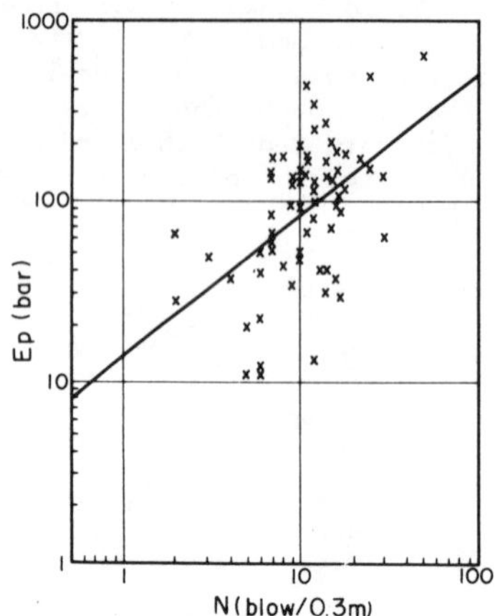

100
10
1
0.1
Pf (bar)
1
10
100
N (blow/0.3m)

Fig. 2.(c) N values-pressuremeter parameter relationships (alluvial gravelly soils)

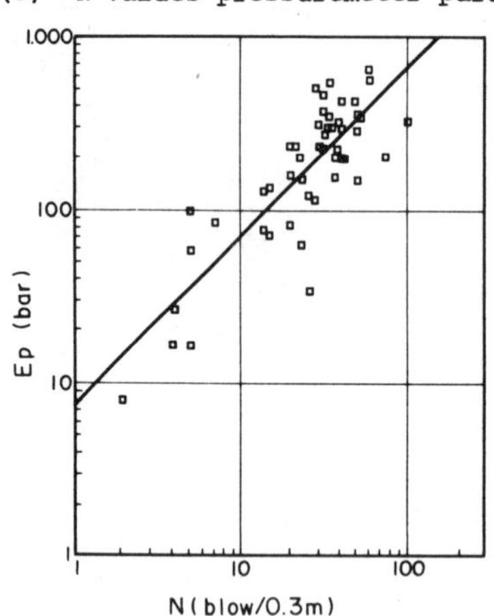

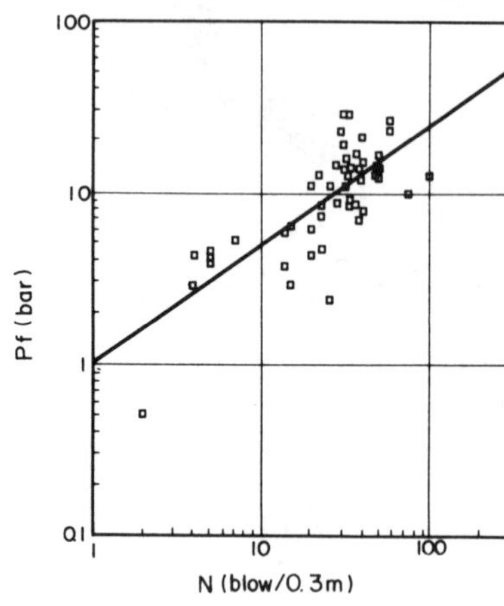

Fig. 2.(d) N values-pressuremeter parameter relationships (alluvial organic soils)

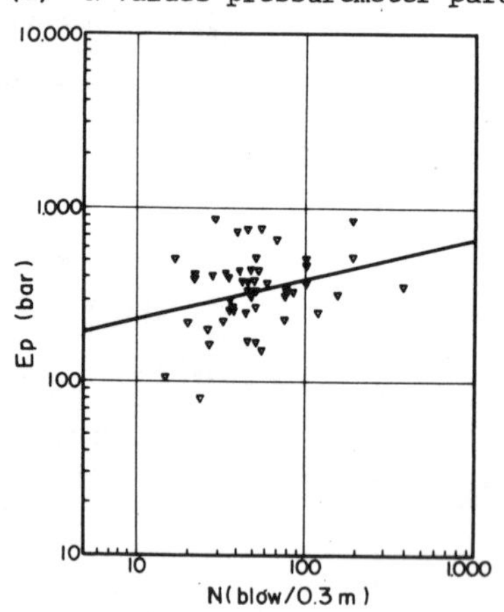

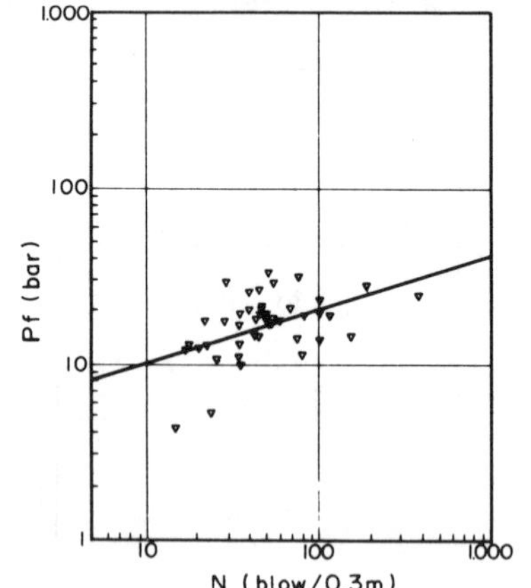

Fig. 2.(e) N values-pressuremeter parameter relationships (diluvial clayey soils)

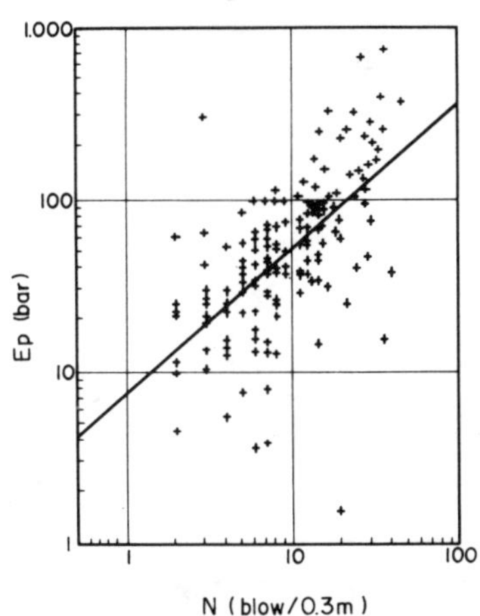

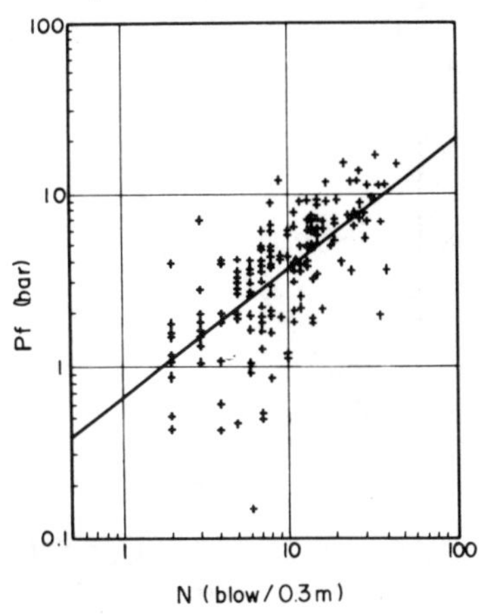

Fig. 2.(f) N values-pressuremeter parameter relationships 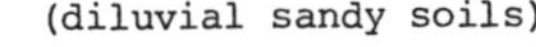 (diluvial sandy soils)

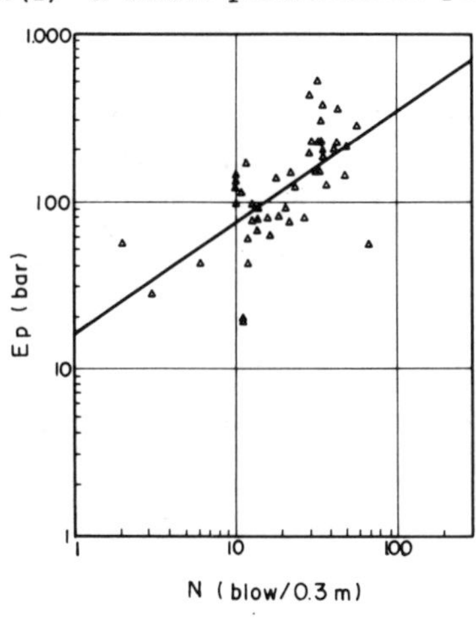

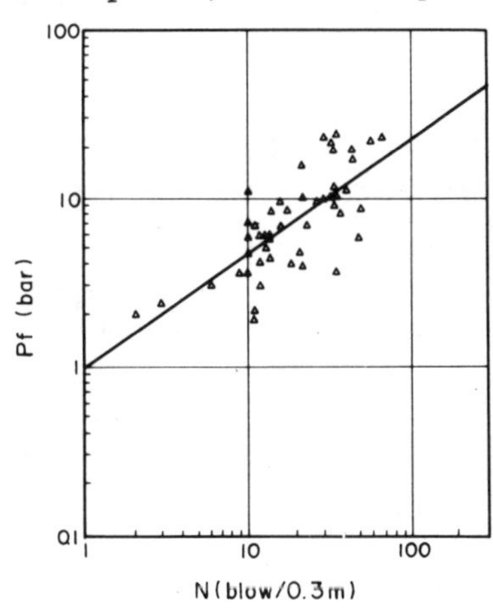

Fig. 2.(g) N values-pressuremeter parameter relationships (diluvial gravelly soils)

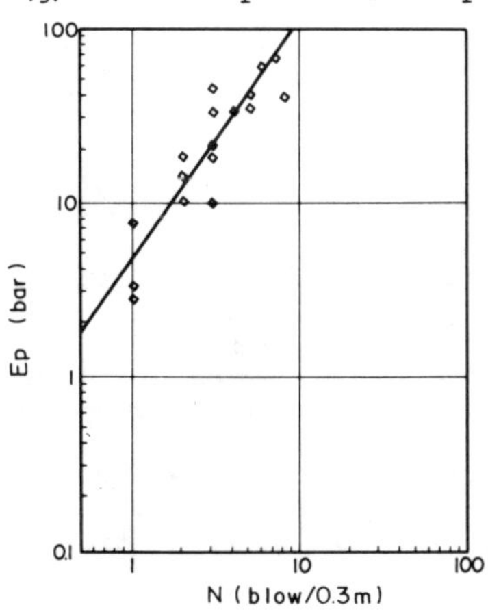

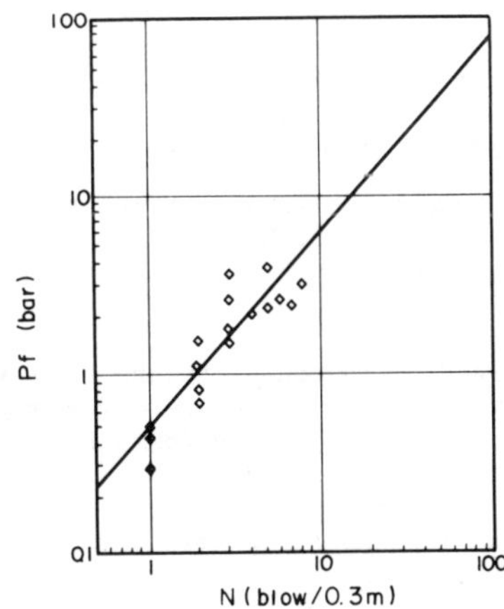

Fig. 2.(h) N values-pressuremeter parameter realtionships (Tertiary mudstone)

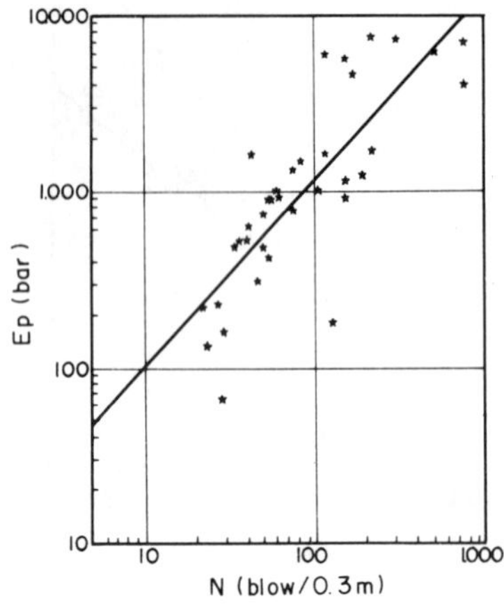

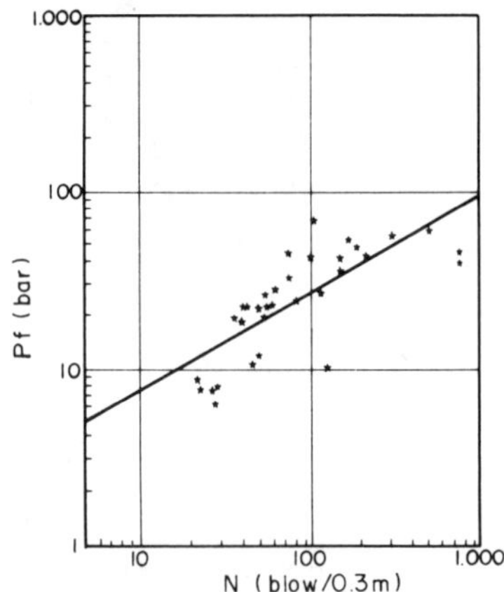

Table 2. Constants a and b

Soil type		Constants for Ep		Constants for P_f	
		a (bar)	b	a (bar)	b
Alluvial	Clayey soils	8.28	0.828	0.668	0.756
	Sandy soils	7.70	0.838	0.658	0.724
	Gravelly soils	16.16	0.663	0.997	0.673
	Orgnaic soils	4.78	1.376	0.550	1.083
Diluvial	Clayey soils	13.96	0.776	0.973	0.707
	Sandy soils	7.35	0.980	0.993	0.695
	Gravelly soils	133.8	0.230	5.08	0.304
Tertiary	Mudstone	9.07	1.058	2.15	0.553
data sets		6.84	0.986	0.669	0.792
Values by R. Yoshinaka		6.78	0.999		

5 CONCLUSION

More than 500 data sets of standard penetration and pressuremeter tests carried out at the same location for various types of soils in Japan were collected and the relationship between the N value of the standard penetration test and the deformation modulus, E_p, and the yielding pressure, P_f, of the pressuremeter test were studied, assuming that the relationship could be approximated by the formula:

$$E_p \text{ or } P_f = aN^b$$

where a and b are constants and take different values for E_p and P_f. The degree of approximation of the formula for all data sets and individual soil types did not differ and no specific advantage could be found in separating soil types in determining a and b.

Furthermore, as the deformation modulus, E_p, and the yielding pressure, P_f , represent different soil characteristics, the degree of approximation of the above formula for the relationship between the N value and the values of E_p and P_f are expected to be different. However, when plotted against the N value, E_p and P_f are scattered to a similar degree.

The values of E_p or P_f estimated from the N values based on the above formula scattered in the range of 1/3-3 of the values actually observed in the pressuremeter tests. This must be taken into consideration when this formula is applied to practical situations.

ACKNOWLEDGEMENT

The authors wish to express their thanks to Mr. D. Flory and Miss N. Miyashita for their editing and typing works.

REFERENCES

Terzaghi, K. & R.B. Peck 1948, Soil mechanics in engineering practice, New York, John Wiley & Sons.

Peck, R.B., W.E. Hanson & T.H. Thornburn 1953, Foundation Engineering, New York, John Wiley & Sons.

Yoshinaka, R. 1968, Coefficient of subgrade reaction, Civil Engineering journal, Vol. 10, No.1, Institute of construction ministry.

Proceedings of the Second European Symposium on Penetration Testing / Amsterdam / 24-27 May 1982

On the standardization of the SPT and cone penetration test

Z.Q.WANG & W.X.LU
Chinese Academy of Building Research, Beijing, China

ABSTRACT

This paper deals with the standardiation problems of the SPT and cone penetration on the basis of the their developments and practical experiences in China. The authors present a general idea that the sufficient effectiveness for qualitative and quantitative measuring of any penetration technique can only be fulfilled with the cancellation of accidental errors to a max. extent and the diminish of systematic errors. Thus, the standardization should be carried out with rationalized testing operation and design of the probe set-up and the effectiveness of its working mechanism. Some research works in this area and several comments from the authors are given for referances of proposed international standards of penetration testing.

1. PROBLEMS RELATING APPLICATIONS AND STANDARDIZATION OF SPT

1.1 Brief review of the application of SPT

Since the thirties the SPT was used as a normalized method for evaluating certain soil properties, it has become a world-wide insitu technique due to its simplicity and effectiveness. The most outstanding development which geotechnical engineers have been much concerned with so far is that SPT has been so almighty as to give more than 10 soil parameters and to make many geotechnical assessment as we want. For example, the following applications of SPT may be the most popular and routine ones:

--- Estimating relative density of sand both qualitatively and quantitatively relating to effective overburden pressure.

--- Estimating internal friction angle of cohesionless soils with respect to D_r.

--- Estimating undrained shear strength of salurated clays.

--- Estimating shear wave velocity Vs and dynamic shear modulus G in soils.

--- Evaluating bearing capacity of sands.

--- Evaluating foundation settlement in sand and its coeffecient of compressibility.

--- Evaluating bearing capacity of a single pile.

--- Evaluating liquefaction potential of sands.

However, it should be noted that there might be some discrepancies between so wide applications of SPT and its original set-up even somewhat stardardized, because there are many procedures still in need of being specified or normalized. In recent years many important comments and suggestions have been published on such problem,(Fletcher 1965, Kovacs et al. 1979, de Mello 1971, Kovacs et al. 1981) undoubtedly these contributions will merit appreciation as to make SPT more rationalized.

1.2 Deficiencies in current application of SPT

It has been widely acknowledged that in the application of SPT, many accidental errors encountered. And in the original equipment and set-up of typical cathead-rope together with donut hammer system, many systematic errors do exist. As a comprehensive investigation of these errors, the following categories in which errors may be classified are:

(1) Errors due to inadequate manipulation;

(2) Errors induced by unsatisfactory performance of the testing equipment;

(3) Deviations due to different procedures in some unstandardized process;

(4) Deviations due to different method for processing and analyzing test data.

The former two items have been already noticed and put into consideration for making better results towards a more standardized SPT. The author won't add any more in these aspects. Never-the-less, the latter two items listed above have not been sufficiently noticed yet, so a short discussion will be given in the following:

According to our experiences and experiments accumulated in China, different procedures in manipulation may cause quite significant differences in number of blow count. Table 1.1 shows such fluctuations through a number of contrasting tests given by investigation teams in Liao Ning Province, China. All data were obtained by randomly adopted method in a common site of a project. The scattering of N-values naturally comprises somewhat non-homogenity of soils of same stratum. Of course, these data are not practically applicable.

1.3 Further considerations of the standardization of SPT

From the background mentioned above, we are convinced that for obtaining more reasonable and effective results of SPT, we should not only standardize the dimensions and set-up of testing equipments, but also specify every testing and data processing procedure which may influence the testing results.

From Table 1.1, we can see a very surprising range of N values obtained by different methods of manipulation for a common soil layer. Because of a lot of various errors introduced likely in the whole process, we can hardly distinguish the most decisive factor influencing the most probable values of N's, but by the author's experiences such a random scattering of N values is caused mainly by different mode of hammering and different method of drilling. Referring to the additional recommandations by the International Commission of the International Society of Soil Mechanics and Foundation Engineering(Arce et al. 1971), the item that drilling mud may be used instead of casing is very meaningful in eliminating accidental errors.

Free fall hammer is apt to eliminate manual errors and unexpected friction in winch rope system which will cause increase of N values, In addition to this consideration, the so called automatic trip hammer can diminish exhausting labor and in turn may further ensure the carefulness of the operator.

The most important factor which may affect N value of SPT may be the length correction of drill rods during penetrating. The influence of the length of drill rod on N-values may be demonstrated in the following aspects:

(1) Increasing of slenderness ratio of the drill rods during driving will be susceptible to form rod buckling which will consequently consume the dynamic energy of hammering and increase the N-value necessary for reaching the same penetration.

(2) Increasing the susceptibility of uneffectiveness of connecting or loosening of rod coupling which will isolate energy transmitting, and as a sequence it will need more blow counts to meet penetration requirement.

(3) Increasing the mass of drill rods which will act as an impacting body during driving. By law of momentum, we can divide the whole set-up into two parts: one is the active part of impact denot-

ing the falling hammer with mass M_1; the other is the passive part of impact denoting the whole system receiving driving action, i.e. the guide pipe anvil, drill rod, split spoon sampler. Let the total mass of the latter be M_2, and V_1, V_2 be the velocities of M_1, M_2 respectively. Then, by the law of elastic impact we have $M_1 V_1 = M_2 V_2$. The behavior of the driving system after impact will depend on the corelation between M_1 and M_2. When the depth of testing is relatively shallow, in other words if the length of the rod is so small as to meet the case of $M_1 > M_2$, the hammer will go downward together with the rod, and $V_1 \approx V_2$ due to resistance of soil to the spoon sampler, Then normal penetration will go on under both dynamic and static loading. However, when $M_1 < M_2$, many troubles will occur: (1) The hammer will jump back upward then fall down several times with a convengent rebounding height; (2) Driving energy will be partially consumed in rebounding of the hammer, then the energy ratio ER_v for velocity or efficiency(Kovacs et al. 1981) will become much smaller as shown below:

$$ER_v = \frac{E_v}{WH} = \frac{M_1\sum_{i=1}^{n}(V_{1i}^2 - V_{1i}'^2)}{2WH} = \frac{M_1\sum_{i=1}^{n}V_{1i}^2(1-e^2)}{2WH} \qquad (1.1)$$

where E_v --- kinetic energy just before impact i.e. the energy or velocity

M --- mass of the falling hammer

V_{1i} --- velocity of the hammer just before impact in ith turn

V'_{1i} --- rebound velocity of the hammer just after impact in ith turn

e --- coefficient of rebound or recovery

W --- weight of the hammer

H --- measured fall height

Therefore the efficiency of driving is too low to reflect the soil characteristics. If $M_1 << M_2$, it will be almost impossible to indicate soil properties by using N-value, because N-value is no more a response paramenter of soil properties undereath. Obviously, there should be a critical length of drill rods corresponding the criteria of $M_1 \approx M_2$.

1.4 Correction of N-value with depth of testing

Even within the above critical limitation of rod length(namely the depth of testing below the ground surface), the number of blow count (N-value) will still be affected by variation of rod length because impact energy loss will be altered. Even factors resulting in loss of energy are more complicated and unexpected, we can explore certain modifications to estimate this influence by experiments rather than simply theoretical assessment. Fig. 1.1 is a reducing correction factor of N-values vesus length of rod which has been adopted by the national current Code of Soil and Foundation Design for Industrial and Civil Building. This can be considered as an overall modification to N-values with respect to length of rod. Engineering practice did prove this correction is effective and important.

In a broader sense, correction of N-values related to length of rod might be a problem of controversy. There has been seemingly a tendency of not making such correction, because most of the empirical formula of N-valueversus soil parameter

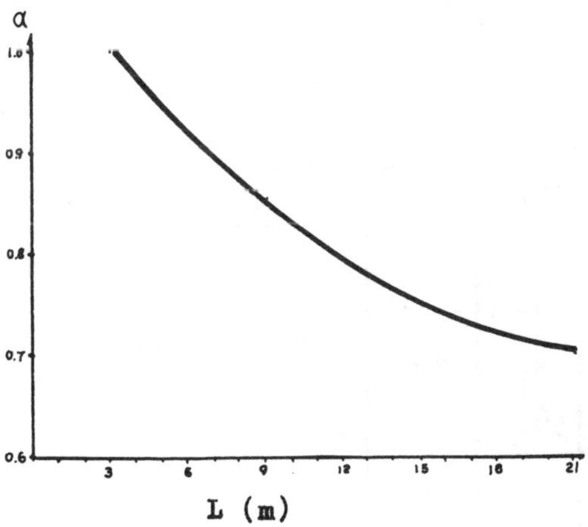

Fig. 1.1 Correction factor α of N-values vesus length of rod(L)

Table 1.1 Results of SPT obtained by various manipulations (in no. of blow count)

methods / soil strata				percussion with mud						rotaray with mud						percussion with casing						remarks
				A			B			B			C			C			B			A. automatic trip hammer B. mechanically controlled winch rope hammer C. manually controlled cathead rope hammer
Depth(m)	Thickness(m)	Soil classif.		no. of bore-holes	range of N	aver. N	no. of bore-holes	range of N	aver. N	no. of bore-holes	range of N	aver. N	no. of bore-holes	range of N	aver. N	no. of bore-holes	range of N	aver. N	no. of bore-holes	range of N	aver. N	
0-2.8	2.8	silt/ sandy clay		1	3	3	1	10	10	1	8	8	3	6-8	7				1	10	10	aver. N = 7.7
2.81-3.85	1.05	silty sand/ sandy loam		4	4-7	6	4	8-10	9	5	6-14	9	3	3-7	5	3	3-14	7.5	3	5-8	6.5	aver. N = 7.1
3.86-14.6	10.75	silty sand	upper	4	4-7	7.8	6	6-13	9	4	9-23	14.5	1	6	6	5	5-11	7.4	4	7-15	10.5	aver. N = 9.2
			lower	4	14-21	17	6	25-53	34	5	38-49	44.5	3	24-31	28	5	15-20	16.5	4	14-25	19.0	aver. N = 26.5
14.61-26.60	12.00	fine sand	upper	4	11-30	18	6	20-50	35	4	20-42	34	3	25-31	27	5	9-23	16.4	4	14-25	17.9	aver. N = 24.7
			lower	4	36-65	44	6	57-94	79	5	47-100	67	3	61-86	76	6	35-82	61.8	3	21-37	27.8	aver. N = 59.3

neglect this effect in the past. But the author wishes to stress that as SPT tends to be a widespread quantitative measuring technique more accurate and reliable N-value is more necessary than ever.

2. PROBLEMS IN THE APPLICATION OF CONE PENETRATION TEST

2.1 Brief review and remarks on current application

There are several kinds of cone penetrometers in the world. What we have widely used in China is an electrical static cone penetrometer which was developed early from 1963 (Wang 1978). The auther wishes to discuss the standardization of cone penetration test in the light of eletrical static cone penetrometer in order to refer to our experiences in China.

The Chinese electrical static cone penetrometer was developed with the background of failing to measure the cone resistance in deep soil layer, especially for soft or loose saturated soils by using the routine double rod apparatus before the sixties. In order to eliminate frictional errors in measuring we began to design a probe(sonde) with strain guage transducer so as to measure cone resistance directly from the cone. The whole system consists of three main parts: probe transducer system, penetrating system, recording and displaying system. The junction part is the probe.

Fig. 2.1 shows the construction of the "single bridge probe". This type of probe is different from the later Fugro one. The main feature of it is the tightness and transducing character. We designed a conic point with a cylindrical sleeve on the cone in order to measure the specific penetration resistance Ps which can be closely related to the soil compressibility and strength at the same time.

The special construction of the probe transducer consists of a hollow cylinder which is deformed by a coaxial pushing element. During penetration, the resistance of soils transmitted by the sleeved cone to the hollow cylinder can be transduced by a strain gauge, and the proper thickness of the hollow cylinder can be chosen to satisfy the sensitivity and measuring range requirement. In practice, the absolute strain observation of the cylinder is not needed but the re-

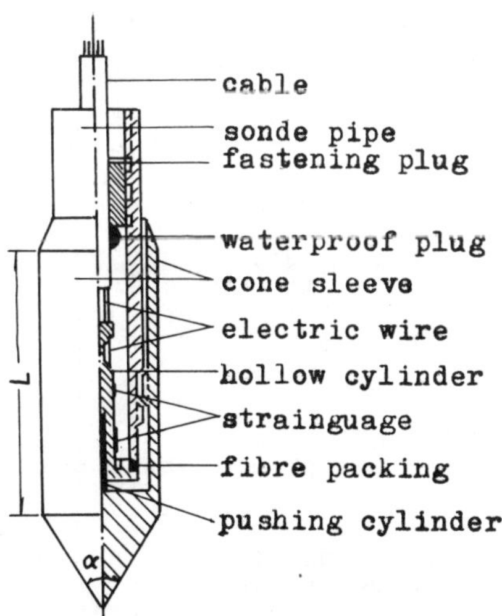

Fig. 2.1 Single bridge probe

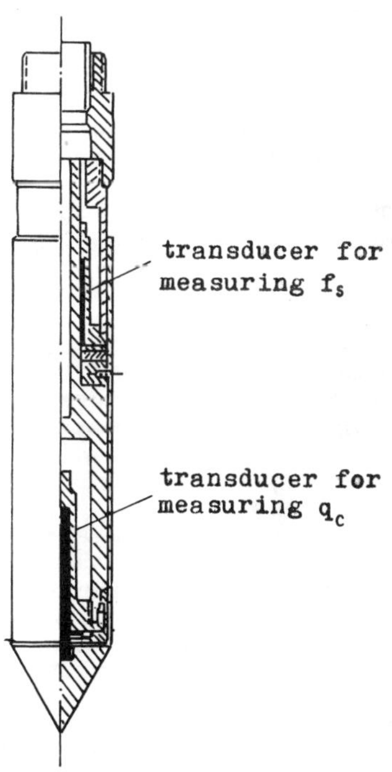

Fig. 2.2 Double bridge probe

lative strain is recorded to calculate the specific penetration resistance by calibration data.

The thightness is guaranteed by fibre packings with epoxy resin in an unloaded gap and not by close contact of smooth surfaces, because the latter way can hardly be proved effective. Therefore such a construction of our prototype can really solve the discrepencies between tightness and transducing. The probe once set up can be used for a long period without cleaning.

Moreover, another type of "double bridge sonde" was developed by using a similar construction and working mechanism to the single bridge one(Fig. 2.2). The main uses of it are to measure the point resistance q_c and the lateral friction f_s for simulating the single pile action.

This probe was normalized in our national Code of Site Investigation for Industrial and Civil Buildings. Many of the soil parameters including compressive modulus, undrained shear strength, consistency of clayey soils, even the bearing capacity and liquefaction potential can be obtained and analyzed by such technique.

2.2 Key problems relating the standardization of cone penetrometer

There are currently several types of cone penetrometers. Generally, these can be devided into two categories by geometry of the probe: (1) A cone with a cylindrical sleeve directly above it; (2) A cone with aninverse conical segment shoulder then a cylindrical sleeve above. These are commonly known already.

People used to apply the measured cine resistance q_c and lateral skin friction regardless of what type of the penetrometer may be. Never-theless, experiments show that the geometry of the probe has significant influence on the measured cone point resistance q_c and lateral skin friction f_s . The key points would be as follows:

(1) For better simulating of a single pile action, the geometry of the probe should fit that of a pile,

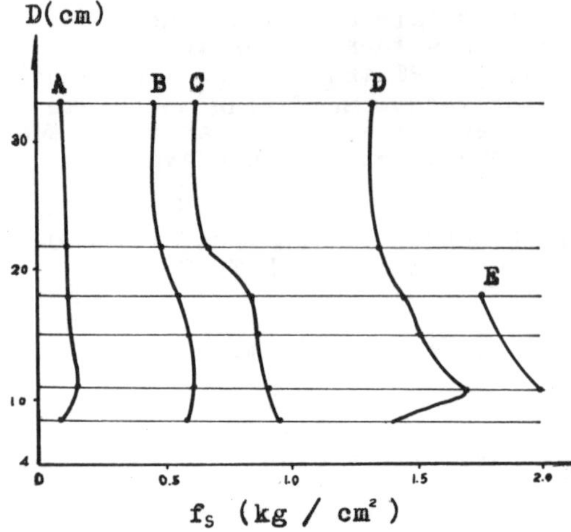

Fig. 2.3 Distance(sleeve center to point) effect on measured f_s
A. soft clay B. clay C. sandy silt D. sand E. tertiary clay

so it would be better to forget any neck or gap between the cone point and the cylindrical sleeve. Otherwise soil particles will insert or penetrate in it causing fictitious skin friction.

(2) For closely following the working mechanism of a single pile, the cone point resistance must be separated from lateral skin friction or vice-versa. Our experiments show that larger friction value f_s will be measured if the sleeve is directly above the cone, because shear failure surface around the cone will squeese the sleeve and cause an increase of skin friction. This effect can be easily realized from Fig. 2.3, where D is the distance from the center of the sleeve to the cone point.

Therefore, the sleeve should be designed with a distance apart from the cone point of the probe just like that of a pile.

2.3 Proposed geometry of the standardized probe for cone penetration test

So far as our experience is concerned, the main geometric factor of an effective probe for measuring both q_c and f_s is its appearance rather than its dimension. Fortunately, the dimensions of the sleeve and the cone for different probes used in the world almost meet together coincidentally so far, and need no more consideration about its standardization. Thus, the key problem may be the geometric arrangement of the probe. Based on the test results such as shown in Fig. 2.3, we are developing new probes as shown in Fig. 2.4(A)(B). One probe(Fig. 2.4 A) is composed of three measuring elements:

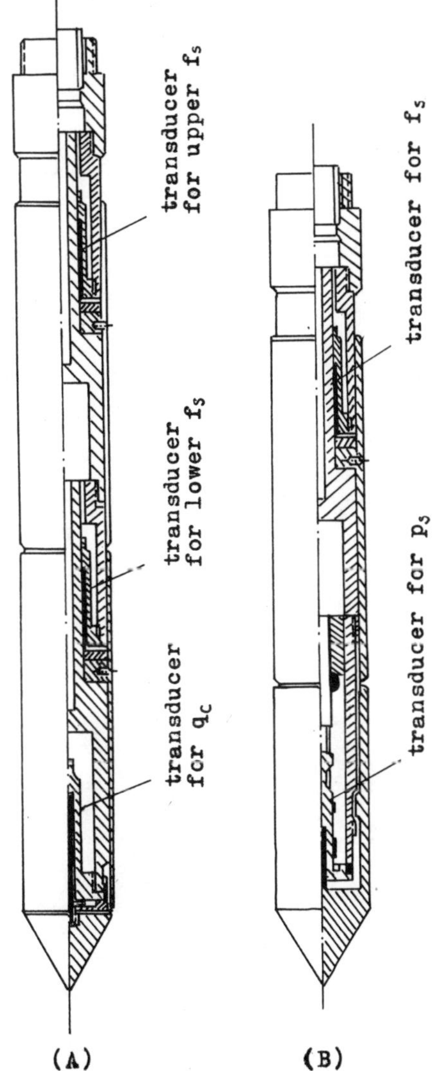

Fig. 2.4 Proposed probes for comprehensive measuring

(1) A cone point for measuring q_c ;

(2) A lower cylindrical sleeve for measuring apparent f_s which might be associated with q_c ;

(3) An upper cylindrical sleeve for measuring true value of f_s which is independent of q_c .

Another alternative(Fig. 2.4 B) of the proposed geometry of the probe is really a composite one of both the single bridge probe(Fig. 2.1) and the double bridge probe (Fig. 2.2). By using such a multiple probe, we can obtain the following parameters;

(1) The specific penetration resistance p_s by which a number of soil parameters can be obtained (Wang 1978, Zhou 1981).

(2) The lateral skin friction f_s can be directly measured with the upper cylindrical sleeve of the probe.

(3) The cone resistance q_c which can be calculated by the formulation:

$$q_c = p_s - \frac{f_s \cdot A_L}{A_c} \qquad (2.1)$$

where A_L is the total frictional surface of the sleeve of the single bridge probe.

A_c --- the crosssectional area of the conical point.

The calculated q_c will be logically reliable due to its physical significance. And from the practical point of view, it should be more reliable, because the measuring system is no more complicated than the original "double bridge probe", and less error could be introduced.

CONCLUSIONS

(1) With regard to the standardization of SPT, one additional aspect should be put into consideration i.e. the specification of manipulation and modification of N-value with length of rod.

(2) As one of the cone penetration tests --- the static cone penetrometer is in need of standardization. The main problem is the geometry of the probe for which the author wishes to recommand a more comprehensive design as a tentative one.

REFERENCES

Arce, C. M., Torres, F. L., and Vercelli, H. J. (1971) "Compared Experiences with the SPT," Proceedings of the 4th Pan American Conference on Soil Mechanics and Foundation Engineering, San Juan, Puerto Rico, Vol. II, pp. 95-104.

Fletcher, G. F. A., (1965) "Standard Penetration Test: Its Uses and Abuses," Journal of the Soil Mechanics and Foundation Engineering, Vol. 91, No. SM4, pp. 67-75.

Kovacs, W. D., Evans, J. C., and Griffith, A. H. (1979) "Towards a More Standardized SPT," Proceedings of the IX International Conference on Soil Mechanics and Foundation Engineering, Tokyo, Japan, Vol.II, Paper 4-18, pp. 269-276

Kovacs, W. D., Salomone, L. A., and Yokel, F. Y. (1981) "Energy Measurement in the Standard Penetration Test" NBS Building Science Series 135

de Mello, V. F. B. (1971) "The Standard Penetration Test," Proceedings of the Fourth Panamerican Conference on Soil Mechanics and Foundation Engineering, ASCE, Vol. 1, pp. 1-86.

Wang, C. C.(Wang, Z. Q.) (1978) "Some Experiences with an Electrical Static Penetrometer", Bulletin of the International Association of Engineering Geology, No.18, pp. 153-156, KREFELD

Zhou, S. G. (1981) "Influence of Fines on Evaluating Liquefaction of Sand by CPT", Proc. International Conference on Recent Advances in Geotechnical Earthquake Engineering and Soil Dynamics, Vol.I, pp.167, St. Louis

Proceedings of the Second European Symposium on Penetration Testing / Amsterdam / 24-27 May 1982

Correlations of penetration test results with in-situ and laboratory test data

J.L.WITHIAM, T.J.SILLER, R.M.BORT & A.J.EGGENBERGER
D'Appolonia Consulting Engineers, Pittsburgh, PA, USA

P.P.CHRISTIANO
Carnegie-Mellon University, Pittsburgh, PA, USA

U.DAYAL
Indian Institute of Technology, Kanpur, India

1 INTRODUCTION

When evaluating data accrued from site exploration and testing, the geotechnical engineer rarely, if ever, is confronted with too much information. Almost always gaps are found in the data which might be filled either with knowledge acquired through experience with similar sites or with supplemental data obtained by additional exploration and testing. Equally often, if the engineer were to be given a second chance to conduct the work, with the aid of hindsight but with the proviso that the total effort (cost) be equal to that associated with the previously conducted program, some changes would be made. The extent of the changes would be a function of the individual engineer's experience with similar sites, the range of variability in data considered acceptable, and the manner in which the information is to be used in analyses. One option the engineer might wish to exercise is the partial substitution of cone penetration tests for standard penetration tests; nearly continuous, relatively reliable cone penetration data are acquired with the penalty of recovering fewer soil samples via split-barrel samplers. Another option might be the inclusion of continuous undisturbed sampling in one or two borings while reducing the number of standard penetration tests and disturbed samples.

While the various options are numerous and would vary among engineers, it might be useful to have, for particular types of sites, a comparison of data acquired through different kinds of field and laboratory tests. Thus, when confronted with a site having similar geology, one may be better able to choose the exploration and testing programs which would furnish a good balance of information. The present paper is intended to provide such data comparisons for a marine site, in which the field exploration and testing program on primarily silty and clayey sands comprised the following: standard penetration tests, cone penetration tests, pressuremeter tests, crosshole seismic tests, and continuous, undisturbed sampling. In addition, laboratory tests included the determination of water contents from disturbed samples, and water contents, dry densities, and confined moduli from oedometer tests using undisturbed samples.

One of the main objectives of the program was the acquisition of data for settlement analysis. Although in the United States settlements of structures founded on sands are often estimated solely on the basis of standard penetration test data, additional data was sought for three reasons: first, large industrial structures are to be constructed on mats, incurring significant stresses at great depths and thereby precluded the use of existing empirical correlations between penetration resistance and settlements of spread footings; second, significant contents of clay and silt within the sands permitted the recovery of undisturbed samples whose compressibility could be tested in the laboratory; third, a rather complex dewatering and excavation sequence, which will partially preload the site, requires that reloading moduli as well as virgin loading moduli be determined. Accordingly this paper describes the relationships among the various data obtained, with a view both toward correlating different data (e.g., standard penetration vs. cone penetration resistance) at nearby points, and toward determining the extent to which some type of information (e.g., large quantities of standard penetration test data) may be of only marginal value compared with other data that could otherwise be acquired.

2 SITE DESCRIPTION

The site is located on the coastal plain of the United States. As indicated by a typical boring log in Figure 1, the soils comprise mainly silty and clayey sands with occasional interbedded lenses of clays and silts. The uppermost 20 to 30 feet, which lie above the groundwater table, are desiccated. At a depth of approximately 70 feet there exists a clayey silt layer, varying in thickness from five to ten feet, which separates two geologically similar formations. Both formations comprise sands varying in density from very loose to dense, and each constitutes a single, statistically homogeneous formation. A third geological formation, extending below a depth of approximately 140 feet, comprises very dense sands and has no significant influence on the foundation behavior.

The large plan dimensions of several structures to be built at the site necessitated the drilling of numerous borings and the implementation of extensive field tests. Figure 2 shows a layout of the field exploration and testing program. It comprised the drilling of borings, including 90 in which standard penetration tests (SPT) were performed and 2 from which continuous undisturbed samples (CS) were recovered; 6 cone penetration tests (CPT); 3 pressuremeter tests (PMT); and 3 crosshole seismic tests (CHT).

3 DATA CORRELATIONS AND COMPARISONS

3.1 Cone Penetration Tests vs. Standard Penetration Tests and Water Contents

Cone penetration resistance measured at six locations were compared with standard penetration resistance in nearby borings and also with water contents measured in SPT samples. The comparisons are shown in Figure 3 and Table 1.

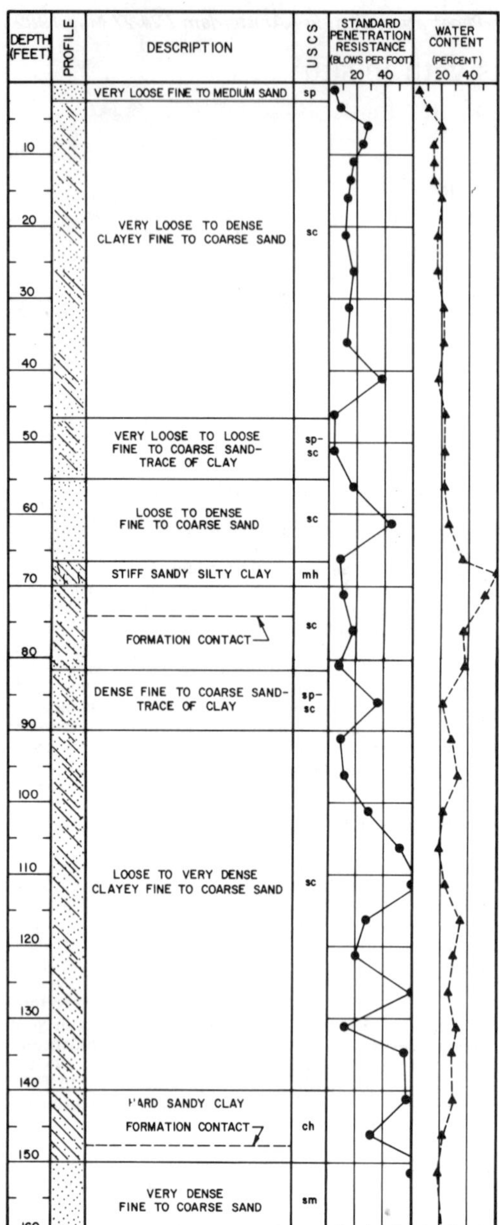

FIGURE 1
TYPICAL BORING LOG

Correlations between cone resistance and standard penetration resistance are highly variable. The standard penetration resistance encountered in B-43, for example, correlates well with nearby cone resistance, especially considering that the SPT data were acquired at five-foot intervals. In other instances, however, major trends differ between the two penetration methods. This may be observed in B-21, for example, at an elevation interval between El. 250 and El. 235, where the cone data suggests that the soil is increasing in stiffness while the SPT data indicates the contrary.

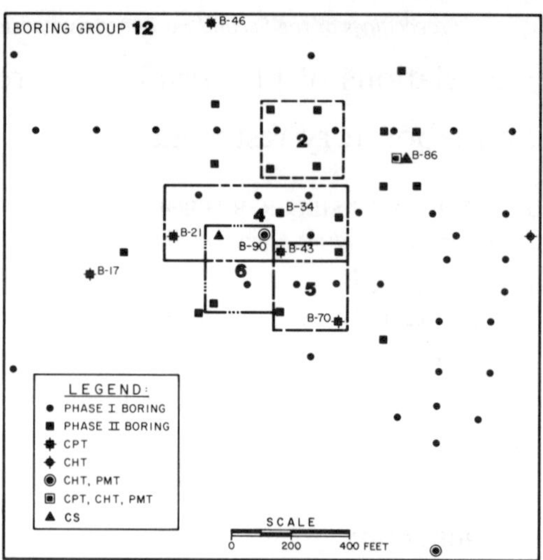

FIGURE 2
BORING PLAN AND GROUPING

In general, good correlation exists between SPT and CPT, the correlation coefficient ranging from 0.53 to 0.78. Note that this correlation represents the entire soil profile, including the zone above the groundwater table and the clayey silt layer near El. 210. When these two zones are excluded from the regression analysis, the correlation improves significantly. For example, the correlation coefficient for B-86 increases from 0.62 to 0.76. The variability associated with individual soundings may be minimized by correlating SPT and CPT resistances averaged over the six borings. The correlation is 0.74 on this basis.

It is also interesting to compare trends in SPT and CPT with those in water content measured from disturbed soil samples. The comparisons shown in Figure 3 indicate a poor correlation between water content and cone resistance on a boring by boring basis. The correlation coefficient between average water content and average SPT and CPT is significantly greater, especially when the zone above the groundwater table is deleted.

3.2 Crosshole Seismic Data vs. Cone Penetration Resistance

Shear wave velocities and attendant shear moduli associated with very small strains may be obtained from crosshole tests. It might be expected, that despite the tendency of crosshole tests to mask localized trends in stiffness, overall trends might be well represented. As shown in Figure 4, however, the trends in shear modulus show rather poor correlation with the trends in cone penetration resistance. Apparently, the "initial shear modulus," describing the soil stiffness at very small strains, is not well correlated with the stiffness associated with the very large strains incurred during penetration of the cone. Thus for the purpose of estimating settlements associated with neither very small nor very large strains, one may be faced with equally unsatisfactory data arising from the two tests described above.

3.3 Comparison of Spatial Averages of Standard Penetration Resistance and Water Content.

It is of value to compare standard penetration resistance and water contents among borings and among groups of borings. Figure 5 compares SPT resistance and water contents among borings B-34, 43, and 90 located within a radius of 65 feet. It may be observed that while the three

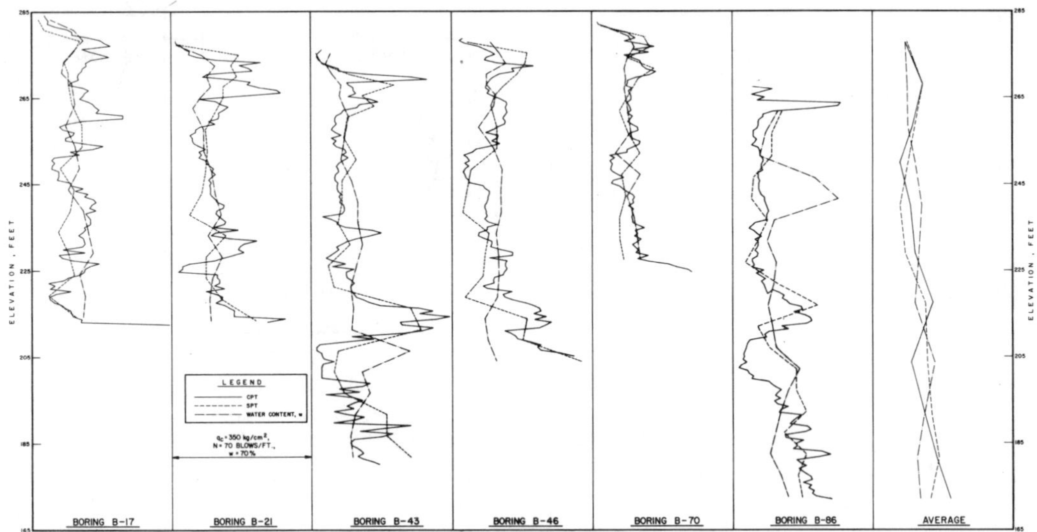

FIGURE 3
CPT-SPT-w PROFILES

TABLE I. CORRELATION COEFFICIENTS

	B-17	B-21	B-43	B-46	B-70	B-86	AVERAGE*
N vs. q_c	0.66	0.53	0.78	0.73	0.65	0.62	0.74
q_c vs. w	0.26	0.44	-0.14	-0.16	0.61	-0.16	-0.37
N vs. w	0.21	0.17	-0.03	-0.20	-0.14	-0.29	-0.49

*Excludes soils above water table and in silt layer near formation contact.

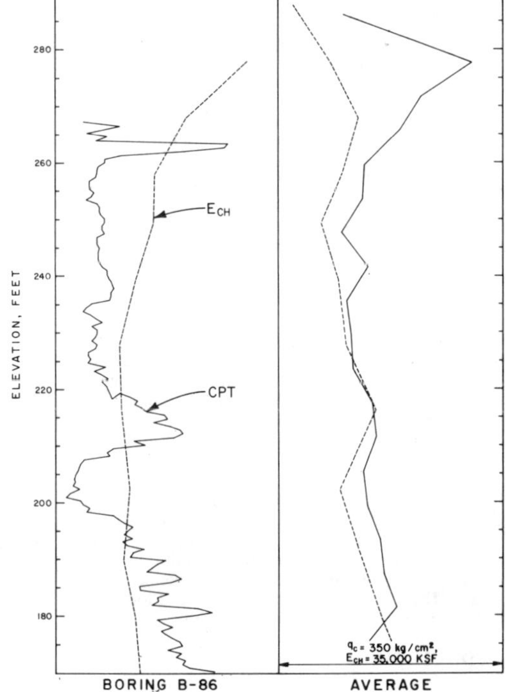

FIGURE 4
CPT—CHT COMPARISON

borings exhibit the same general trends, there is rather wide variation in blowcounts. The average of the standard deviations is 10 blows/foot. Part of the reason for differences from one boring to another is, or course, due to the natural variation of the subsoil stratigraphy. But it may also be seen in Figure 5 that the variation among water contents is considerably less, the average standard deviation being 6.0%.

Recognizing that the settlements of large mat foundations would be most strongly correlated with average spatial properties rather than properties indicated by individual borings, one might wish to know how much information is required to attain a valid statistical representation of soil properties over different regions of the site. That is, when is the acquired data of sufficient volume to obviate the acquisition of additional data? To these ends the boring layout shown in Figure 2 has been divided into 12 groups; several groups represent clusters of borings beneath proposed structures, while others simply were constructed to include borings in the same general area.

Except for Group 12, which comprises all borings, Group 4 contains the largest number of borings (11). Standard penetration resistance averaged within depth intervals of approximately ten feet, and the corresponding standard deviation in penetration resistance are shown in Figure 6a; shown in Figure 6b are water content data. The same general trends appear in each data set. However, the water content data shows less scatter than does the penetration data. It may be seen further that the subset of data associated with borings B-34, 43, and 90 shown in Figure 5, indicate the same trends as the data shown in Figure 6, but have greater scatter. Furthermore, the statistical descriptions of penetration resistance and water contents corresponding to Group 5 correlate very well with those for Group 6; in addition the data for each group correlates well with the data associated with Group 4 (which contains part of Groups 5 and 6 as subsets).

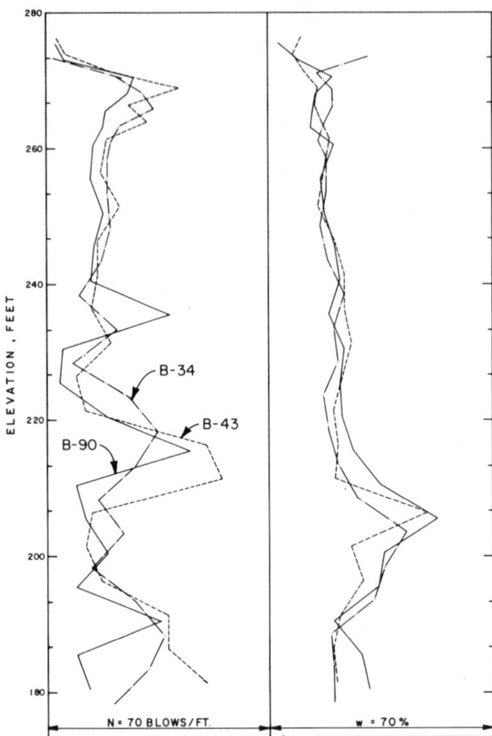

FIGURE 5
SPT—w COMPARISON
WITHIN THREE-BORING CLUSTER

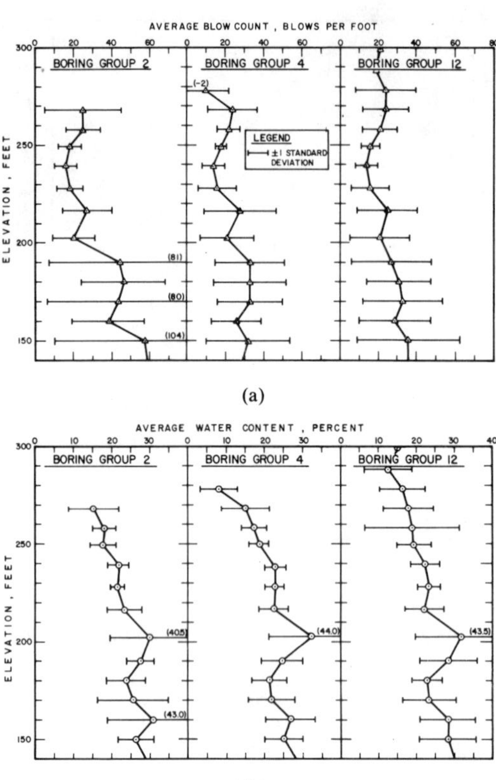

(a)

(b)

FIGURE 6
VARIABILITY OF SPT AND w
PROFILES BY BORING GROUP

There exists very good correlation between data associated with Groups 2 and 4 and, in fact, among most other groups of borings. Moreover, as shown in Figure 6, for average penetration resistence and for water content the correlation is very good between both Groups 2 and 4 and between either of these groups and Group 12, which comprises all borings. It may therefore be inferred from these correlations that if data were to be acquired for the purpose of statistically characterizing the present site (at least with respect to penetration resistance and water content), only a relatively small subset of borings (as included in Group 4, for example) would have had to be drilled.

The same inference may be drawn from Figure 7, which shows 95 percent confidence levels in mean standard penetration resistance and mean water contents as functions of sample numbers averaged over depth intervals of approximately 10 feet. Figure 7 was constructed by tabulating mean values of penetration resistance and water of sample numbers averaged over depth intervals of approximately 10 feet. Figure 7 was constructed by tabulating mean values of penetration resistance and water contents within radii of 200, 300, and 500 feet centered about boring B-90. It may be seen in Figure 7 that as the radius increases and more data are acquired, the 95 percent confidence interval becomes smaller. The interval is narrowest for the upper layers which were more frequently sampled, and is widest for the deeper layers which were sampled the least. It is evident, however, that beyond a certain number of data points included in the process (e.g., 15 to 20 points), little additional benefit is gained by including more points.

3.4 Pressuremeter versus CPT and SPT

Attempts to relate PMT results with CPT and SPT data on a boring by boring basis were unsuccessful. However, an alternate procedure suggested by Baguelin, et al. (1978), relating cone resistance and the pressuremeter limit pressure P_l, found a good correlation of 0.84. A similar comparison with N values from SPT's provided an excellent correlation of 0.96

3.5 Comparisons with Laboratory Tests

Oedometer tests using undisturbed samples were conducted to provide confined virgin and unload-reload moduli for settlement analyses. These data were then compared with moduli provided from PMT's, CPT's and CHT's. The results are summarized in Figure 8 with the exception of CHT moduli which extended beyond the modulus scale of the figure.

In general, moduli derived from PMT's (Ladanyi, 1972) and CPT's (Schmertmann, 1970) show good agreement with virgin moduli from oedometer tests and provide indirect evidence of the normally consolidated soils within the study area. Reload moduli are typically 3 to 5 times greater than these virgin moduli while moduli based on CHT's are approximately 50 to 100 times greater.

4 CONCLUSIONS

The main objective of this paper was to present comparisons among data acquired through various field and laboratory tests. For the site considered herein, the following conclusions are reached:

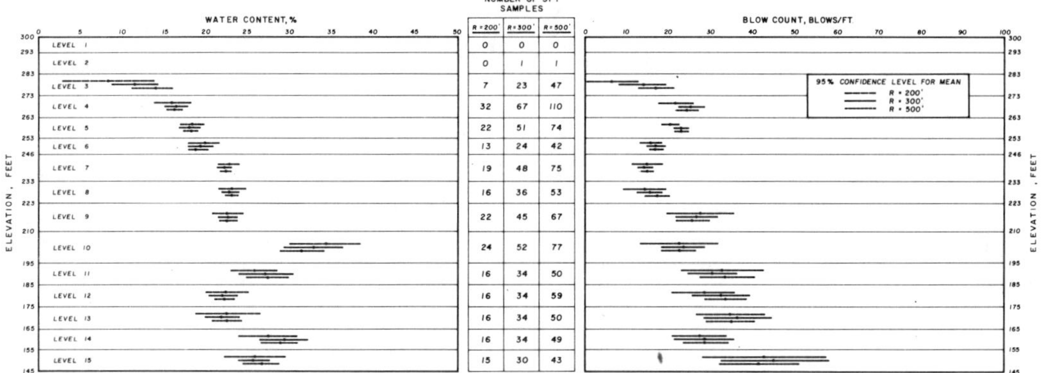

FIGURE 7
95% CONFIDENCE LEVELS FOR MEAN SPT AND w PROFILES

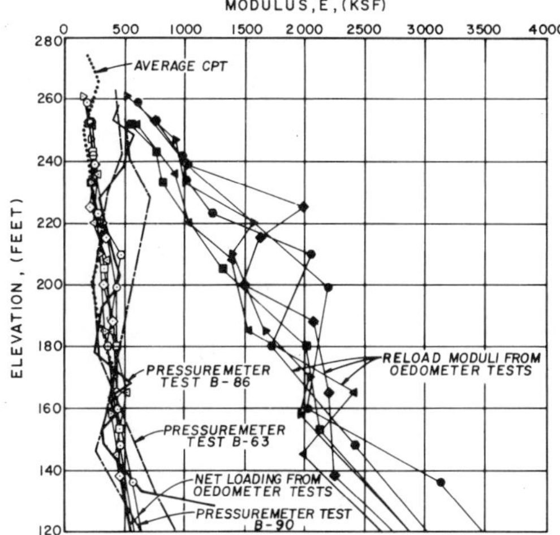

FIGURE 8
MODULUS COMPARISON FROM
CPT, PMT AND LABORATORY TESTS

1. Individual CPT and SPT soundings may or may not correlate well when each are conducted at the same location. Both CPT and SPT appear to correlate poorly with water content (even below the groundwater table) for individual soundings. SPT, CPT, and water content data averaged over several soundings (6 in the present case) show reasonably good correlation.

2. At any prescribed elevation, statistical distributions of SPT resistance and water content among different groups of borings are nearly the same, provided at least 5 or 6 borings are included in a group. The standard deviation reduces as more data are considered but once approximatly 20 data points have been grouped, additional data is of negligible value in providing a useful statistical description.

3. When data are combined for a sufficient number of borings (on the order of 10 to 20), the distribution of water content exhibits a smaller scatter than does that of SPT resistance and probably can be used to predict soil density.

4. Correlation between oedometer virgin moduli and average PMT and CPT moduli are good. In general these moduli which are representative of a normally consolidated soil deposit are approximately 3 to 5 times less than unload-reload moduli and 50 to 100 times less than small strain elastic moduli associated with seismic crosshole tests.

5 REFERENCES

Baguelin, F., J. F. Jezequel, D. H. Shields, 1978, The Pressuremeter and Foundation Engineering, Clausthal, Germany, Trans Tech Publications.

Ladanyi, B., 1972, In Situ Determination of Undrained Stress Strain Behavior of Sensitive Clays with the Pressuremeter, Can. Geo. J., 9:313-319.

Schmertmann, J. H., 1970, Static Cone to Compute Static Settlement Over Sand, J. Soil Mech. and Found. Div., ASCE, 96:pp. 1011-1043.

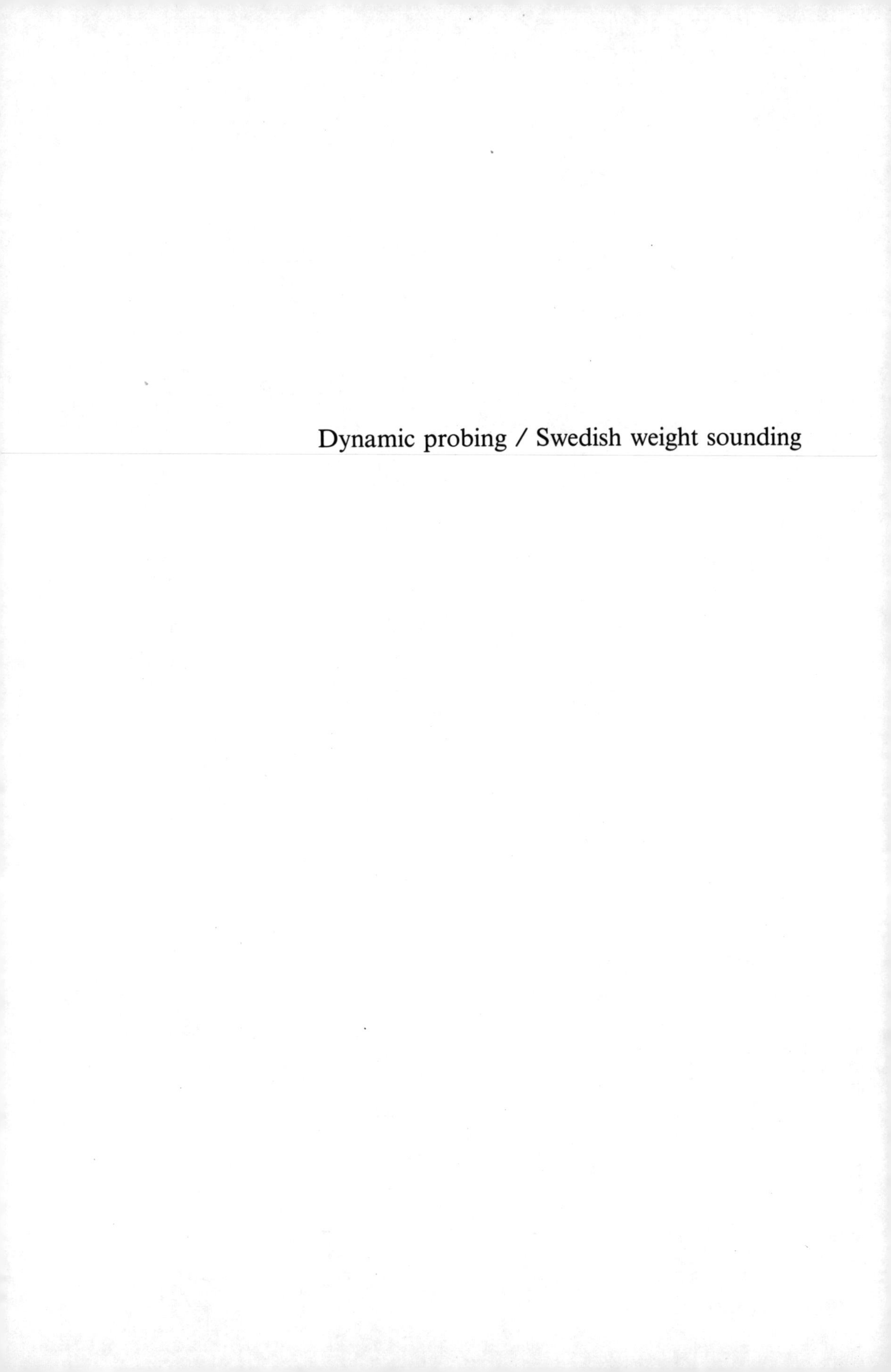

Dynamic probing / Swedish weight sounding

Proceedings of the Second European Symposium on Penetration Testing / Amsterdam / 24-27 May 1982

Dynamic penetration testing
State-of-the-art report

K.-J.MELZER
Battelle-Institut e.V., Frankfurt a/Main, Germany

U.SMOLTCZYK
Institut für Grundbau und Bodenmechanik, Universität Stuttgart, Germany

1 INTRODUCTION

The approach to any state-of-the-art report is necessarily influenced by the following tasks and problems:

- A qualitative view of the current knowledge is to be given.
- Future research and development in problem areas are to be outlined.
- It is to be attempted to present all possible approaches to the topic, and not ignore all except the one "right" solution; many approaches that appear contradictory may still be "right".
- Current knowledge undoubtly is relative and depends on the person, the time and the place. For example, some important work may not be mentioned partially from ignorance and partially because of the subjective estimate as to whether or not it matches the topic.

With these considerations in mind, the authors set the scope of the report by selecting a time frame ranging from the First European Symposium on Penetration Testing (ESOPT I) in Stockholm in 1974, to about the end of 1981. The main geographic area covered is Europe.

The report is diveded into six parts. In the first part, a summarizing survey is given of the state-of-the-art related to dynamic penetration testing at the ESOPT I. In the following parts, major advancements - if any - will be discussed under the following headings: Parameters influencing the penetration resistance; Field application; Empirical prediction of engineering properties; Theoretical considerations; Standardization.

In order to ensure that most of the recent advancements are covered, a circular letter was distributed to the state-of-the-art reporters of ESOPT I and/or their associates in about 20 European countries. This letter contained the key question: "Are there any changes in the content and/or philosophy of your state-of-the-art report of 1974, and what are the recent advancements?"We received eleven answers, which seems to be encouraging. Nevertheless, the result was somewhat discouraging as regards the advancements we would have expected. A similar picture emerged from our search of common literature data bases.

2 SUMMARY, ESOPT I, 1974

The state-of-the-art reports of ESOPT I were reviewed from the following angles in order to arrive at a qualitative picture for the European countries under consideration:

- Penetrometers in use;
- Standardization of equipment(at least one type);
- Mode of interpreting test results, viz.
  -- qualitatively (e.g. identification of layers),
  -- quantitatively (e.g. ranges or relations of engineering soil properties as a function of penetration resistance).

Only the penetrometers that are used for continuous dynamic penetration testing are considered. Because of the great variety of penetrometers in use, the following arbitrary classification according to the hammer masses applied was chosen:

Type	Abbr.	Mass, kg
Light	DPL	≤ 10
Medium	DPM	$> 10 < 40$
Heavy	DPH	$\geq 40 < 60$
Super-heavy	DPSH	> 60

The results are summarized in Table 1, from which the following general conclusions can be drawn for the countries considered:

- A large number of different dynamic penetrometers have been in use, in most cases for a considerable time;
- The degree of standardization appears to be low (only seven out of 20 countries have standardized one or more types of equipment);
- Results are mainly interpreted qualitatively;
- The engineering properties of soils, e.g. relative density, shear strength, compressibility, bearing capacity, are primarily derived from results obtained in cohesionless soils.

The question that arises is: what has been accomplished since then?

Table 1: Qualitative summary of the state-of-the-art in 1974 (ESOPT I) and progress until 1981

No.	Country	Penetrometers in Use				Standard	Interpretation of Test Results				
							Qualitatively	Eng. Propert. of Soils		Bearing Capacity	
		DPL	DPM	DPH	DPSH			Cohesionless	Cohesive	Shall.Found.	Piles
1	Belgium	x	-	x	-	-	x	-	-	-	-
2	Bulgaria	-	x	x	x	x	x	1,2	1,2	1	3
3	C.S.S.R.	x	-	-	x	x	x	1	-	1	3
4	Denmark	x	-	x	-	-	x	-	-	-	3
5	Finland	-	-	-	x	x	x	1	-	3	1
6	France	x	x	x	x	-	x	1,2	1	-	1
7	F.R.G.	x	x	x	-	x	x	1,2	-	-	-
8	G.D.R.	x	-	x	x	-	x	1,2	1,2	-	-
9	Greece	-	-	x	-	-	x	1	1	-	-
10	Hungary	-	-	x	-	x	x	1	-	-	-
11	Italy	x	-	-	x	-	x	1,2	-	1	-
12	Norway	-	-	-	x	x	x	-	-	-	1
13	Poland	x	-	-	x	-	x	1	1	-	-
14	Portugal	x	-	-	-	-	x	-	-	-	-
15	Spain	-	-	-	x	-	x	-	-	1	1
16	Sweden	-	-	-	x	x	x	3	-	-	-
17	Switzerland	x	x	x	x	-	x	3	3	-	-
18	Turkey 1)										
19	U.K.	-	-	-	x	-	x	3	-	-	-
20	U.S.S.R.		x		x	-	x	1,2	-	-	3

1 = Ranges (estimates) 3 = Not specified (or rarely) ▭ = Progress since 1974

2 = Relations 1) = No definite statement possible

3 PARAMETERS INFLUENCING THE PENETRATION RESISTANCE

A distinction is generally made between two groups of parameters influencing the penetration resistance:

Group I: Soil related parameters such as soil type, grain characteristics, degree of saturation, density, consistency, shear characteristics, compressibility, degree of consolidation, skin friction, etc. These parameters will be discussed mainly in chapter 5.
Group II: Equipment-related parameters such as cone geometry, rod type, anvil (weight, cushion), hammer (geometry, weight, release mechanism, height of fall, penetration depth), test procedure (number of blows per minute, interruptions, etc.).
These parameters are dealt with in this chapter.

The importance of unterstanding and considering the equipment-related parameters has considerably increased during the last ten years, mainly for two reasons:
- The reliability of the results of dynamic penetration tests had to be improved, compared with static penetration testing;
- Establishing relations between the results of various dynamic tests and those of the static test has become more and more important with the increasing drive for standardization on the national and international levels.

Baudrillard (1974) has tackled the old controversy that dynamic penetration test results are not reliable and cannot be used for the design of foundations. He proves that with properly designed equipment, with an adequate procedure, and with full understanding of the mechanism of driving a cone into the soil, such dynamic penetration testing allows measurements to be made that are as reliable as those performed with the best static equipment. This is only one example that stands for various others (Pfister, 1974; Schultze, 1974 b, 1976; Meardi, 1974; Uriel, 1974), not to mention the attempts to modify the basic philosophy of the dynamic test by combining it with a vane or static test (e.g. Martin, 1974; Sherwood/Child, 1974; Bergdahl/Möller, 1981) or a sleeve jacket (Bergdahl, 1979c).

An essential contribution to the basic unterstanding of the interaction of the various equipment-related parameters is the research by Krämer (1977, 1980, 1981). Using the dimensional analysis approach, he developed a method to describe the interaction of the group-II parameters considering soil type and density. The results enable the user to choose the proper equipment for a given job. For instance, the maximum depths down to which various dynamic penetrometers give reliable results were evaluated: DPL = 6-10 m, DPH = 14-25 m, DPS = 40 m. Furthermore, relations can be established between the results of dynamic penetrometers having different group-II characteristics.

Other authors still attempt to express the dynamic penetration resistance using pile driving formulas, the most familiar being the Dutch formula (Sanglerat, 1974; Lareal, 1974; Puech, 1974; Delmarcelle, 1981). Also the approach to use the work during the penetration test as a means to correlate the results of various penetrometers with each other seems to be successful (Schultze, 1974a; Teferra, 1975; Biedermann, 1979; Borowczyk/Frankowski, 1981).

Another approach that is still made is to correlate directly the number of blows of various penetrometers under given conditions (e.g. Tammirinne, 1974; Biedermann, 1979; Faraco, 1981). However, one has to consider that these relations may strongly depend on the soil conditions of the geographic areas for which they were established.
Let us close this chapter with an example that one of our colleagues conveyed to us (Escario, 1981). The author normally uses the "Borro's test" (4x4 cm square point; ⌀ 32 mm rods, 65 kg hammer, height of fall: 0.5 m), which is the most popular test in Spain. On one occasion, he used the DPA equipment and the static penetrometer recommended by ISSMFE along with his regular dynamic penetrometer. The soil was a poorly compacted clay in the upper meters, followed by a natural clay. The reason for using the two additional tests was to check whether it was possible to get a better definition of the different layers. The answer was somewhat sad in that there was no difference in the results of the two dynamic penetrometer tests (number of blows/20 cm penetration), one of them being a dynamic test according to our new standards; even the static cone penetrometer test did not permit better conclusions. This may be a coincidence; but it is by no means encouraging people to turn to new standards without first taking a closer look at what they are used to.

4 FIELD APPLICATION

The kinds of subsoil exploration for which dynamic penetration tests are mostly used can be divided into the following three groups:

a) Preliminary field investigations
b) Final field investigations
c) Construction supervision

Which type of penetrometer should be used for these investigations is strongly influenced by the following factors:

- Topography (accessibility of the area of investigation)
- Geology (general information about soil types etc.)
- Necessary depth of penetration
- Necessity of key penetration tests in connection with borings.

Considering the tasks and influencing factors mentioned above, the process of selecting the right equipment has to be optimized under economic and technical aspects. Krämer (1980, 1981) and Bergdahl (1979 b, 1981 b) have given at least examples for this process. However, specific guidelines that would be valid worldwide cannot be given. Optimization can only be regional.

Usually one will start the investigations (especially those of groups a and b) with the simplest equipment available which is suitable for the specific job in hand. General rules concerning, for example, the best penetration equipment depths (DPL: 6 to 10 m; DPH: 14-25 m; DPSH: 40 m) are serving as guidelines. Additional influences on the results of penetration tests, such as skin friction, groundwater level, critical depth, etc. are being considered when selecting the equipment which is first to be used. After first results of the site investigation are available, the use of other, maybe more sophisticated, equipment which is more suitable for the problems encountered will be taken into consideration.

For construction supervision, especially the lighter equipment will be preferred. For instance in the area of road construction, the DPL is mostly used; its results correlate well with the CBR values (e.g. Kindermans, 1976; Centre de Recherches Routières, 1978).

Another aspect should be taken into account which has become increasingly important in recent years (e.g. Schultze, 1975; Biedermann, 1979): the necessity of applying statistics and probability calculations to results of field and laboratory tests. This already influences the design even of preliminary field investigations (e.g. distance of penetration test points). Even more important is that the test results - which under normal circumstances are subject to considerable scattering - should be analyzed using statistical approaches; however, care must here be taken that basic physical principles are not violated. Crespellani/ Loi (1978) have shown a method of analyzing penetration resistance-depth diagrams statistically; however, there is the risk that valuable detailed information may get lost when using curve-fitting filtering techniques.

5 EMPIRICAL PREDICTION OF ENGINEERING PROPERTIES

It has always been the desire of engineers and researchers to extend the use of penetration test results beyond the qualitative interpretation towards the determination of engineering properties. Attempts have been made to correlate penetration test results empirically especially with the following properties: relative density or dry density (cohesionless soils), consistency (cohesive soils), shear strength parameters, compressibility, bearing capacity of shallow foundations and piles. We will try to summarize the progress in these areas in the following paragraphs (see also Tabele 1).

5.1 Relative density

The first step to evaluate the relative density of soils on the basis of dynamic penetration test results still is to establish correlations between ranges of the number of blows and the relative density. Numerous examples have been published in the past and frequently summarized (Schultze, 1974a; Wolski, 1974; Biedermann, 1979). More recently, research efforts in Sweden have led to such relations, which also indicated cases where the number of blows should be corrected (Bergdhal, 1979b; 1981a).

The second step is to establish direct relations between the relative density and the number of blows. Besides the relative density (e.g. Hausner/Sperling, 1974), the dry density is used in some instances and correlated with the penetration resistance determined from pile driving formulas (e.g. Tammirinne, 1974). However, when establishing direct relations one must possibly consider additional soil parameters such as the grain size distribution, soil compactibility, overburden pressure, and critical depth. These dependencies, which have been well known for more than 10 years now (e.g. Melzer, 1968; Schultze, 1974a, 1976; Puech et al., 1974), have been confirmed (Teferra, 1975, 1976a).

Existing resarch material has been thoroughly evaluated (Stenzel/Melzer, 1978) in order to derive relations between the relative density and the number of blows (valid below the critical depth) of the DPL, DPH and the SPT for sand ($U \leqq 3$) and sand-gravel mixtures ($U \geqq 3$) (conditions: above and below the groundwater level); these relations have been introduced into the German DIN 4094 standard (1980). Almost at the same time, Polish researchers (Borowczyk/Frankowski, 1981) established similar relations for sand ($U \leqq 3$) to evaluate the relative density from results achieved with the DPL, DPB and SPT. The relations for DPL and SPT from both sources are shown in Fig. 1.

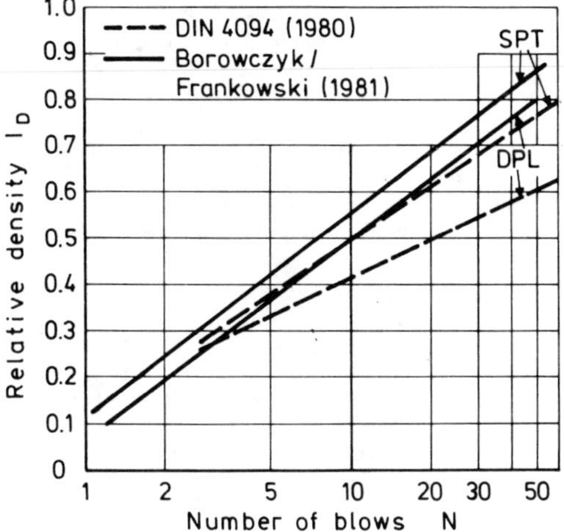

Fig. 1: Relations between relative density and number of blows for sands

5.2 Consistency

Only very few correlations are known between the relative consistency of cohesive soils and the number of blows of dynamic penetrometers (Hausner/Sperling, 1974; Wolski, 1974; Rollberg, 1976). As far as we know, nothing essentially new has been published since ESOPT I. This ist not very surprising because of the strong effects of porewater pressure and skin friction on the dynamic penetration resistance in cohesive soils which make the interpretation of test results more difficult than in cohesionless soils.

Before this background, the research efforts by Schormann (1979) and Biedermann (1979, 1980) have to be mentioned as an essential step forward towards the evaluation of the properties of cohesive soils on the basis of the results of dynamic penetration tests (DPL, SPT). Both authors concentrated on one specific soil type, commonly known as "Rhineland Silt". They did not try, however, to correlate the relative consistency but used terms like plasticity index, degree of saturation, Atterberg limits etc. Their efforts were successful to the point that relations are now available with which it is possible to determine compressibility and shear strength parameters from the results of dynamic penetration tests for this specific soil. These attempts confirm Baudrillard's thesis of 1974 (see chapter 3).

5.3 Shear strength parameters

Basically, three groups of methods can be defined for the determination of the shear parameters friction angle and cohesion.

Group I: The shear parameters are directly correlated with the results of penetration test results.

Group II: The shear parameters are indirectly determined on the basis of relations between other soil properties (e.g. the relative density) and penetration test results.

Group III: The shear parameters are derived from penetration test results, using theoretical methods.

Only very few attempts with group-I methods have been reported for dynamic penetration test results within the time frame covered here (Baudrillard, 1974; Pfister, 1974). In both cases, the friction angle for cohesionless soils and the undrained shear strength for cohesive soils was successfully correlated with a dynamic penetration resistance driving formula.

The indirect methods according to group II still seem to be most successful, especially for determining the friction angle on the basis of dynamic penetration test results. For cohesionless soils, the relative density (or void ratio) is derived from the test results, and the friction angle is then determinded by means of the relative density (Teferra, 1975; Schultze, 1976). Similar methods are applied in the case of cohesive soils, as was mentioned above (Schormann, 1979; Biedermann, 1979, 1980).

In group III primarily the results of static cone penetration tests are used to derive the friction angle theoretically. Mitchell/Lunne (1978) compared a number of theoretical and empirical prediction methods. Further attempts at modelling the cone-soil interaction for the static and dynamic cases are described in chapter 6.

One essential point should be emphasized in this context: as long as there is no theoretical method available to determine shear parameters from dynamic penetration tests, direct correlations between the results of static and dynamic penetration tests continue to be important, (e.g. Puech et al. 1974; Schultze, 1976; Biedermann, 1979).

5.4. Compressibility

The approaches to determine the compressibility of soil can be divided into two groups:

Group I: Penetration test results are directly correlated with compressibility indices (e.g. modulus of compressibility).

Group II: The compressiblity is indirectly determined on the basis of relations between other soil properties (e.g. the relative density or void ratio) and penetration test results.

As regards group I, one way taken is to correlate the results of pressuremeter tests with the number of blows required for tests with the DPL and DPH according to DIN 4094; various soil types - especially silt, however - were investigated in this way (Schultze/Biedermann, 1977; Biedermann, 1980). The second way, which was first demonstrated by Sherif (1973), is to determine the modulus of compressibility by observation of the settlements of actual structures and to correlate these values with the results of penetration tests performed during subsoil explorations on the corresponding construction sites (Schultze, 1974a,c; Sievering, 1980). Various soils (cohesive, cohesionless) and types of foundation were investigated in this way. The penetrometer types used were the SPT and the CPT. However, there are possibilities of relating the results to the resistance of other dynamic penetrometers. Wennestrand (1979) confirmed the usefulness of this approach by predicting the settlements of the shallow foundation of a bridge on fine sand; among other methods he used the results of tests with a DPA. Predicted and measured settlements compared well.

As for Group II, Jänke (1975) and Teferra (1976b) established relations to derive the coefficients v and w in the equation (constrained modulus) $E_s = v \cdot \sigma^w$ from the results of dynamic penetration tests in cohesionless soils. Schormann (1979) used a similar approach to evaluate the compressibility of silt, using results from DPL tests.

5.5. Bearing capacity of shallow foundations

Reports about relations to determine the bearing capacity of shallow foundations directly from the results of dynamic penetration tests have become scarce since 1974. Only Sweden reported some progress by giving ranges for ground pressures of spread footings as function of dynamic penetration test results (Bergdahl, 1979b). Most of the time, the bearing capacity is determined indirectly through relative density or shear strength predictions.

5.6. Bearing capacitiy of piles

Because of the similarity of penetrometers and piles, nummerous attempts have been made to determine the pile bearing capacity on the basis of penetrometer test results. Most of the commonly known methods are described in the Proceedings of ESOPT I (1974). The most comprehensive contribution concerning this problem area that was made since then is the research effort presented by Rollberg (1976; Biedermann, 1979). The basis for his work were the results of 248 pile loading tests and corresponding static and dynamic penetrometer tests. Two essential improvements in the evaluation of the corresponding test results have been made by Rollberg:

- He provided an exact definition of the failure load (ultimate bearing capacity) from pile loading test results;
- Not the single penetration resistance values are used for interpretation, but the integral of the penetration versus depth curve is applied.

Using statistical approaches, Rollberg presents relations for the determination of pile failure loads, permanent settlement and pile driving work as a function of static and dynamic penetrometer test results. An independent check calculation of three test cases (Smoltczyk/v.Koten/Hilmer,1978) confirms this method so far (Rollberg, 1978).

6 THEORETICAL CONSIDERATIONS

In this chapter we shall deal with those investigations which try to explain the interaction of a dynamic penetrometer and its surrounding soil by using theoretical analyses.

It might well be questioned if a sounding procedure of this kind is at all apt to be interpreted by any type of purely analytical treatment since most of the soil successions which we have to investigate do not fulfill the requirements which would enable us to define a sufficiently simple medium for this purpose. "Simple", however, does not mean homogeneous. Stratification can well be accounted for by modern theories but can also yield an improper simplification as may be indicated by the example of fig.2.

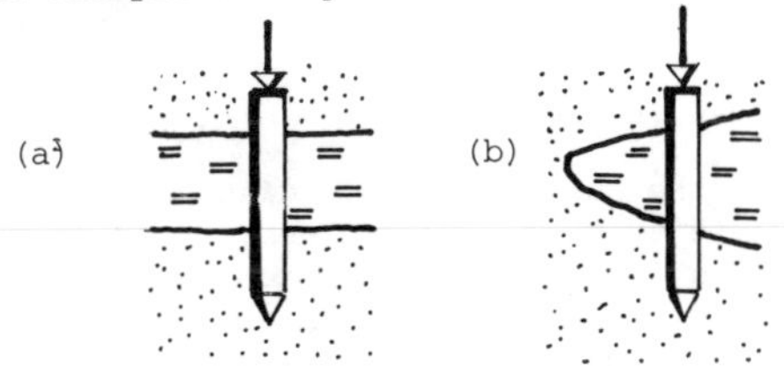

Fig.2. Inadequate simplification (a) of inhomogeneous soil (b)

The main obstacle is the fact that the interaction of the sounding cone with the contacting soil has both a local and a long distance influence on the machanical behaviour of the cone. The particular local strength condition due to any small embedment of alternate material may have the same significance to the rate of penetration as have the boundary conditions of the general soil succession which are still at some distance.

But keeping this in mind, we still do believe in the value of deductive considerations of idealized systems. They can, at least, help to reduce the wide variety of influencing parameters to those which remain of paramount importance compared to others of minor significance. They also may help us in the problem of scaling and in evaluating the influence of the geometry of the device.

The axisymmetric problem of a cone-pointed rod driven into a continuous body refers both to the penetrometer and the driven pile. A closely related problem is that of a projectile shot into ground.

We shall try to give a review on the results of analyses published so far which might be used here.

Static analyses

The earlier static approaches considered soil to be a non-elastic, ideal-plastic body of Coulomb behaviour at failure.

Nowatzki (1971) and Nowatzki/Karafiath (1972) presented an analysis of the penetration of a rigid cone into a Coulomb soil by a numerical computation of the field of stress characteristics for an axisymmetric state of stress following the Sokolovski procedure. For the intermediate principal stress the assumption (1) $\sigma_2 = \sigma_3$ was made. Further, assumption (2) considered the soil above the cone to act as an external surcharge only, overburden shear being disregarded.

To prove the theoretical findings, penetration tests by alu-cones were made in sand. The intention of evaluating the shear parameter ϕ from the "cone index" (that is the stress measured at 6" penetration depth in the problem area of vehicle mobility), however, seems to be rather questionable for the assumptions cited above. In fact, the initial state of stress in a horizontal plane does not at all remain hydrostatic during the penetration test and the shear strength contribution of the soil above the cone should be substantial.

Obviously without knowing about these papers, Yong/Chen (1976) repeated the same type of calculus but assuming (i) a passive Rankine zone adjacent to the cone, and (ii) $\sigma_2 = \sigma_1$. Both these assumptions seem to be strongly misleading.

Besides, the fulfilment of boundary conditions along the failure surface was only possible by introducing a stress jump condition which indicates an improper subdividing of the plastified region. This analysis, therefore, might not be used without further clarification by its authors.

In a recent paper, Karafiath (1980) tried to adapt the theory of plasticitiy to the progressing indentation of a cone. The process of penetration is thought of as if it occurred in finite, small increments. The stress state produced in the soil by a previous stage of penetration constitutes the boundary conditions for the next increment at the upper level of the cone. Like Nowatzki did, assumption (1) is used to solve the differential equations without regarding the rates of displacement. The author defends this assumption by reference to the triaxial test and to the tests of Bennett/Gisbourne (1971). The triaxial test, however, is no useful argument since the hydrostatic state of horizontal stress is a boundary condition imposed by the liquid around the sample which notably deviates from the situation in a soil continuum. On the other hand, the model tests of Benett/Gisbourne as well cannot prove the validity for the simple reason that just three normal stress components were measured and again an assumption had to be made in order to get any information on the relative position of the intermediate principal stress to the minor principal stress. The only conclusion possible might be drawn by saying that probably $\sigma_2 \rightarrow \sigma_3$ is closer to reality than $\sigma_2 \rightarrow \sigma_1$.

Although Karafiath presents some arguments for disregarding the displacements prior to the state of failure, the conventional theory of ideal plasticity used to compute stress fields under simplified assumptions cannot be looked upon as being an adequate approach to analyse the continuous penetration of a cone into a real soil, shear displacements being of paramount importance.

In both cohesive and non-cohesive soils, published model test data indicate continuous displacement fields of the confined plasticitiy type in front of the tip and laterally, see e.g. Roy/Michaud/Tavenas (1974).

Reference ist also made to the photo given by Malyshev/ Lavisin (1974). It was this kind of observation which at the last symposium yielded the statement that in no model test any Prandtl type failure pattern had been observed. It is no point to compare this conclusion with the model test results for flat foundations where also such failure patterns can but seldom be obtained: for shallow foundations the unconfined plastic failure truly has a physical meaning and therefore may reasonably be adopted as approaching reality. Contrary to this, unconfined plastic failure has no physical relevance to the penetrometer problem and should thus not be considered further.

A more refinded approach considers the soil as an ideal-elastic, ideal-plastic body of Coulomb behaviour at failure. Rohani/Baladi (1981), dealing with the problem of dynamic soil-track interactions made an investigation on the resistance of soil to the penetration of a right-circular cone. Their calculation was based on a solution published by Vesic (1972) for the expanding spherical cavity in an ideal-elastic, ideal-plastic (in terms of ϕ and c) soil continuum. The basic premise of their model was the cone penetration process to be viewed as the expansion of a downward series of such cavities. The characteristic soil parameter ist the "stiffness ratio" $\bar{G}/(+ q.\tan\phi)$ with q being the surcharge and $\bar{G}$ a shear modulus modified from G to $\bar{G}$ in order to consider the free-surface effect. This effect was found to be significant for depths up to six times the cone height.

According to this paper, the shear modulus is of minor influence on the cone index CI for cohesive soil and for non-cohesive soil with angles of friction less than 30°. On the other hand, the increase of CI with depth was more than linearily proportional. This indicates the consequence - see the stiffness ratio - of such approach: for any cavity to be expanded by the cone at a deeper level the failure condition allows for an elevated limit stress level in direct proportion to the depth increment. If penetrometers are used close to the surface as done by Rohani/Baladi this may be neglected. For our problem, however, it certainly cannot be omitted. For deeper going penetrometers we need to consider the growing influence of plastic volumetric strains compared to that of shear strains.

Dynamic analyses

Since the subject of this discussion is the dynamic type of sounding device, an analysis should be looked for which implements the acceleration forces in the mathematical model. We may, for example, make use of investigations dealing with projectile penetration into soil. On this topic the last state-of-the-art report seems to have been presented by McNeil in 1972.

Thigpen (1974) published this kind of analysis by regarding the soil (the "target") as an ideal-elastic, ideal-plastic body with a plastic condition of the v.Mises type and with a normality flow rule. The projectile is seen to slide without mantle friction. The initial conditions are specified in terms of velocities and stresses at t = 0, stresses being zero at the moment of impact. The numerical computations were done for time steps. Fig.3a and b are taken from this paper to demonstrate the results.

They indicate the region of plastic flow to extend to about 2 projectile radii in front of the projectile and to about 9 radii laterally at maximum. Following this, shear stresses are concentrated to a rather small region around the projectile.

The results were compared to experimental data and a good agreement was found with the deceleration data whilst the penetration depth as predicted by the theory was only 44% of the measured one.

In view of the simple bilinear material equation used this is a very good agreement. It may, however, be due to the type of material which was a very soft rock with only cohesion for shear strength.

A further simplification of this kind of analysis was done by Yankelevsky/Adin (1980) for frictionless cohesive soil. In this, the continuum is replaced by a stack of discs, fig.4.

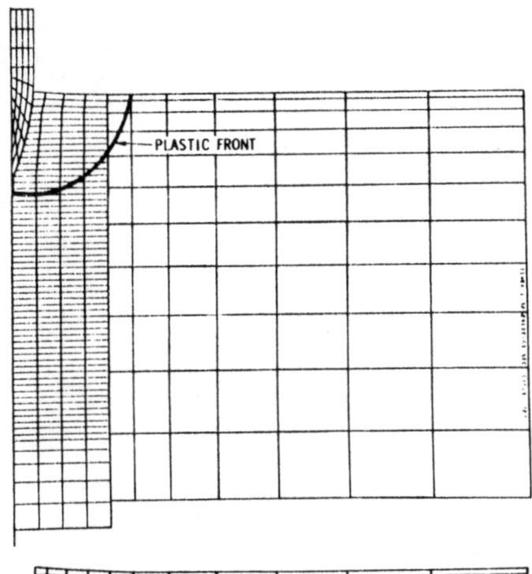

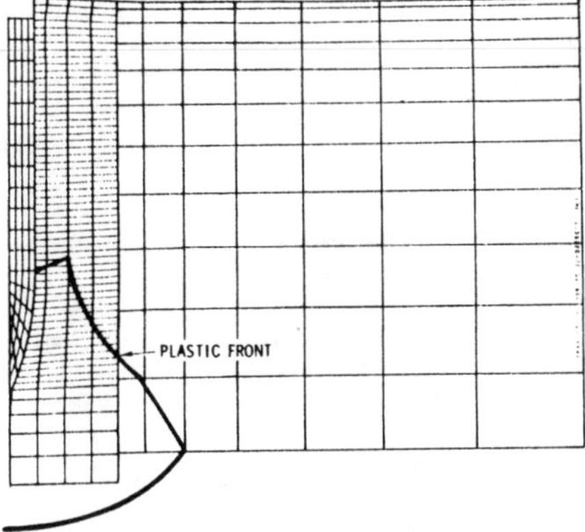

Fig.3 Projectile penetration into soft rock (Thigpen, 1974).
a - t = 1.52 ms, b - t = 13.92 ms.

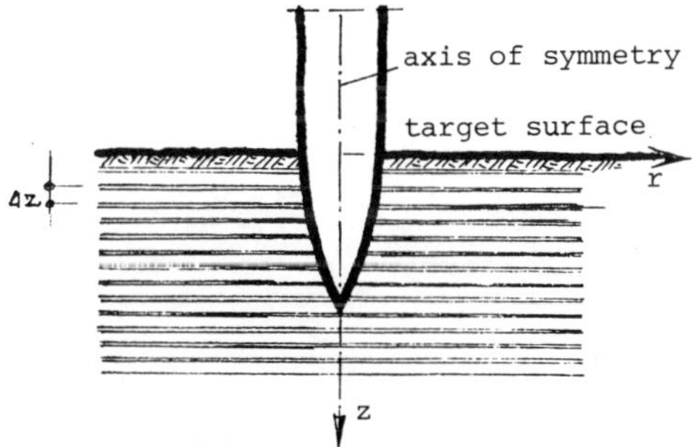

Fig.4 Disc model of soil penetrated by a projectile

The overall disc response is expressed by a single parameter $\varepsilon = \varepsilon_1 + \varepsilon_2 + \varepsilon_3$, the interaction pressure on the internal boundary being related to the radial displacement, velocity and acceleration. The plastic shock wave front has an instantaneous radius

$$h(t) = \frac{u(t)}{\sqrt{1-\rho_0/\rho}} = \frac{u(t)}{\sqrt{\varepsilon}}$$

(u - displacement; ρ_0- initial and ρ - actual density) which strongly depends on the volume change ε.

The computer time for this simpler one-dimensional approach, of course, is much less than for the earlier two-dimensional one. It is remarkable to note that the disregarding of the soil response in front of the penetrator tip here (as demonstrated by Thigpen) did obviously not much affect the results of the purely lateral displacement approach.

It is therefore a promising theoretical tool and further research work should be motivated to (i) extend it to frictional materials, and (ii) to consider the volume decrease provided by the crushing of soil grains within the immediate surroundings of the penetrometer.

The first aspect is included in a recent analysis done by Chow (1981) who investigated the dynamic behaviour of driven piles subjected to low frequency wave loading, and impact loading. He found that the finite element model gave a rather damped response. For constant energy input from a hammer, the one-dimensional and three-dimensional analyses diverged with increasing soil strength (one-dimensional approach means using the wave equation model). Applying these results to our problem would mean that layers which response by a high number of blows should be evaluated with due respect to their spatial reaction. Much more research work seems to be needed in this field, especially in verifying the soil models used so far.

7 STANDARDIZATION

"The dynamic penetrometer test: A standard which is not standardized": under this title a paper was published in 1970 (Ireland et al, 1970). Does it still hold true? The large variety of dynamic penetrometers in use, as shown by Zweck/Bergdahl (1974) and also by Begemann (1974), aggravates the problem of standardizing this equipment, at least from the internationl point of view. On a national level, 7 out of the 20 countries listed in Table 1 already had standards in 1974. As far as we know, Poland and the U.S.S.R. have established standards since then. Denmark and Belgium are following the recommendations of the ISSMFE Subcommittee on Standardization of Penetration Testing. Greece is using the German standard DIN 4094. Finland, Sweden and the FRG have improved their standards. Thus, more than fifty percent of the European countries are using standards (Table 1).

On an international level, the ISSMFE Subcommittee on Standardization of Penetration Testing ("in Europe" until 1981) drafted a standard for dynamic penetration testing following the Conference on the Standardization of Penetration Testing in Europe in The Hague in 1976. This standard essentially considers the Swedish equipment, procedures A and B (Dahlberg, 1974; Bergdahl, 1979a). At the 9th ICSMFE in Tokyo 1977 this standard was adopted by the Executive Committee and since then has been distributed worldwide. In Brighton, 1979, the Subcommittee found that it was desirable to adopt the Recommended Standard as International Standard. At the 10th ICSMFE in Stockholm 1971, a Recommended Standard for a light penetrometers test (essentially complying with the German standard DIN 4094) was approved of by the ISSMFE Executive Committee.

At this point, some general remarks about the push for international standardization of dynamic penetration testing, may be allowed.

The dynamic penetrometer is probably the oldest of its kind. As a consequence, a large variety of equipment is used in different countries, which is in some cases defined by national standards, in others not. Any international standard defining only one or two types of this equipment will give rise to severe difficulties, e.g.

- Existing national standards will be overruled;
- Private industry and consultants as well as research institutes will be forced to buy new equipment, which is inconceivable at the present economic situation.

Therefore, international standardization should be pursued rather cautiously. Instead of standardizing one or two penetrometers only, future efforts should be directed towards standardizing the relations between the results of various penetrometers that are used worldwide. Therefore the recommendation of the ISSMFE Subcommittee, to conduct side-by-side tests with the penetrometer normally used and one of the "standard" penetrometers appears to be very sensible. However, the recommendation of the same committee that "papers to international conferences or journals presenting results from penetration tests should also include results from at least one recommended standard penetration testing method" (or the paper being rejected?) should be reconsidered. This will not encourage the use of "standard" penetrometers. As a matter of fact, it will promote the publication of papers by a very few privileged institutions only, thus reducing the number of papers concerning the subject area even further.

8 CONCLUSIONS

Based on the results of our state-of-the-art report, the following main conclusions can be drawn (see also Table 1):

- There has been no drastic change in the basic philosophy concerning dynamic penetration testing in the various countries considered since 1974.
- Generally speaking the application of dynamic penetrometers in the field of subsoil exploration has not increased. Also the number of different types of penetrometer used in the individual countries has remained more or less constant.
- The progress in standardization we have seen since 1974 has increased; more than half the European countries considered are following some standard or other. Future standardization efforts should be pursued rather cautiously.
- Attempts to interpret the results of dynamic penetration tests more quantitatively are concentrated mainly on a certain group of countries where this research has become a necessary tradition. As regards the determination engineering properties of cohesionless soils, the efforts have been expanded. With respect to the determination of cohesive soils and pile bearing capacity, few essential steps forward have been made. Future research efforts should be made in this direction.
- Attempts to improve the understanding of the influence of equipment mechanisms on the test results as well as of the dynamic process of cone-soil interaction have increased especially with a view to improving the quantitative interpretation of the test results. We trust to see more intensive research efforts - both empirical and theoretical - in this field in future.

9 ACKNOWLEDGEMENT

The authors wish to thank their European colleagues who contributed to this state-of-the-art report by providing information about the present situation in the area of dynamic penetration testing in their countries.

10 REFERENCES

Baudrillard,J. 1974, New Development in Dynamic Penetration Testing. Proc.ESOPT I, vol.2.2: 25-34, Stockholm.

Bennett,D.H.& Gisbourne,R. 1971, The penetration resistance of dry sand. Proc. Roscoe Memorial Symp.

Bergdahl,U. 1979a, European Standard on Penetration Testing - A Necessity. Proc. Symp."Sondierungen und In-Situ-Messungen": 87-98, Bundesversuchs-u.Forschungsanstalt Arsenal, Geotechnisches Institut, Wien.

Bergdahl,U. 1979b, The Use of Penetrometers in Sweden. Proc.Symp."Sondierungen und In-Situ-Massungen": 99-122, Wien.

Bergdahl,U. 1979c, Development of the Dynamic Probing Test Method. Proc.7.ECSMFE, vol.2: 201-206, Brighton.

Bergdahl,U. 1981a, Reduction of N-Values from SPT-Tests? Proc.10.ICSMFE, vol.4, Stockholm (to be published).

Bergdahl,U. 1981b, Current Practice in Planning and Carrying out Site Investigations in Sweden. Report to 10.ICSMFE, Subcommittee on Site Investigations.

Bergdahl,U.& Möller,B. 1981, The Static-Dynamic Penetrometer. Proc.10.ICSMFE, vol.2: 439-444, Stockholm.

Bergmann,H. 1974, General Report: Central and Western Europe. Proc.ESOPT I, vol.2: 29-39, Stockholm.

Biedermann,B. 1979, Dynamische und statische Sonden und ihre praktische Bedeutung in der Bodenmechanik; State-of-the-Art-Report. Proc.Symp."Sondierungen und In-Situ-Messungen",Bundesversuchs- und Forschungsanstalt Arsenal, Wien.

Biedermann,B. 1980, Vergleichende Untersuchungen mit der Seitendrucksonde im Schluff. Diss.TH Aachen.

Borowczyk,M.& Frankowski,Z.-B. 1981, Dynamic and Static Sounding Results Interpretation. Proc.10.ICSMFE, vol.2: 451-454, Stockholm.

Centre de Recherches Routières 1978, Mode opératoire, estimation rapide de la portance des sols à l'aide d'une sonde de battage légère type C.R.R..Méthode de Mesure C.R.R.,MF.39, Bruxelles.

Chow,Y.K. 1981, Dynamic Behaviour of Piles. Ph.D.Thesis Univ.of Manchester.

Crespellani,T.& Loi,A. 1978, Analisi numerico-statistica di penetrometriche su vasta scala. Riv.Ital.Geotechn.12: 78-100.

Dahlberg,R.& Bergdahl,U. 1974, Investigations on the Swedish Ram-Sounding Method. Proc.ESOPT I,vol.2: 93-101, Stockholm.

Delmarcelle,A. 1981, Corrélation entre deux sondes de battage légères: Sonde C.R.R. et sonde Van Vuuren modifiée. La Technique Routière 2: 36-44.

Escario,V. 1981, Personal Communication.

DIN 4094 Teil 2, Baugrund. Ramm- und Drucksondiergeräte. Anwendung und Auswertung (Vornorm). DIN Berlin.

Finish Geotechnical Society 1980, Kairausopas I: Painokairaus, Tärykairaus, Heijarikairaus. Rakentajain Kustannus Oy, Helsinki.

Folque,J. 1974, Penetration Testing in Portugal; State-of-the-Art-Report. Proc. ESOPT I,vol.1: 105-106, Stockholm.

Hausner,H.& Sperling,G. 1974, Experiences Made with the Penetrometer of VEB Baugrund Berlin. Proc.ESOPT I, vol.2: 163-167, Stockholm.

Ireland,H.O.,Moretto,O.&Vargas,M. 1970, The Dynamic Penetrometer Test: A Standard Which Is Not Standardized. Géotechnique 20:185-192.

Jänke,S. 1975, Auswertung von Sondierungen zur Ermittlung der Setzung von Flächengründungen in nichtbindigem Untergrund. Baumaschine &Bautechnik 22: 121-126.

Karafiath,L.L. 1980, Cone Resistance in Sand by Incremental Method. J.Terramechanics 17: 101-113.

Kindermans,J.-M. 1976, La sonde de battage légère en construction routière; étude d'information et expérimentation. Centre de Recherches Routières,CR5, Bruxelles.

Krämer,H.-J. 1977, Vergleichende Untersuchungen von Rammsonden mit verschiedenen Fallgewichten und Spitzenquerschnitten. Baumaschine&Bautechnik 24: 237-253.

Krämer,H.-J. 1980, Geräteseitige Einflußparameter bei Ramm- und Drucksondierungen und ihre Auswirkungen auf den Eindringwiderstand. Diss.Univ.Karlsruhe.

Krämer,H.-J. 1981, Die Auswirkungen geräteseitiger Einflußparameter auf den Sondierwiderstand. Baum.&Bautech.28: 102-116.

Lareal,P.,Sanglerat,G.& Gielly,J. 1974, Comparison of Penetration Test Data Obtained by Different Static or Dynamic Penetrometers. Proc.ESOPT I,vol.2.2: 229-236, Stockholm.

Malyshev,M.V.&Lavisin,A.A. 1974, Certain Results Obtained in Cone Penetration of a Sand Base.Proc.ESOPT I,2.2:237-239.

Martin,D. 1974, Tower Foundation Design by Means of the"Penevane".Proc.ESOPT I, 2.2: 253-257, Stockholm.

McNeil,R.L. 1972, State-of-the-Art Paper. Proc.Conf.Rapid Penetration of Terrestrial Materials, J.Colp,ed.,Texas A.&M.University, College Station: 11-125.

Meardi,G.&Meardi,P. 1974, Employment of the Dynamical Conical Penetrometer in Italy with External Protecting Casing. Proc. ESOPT I, 2.2: 259-261, Stockholm.

Melzer,K.-J. 1968, Sondenuntersuchungen in Sand. Mitt.VGB 43, Aachen.

Mitchell,J.K.& Lunne,T.A. 1978, Cone Resistance as Measure of Sand Strength.Proc. ASCE 104, GT7: 995-1012.

Nowatzki,E.A. 1971, A Theoretical Assessment of the SPT. Proc.4.Panameric.CSMFE San Juan, Puerto Rico,vol.II: 45-61.

Nowatzki,E.A.& Karafiath,L.L. 1972, Effect of cone angle on penetration resistance. Highway Research Record no.405: 51-59.

Pfister,P. 1974, Sermes Dynamic Penetration Test: Its Use in Sandy Soils. Proc. ESOPT I, vol.2.2: 301-306,Stockholm.

Puech,A. et al. 1974, Contribution to the Study of Static and Dynamic Penetrometers.Proc.ESOPT I,2.2: 307-312,Stockholm.

Rohani,B.& Baladi,G.Y. 1981, Correlation of Mobility Cone Index with Fundamental Engineering Properties of Soil. Final Report, Structures Lab.,US Army Eng.Waterways Exper.Stat., Miscellan.Pap. SL-81-4.

Rollberg,D. 1976, Bestimmung des Verhaltens von Pfählen aus Sondier- und Rammergebnissen. Forschungsberichte aus Bodenmech.u.Grundbau FGB 4, Aachen.

Rollberg,D. 1978, Diskussionsbeitrag zu Smoltczyk/van Koten/Hilmer. Baumaschine und Bautechnik 25: 388-389.

Roy,M. et al. 1974, The Interpretation of Static Cone Penetration Tests in Sensitive Clays. Proc.ESOPT I, vol.2.2:323-330.

Sanglerat,G. 1974, Penetration Testing in France; State-of-the-Art-Report. Proc. ESOPT I, vol.1: 47-58,Stockholm.

Schormann,K. 1979, Baugrunduntersuchungen des rheinischen Schluffs durch leichte Rammsondierungen. Forschungsberichte Nordrhein-Westfalen Nr.2825, Westdeutscher Verlag, Opladen.

Schultze,E. 1974a, Examples of Evaluating the Results from Sounding Tests. Proc. ESOPT I,vol.2.2: 353-359, Stockholm.

Schultze,E. 1974b, Interpretations of Dynamic Penetration Tests (Group Discussion).Proc.ESOPT I,2.2: 73-79.

Schultze,E. 1974c, Bericht über "Conference on the Settlements of Structures" (Cambridge 1974). Vorträge Baugrundtagung Frankfurt: 215-248, DGEG Essen.

Schultze,E. 1975, Some Aspects Concerning the Application of Statistics and Probability to Foundation Structures. Proc.2.IC Applications Statistics and Probability in Soil & Struct.Engng., vol.II: 457-494, Aachen.

Schultze,E. 1976, Die praktische Anwendbarkeit verschiedener Sondenuntersuchungen. Report at Techn.Univ.Graz (unpublished).

Schultze,E.& Biedermann,B. 1977, Pressuremeter, Penetrometer and Oedometer Tests. Proc.9.ICSMFE, vol.3: 501-503, Tokyo.

Sherif,G. 1973, Setzungsmessungen an Industrie-und Hochbauten und ihre Auswertung. Mitt.VGB 57, Aachen.

Sherwood,D.E.& Child,G.H. 1974, A Static-Dynamic Sounding Technique. Proc.ESOPT I, vol.2.2: 361-366, Stockholm.

Sievering,W. 1980, Die Zuverlässigkeit von Setzungsberechnungen. Forschungsberrichte aus Bodenmech.& Grundbau FGB 6,Aachen.

Smoltczyk,U./van Koten,H./Hilmer,K. 1978, Dynamische Untersuchung von Pfählen. Baumaschine & Bautechnik 25: 65-71.

Stenzel,G.& Melzer,K.-J. 1978, Bodenuntersuchungen durch Sondierungen nach DIN 4094. Tiefbau 28: 155-160; 240-244.

Tammirinne,M. 1974, Factors Effecting the Dynamic Penetration Resistance. Proc. ESOPT I, vol.2.2: 383-391,Stockholm.

Teferra,A. 1975, Beziehungen zwischen Reibungswinkel, Lagerungsdichte und Sondierwiderständen nichtbindiger Böden mit verschiedener Kornverteilung. Forschungsberichte aus Bodenmech.& Grundbau FGB 1, Aachen.

Teferra,A. 1976a, Bestimmung der Lagerungsdichte aus Sondierungen. Der Bauingenieur 51: 329-331.

Teferra,A. 1976b, Beitrag zur mittelbaren Bestimmung des Steifemoduls aus Sondierungen in nichtbindigen Böden. Die Bautechnik 53: 306-311.

Thigpen,L. 1974, Projectile Penetration of Ealstic-Plastic Earth Media. J.Geotechn. Engng.Div.,Proc.ASCE 100: 279-293.

Uriel,A.O. et al. 1974, Test Results Concerning the Influence of the Cone Angle in the Dynamic Penetration Resistance. Proc.ESOPT I,2.2: 401-406,Stockholm.

Vesic,A.S. 1972, Expansion of Cavities in Infinite Soil Mass. J.SMF Div.,Proc. ASCE 98:265-290.

Wennerstrand,J. 1979, Comparison of Predicted Settlements for a Fine Sand. Proc. 7.ECSMFE,vol.2: 295-298, Brighton.

Wolski,W. 1974, Penetration Testing in Poland; State-of-the-Art-Report. Proc. ESOPT I, vol.1:97-104,Stockholm.

Yankelevski,D.Z.& Adin,M.A. 1980, A Simplified Analytical Method for Soil Penetration Analysis. Int.J.Num.Analyt. Methods Geomech. 4: 233-254.

Yong,R.N.& Chen,Ch.-K. 1976, Cone Penetration of Granular and Cohesive Soils. J.Engng.Mech.Div.,Proc.ASCE 102:345-363.

Zweck,H.& Bergdahl,U. 1974, Summary of Discussion Group 2: Standardization and Future Development. Proc.ESOPT I,2.1:53-55.

Proceedings of the Second European Symposium on Penetration Testing / Amsterdam / 24-27 May 1982

The weight sounding test (WST)
State-of-the-art report

B.B.BROMS
Royal Institute of Technology, Stockholm, Sweden

U.BERGDAHL
Swedish Geotechnical Institute, Linköping

1 INTRODUCTION

The most common penetration testing method in the Scandinavian countries and in Finland is by far the weight sounding method (Aas, 1969, Broms, 1974). In, for example, Denmark about 90 to 95% of all penetration tests are weight soundings (Hansen, 1967). It is estimated that roughly around twenty thousand weight soundings are carried out each year.

The weight penetrometer consists of a screw shaped point, rods with 22 mm diameter and a number of weights (5, 10, 10, 25, 25 and 25 kg) as illustrated in Fig. 1. The point is manufactured of a square steel bar which is twisted one turn to the left.

The load on the penetrometer is gradually increased to 0.05, 0.15, 0.25, 0.50, 0.75 nad 1.0 kN by increasing the number of weights. The load on the penetrometer is adjusted to keep the penetration rate constant, about 20 mm/s (1.2 m/min). When the penetrometer does not penetrate any further when loaded to 1.0 kN it is roatated and the number of halfturns every 0.2 m of penetration is recorded (halfturns/0.2 m) Fig. 2.

When the weight penetrometer was first developed it was rotated by hand. Today the weight penetrometer is usually rotated by machine to reduce the time for a test.

A gasoline driven engine type Borros, Fig. 3, is often used for the rotation of the rods and the number of halfturns every 0.2 m (N_{wst}) is registered automatically by a counter. The penetrometer is usually pushed down by the two operators and the applied load is read on a separate force indicator (dynamometer). Today it has become more common to use a crawler-mounted boring rig also for weight sounding tests e.g. the one manufactured by Geotech, Fig.4.

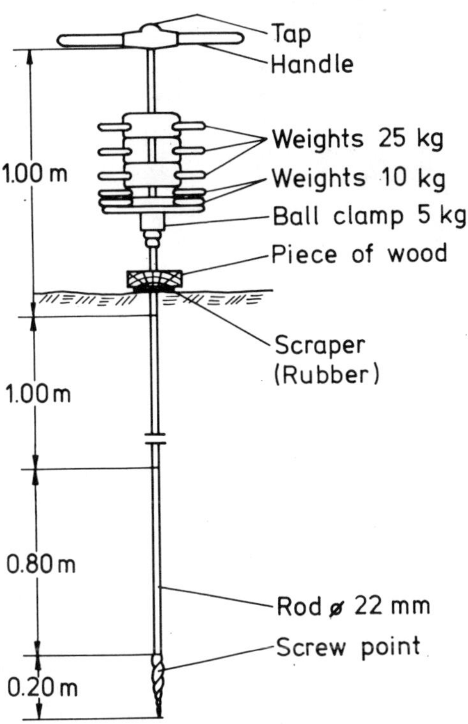

Fig. 1 Details of the manually operated weight penetrometer.

If an engine is used it becomes much more difficult to determine the penetrated soil type (clay, silt, sand or gravel) from the sounds and the vibrations of the penetrometer rods during the rotation. The vibrations from the engine can also affect the results. Investigations in Sweden (Bergdahl, 1969) and in Norway (Senneset, 1974) indicated, however, that the effects normally are relatively small.

The weight sounding method was standardized in Sweden in 1964 and 1970, in Finland in 1968 and in Norway in 1973. The proposed European standard, approved at the meeting of the Executive Committee of the ISSMFE in Tokyo in 1977, corresponds closely to the different standards in the Scandinavian countries.

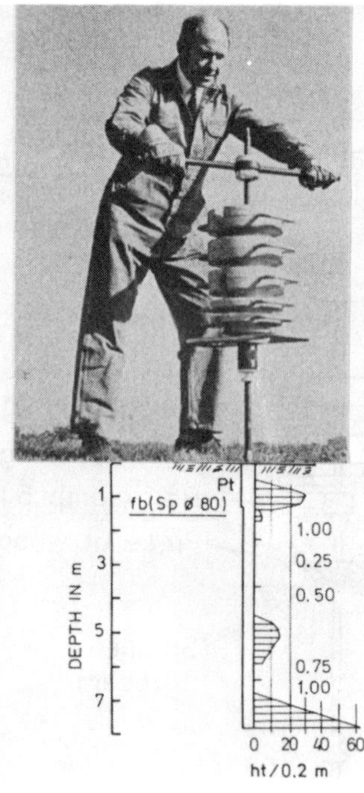

Fig. 2 Manually operated weight penetrometer with an example of test results.

2 DEVELOPMENT OF THE WEIGHT SOUNDING METHOD

The weight sounding test is an old method. Already at the end of the 19th century it was common in Sweden to push or hammer down square or circular steel rods with 15 to 30 mm diameter to determine the depth of the soft alluvium above underlying till or bedrock.

The first modern weight sounding device was described by O. Olsson in 1915 in a manual published by the Swedish State Railway. One metre long steel rods with 15 mm diameter were used which were provided with a twisted square steel point. The penetration resistance was expressed in the number of men required to push down the penetrometer. If it was not possible to push down the penetrometer by hand a 90 kg weight was added and the penetrometer was rotated. The penextration every 25 halfturns was recorded.

Fig. 3 The Borro power unit in use for rotating the weight penetrometer.

Fig. 4 The Geotech crawler is often used for weight sounding tests.

This early weight penetrometer was modified by the Swedish Geotechnical Commission (1917, 1922). The force required to push down the penetrometer was measured using a series of weights. The soil was classified with respect to the force required to push the penetrometer. A soil could for example be classified as a "50 kg clay" when a force of 0.5 kN (50 kg) was required. When the penetrometer did not go down any further when loaded to 1 kN (100 kg) it was rotated and the penetration every 25 halfturns was measured.

The weight sounding method was used extensively by the Geotechnical Commission, 1922 to determine the layer sequence and the thickness of the different strata during investigations of the stability of railway embankments. Several large landslides had occured at that time in Sweden e.g. at Lake Aspen (1913) and at Getå (1918) where 41 persons lost their lives.

Since the Recommendations of the Geotechnical Commission the development had been concentrated to an increasing penetrability of the weight penetrometer by increasing the strength of the different parts. The penetrability has also been increased by the mechanization of the turning of the penetrometer and to combine the weight sounding with percussion drilling in order to penetrate dense strata.

There are also extensive investigations on the interpretation of the test results for different application. The investigations also pointed out the necessity of standardization of the penetrometer and the test procedure.

3 FACTORS AFFECTING THE PENETRATION RESISTANCE

The need of standardization of different sounding methods was discussed intensively in Sweden in the 1950th. The Swedish Geotechnical Society (SGF) therefore appointed in 1958 a committee on penetration testing for standardization and mechanization of existing penetration testing methods.

The committee carried out a large number of tests to study the influence of different factors on the penetration resistance. Such factors as the wear of the penetrometer point, straightness of the rods, overburden pressure, method of rotation and friction along the rods were investigated.

Senneset (1973) investigated for example the effect of the shape and the wear of the penetrometer point in different soils. (The penetrometer point is manufactured of a square steel bar which is rotated one turn to the left. Sometimes the twist was less than one turn). Senneset (1973) found that in soft clay the twist (½, 3/4 or 1 turn) did not significantly affect the results.

The penetration resistance was found by Senneset (1973) to be reduced in clay for a worn point compared with a new point. In a silt the difference was small. In dense sand the penetration resistance was higher for the worn point than for a new point when the depth was large. Close to the surface, however, the penetration resistance was lower for the worn point.

Similar results have been reported by Bergdahl (1969). He found that the penetration resistance in a sand was somewhat lower close to the ground surface for a worn point than for a new point. Below 7 m the penetration resistance was about 25% higher for a worn point compared to a new point. In a silt the resistance was slightly reduced when a worn point was used.

A change of the penetration resistance was observed when the penetrometer was rotated by hand or by machine. An increase of 60 to 70% has e.g. been reported by Senneset (1973) for a clay when the weight penetrometer was rotated by a petrol driven engine compared to the resistance (halfturns/0.2 m) when the penetrometer was rotated by hand. Bergdahl (1969) found that for a soft clay from Sweden the increase was 30%. Natukka (1969) has reported for a clay in Finland an increase of the penetration resistance of 14%. For a silt the penetration resistance was reduced by 7 to 17% when the penetrometer was rotated by machine.

Senneset (1973) has also investigated the effect of the straightness of the rods on the measured penetration resistance. The effect was small except in a clay close to the ground surface. The penetration resistance was larger for the bent rods than for the straight rods. For large depths the penetration resistance was slightly smaller for the bent rods than for the straight rods.

Also the speed of rotation of the rods can affect the results. For a soft clay Bergdahl (1969) found that the penetration resistance increased 22% when the speed of rotation was 125 rpm compared to a manually rotated penetrometer. When the speed was 50 rpm the increase was 16%. For a fine sand the difference was smaller. For sand and sandy gravel the increase was 10%. The speed of rotation should therefore be between 15 and 40 rpm as stated in the recommended standard. The recommended rate is 30 rpm and the rate should not exceed 50 rpm.

The effect of the diameter of the sounding rods on the penetration resistance has been investigated by Bergdahl (1969). In clay the effect was found to be small while in a sand the penetration resistance was 7 to 8% larger for a 22 mm rod than for a 19 mm rod, due to the difference in surface area. In silt the difference was small.

4 RELATIVE DENSITY

Weight soundings are used in Sweden and Finland to get an indication of the relative density of cohesionless soils and the degree of compaction (Hellman et al, 1979). The increase of the penetration resistance with depth close to the surface is then utilized above the critical depth. Below the critical depth the penetration resistance can be used. However, the penetration resistance is besides the relative density also affected by the particle size and the gradation of the soil. It is therefore necessary to calibrate each material separately if the relative density should be determined. However, the friction along the uncased rods can affect the results and the relative density can be overestimated. According to the Swedish Building Code (SBN, 1980) sand is classified as dense when the penetration resistance exceeds 15 halfturns/0.2 m and as medium to loose between 1 and 15 halfturns/0.2 m penetration.

A somewhat different classification is used by the Swedish Road Administration (1976) where sand is classified as dense when N_{wst} > 30 halfturns/0.2 m penetration.

In order to compare the results from weight soundings with other methods used internationally Bergdahl and Sundqvist (1974) and Helenelund (1966) have proposed the following classification.

Classification	Sweden (Bergdahl and Sundqvist 1974) Halfturns/0.2 m	Finland (Helenelund 1966) Halfturns/0.2 m
Very loose	<8	<10
Loose	8-20	10-30
Medium	20-60	30-60
Dense	60-100	60-100
Very dense	>100	>100

The penetration resistance from weight soundings is classified in Norway as very low, low, medium, dense and very dense when the number of halfturns/0.2 m penetration is 0, <7, 7-25, 25-50 and >50, respectively (Senneset, 1974). These values are somewhat lower than those proposed by Bergdahl and Sundqvist (1974) and by Helenelund (1966).

Test data indicate that the penetration resistance of cohesionsless soils is not affected when the material is either dry or saturated. Bergdahl (1973) has for example reported from laboratory tests that the penetration resistance was about the same at the same relative density when the material was either dry or saturated. The penetration resistance is, however, affected by the degree of saturation due to the false cohesion as found by Bergdahl (1973) and the relative density can be overestimated.

The size and gradation of the soil also affect the results as pointed out by Bergdahl (1973). An investigation in Finland (Gardemeister et al, 1974) has also indicated that the particle size affects the penetration resistance. At a given dry unit weight the penetration resistance was found to increase with decreasing particle size. Also the water content affected the results as well as the location of the ground water level (Tammirinne, 1973).

An interesting series of weight sounding tests have been carried out by Dahlberg (1974). He investigated the effect of the overburden pressure on the penetration resistance of a fine sand. These tests were made in an excavation where the depth was increased gradually and the overburden pressure was reduced. The investigation indicated that the penetration resistance was affected down to a depth of 0.9 to 2.6 m below the bottom of the excavation where the effective overburden pressure was 25 to 30 kPa. Below this depth the penetration resistance did not change. These tests indicated that the penetration resistance at weight soundings was very much governed by the lateral pressure in the ground which is uneffected by a reduction of the overburden pressure except close to the bottom in a small excavated area.

5 COMPRESSIBILITY

The weight sounding method can also be used to estimate the settlement of spread footings and rafts. The compression modulus (M) is according to Helenelund (1966) approximately equal to 10 to 20 MPa at 10 to 30 halfturns/0.2 m, 20 to 50 MPa at 30 to 60 halfturns/0.2 m and 50 to 80 MPa at 60 to 100 halfturns/0.2 m.

6 SHEAR STRENGTH AND SENSITIVITY

It is normally possible to separate with weight sounding the stiff surface crust of a deep clay layer from the underlying soft clay. The penetrometer must normally be rotated in the stiff surface crust while in the underlying soft clay the penetrometer often can be pushed down with a force less than 1 kN.

The friction along the penetrometer rods is normally low in the Scandinacian countries due to the often high sensitivity of the soil. The sensitivity is normally 10 to 20 while in quick clay it is higher than 50.

In areas where the sensitivity of the clay is low and where sand and gravel layers occur close to the surface the skin friction resistance along the shaft of the penetrometer rods can be large and influence the results.

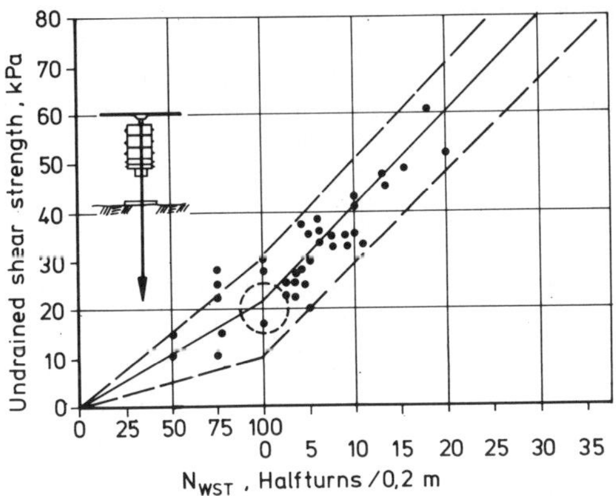

Fig. 5 Relationship between undrained strength and weight sounding resistance N_{wst} (Fukuoka 1974).

Several attempts have been made to correlate for clays the results from weight soundings with the undrained shear strength. However, the proposed relationships are very uncertain. Fukuoka (1974) has correlated the penetration resistance with the unconfined compression strength (Fig. 5). It can be seen that the penetration resistance increased with increasing shear strength. In the case the penetration resistance is less than 1 kN the undrained shear strength of the clay will not exceed 20 to 30 kPa. Similar results have been reported by Saarelainen (1979) for soft clays from Helsinki in Finland. When the penetration resistance was 0.5 kN the average undrained shear strength was about 13 kPa. The scatter of the results was large particularly close to the ground surface.

Möller (1980) has utilized the increase of the penetration resistance with depth as an indication of the sensitivity of clays. In the case the penetration resistance is about constant the sensitivity of the clay is high because the shaft resistance is small. When the sensitivity is low the shaft resistance will be high and the penetration resistance will increase with depth (Fig.6).

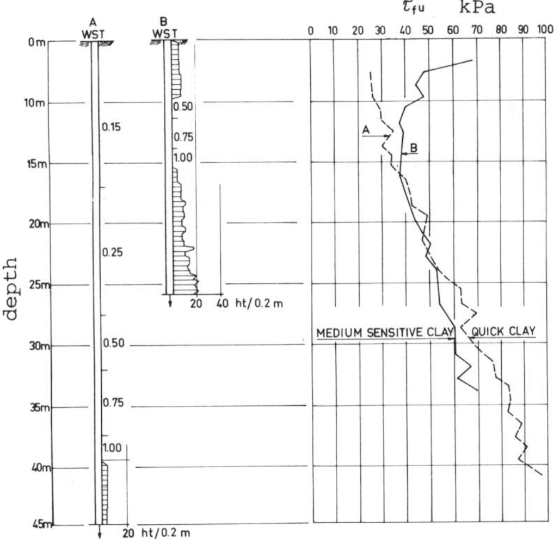

Fig. 6 Due to the effect of the sensitivity on the penetration resistance it can be impossible to estimate the shear strength of clays from weight sounding tests.

7 COMPARISONS WITH OTHER PENETRATION TESTING METHODS

Comparisons with the weight sounding method and the standard penetration test (SPT) have been carried out in Japan where the weight sounding method is frequently used. The following relationships have been reported for cohesionless soils (Muromachi et al, 1974):

$$N_{SPT} = 1.07\, N_{wst}^{0.755} \qquad \text{(Ueda)}$$

$$N_{spt} = 0.42\ N_{wst} \quad \text{(Miki)}$$

$$N_{SPT} = 0.34\ N_{wst} + 2 \quad \text{(Inada)}$$

where N_{SPT} is the standard penetration resistance (blows/ft) and N_{wst} is the penetration resistance (halfturns/0.2 m) from the weight soundings. The difference between the different expressions is relatively small at $N_{wst} > 10$.

In Fig. 7 is shown a comparison between the results from weight soundings and standard penetration tests carried out in a preloaded deposit of medium to coarse sand. The relationship $N_{SPT} = 0.37\ N_{wst} + 1.70$ corresponds closely to that proposed by Inada mentioned above.

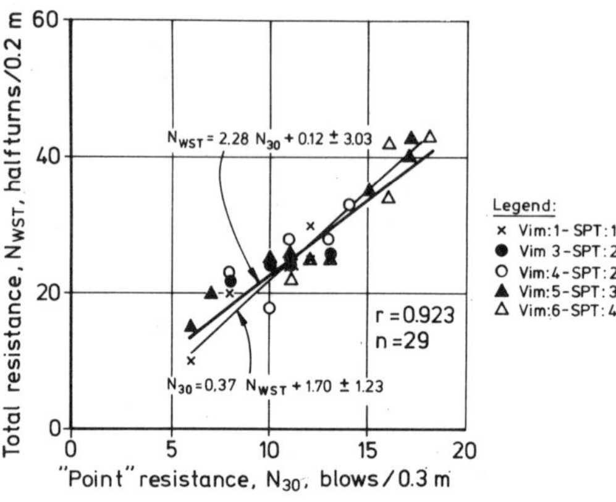

Fig. 7 Relationship between weight sounding resistance N_{wst} and SPT-values (Dahlberg, 1975).

Tammirinne (1974) has compared the penetration resistances of weight soundings and cone penetration tests (CPT) in coarse and fine sand using the Dutch cone penetrometer. For coarse sand the following relationship was obtained

$$q_c\ \text{(MPa)} = 0.5\ N_{wst}\ \text{(halfturns/0.2 m)}$$

For fine sand

$$q_c\ \text{(MPa)} = 0.2\ N_{wst}\ \text{(halfturns/0.2 m)}$$

where q_c is the cone penetration resistance.

Similar results have been reported by Bergdahl (1973). The proposed relationships correspond closely to those used in Sweden for the classification of cohesionsless soils.

8 SOIL EXPLORATION

Weight soundings can be used in almost all soils. Its main application is in soft clay and silt and in loose to medium dense sand. It can be used in inaccessable areas where it is difficult to bring in the relatively heavy units required for e.g. cone or standard penetration test or for ram soundings.

Weight soundings are used in Sweden primarily in the exploratory phase of an investigation to determine the stratification, location and thickness of the different strata, the depth to "firm bottom", the depth where soil samples should be taken and the most suitable location of e.g. field vane tests and other penetration tests, e.g. cone penetration tests or ram soundings, Fig. 8. Weight soundings can also be used to estimate the length and the bearing capacity of point bearing and friction piles as well as the bearing capacity and settlement of spread footings and rafts on cohesionless soils (sand and gravel). Weight soundings have been used as well to check the relative density of compacted fills and the digability of clay, silt, sand and gravel (Korhonen and Gardemeister, 1972).

The vibrations and noise in the sounding rods during a penetration test can also be useful to identify the soil type (clay, silt sand or gravel). A scraping noise is an indication of sand. In gravel or in stony soils a screeching noise can often be heard. However, this advantage cannot be used when machines are used for the rotation of the rod. When the penetrometer point strikes a stone or a boulder the point may catch a corner which gives the penetrometer a jerk.

The accuracy of the method depends to a large extent on the experience of the driller. He can normally determine the soil conditions (clay, silt, sand, gravel or till) from the sounds and vibrations in the sounding rods during the rotation of the rods. If a fill or a sand or gravel layer is located close to the ground surface the interpretation of the results becomes much more difficult because of the friction along the rods which affects the penetration resistance. Predrilling or precoring will than be required.

A predrilling or precoring (50 to 100 mm) should be used when e.g. a hard layer is located close to the surface since the friction in this layer may affect the results. Casing is sometimes used when the depth exceeds 20 to 25 m (Natukka, 1969), to reduce the friction along the rods or

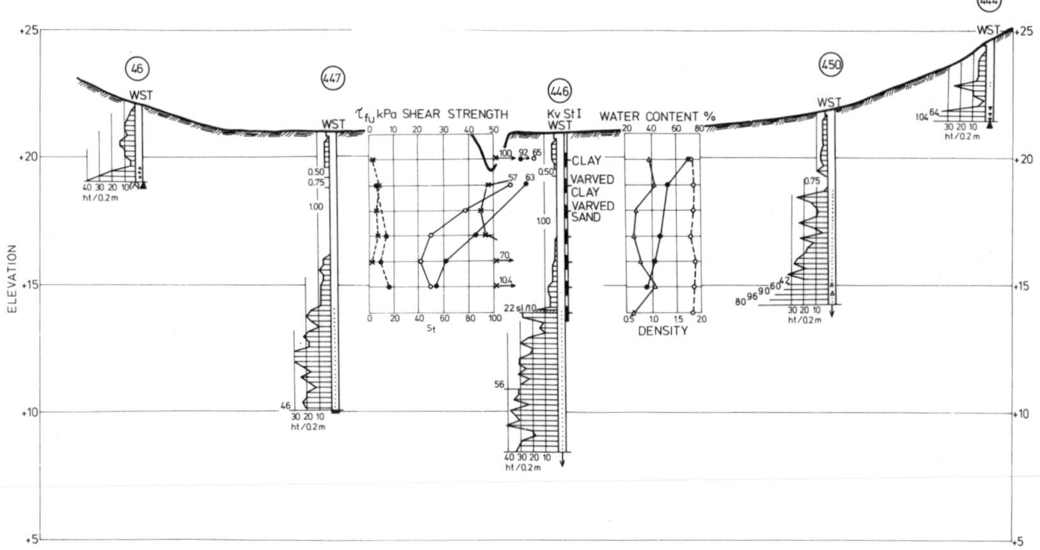

Fig. 8 The different soil layers in a clay filled valley can be indicated by a number of weight sounding tests. The investigation is often completed by field vane tests, piston sampling and laboratory tests.

when the density in cohesionless soils at greater depths (>10 m) should be estimated.

A weight sounding test is normally terminated when a certain predetermined penetration resistance or depth is reached. It is frequently prescribed that a weight sounding test should be interrupted when the penetrometer does not go down noticable when rotated or after 10 blows by a 3 kg sledge hammer. As an alternative two 25 kg weights can be used which are dropped from a height of 0.20 m. When neither of these criteria can be satisfied the test is interrupted when the penetration resistance increases with depth and the penetration resistance for two consecutive 0.2 m intervals exceeds 40 halfturns/0.2 m and the penetration is less than 10 mm/blow at 5 blows by a sledge hammer or by the two 25 kilogram weights. If the penetration exceeds 50 mm/5 blows or the penetration resistance decreases with depth the test should be continued. The refusal may be only temporary and caused by a stone.

The maximum depth of all weight soundings should be recorded even if the penetration test has been interrupted close to the ground surface. If the depth of refusal varies then there are probably stones or boulders in the soil which may affect for example the installation of piles or sheet piles.

Large variations in penetration resistance have sometimes been observed as pointed out by Mortensen (1973). These variations have in most cases been caused by errors in the indicated location of the soundings in a plan view or in the indicated level of the ground surface. These errors are of course possible to all soil investigations. In some cases the deviations have been caused by excessive wear of the penetrometer point. The difference in penetration resistance can be large in particularly cohesionless soils (sand and gravel) and the relative density can be overestimated if a worn point is used. In extreme cases it may be necessary to replace the point after 5 to 25 soundings in dense or cemented soils.

The weight sounding is a flexible and in expensive investigation method. Because of the relatively low cost it is possible to carry out a large number of sounding for every boring. The costs today (1981) in Sweden is about 50$/sounding for a 10 m deep hole in soft clay or loose to medium dense sand if a tractor is used.

Soil investigations and weight soundings are in many cases carried out at a fixed price for each borehole or per metre of penetration regardsless of the soil conditions. The disadvantage with this method is that it is often difficult to make any changes in the boring program to fit the actual conditions in the field. The advantage with this system is that the owner knows in advance and can control the total cost of a certain investigation.

Shallow Foundations

The design of spread footings for structures is in Sweden primarily based on the results of weight soundings. The weight soundings are normally supplemented by screw borings at which disturbed but representative samples are obtained so that the soil in the different strata can be classified. However, there are examples where shallow foundations have been designed on the basis of soundings alone. In one case a building was founded just above a peat layer with excessive settlements as a result (Mortensen, 1973). It was not possible to detect the peat layers below the foundation level from the weight soundings alone due to the high friction resistance along the penetrometer rods.

Different rules of thumb based on local experience exist. For example in Sweden bridge abutments can be founded on spread footings on sand when the penetration resistance (N_{wst}) exceeds 10 halfturns/0.2 m. In Denmark it is estimated that a two storey building can be founded on spread footings in boulder clay when the resistance is 10 to 15 halfturns/0.2 m.

Weight soundings are normally used in Sweden to estimate the allowable bearing capacity of spread footings on gravel, sand or silt. According to the Swedish Building Code (SBN 1980) sand and silt are classified as dense when the penetration resistance exceeds 15 halfturns/0.2 m and as medium to loose when the penetration resistance is between 1 and 15 halfturns/0.2 m penetration as mentioned previously. Spread footings and rafts are not used when the penetration resistance is less than 1 halfturn/0.2 m penetration, because of the large expected settlements. The Swedish Road Administration (1976) classifies sand and gravel as dense when the penetration resistance N_{wst} exceeds 30 halfturns/0.2 m and as medium dense between 10 and 30 halfturns/0.2 m. The allowable bearing capacity according to the Swedish Building Code is calculated using a special bearing capacity formula according to which the soil pressure increases linearly with increasing size of the footing. In order to keep the settlements within acceptable limits the allowable soil pressure is limited. Thus the maximum allowable soil pressure is 0.5 and 0.3 MPa for medium to coarse sand when the soil is dense and loose, respectively. The corresponding maximum soil pressures for a fine sand is 0.4 and 0.2 MPa, respectively. In gravel the maximum soil pressure is 0.6 MPa. According to the Swedish Road Administration the allowable soil pressure is calculated in a similar way where the factor of safety F_ρ = 1.5. The angle of internal friction is determined from the following table:

Soil type	Penetration resistance halfturns/0.2 m 10-30	>30
Fine sand	31°	35°
Medium to coarse sand	35°	38°
Gravel	38°	42°

The allowable soil pressure according to these recommendations must not be used without a settlement calculation.

Deep Foundations

The results from weight soundings are also used to predict the driving resistance of piles as illustrated in Fig. 9 (Helenelund, 1974). It can be seen that the driving resistance of precast concrete piles (0.25 x 0.25 m) driven by a 3 to 4 tons drop hammer increased in general with increasing penetration resistance. However, the scatter of the results is large particularly for glacial till.

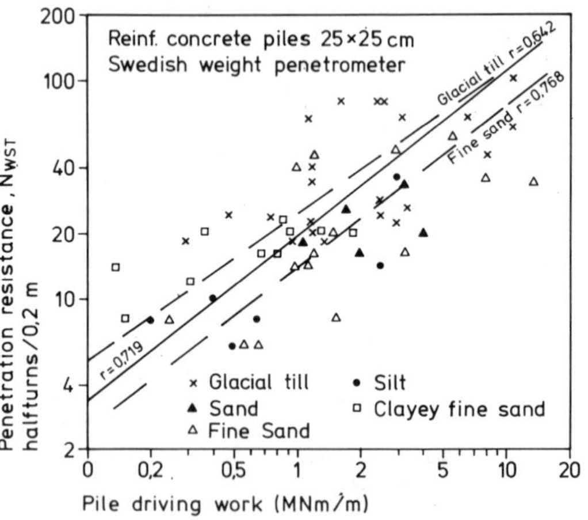

Fig. 9 Driving resistance of precast concrete piles in relation to the weight sounding resistance in different soils (Helenelund 1974).

The driving resistance of fine sand and silt is often high probably due to the high negative pore water pressures that develop during the driving when the relative density is high. In glacial till stones and boulders in the soil have often a large influence on the penetration resistance as determined by weight soundings as well as on the driving resistance of timber and concrete piles and of sheet piles. It is therefore not surprising that the scatter of the test results is large.

Experience in Finland (Eklund, 1970) indicates that the length of precast concrete piles corresponds approximately to the maximum depth that can be reached with weight soundings. In Sweden it is often assumed that the length of end-bearing piles will correspond to the penetrated depth plus 1 or 2 m. However, dynamic probing is a more accurate way of determining the length of end-bearing piles. In Denmark (Winkel, 1969) it has been found that point bearing piles in boulder clay can be driven 0.5 to 1.0 m deeper than the maximum depth of a weight sounding test at a penetration resistance of 60 to 80 halfturns/0.2 m (N_{wst}).

Results from weight soundings can also be used to estimate the bearing capacity of friction concrete and timber piles in sand. In Fig.10 is shown the results from a number of pile load tests in sand (Norwegian Pile Committee, 1973). The ultimate bearing capacity of the piles divided with the total surface area of the piles (Q_{ult}/A_f) neglecting the point resistance has been plotted as a function of the penetration resistance from weight soundings. It can be seen that the bearing capacity of timber piles is normally higher than that of concrete piles at the same penetration resistance due to the taper of the timber piles. The ultimate bearing capacity per unit shaft area for timber piles increases in general with increasing penetration resistance of the soil from about 20 kPa when the penetration resistance is low ($N_{wst} < 4$ to 60 to 80 kPa when the relative density is high ($N_{wst} > 24$). The results are applicable to piles with a length of 12 to 20 m.

According to the Recommendations on design of bored piles published by the Swedish Commission on Pile Research the angle of internal friction in cohesionless soils can be determined within the range 15-50 halfturns/0.2 m. The evaluation for silt, sand and gravel is summarized in the following table.

Soil Type	Penetration resistance halfturns/0.2 m			
	15	20	30	50
Silt	-	-	-	27°
Sand	31	32	34	38
Gravel	32	34	37	40

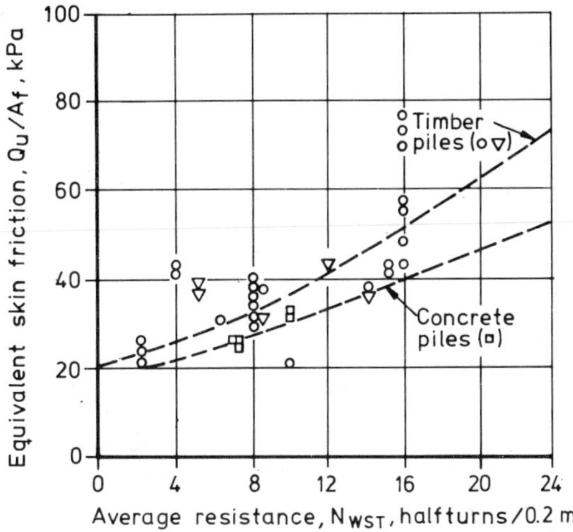

Fig. 10 Relationship between the equivalent skin friction resistance and the average penetration resistance from weight soundings. (Norwegian Pile Committee, 1973.)

Compaction Control

Weight soundings are frequently used in Sweden to check the increase in relative density of fine and medium sand below the ground water level where other methods cannot be used or are difficult to apply. Lagging and Eresund (1974) have for example described a case where weight soundings were used to check the uniformity of a hydraulically placed fill. The thickness of the fill varied between 8.3 and 9.0 m. The surface was compacted by a 3 tons vibratory roller.

The penetration resistance of different soils can vary considerably at the same relative density due to a variation in grain size or gradation of the material as reported for example by Bergdahl and Tammirinne (1973). Therefore the weight penetrometer must be calibrated for each particular soil before it can be used. The different control methods have been compared by e.g. Hellman et al (1979).

Weight soundings have also been used to check the compaction caused by pile driving using short tapered timber or concrete piles (compaction piles). This method has been used in Sweden to compact sand and gravel for e.g. bridge abutments as reported by Sandegren (1974). The length of the piles was 6 to 8 m. The compaction when the spacing of the 235 x 235 mm precast concrete piles was 1,5 was small. The compaction increased considerably when the number of piles was doubled.

SUMMARY AND CONCLUSIONS

The most common penetration testing method in Sweden, Finland, Norway and Denmark is by far the weight sounding test (WST). It is estimated that more than 20.000 weight soundings are carried out anually (1981). Weight sounding methods can be used in almost all soil types. It is an inexpensive method which can be used in inaccessible areas which are difficult to reach with other types of penetrometers. The main application is in soft to medium stiff clay and silt and in loose to medium dense sand. The main limitation of the method is that layers of dense sand or gravel or layers of glacial till are difficult to penetrate. Predrilling may then be required. Weight soundings are primarily used during the exploratory phase of a soil investigation to determine the depth and thickness of the different strata. The method has also been used to estimate the length and bearing capacity of friction and point bearing piles in cohesionless soils as well as the allowable bearing capacity and settlements of spread footings and rafts. Weight soundings can also be used to check the compaction of fills and the diggability of soil

REFERENCES

Aas, G., 1969. Sonderingsmetoder i Norge (Penetration testing methods in Norway). Nordic Meeting on Penetration Tests in Stockholm Oct 5-6, 1967. Swedish Geotechnical Institute, Report No.31, pp 19-22.

Andresen, A. and Rygg, N., 1975a. Rotary Pressure Sounding, Proc, European Symp. on Penetration Testing, Vol.2:2, Stockholm. pp 15-18.

Andresen, A. and Rygg, N., 1975b. Borrigg og metoder for vegtrace undersökelser (Boring Equipment and Methods for Road Investigations) Nordisk Geoteknikermöde i Köpenhavn, pp 357-366.

Bergdahl, U., 1969. Resultat av försök med viktsond. (Results from weight soundings). Scandinavian meeting on Penetration Tests, Stockholm Oct. 5-6, 1967. Swedish Geotechnical Institute, Report No.21, pp 51-59.

Bergdahl, U., 1973. Sondering i friktionsmaterial - Resultat av laboratorieförsök (Penetration tests in Cohesionless soils - Results from Laboratory Tests). Nordiskt Sonderingsmöte i Otnäs den 5-6 maj 1971. Finnish Geotechnical Society, pp 6-24.

Bergdahl, U., and Sundqvist, O., 1974. Geotechnical Investigations at the Demonstration Site, Borros Equipment. Proc. European Symposium on Penetration Testing, Vol.2, pp 193-198.

Broms. B., 1974. General Report, Scandinavia, Proc. European Symposium on Penetration Testing, ESOPT, Vol.2.1, pp 14-23.

Dahlberg, G., 1913. Skred och sättningar eller s k rad vid järnvägsbyggnader och liknande arbeten.

Dahlberg, R., 1973. Bestämning av friktionsjordars kompressionsegenskaper med hjälp av sondering. (Determination of the Deformation Properties of Cohesionless Soils with Penetration Tests). Nordiskt sonderingsmöte i Otnäs den 5-6 maj 1971, pp 25-48. Finnish Geotechnical Society.

Dahlberg, R., 1974a. Penetration Testing in Sweden, Proc. European Sympisium on Penetration Testing (ESOPT), Vol.1, pp 115-131.

Dahlberg, R., 1974b. A Comparison Between the Results from Swedish Penetrometers and Standard Penetration Test. Results in Sand, Proc. European Symposium on Penetration Testing, ESOPT, Vol.2.2, pp 67-68.

Dahlberg, R., 1974c. The effect of overburden pressure on the penetration resistance in a preloaded natural fine sand deposit. Proc. European Symposium on Penetration Testing, ESOPT, Vol.2.2, pp 89-91.

Gardemeister, R., and Tammirinne, M., 1974. Penetration Testing in Finland. Proc. European Symposium on Penetration Testing (ESOPT), Vol.1, pp 35-46.

Hansen, Å., 1969. Sonderingsmetoder i Danmark (Penetration testing methods in Denmark). Nordic Meeting on Penetration Tests in Stockholm Oct. 5-6, 1967. Swedish Geotechnical Institute, Report No.31, p 13.

Hartmark, H., 1969. Norska erfarenhetern av dreiesondering (Experience in Norway with Weight Soundings). Nordic Meeting on Penetration Testing in Stockholm Oct. 5-6

1967. Swedish Geotechnical Institute, Reprints and Preliminary Reports No.31, pp 45-49.

Helenelund, K.V., 1066. Kitkamaalajien Kantavuusominaissukusta Valtion (On the bearing capacity of frictional soils) Teknillinen Tutkimus laitos, Tideotus. Sarja III, Rakennus 97, Helsinki.

Helenelund, K.V. 1974. Prediction of Pile Driving Resistance from Penetration Tests. Proc. European Symposium on Penetration Testing, Vol.2.2, pp 169-175.

Hellman, O., Pramborg, B.O. and Svensson, G. 1979. Kontroll av packad friktionsjord. Kontrollmetoder för bestämning av deformations- och brottbärighetsegenskaper (Control of the Compaction of Cohesionless Soils. Methods to check the deformation properties and bearing capacity). Swedish Council for Building Research, Report R 102:1979, Stockholm, 146 pp.

Korhonen, K-H. and Gardemeister, R., 1972. Ett nytt system för klassificering av schaktbarhet. (A new classification system with respect to diggability). Väg- och Vattenbyggaren nr 3. Also Finnish Geotechnical Society, Nordiskt Sonderingsmöte i Otnäs den 5-6 maj 1971. pp 139-145.

Lagging, L.B. and Eresund, S., 1974. Test Loading at a Hydraulic Sand Fill. Proc. European Symposium on Penetration Testing, Vol, 2.2, pp 221-227.

Mortensen, K., 1973. Der begås mange fejl ved utførelse og anvendelse af drejesonderinger (Many mistakes are made when weight soundings are used). Nordiskt Sonderingsmöte Otnäs den 5-6 maj 1971. pp 131-133. Also Ingeniørens Ugeblad No.23, June 3, 1966. Finnish Geotechnical Society.

Muromachi, T., Oguro, I. and Miyashita, T., 1974. Penetration testing in Japan, Proc. European Symposium on Penetration Testing (ESOPT), Vol.1, pp 193-200.

Möller, B., 1980. Bedömning av lerors sensitivitet ur vikt- och trycksonderingsresultat (Estimate of the sensitivity of clays from weight and ram soundings). Swedish Geotechnical Institute, Report 1-206/79 15 pp.

Natukka, A., 1969. Finska Sonderingskommiténs rekommendationer för viktsonderingsstandard (Recommendations for Weight Soundings by the Finnish Penetration Committee on Testing.) Nordic Meeting on Penetration Testing in Stockholm Oct. 5-6. 1967. Swedish Geotechnical Institute, Reprints and Preliminary Reports No.31, pp 39-43.

Nordiskt sonderingsmöte i Otnäs 5-6 maj, 1971. Föredrag och diskussioner, Finlands Geotekniska Förening, 1973.

Norwegian Geotechnical Society, Sounding Committee, Dreiesondering - Forslag till Standard (Weight Sounding - Proposed Standard), Oslo, 1972.

Norwegian Pile Committee, 1973. Veiledning ved pelefundamentering (Guide for Pile Foundations) Norwegian Geotechnical Institute, Veiledning nr 1, (Guide No.1) 108 pp.

Olsson, H., 1915. Banlära. Järnvägars byggnad och underhåll (Railway Construction Guide, Construction and Maintenance of Railways). Stockholm, Sweden, Vol.1, 512 pp, Vol.2, 421 pp.

Pitkänen, R., 1971. Daalujen tunketuminen eri maalajiolo suhteissa sekä korrelaatio heijariä ja painokairangvastusten kanassa. M. Sc Thesis, Helsinki University of Technology,

Sandegren, E., 1974. Swedish Weight Sounding as an Aid to Control Compaction of Loose Sand by mean of Piles. Proc. European Symposium on Penetration Testing, Vol.2.2, pp 331-336.

Saarelainen, S. 1979. Grovt sättningsestimat på grundval av viktsonderingsmotstånd. (A rough estimate of settlements from weight soundings) Nordiska Geoteknikermötet 1979 NGM -79, Finnish Geotechnical Society, pp 718-727, Esbo, Finland.

SBN, 1980. Svensk Byggnorm (Swedish Building Code) National Board of Physical Planning and Building) Liberförlag, Stockholm.

Senneset, K., 1973. En undersøkelse vedrørende dreieborutstyr (An investigation of weight soundings equipment) Nordiskt sonderingsmöte i Otnäs den 5-6 maj 1971. Finnish Geotechnical Society.

Senneset, K., 1974. Penetration Testing in Norway, Proc. European Symposium on Penetration Testing (ESOPT) Vol.1, pp 85-95.

Svensson, M., 1899. Beteckningssätt för lös mark å undersökningsprofiler (The recording of Soft Soils in Boring Profiles) Tekn, Tidskrift Vol.29, No.3, pp 55-56.

Swedish Committee on Pile Research, 1979. Recommendations on Bored Piles, Design construction and supervision. Report No. 58, Stockholm.

Swedish Geotechnical Commission, 1917. Vägledningar vid jordborrningar för järnvägsändamål (Guide to Soil Borings for Railways). Stat. Järnv. Geotekn. Medd. No.1, 37 pp.

Swedish Geotechnical Commission, 1922. Statens Järnvägars Geotekniska Kommission. 1914-22. Slutbetänkande (Geotechnical Commission on the Swedish State Railways, Final Report). Stat. Järnv. Geotekn. Medd. No.2, Stockholm, 180 pp.

Swedish Geotechnical Commission, 1922. Statens järnvägsrs geotekniska kommission. 1914-22. Slutbetänkande (Geotechnical Commission of the Swedish State Railways,

Final report). Stat. Järnv. Geot. Medd. No.2, Stockholm, 180 pp.

Swedish Road Board, 1976. Bronormer, Statens Vägverk TB 103, 1976-09 Stockholm, 200 pp.

Tammirinne, M., 1972. Karakearakeisten maalajien ja karkeiden silttimaalajien tiiveyden määvittäminen kairausvastuksen perus teella. Valtian taknillinen tut kimuskeskus. Rakennus - ja yhdryskintateknikka) Jalkaisu 2. Helsinki.

Tammirinne, M., 1973. Bestämning av torrvolymvikt i grus, sand och grov silt på grund av sonderingsmotstånd (determination of the Dry Unit Weight of Gravel, sand and coarse silt from the penetration resistance) Nordiskt Sonderingsmöte i Otnäs den 5-6 maj 1971, pp 49-62. Finnish Geotechnical Society.

Tammirinne, M., 1974. Relation between Swedish Weight Sounding and Static Penetration Test, Resistance of Two Sands, Proc. European Symposium on Penetration Testing ESOPT, Vol.2.1, pp 154-156.

Winkel, C.T., 1969. Erfaringer vedrørende sondeboret i Danmark (Experience with soundings in Denmark).Nordic Meeting on Penetration Testing in Stockholm, Oct. 5-6, 1967, Swedish Geotechnical Institute, Reprints and Preliminary Reports No.21, pp 35-37.

Proceedings of the Second European Symposium on Penetration Testing / Amsterdam / 24-27 May 1982

Two decades of dynamic cone penetration testing in India

V.S.AGGARWAL
Central Building Research Institute, Roorkee (UP), India

SUMMARY

The studies on dynamic cone penetration during last two decades in India have been presented in the paper, with particular reference to the test standardisation and prediction of allowable soil pressure in non cohesive soil deposit. The extensive experience with the test has demonstrated that the dynamic probing can be done successfully upto 30m depth besides the method is an economical alternative to other existing methods of soil exploration.

INTRODUCTION

Penetration tests are extensively used as a quick and easy means of sub-soil exploration. It is expected to give qualitative and relative information for site selection etc. In view of high cost, more time and difficulties of performing the standard penetration test below water table led to the development of simple method of dynamic cone penetration test. Various types of cones, rods and drop weight have been used by various workers, Meyerhof (1956), Palmer & Stuart (1957), Rodin (1961), Alam Singh and Punmia (1966), Schultze & Knausenberger (1957), Desai (1968). In European countries mostly 51mm dia. cone is in use and its correlations with Standard Penetration Tests (SPT) have been given, Meyerhof (1956). This test is becoming popular in India too during the last about two decades because it is simple, economical and less time consuming. These tests are of two types (i) cased, (ii) uncased. Conducting a dynamic cone penetration test in a bore hole has no advantage over SPT but test in uncased borehole has an advantage over other test as it gives a continuous record about the homogenity of the soil and its consistency and no bore hole is required. The main disadvantage of the test is that no soil samples can be obtained.

HISTORIC DEVELOPMENTS & TEST PROCEDURE

In India Khanna et.al (1953) for the first time adopted Dynamic Cone Penetration Test (DCPT) during the soil investigation of Hirakud Dam. They used 51mm dia. cone with 60° apex angle which was penetrated under two methods (i) 63.5kg hammer falling from 75cm of height and (ii) 20.4 kg hammer falling from 45cm height. Number of blows were recorded for every 30cm penetration of cone. The test were carried out in two seasons, wet and dry. They pushed the cone first upto 3m and later on a borehole was sunk and a casing was lowered.

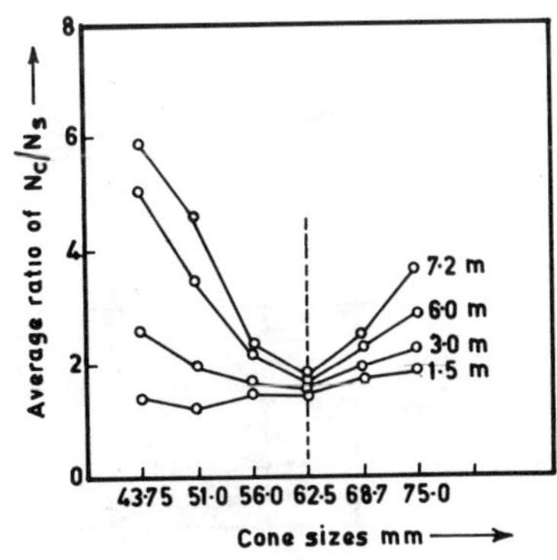

FIG. 1. Standardisation of cone size

Cone was again pushed below the casing. This method was repeated.

Cone Penetration Test in Uncased Boreholes

It was found that the method of driving cone with bentonite slurry is too cumbersome so it was thought that a dry uncased cone should be used. First question which came to the mind was the standardisation of the size of cone. This project was taken up in CBRI and cones of different sizes varying from 43.75 to 75.0mm were used in the study. On the basis of an exhaustive study Mohan et.al (1970), it was found that a cone of 62.5mm dia. gives the best correlation with SPT (Fig 1).

Method of Deep Probing

In case investigation is to be carried out beyond 9m depth then dynamic cone penetration test has to be carried out either by circulating 5 percent bentonite slurry or the cone can be advanced by making a bore hole upto the depth cone has already been pushed in, Fig 2. In this case the cone is advanced with the help of impact blows upto 9m and N_c values are recorded at every 30cm depth. Beyond this depth to avoid friction on drill rod, first a bore hole is advanced by about 50cm with the help of sand boiler and this is followed by cone penetration test and N_c values are again recorded upto the max. no. of blows of 50. Again, a bore hole is sunk as discussed earlier and the penetration test is conducted. The method is repeated upto 30m.

Desai (1969) used it during sub soil exploration and has established the correlation Nc/Ns which varies from 0.9 to 3.5 for a depth range of 1.5 to 3.5m in saturated silty fine sand. He has also found out that it does not reflect the density only but affected by grain size too.

Sikka (1965) studied the effect of shape of the cone on penetration resistance on model tests in the laboratory. It was found that the shape has no effect on penetration resistance. Similar findings have been reported by Virendra Kumar and Jain (1965) but Narahari et.al (1965) have reported on the basis of work carried out that the penetration resistance increases with the increase in size of cone.

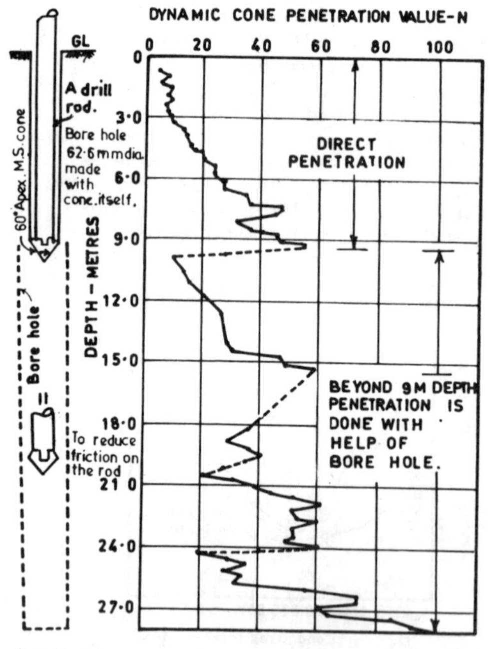

FIG.2. DYNAMIC PENETRATION CONE TEST UPTO 30(m)

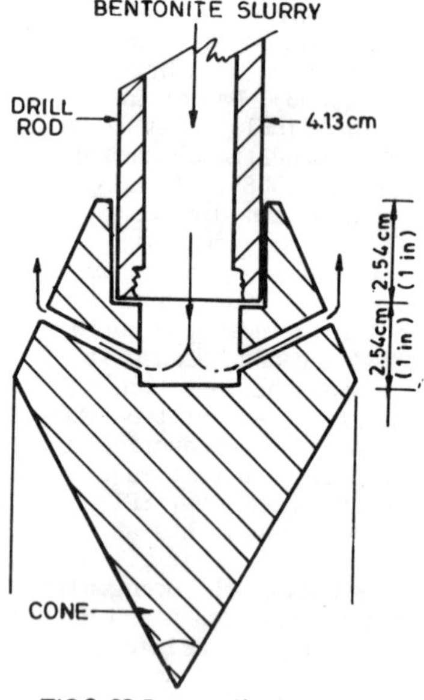

FIG.3. 62·5 mm dia cone.

FURTHER DEVELOPMENTS

Dynamic Cone Penetration Tests with Bentonite

Dynamic cone penetration test was in vogue in European countries in addition to standard penetration test. The cone is pushed either under a dynamic load of 63.5 kg falling from a height of 75cm or the test is performed in a bore hole as in case of Standard Penetration Test. However, it was experienced that due to development of high friction along the drill rod in case of direct driving of the cone, the number of blows increases tremendously. To over come this friction, penetration of cone by circulating bentonite slurry through the cone. In view of this Sengupta and Aggarwal (1968) tried the penetration of 62.5mm (Fig 3) cone by circulating slurry to overcome the friction along the side of drill rods.

A vane was attached at the top of the cone which is rotated off and on to avoid accumulation of sand particles over the cone and to keep the slurry flowing. Complete set up to carry out the test is shown in Fig 4.

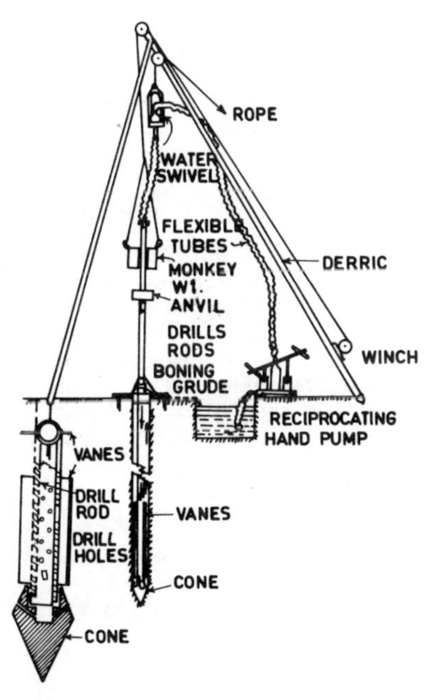

FIG.4 . Conepenetration test for soil exploration.

Bentonite slurry from the tank is passed through the swievel and 'A' rods to the cone with vents by means of reciprocating pump. The slurry after passing through the cone comes to the ground along the drill rod and thus eliminates the friction. The work was carried out at many sites on alluvial deposits. A correlation of Nc = Ns was established which indicates the complete removal of skin friction along the driven rod.

Light Weight Penetrometer

Alam Singh (1973) made use of light weight dynamic penetrometer developed by him earlier for the study of the following factors on model scale in laboratory. Penetration resistance, overburden pressure and relative density. The penetrometer had a base diameter of 35mm, vertical angle 60°, and a 10 kg weight falling through a height of 50cm was used and the N values were recorded for 20cm penetration of the cone. The tests were carried out at different relative densities and overburden pressures. He arrived at the conclusion that penetration resistance increases with an increase in density index and overburden pressures. The penetrometer is very handy and light to operate and transport. It gives fairly accurate results for preliminary investigations for shallow foundations of buildings, exploring subgrades of high-ways, railways in sandy regions.

STATISTICAL CORRELATIONS

Correlations have been developed with SPT because it is the only reliable and dependable field test with which different soil parameters have been correlated earlier. In view of this fact, the (DCPT) Nc values were correlated with the NSPT and is expressed by equations 1,2,3.

$N_c = 1.5\ N_s$ upto 4m depth ...1

$N_c = 1.75\ N_s$ from 4 to 9m depth ...2

Nc = Ns When cone penetrated by circulating bentonite slurry ...3

Where N_c = Number of blows of cone

N_s = Number of blows of SPT

The above correlations have been introduced in IS:4968 (Part I and 2) 1968 also.

SAFE BEARING PRESSURE FOR SHALLOW FOUNDATIONS

Main function of these tests is to determine the allowable soil pressure for the design of footing for the structure so far all the correlations are indirect that is number of blows of cone (N_c) are converted into number of blows SPT (N_s) and then the other parameters such as relative density, angle of internal friction etc. and the bearing capacity were determined. This was the major issue which was rattling in the minds of research workers that a direct correlation of dynamic cone penetration should be developed to determine the bearing capacity of soil. Desai (1970) developed charts which gave direct value of allowable bearing pressure and relative density for different values of surcharges from 50.8mm diameter dynamic cone test. Dinesh Mohan et.al (1970) made use of 62.5mm diameter cone to determine a direct value of allowable pressure by carrying out plate load test and SPT at the same depths. Charts were developed (Fig 5) and the values

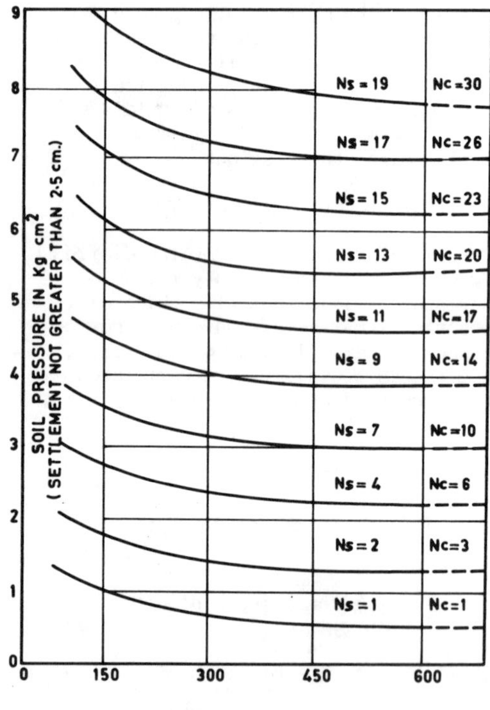

FIG.5. CBRI allowable pressure chart for Ns and Nc.

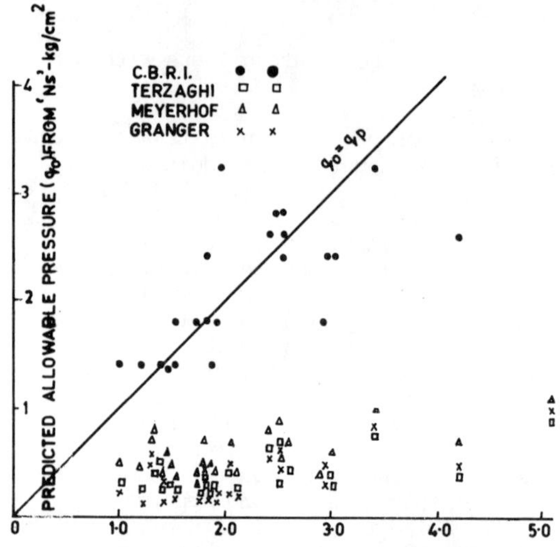

FIG.6. Comparison of observed and predicted allowable pressure.

of allowable pressure obtained from the existing methods were compared with those obtained from the charts developed by CBRI (Fig 6).

The comparison of values obtained is given in Table I. It has been found that the values obtained from CBRI charts varies from -25 to + 30 per cent whereas in other cases, the decrease in allowable pressure is exhorbitant, thus leading to uneconomical designs.

DYNAMIC CONE TEST FOR DIFFICULT SOILS

Narahari et.al (1967) made use of the dynamic cone penetration test in gravelly soils. First they carried out tests in the laboratory by compacting different sizes of gravels and penetrated different sizes of cones. They established the following equation to determine ultimate bearing capacity

$$q_o = \frac{K.N'}{D_c} B_f$$

where q_o = UBC (surface)

N' = cone penetration value

$D_c = 8\ B_c$

B_c = diameter of cone

B_f = width of footing

K = constant of proportionality

= 2.5 for 62.5mm cone

1.5 for 75mm cone.

Later on dynamic cone test was used very exhaustively in the areas which were predominantly covered by gravels.

Based on the field studies the equation 4 was modified as detailed below (Mohan et.al 1971)

$$q_a = \frac{1}{2.54} \; \frac{N''_c \, S_a}{D_c . B_f} \qquad \ldots 5$$

where q_a is the allowable soil pressure in t/m^2, N''_c is the cumulative number of blows, S_a is the settlement (cm) B_f is the width of the footing (m) and D_c is the critical depth in meter.

The critical depth is determined from the cumulative number of blows (N''_c) and D_c/B_c ratio relationship. D_c/B_c is read from the (Fig 7) at a point C where the slope of curve changes. From this the critical depth is determined.

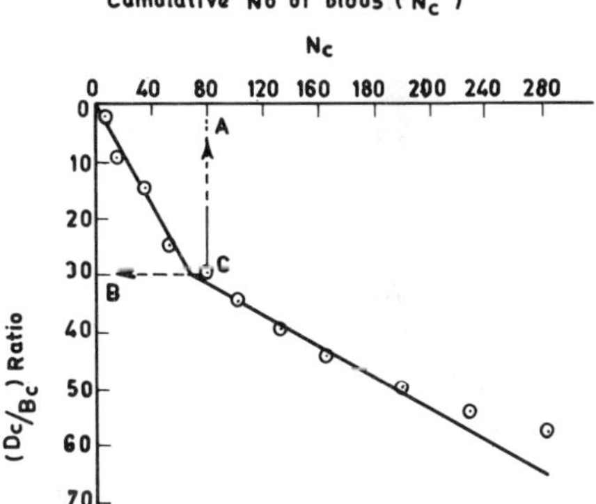

FIG. 7. Computation of critical depth.

DISCUSSION & CONCLUSIONS

Standard penetration test is carried out in a bore hole. Sinking of bore hole upto 4-5m deep in a day is possible when it is shallow but with the increase in depth, the progress of sinking of bore hole increases. Thus sinking of bore hole is expensive and time consuming On the other hand, dynamic cone penetration tests can be performed in about

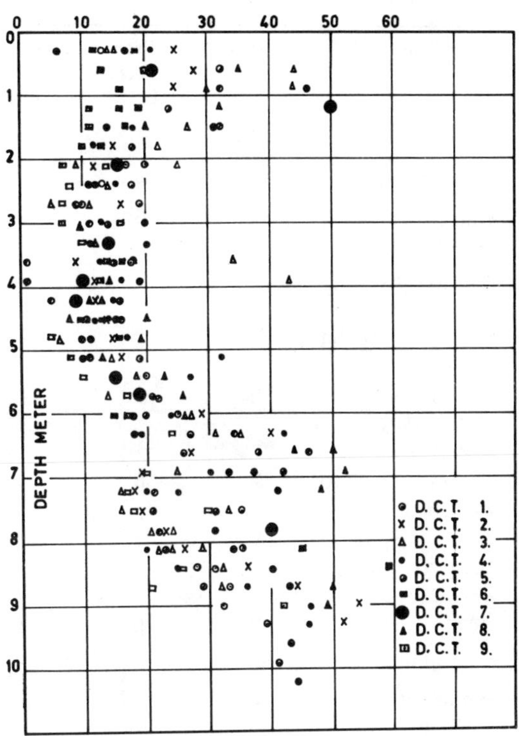

FIG. 8. Dynamic cone penetration test at Maharani Bagh New Delhi.

4 hours upto a depth of about 10m which saves time, labour and money.

On the basis of the dynamic cone penetration tests the number of boreholes can be avoided or even decreased to. Recently, one site at Delhi was explored by means of dynamic cone penetration tests and the recommendations for allowable pressures was made. The results of dynamic cone penetration tests indicate the uniformity in strata (Fig 8). This method of exploring prove to be very economical and less time consuming.

Penetration of cone by using bentonite slurry is very simple as compared to standard penetration tests. The only draw back of this test is that it does not provide with the soil sample for identification.

Use of dynamic cone penetration tests in difficult soils such as gravels is

Table I - Comparison of Allowable Pressures

Location	Classi-fication	'Ns' from B.H.	Corrected 'Ns' at surface	Allowable pressure for 7.5mm settlement PLT kg/cm2
1	2	3	4	5
Church Building Lucknow	SM	6	2	1.47
Indraprastha New Delhi	SM	7	4	1.80
Delhi Polytechnic Delhi	SM-ML	5	2	1.87
Indian Law Institute, New Delhi	SM	4	2	1.40
Medical Enclave, Amritsar	SM	6	4	3.00
Girls Hostel Medical College, Amritsar	SM	7	4	2.96
ITRC, Lucknow	SM	5	3	1.80
-do-	SM	5	2	1.50
Boiler site, Faridabad	SM-ML	10	7	3.40
Tractor Factory Site, Ghaziabad	SM	8	5	2.40
-do-	SM	9	6	2.50
New Hostel, CBRI	SM	4	3	1.87
CBRI site	SM	18	18	8.50
-do-	SM	11	10	5.10
-do-	SM	4	3	1.75

gaining importance because no other simple method to explore such deposits is available. Due to shortage of time the use of dynamic cone penetration tests was made recently to explore a site at Dehradun which was predominant with gravels. The allowable pressure was determined by using empirical co-relation and the value thus determined was in good agreement with the value obtained from plate load tests.

The chart developed at the CBRI (Fig 5) may be used to predict the allowable soil pressure for 25mm deformation, however, this may be extrapolated directly for 50mm deformation in accordance with IS: 1904 (1978).

ACKNOWLEDGEMENT

Help of Dr R.K. Bhandari, Head, Soils Div and shri B.G. Rao is gratefully acknowledged. The paper is being published with the permission of Director, CBRI, Roorkee.

REFERENCES

Aggarwal V.S and Tolia D.S. (1980) Penetration Tests for Shallow Depths Proc. of Regional Symp on Application of SM&FE, West Bengal, 21-22 Dec.

Alam Singh and Punmia, B.C. (1966) Progress report in the Penetration Resistance of Soil. Jr of INS of SM&FE 1:114.

Table I contd.

Allowable pressure for 3m footing (kg/cm^2)				Per cent differences w.r.t. PLT			
Terzaghi	Meyerhof	CBRI	Granger	Terzaghi	Meyerhof	CBRI	Granger
6	7	8	9	10	11	12	13
0.3	0.6	1.4	0.31	-80	-59	-5	-79
0.4	0.7	2.4	0.45	-78	-61	+33	-75
0.3	0.5	1.4	0.22	-84	-73	-25	-88
0.25	0.4	1.4	0.14	-82	-72	0	-90
0.3	0.6	2.4	0.31	-90	-80	-20	-90
0.4	0.7	2.4	0.45	-86	-76	-19	-85
0.3	0.5	1.8	0.22	-83	-72	0	-88
0.3	0.5	1.4	0.22	-80	-67	-7	-85
0.75	1.0	3.2	0.80	-78	-71	-6	-76
0.65	0.8	2.6	0.56	-73	-67	+8	-76
0.70	0.9	2.8	0.67	-72	-64	+12	-73
0.25	0.4	1.8	0.14	-87	-78	-4	-92
1.8	1.8	7.8	1.65	-79	-79	-8	-80
0.9	1.1	4.5	0.94	-82	-78	-12	-81
0.25	0.4	1.8	0.14	-86	-78	+3	-92

Aggarwal V.S. and Tolia D.S. (1969) Discussions on Correlation of Dynamic Cone Tests and SPT. Limitations of use of Penetration Allowable Pressure Curves for Shallow Foundations. Jr of INS of SM&FE Vol 8, No. 3 pp 304-307.

Alam Singh and Punmia, B.C. (1966) Progress report on Research in the Penetration Resistance of Soil. Jr. INS of SM&FE, pp 1-114.

Alam Singh and Sharma Damodor (1973) Development of a light weight Dynamic Penetrometer. Jr of IGS Vol 3, No. 1

Desai M.D. (1967) Discussion on Dynamic Cone Test Jr. of INS of SM&FE p. 363.

Desai M.D. and Roy M.B. (1968) Correlation of dynamic cone and Standard Penetration Test Jr. INS of SM&FE, 7:311

Desai, M.D. (1968) Limitation of Use of Penetration allowable pressure curves for shallow foundations Jr. of INS of SM&FE 7:427.

Desai M.D. (1968) General report and Discussions at the Symposium on Earth Rockfill Dams at Talwara, Punjab.

Dinesh Mohan, Aggarwal, V.S. and Tolia D.S. (1970) Correlation of Cone Size in the Dynamic Cone Penetration Test with the Standard Penetration Test. Geotechnique 20(3) 315-319.

Dinesh Mohan and Sengupta D.P. (1970) The Dynamic Cone Penetration Test Jr. of American Society of Civil Engineerig 40(2): 49-50.

Dinesh Mohan, Aggarwal V.S., Tolia D.S. (1971) Bearing Capacity from Dynamic Cone Penetration Test. Jr. of IGS Vol 1 No. 2.

Dinesh Mohan, Narahari D.R. Rao B.G.(1971) Field and Laboratory Tests on Gravel and Boulder deposits. Proc. 4th ARC on SM&FE Vol. I, pp. 49-55.

Dinesh Mohan, Narahari D.R. and Rao B.G. (1971) Field and Laboratory tests on Gravel and Boulder Soils. Proc of 4th ARC on SM&FE Bangkok, 1:49-55.

Handa C.L. Dr Pais and Chetty (1965) Foundation Studies for Yamuna Barrage at Delhi Proc 6th Int. Conf of SM&FE, Canada Vol. II, P. 65.

Jain G.S., Sengupta, D.R. and Aggarwal V.S. (1968) Dynamic Cone Penetration Tests for Site Investigations. Proc of Symp. on Earth Rock Fill Dams INS of SM&FE, 1:287.

Khanna P.L., Varghese and Hoon R.C. (1953) Bearing Pressure and Penetration on Typical Soil Strata in the Region of Hirakud Dam Project. Proc of 3rd Int Conf. of SM&FE, Switzerland, Vol. 1. pp. 246.

Meyerhof, G.G. (1956) Penetration tests and Bearing Capacity of Cohesionless soils. J. Soil Mech. Fdn. Div. ASCE, 82, 866.

Narahari, D.R., Rao B.G. and Makol R.L. (1965) Cone Penetration Tests in dry Sand of Varying Densities. Jr. INS of SM&FE 4: 444-454.

Narahari D.R., Rao, B.G. and Makol R.L. (1966) Effect of borehole size on the results of Cone Penetration Test. Jr. of INS of SM&FE, 5(1) 83-95.

Narahari D.R., Rao B.G. and Jain R.C. (1967) Dynamic Cone Penetration Tests in Gravels and Gravelly Soils. Proc of Symp. on Site Investigations for foundations, Roorkee 1:16.

Palmer, D.J. and Stuart, S.G. (1957) Some Observations on Penetration Test and a Correlation of the Test with a new Penetrometer. Proc. 4th Int. Conf on SM&FE, London 1:231.

Patel J.A. and Dinesh Mohan (1961) Penetration tests in Soils. Ist ARC on SM&FE, New Delhi 1(22): 1-9.

Rodin S. (1961) Experience with Penetrations with Particular reference to Standard Penetration Tests. Proc. 5th Int. Conf. SM&FE, Paris 1, 517.

Sengupta D.P. and Aggarwal V.S. (1966) A Study on Dynamic Cone Penetration Test Jr. of INS of SM&FE 5(2) 207.

Saha, H.L. (1961) Cone Penetrometer Tests and its Shortcomings with reference to Founding 130KV Transmission Tower in W. Bengal. Proc. of Symp. on Foundation Engg, Bangalore, P. 1-7.

Schultze, E. and Knausenberger, H. (1957) Experience with Penetrometers. Proc. 4th Int. Conf SM&FE, London, 1,249.

Sengupta D.P. (1967) Cone Penetration Tests for Soil Exploration in Calcutta Region. Jr. of Institution of Engineers (India) Civil Engg Div. 48(2) 193-208.

Sikka D.V. (1968) Model Experiments to Determine Effect of Shape and Size of Penetrometer on Dynamic Penetration Resistance of Sand. Jr. INS of SM&FE 7(1) 61-76.

Sridhava, S. (1971) Discussions on paper Effect of Surcharge on Dynamic Cone Penetration test by M.D. Desai, Jr. IGS, Vol. 1 No. 2.

UP, I.R.I. Report (1968) Note on the Use of 62.5mm and 50.8mm diameter Dynamic Cone for Penetration Tests. Fundamental basic research report for 24th Zonal meeting Bhopal, India.

Proceedings of the Second European Symposium on Penetration Testing / Amsterdam / 24-27 May 1982

Pile driving, a dynamic penetration test?

F.B.J.BARENDS & H.L.KONING
Delft Soil Mechanics Laboratory, Netherlands

1. INTRODUCTION

In the period of 1976 to 1981 a national project was undertaken to promote the exchange of experience and knowledge about the pile-driving process with the emphasis on improvement and application of existing computer models. Under the auspices of CIAD (a non-profit organisation which promotes the computer use in engineering practice) a study team drawn from different organizations worked with a great spirit to reach this goal. A wide variety of activities were put forth:

- inventarization of driving-problems, relevant literature, available models;
- organization of pile-driving test including soil instrumentation;
- soil investigation using many methods;
- evaluation of measurements, prediction and calibration of computer models;
- performance of static and dynamic bearing test;
- normalization of the use of computer models.

The completion of these activities has been made possible by a grant from PRONORM (Programming of normalization) and by the active contribution of the team members and the facilities of their organizations, to wit:

- ABT, consultant;
- Ballast-Nedam Group, contractor;
- Bos-Kalis Westminster Group, contractor;
- Municipality Works Rotterdam, government;
- Holland Beton Group, contractor;
- Delft Soil Mechanics Laboratory, institute
- Nederhorst Ground Techniques, contractor;
- Rijkswaterstaat DIV, government;
- Rutten and Kruisman, consultant;
- Shel IPM, company;
- Volker-Stevin, contractor.

The present article reviews some of the most significant findings and conclusions which are embodied in a massive detailed report recently completed (CIAD, 1981).

2 INVENTARIZATION AND EVALUATION

Pile-driving is an art mainly based on experience. The fundamentals are merely emperical. Characteristic phenomena are hardly understood or only in a qualitative sense. Since the recent development and the scale increase in pile-driving the need for more comprehension of the driving process became apparant.

In the first stage of the study an extensive inventarization was completed about experienced pile-driving problems, relevant literature and available computer models.

Pile-driving problems are collected and supplemented on the base of practical experience of the team members and from literature. The information has been ordered regarding soil, pile and equipment. A survey is presented in table I.

pile system	wood	concrete	steel	made in situ	special systems
year prod. %	57.8	34.0	0.3	4.3	3.6
total bearing capacity %	13.0	64.0	1.0	12.5	9.5
problem chance	0.03	0.02	0.01	0.05	-
cathegory of problems %					
regulations	15	5	5	15	
experience	10	5	15	40	
equipment	20	20	10	5	
pile system	-	-	-	5	
density	10	5	-	-	
tension	-	20	5	-	
pressure	20	5	-	-	
vibration	-	5	15	5	
obstacles	-	5	5	5	
pore pressure	-	10	15	20	
soil expl.	25	10	30	5	

Table I Pile-driving problems

The absolute chance on problems is not large, but regarding the enormous number of piles driven a serious effort to decrease this value is justified. Concerning special pile systems no data are given, as pile-driving problems encountered in this field of engineering are not covered by the traditional experience. Particularly in new fields of engineering a reliable prediction of the pile-driving process is valuable.

Beside a long list of articles published in the international technical literature also previous studies have been consulted. They are evaluated and compiled in intermediate reports (CIAD, 1978; CIADSCRIPT-3, 1981). In-situ pile-driving tests adequately instrumented are scarce.

Assessment of the pile-driving process by means of a mathematical model i.e the one-dimensional wave equation analysis, is common practice nowadays. Several computer models have been evaluated regarding documentation, user-friendliness and applicability. This was achieved by predicting the measured pile-driving behaviour of two in-situ tests performed already before. Only soil exploration data, pile system and type of equipment were provided.

Different computer codes have been used and several team members applied the same code independently.

Comparison of the various computer results with the measurements revealed a divergence too large. A calibration was undertaken to highlight the sensitivity of various input parameters. Apparantly, more than one possible set of values gave satisfactory results. Yet unmeasurable soil properties (added mass) seemed to be essential. In conclusion, prediction using computer models can be advised only in a relative sense.

Computer codes based on the one-dimensional wave equation analysis show two major imperfections:

- the formulation of the driving energy transmission to the pile is inadequate in many cases;
- the surrounding soil is incorporated in an unsophisticated manner.

Moreover, to select model parameters from factually improper data requires a thorough experience to make a computation useful. Prediction is dependent on the user's experience and insight.

To improve this quandary a pile-driving test on three different pile types was set up including dynamic and static bearing capacity tests and an extensive soil exploration applying a great variety of methods, before and after. The driving tests are unique, since instrumentation was placed in the surrounding soil (see figure 1). The driving test was carried out by the Public Work Department of Amsterdam (1979). All recordings of this test are stored at DSML, where they are available.

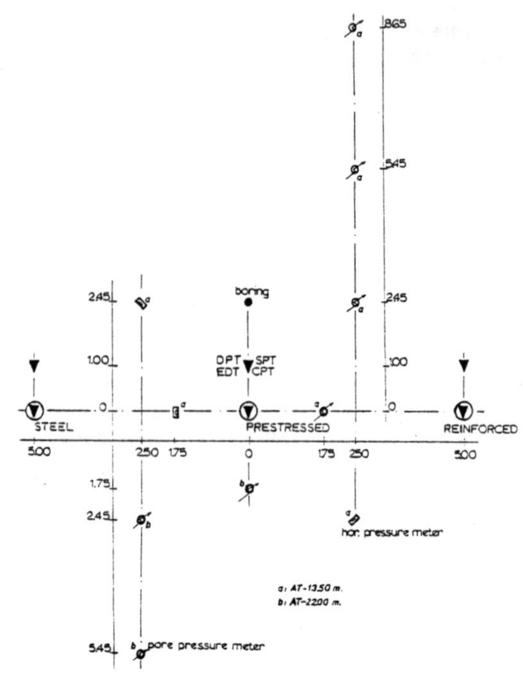

Figure 1 Plan of the locations.

3 SOIL EXPLORATION

Soil investigation is indispensable to determine soil material properties, which in turn provide the required input to a pile-driving analysis or a bearing capacity calculation. As already shown in table I lack of soil exploration or its quality is a common cause of problems during pile-driving.

In relation to the driving test a large soil exploration programme was executed with a two-fold purpose. First, to interrelate the different methods and to avoid a late discovery of essential soil data missing. Second, to find out whether the pile-driving data, such as a blow-count diagram, is useful to scan a soil stratum, and whether it provides informations to estimate the static bearing capacity of the pile.

In this respect the following site investigations have been carried out (see fig. 1):

- CPT, cone penetration test before and after pile-driving;
- SPT, standard penetration test including

boring (british standard);
- DPT, dynamic penetration test;
- EDT, electrical density measurement including cone and friction resistance;
- Ackerman boring.

The following laboratory tests have been carried out:
- grain-size distribution and silt analysis;
- permeability tests;
- cell tests.

In general it can be stated that the different in-situ methods recorded comparative results. However, a significant deviation between CPT, SPT and DPT values was noticed in the deeper clay layer at 16 to 21 meter depth (see figure 2).

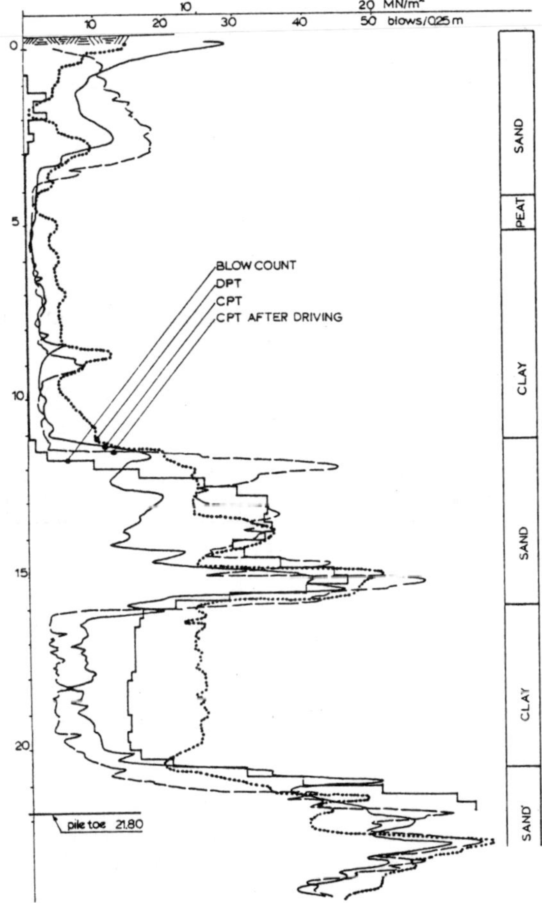

Figure 2 Measurements around the prestressed concrete pile.

The laboratory tests showed a strong influence of the silt fraction to the permeability demonstrated in table II.

Sample	% less than 16 μm	D_{50} (μm)	permeability (m/s)*10^{-6}
17	1.52	310.0	130 - 230
4	2.66	160.0	60 - 130
6	3.42	140.0	20 - 55
12	11.09	210.0	17 - 40
10	21.90	35.0	4 - 12

Table II Silt fraction versus permeability.

Although the soil exploration was rather extensive, no direct information was available concerning the dynamic soil properties. The required dynamic parameters have been empirically determined from the static soil data. In this regard the development of practical methods to determine dynamic soil properties for pile-driving analysis is strongly recommended.

4 PILE-DRIVING TEST, EVALUATION OF MEASUREMENTS

Three different pile systems (prefab concrete, prestressed concrete, steel) have been instrumented with strain-gauges and piezo-electric accelerometers at three levels, viz at the top, in the middle and at the toe. At the site the soil stratum contains two sand layers (see fig. 2). In these sand layers pore water and total pressure meters capable to register high frequencies have been installed (see fig.1). The piles were driven into the deep sand layer with a Delmag D36-02 hammer and using a Hitachi 180 GLS frame. The driving process has been recorded contineously. The blow-count value has been determined. Before and after the test a CPT measurement has been carried out. Results are presented in figure 2.

The maximum stresses in the piles determined from the strain measurements are comprised in table III. The maximum compression stresses occur in the pile top, whereas the maximum tensile stresses in the concrete piles are measured in the middle. An erroneous support before driving slightly cracked the prefab pile nearby the toe (0.2 mm cracks). This has probably caused a tensile stress wave different in character from the one in the prestressed concrete pile. Good support during storage is important.

pile type	compression stress (MN/m^2)	tensile stress (NM/m^2)
prefab (Ø40)	35.6	4.3
prestressed (Ø40)	40.1	4.8
steel (2 Larssen II S)	278.0	52.0

Table III Extreme dynamic stresses.

Maximum tensile stresses in the steel pile occur at the pile top. Accelerations measured in the steel pile are smaller than in the concrete piles.

During driving the cap-cushion became stiffer due to densification. This is clearly shown by the steady increase of the maximum compression stress as the pile went down.

The pile-driving test is quite unique, since the presence of pore water and total pressure meters in the nearby soil. The measurements in the soil reveal a significant pore pressure increase. A higher frequency followed by a lower one could be distinguished. A reasonable static overburden pore pressure was noticed as well.

In figure 3 some of the results are presented. The measurements gave an evident prove that the pore water plays a role in the driving process.

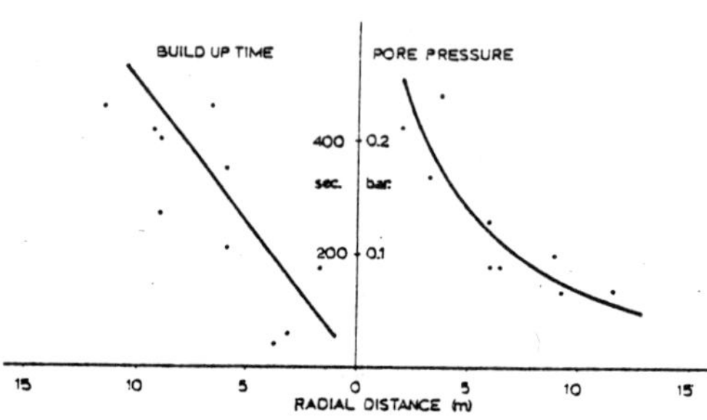

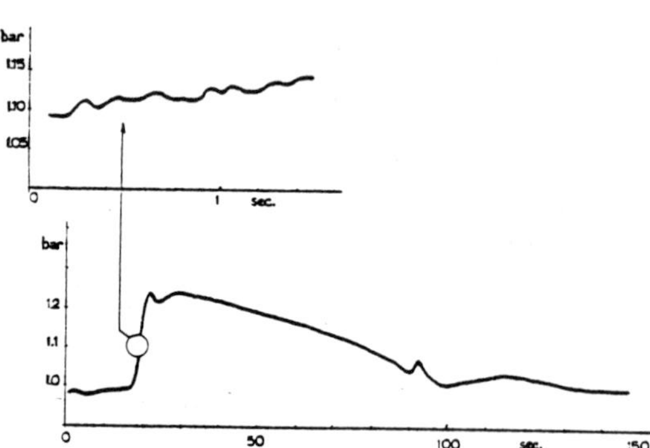

Figure 3 Dynamic pore pressure generation in the upper sand layer at 13.5 meter depth.

5 STATIC AND DYNAMIC BEARING CAPACITY TESTS

The purpose of pile-driving is to economically install a pile foundation, which will provide the required bearing capacity (and stiffness). Finally, it is this bearing capacity in which one is interested.

The traditional method to determine the bearing capacity is by application of some pile-driving formula based on the final blow-count, the driving energy and the weight of the ram and the pile. Such pile-driving formulas abound and several of them are applied to the prestressed concrete pile.

Also other indirect methods to determine the bearing capacity are employed. A dynamic test, in which a steel hammer of 1200 kg is dropped from various heights, has been carried oud on all three piles. For the prestressed concrete pile and the steel pile the ultimate capacity was not achieved, because of cracks at the top and the relatively low impedance respectively.

Finally the bearing capacity of the prestressed concrete pile has been estimated from the results of the cone penetration test. All results obtained are presented in table IV. The scatter in the results of the pile-driving formulas is large, which is a well known fact.

	resistance (MN)		
	side	toe	total
pile-driving formulas			
Eytelwein (Dutch)			4.86
Engineering News			2.54
Mod.Eng. News			0.92
Hiley			2.72
Delmag formula			1.76
dynamic test in-situ			
reinforced	0.75	2.73	3.48
prestressed	0.78	1.97	2.75*
steel	2.91	-	2.91*
prediction			
CPT-method	1.77	1.76	3.53
Nottingham	1.86	1.76	3.62

* ultimate capacity not achieved

Tabel IV Bearing capacity determined indirectly.

To know the truth a static bearing capacity test has been carried out on the prestressed concrete pile. The pile, which was provided with strain-gauges at three levels, was loaded stepwisely; every step has been maintained during four hours. The results are compiled in table V.

	Resistance (MN) side	toe	total	Top displacement (mm)
static test on prestressed concrete pile	0.42	0.07	0.49	0.93
	0.81	0.18	0.99	2.31
	1.15	0.34	1.49	4.09
	1.49	0.52	2.01	5.96
	1.67	0.82	2.49	8.94
	1.70	1.28	2.98	14.52
	1.78	1.74	3.52	55.01
interpretation van der Veen			3.50	} ultimate load
DSML (hyperbolic)			3.80	

Tabel V Bearing capacity measured

The ultimate static bearing capacity measured agrees fairly well with the calculated values of the prediction methods, viz the CPT-method and the method of Nottingham.

Using the installed strain-gauges the measured bearing capacity could be divided into two contributions, the side friction and the toe resistance. It is remarkable that the ratio of both contributions varied significantly with the applied load. Only after the ultimate side friction has been mobilized, the toe resistance increases.

The dynamic and static ultimate bearing capacity are fairly equal, i.e. 3.48 and 3.52 MN respectively. Also the predicted load-settlement behaviour on the base of the dynamic test (not shown here) is in a good agreement with the static measurements.

No coherence was found however in the distribution of the side and toe resistance determined by the dynamic and static methods. The ratio between the ultimate side friction and the ultimate toe resistance is about 1.0 according to the static test (table V) and the predictions (table IV), whereas this ratio is in the order of 0.25 for the reinforced pile according to the dynamic test. In this respect the prediction by the CPT-method or according to Nottingham is superior to the dynamic test.

6 PILE-DRIVING PREDICTION BY COMPUTER

In the past decennia the scale enlargement and the technical improvement of driving equipment and pile systems is a fact. The choice of the best means for a specific case is becoming more critical.

One can wonder whether the pile-driving process is predictable anyway. Many factors play a role, such as the labor skill, the quality of the material and so on, (see table I). Detailed information is required about the different driving systems including the cap-cushion. Dynamic properties of the soil-pile system must be determined.

Regarding the driving equipment and the manufacturing of the pile system the control is in hand, but the usual soil exploration if available is often not sufficient to definitively recommend a certain production method.

Mathematical models offer an opportunity to highlight the sensitivity of the variety of soil properties. They represent a modern utensil to evaluate different possibilities of pile systems, driving equipment and production rates beforehand. The mathematical formulation of the thermo-dynamic process of pile-driving has recently be extended to Diesel hammers and Hydro-blocks. The calibration of the corresponding computer codes by in situ tests has started. Yet much work in this field has to be done.

Most contemporary prediction models contain a rather simple soil-pile interaction, the classic nonlinear spring-dashpot, in which some essential factors such as the soil mass and the pore pressure generation are not included or in a (very) empirical way. This situation prohibits a further refinement on the base of the model itself.

A wide range of acceptable values for the different model parameters has been tested, verifying the measurements of a pile-driving test. The results reveal a significant influence from the impact velocity and the cap-cushion stiffness on the compression stress. The maximum tensile stress decreases if the elastic part of the pile toe displacement is assumed larger. Other effects have been found as well (CIAD, 1981). For an acceptable set of parameter values the scatter in the calculation results with respect to the measured values could be minimized.

To assist the engineer in selecting proper empirical parameter values for a specific pile-driving simulation a norm has been defined, which describes the range of the parameter values as they have been found during the study of the CIAD team 'pile-driving'. This norm is published in CIAD-SCRIPT-3 (1981). The use of this norm will make the computation results realistic in a sense that an accuracy can be achieved of about 50% in tensile stress, 10% in compression stress and 30% in settlement.

In conclusion, computer models can be applied to predict the ultimate driving stresses and the production rate (penetration per blow) for different driving hammers and various pile systems provided that the required dynamic soil behaviour parameters can be estimated with a reasonable accuracy.

7 PILE-DRIVING, A DYNAMIC PENETRATION TEST?

The correlation between the dynamic and the static behaviour of a pile, if it eventually exists, has attracted the attention contineously. In literature many trials to find such a correlation are reported. Hundreds of measured piles have been considered. Yet no useful relation could be derived.

This study proves once again that such a correlation probably does not exist. Compiling the findings concerning this aspect yields the following.

- Pile-driving causes dynamic excess pore pressures. This phenomenon has been measured in sand layers. It can be expected likewise in less permeable soil layers. On the contrary static behaviour of a pile is not affected by such temporary excess pressures.
- It would be obvious to correlate the driving behaviour of a pile with a dynamic bearing capacity test. However, the loading during pile-driving is much larger then applied in a dynamic bearing capacity test.
- The distribution of side and toe resistance appears to differ significantly for a dynamic and a static bearing capacity test, and it is a function of the ratio: actual loading versus ultimate bearing capacity.
- The area of soil mobilized due to the pile behaviour in a static or a dynamic loading environment is principally different with respect to the area size and the nature of the response.

Hence, a direct correlation between the pile-driving resistance and the static bearing capacity does not exist, unless maybe in a purely empirical form valid for a specific situation.

Another way of looking at a pile-driving performance can be as a dynamic penetration tests, i.e. exploration of the subsoil stratification with a large object. Usually the blow-count diagram is quite conformal to the conventional sounding graphs (see figure 2), and in principal pile-driving is a way to detect the geological profile regarding its mechanical properties. Some restrictions are to be mentioned.

- Normalization is not available. Therefore, at least one standard sounding has to be carried out for calibration.
- Pile-driving is rough compared to a normal sounding. Thin layers are not detected and the accuracy of positioning the separation of different layers can be beyond one meter depending on the pile diameter.
- The pile penetration can cause a drastical change in the local soil properties. This fact might well explain that the blow-count diagram produces consequently higher values in the second clay layer at about 18 m depth.

In conclusion, pile-driving is in fact a dynamic penetration test. It can detect deviations in the geological profile. This information can be used to estimate proper input-data for common bearing capacity calculations. In this manner an indirect correlation between the dynamic and static behaviour of a pile exists. Its use for a final computation of a foundation is limited.

REFERENCES

CIAD, 1978, Report of the activities in 1977 (in Dutch), Interimreport CIAD project team 'pile-driving', CIAD, Zoetermeer, 180p.

CIADSCRIPT-3, 1981, Pile-driving by computer (in Dutch), Brochure CIAD project team 'pile-driving', CIAD, Zoetermeer, 55p.

CIAD, 1981, Pile-driving test, bearing capacity test and computer programs (in Dutch), Final report CIAD project team 'pile-driving', CIAD, Zoetermeer, 600p.

The reports can be obtained from CIAD, P.O. Box 74/NL-2700 AB Zoetermeer, The Netherlands.

Proceedings of the Second European Symposium on Penetration Testing / Amsterdam / 24-27 May 1982

Calculation of settlements on sands from field test results

U.BERGDAHL & E.OTTOSSON
Swedish Geotechnical Institute, Linköping, Sweden

1. INTRODUCTION

The Swedish Geotechnical Institute and the Swedish Road Administration have carried out investigations during recent years of different methods of calculating settlements for bridge foundations on spread footings in cohesionless soils. So far, the investigations involve four bridges where different kinds of field investigations and methods of calculation have been used and the results compared with measured settlements. The aim of this investigation is to find the different methods of determining the settlements and to evaluate the accuracy of these methods. The first investigation was reported by Wennerstrand (1979) and the second by Bergdahl and Eriksson (1980). This paper summariezes the results from the subsequent two investigations.

2. TEST SITES

The third test site is situated at the western shore of the Muonio river, on the border between Sweden and Finland in the north, close to the village of Kolari. A new bridge has recently been built across the river and the investigations for this study have been performed for the western abutment of this bridge.

The fourth test site is situated in the valley of the Korsträsk creek, close to Älvsbyn, in the northern part of Sweden. A new bridge across the valley was constructed here in the summer of 1981. The investigations were performed on the northern pier of the bridge, which is situated in the slope of the valley.

3. FIELD AND LABORATORY INVESTIGATIONS

Extensive site investigations have been carried out at both test sites: field as well as laboratory investigations.

The field investigations contained:

- Cone penetration tests in accordance with the recommended European standard.
- Weight sounding tests in accordance with the recommended European standard.
- Dynamic probing in accordance with Swedish geotechnical HfA standard.
- Pressuremeter tests with Menard standard probes.
- Piston sampling in accordance with the Swedish standard.
- Determination of groundwater levels with open stand pipes fitted with filter points.

The HfA type dynamic probing is somewhat different from the recommended European DPB standard. Thus, the height of free fall for the HfA-test is 0.50 m and the cross-sectional area of the point is 16 cm^2. In this dynamic probing, the skin friction along the rods is separated from the point resistance by means of a slip coupling close to the point, Bergdahl et al (1981). The HfA-test results can be converted into DPB test values, since the resistance values, q_d, according to the recommended European standard, are almost identical. Investigations also show that the blow-count in blows /0.2 m for the HfA-test is the same as the N_{30}-values for SPT-test. This correlation has been used for the calculations based on SPT-tests.

The pressuremeter tests at Kolari were made with slotted casings and the Menard standard probe 44 mm dia. Due to the curvature of the volumetric strain-pressure curves from these tests, it was difficult to evaluate an accurate pressuremeter modulus. The pressuremeter modulus has therefore been taken as 9.5 times the net limit pressure at each test level. This is an average of the range 7-12 given by Baguelin et al (1978). At Älvsbyn, the pressuremeter tests were carried out in prebored holes made with screw auger and stabilized by bentonite slurry.

The laboratory investigations consisted of soil classification and determination of the natural water content. For typical samples, the compressibility of the soil was also determined, by oedometer tests. For the Kolari test site, these tests were carried out on recompacted samples for which a medium dense state has been used, $\gamma_d = 16.7$ kN/m^3 for the upper part of the profile and $\gamma_d = 15.2$ kN/m^3 for the bottom part. For the Älvsbyn site, the tests were carried out on the almost undisturbed samples taken with a thin-walled piston sampler.

4. SOIL PROFILES

The results of the site investigations are summariezed in Fig. 1-2.

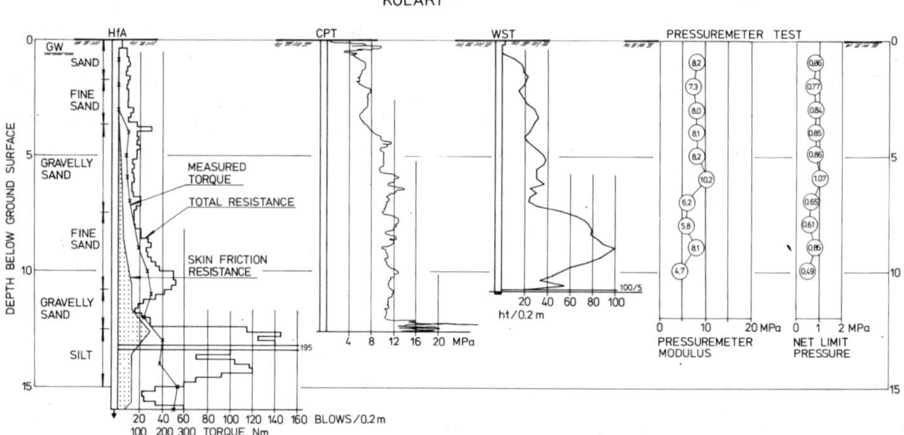

Fig.1. Results of field investigations at the Kolari test site.

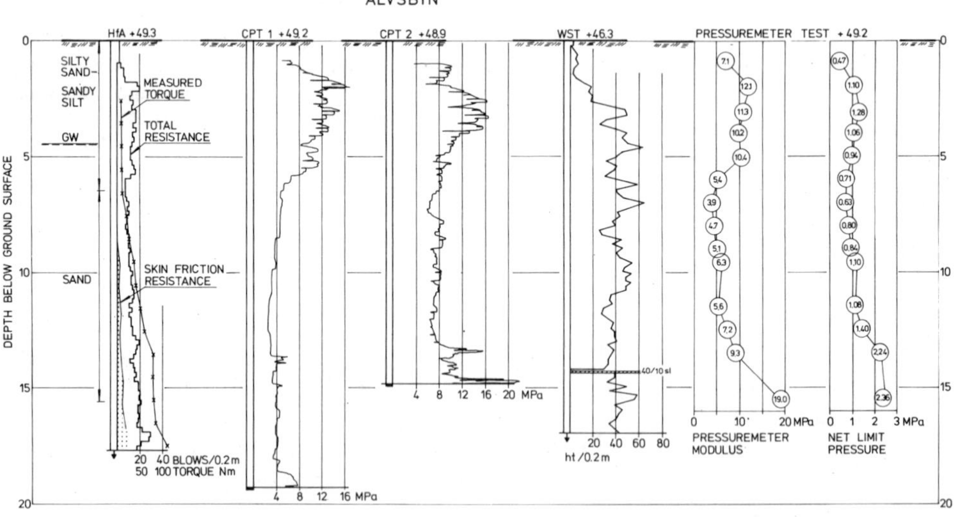

Fig.2. Results of field investigations at the Älvsbyn test site.

The investigations at the Kolari test site indicate a medium dense sand to a depth of about 4 m and below that, a medium dense to dense sand down to a depth of about 12 m. From 3.8-7.8 and from 10.8-12.5 the sand layers contain varying amounts of gravel. A dense to very dense silt was found below a depth of 12 m. During the investigation, the groundwater table was 0.5 m below the ground surface.

The investigations at the Älsbyn test site indicate a medium dense to dense silt and sand to a depth of about 4 m. Below that level, there is a medium dense sand to a depth of about 19 m (cone penetrometer test, CPT 1). Two CPT-tests have been carried out at this test site and the test results are quite different, especially in the layers below a depth of 4 m. Thus, test CPT 1 indicated a loose state of these sand layers. The reason for this difference is probably small differences in the grain size distribution of the soil at the points investigated, which are only six metres apart. In the settlement calculations, an average of those two diagrams has been used. During the investigation the groundwater table was situated 5 m below the ground surface.

5. FOUNDATIONS

The western abutment of the Kolari bridge was founded on a concrete slab (9.5x10.1 m) at a depth of 4.5 m below the average ground surface level (Fig. 3). An embankment, 7.5 m above the original ground surface, was constructed behind the abutment.

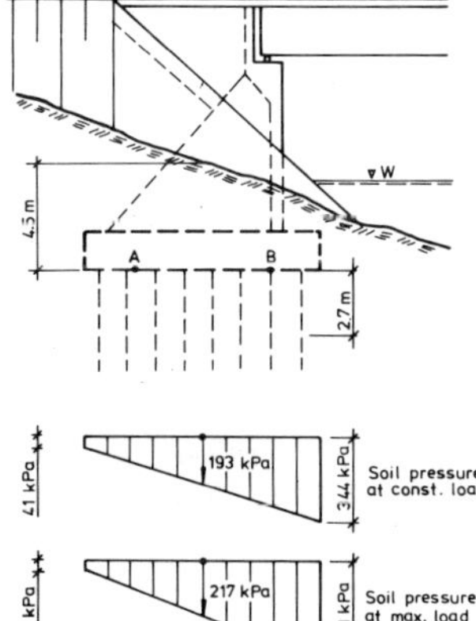

Fig.3. Cross section of the abutment at the Kolari test site, showing calculated soil pressures.

Due to the height of the embankment, the soil pressure is much higher at the water side than to the rear of the abutment (41 kPa-344 kPa under constant load). On average, the soil pressure is 193 kPa at constant load and 217 kPa at maximum load (including traffic load).

During the construction of the abutment, the relative density of the soil layers beneath the foundation level was checked by weight sounding tests. It was then observed that the sand layers close to the bottom of the excavation was loose. It was therefore decided to drive 4 m long timber piles, with a tip diameter of 125 mm and at a spacing of 1.2 m, beneath the entire concrete slab to compact the sand layer to a depth of about 4 m.

The pier of the bridge at Älvsbyn was also founded on a concrete slab (5x8.5 m) at a depth of 2.5 m below the average ground level. The soil pressure was more uniform under this foundation compared to that of the Kolari bridge (Fig. 4). The calculated soil pressure at constant load is on average 183 kPa and the minimum and maximum pressures due to traffic are 195 and 265 kPa respectively.

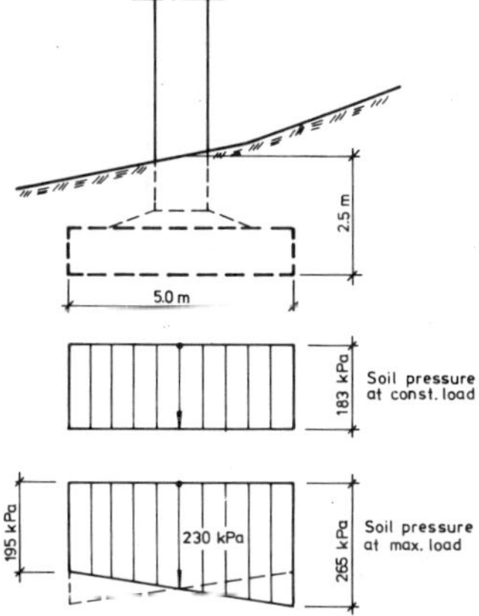

Fig.4. Cross section of the pier at the Älvsbyn test site, showing calculated soil pressures.

6. SETTLEMENT CALCULATIONS

The settlement calculations have been performed according to different methods described in the geotechnical literature. Methods based on both static and dynamic penetration test methods have been used, as well as the method proposed by Menard for the pressuremester test results. The oedometer test results have also been used for settlement calculations. The following methods of calculation have been used:

i For CPT tests: De Béer (1965) and Schmertmann (1978).

ii For dynamic probing: Schultze & Sherif (1973) and Parry (1977).

iii For the pressuremeter tests: Menard (1975).

iv For the oedometer tests, the compressibility of the soil has been evaluated from the test results and the stress distribution has been assessed from the elastic theory.

Settlement calculations were made for the soil pressure both at constant load and at the maximum load. However, as the settlement measurements were made after the concrete slabs had been cast, the weight of this concrete has been excluded. In those methods where the stress distribution can be considered, the settlements have been calculated for the stresses in the characteristic points of the foundation. In other cases, the mean soil pressure of the foundation has been applied.

At Kolari, where a number of piles were driven in the bottom of the excavation, the settlement calculations have been performed for two separate cases, one for the actual footing at the design level and one for an imaginary concrete slab at the lower third point of the pile length, 2.7 m below the actual foundation level.

The results of the settlement calculations have been summariezed in TABLES 1 and 2.

The calculations for the Kolari test site gave settlements within a wide range: 5-44 mm. The lowest values are obtained from methods of calculation based on dynamic probing, while the highest values are obtained from the Menard method, based on pressuremeter test results. The average settlement for all methods, which, according to Jordan (1977) could be the most accurate estimate, is between 18 and 26 mm. The calculated settlements may be compared with the measured settlements, which, one year after construction, are 20 mm at the rear and 24 mm at the river side of the concrete slab. As can be seen from the time-settlement curves (Fig. 5) the final settlement values, which average about 25 mm, have not been reached.

TABLE 1. Settlement calculations for the Kolari test site.

Caln. method	Cald.settlm. mm. At design level		Cald.settlm. mm 2.7 m below design level	
	Constant load	Max load	Constant load	Max load
De Béer (1965) average	32	36	25	29
point A	27	30	20	22
point B	38	42	31	35
Schmertmann (1978) 1 year	21	26	12	15
10 years	25	30	14	17
Parry (1977)	7	8	5	6
Schultze-Sherif (1973)	12	14	8	9
Menard (1975)	41	44	35	41
Oedometer tests	25	29	21	24
Average of all methods	23	26	18	21

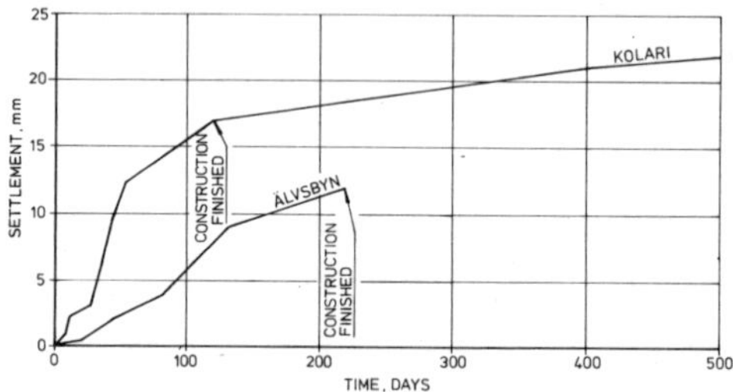

Fig.5. Measured average settlements of the foundations at Kolari and Älvsbyn.

The calculated settlements for the bridge pier at Älvsbyn vary between 15 and 60 mm. In this case, the lowest values are obtained using the Menard method and the highest using the De Béer method.

TABLE 2. Settlement calculations for the Älvsbyn test site.

Caln. method	Calculated settlements mm Constant load	Maximum load
De Béer (1965	45	60
Schmertmann (1978)		
1 year	29	44
10 years	34	51
Parry (1977)	21	27
Schultze-Sherif(1973)	16	20
Menard (1975)	15	19
Oedometer tests	23	30
Average of all methods	25	33

The average settlement for all methods is between 25 and 33 mm, depending on which load is considered. The calculated settlement for the foundation at Älvsbyn may be compared with the average measured settlements at the time of completion of the structure (Fig. 5). From the time-settlement curve, it is possible to extrapolate a final settlement for this foundation to about 18 mm, when the traffic load is not considered.

7. DISCUSSION OF THE TEST RESULTS

A comparison of the calculated and the measured settlements for the test sites described above indicates that the scatter in calculated settlements is large. It appears that the De Béer method overestimates the settlements, for which the Schmertmann method gives reasonable results, except in the Älvsbyn case, where there is some overestimation. This may be due to the large differences in penetration resistances for two cone penetration tests.

The methods of calculation based on the SPT-tests gave reasonable results at Älvsbyn, but underestimated the settlements at Kolari. This may be due to the assumed relationship between the blow-count of the SPT-test and the Swedish standard dynamic probing, HfA. This relationship may differ from one soil type to another.

The results of the settlement calculations based on pressuremeter tests and oedometer test also gave good agreement with measured values, except for the Menard method at Kolari. However, this may be explained by the abovementioned difficulties of evaluating an accurate pressuremeter modulus.

8. CONCLUSIONS

It can be concluded from the investigations revued in this paper and the two earlier investigations that it is possible to obtain reasonable assessments of settlements for spread footings in cohesionless soils, using penetration test results as well as pressuremeter and oedometer test results. However, the scatter in the calculated settlements may be large, which may be due to the behaviour of the test equipment in different types of soil. Possibly the calculation methods can be revised in the future to fit the reality better. In the meantime, it is advisable not to use only one method of settlement calculation and, equally, not to use methods based on the same type of investigation. In the future this investigation will be continued with full scale plate load tests. Later on general recommendations for settlement calculations in cohesionless soils can probably be given.

9. REFERENCES

Baguelin, F., Jézéquel, J.F. and Schields, D.H. (1978). The pressuremeter and foundation engineering. Trans Tech Publication.

Bergdahl, U., Möller, B. (1981). The Static-Dynamic Penetrometer, Proc. Xth ICSMFE, Stockholm, Vol.2, pp. 439-333.

Bergdahl, U., Eriksson, U. (1980). Metoder för in situ bestämning av jords deformationsegenskaper. Försök vid Åstorp. Swedish Geotechnical Institute, Varia No 52.

De Beer, E.E. (1965). Bearing capacity and settlement of shallow foundations on sand. Proc. Symp. Bear.Cap.Settl.Found.Duke Univ. 1965, Lecture 2, pp.15-33.

Jordan, E.E. (1977). Settlement in sand - methods of calculating and factors affecting. Ground Engineering 10.

Menard, L. (1975). The interpretation of pressuremeter test results. Sols-Soils, No 26.

Parry, R. (1977). Estimating bearing capacity in sand from SPT values. ASCE, GT9, 1014-1019.

Schmertmann, J., Hartman, J,P., and Brown P. (1978). Improved strain influence factor diagrams. ASCE, GT8, 1131-1135.

Schultze, E., & Sherif, G. (1973). Prediction of settlements from evaluated settlement observations for sand. Proc. 9th ICSMFE, Moscow (1.3), 225-230.

Wennerstrand, J.(1979). Comparison of predicted settlements for a fine sand. Proc. VII ECSMFE, Brighton, Vol.2, pp.295-298.

Proceedings of the Second European Symposium on Penetration Testing / Amsterdam / 24-27 May 1982

Statistical comparison between the results of dynamic and static penetrometers for Rhineland silt

B.BIEDERMANN
Consulting Office for Soil Mechanics & Foundation Engineering, Würzburg, Germany

1 INTRODUCTION

Normally between the results of dy= namic and static penetration tests only empirically equations are pos= sible because of the difficulty to introduce all the parameters influ= encing the sounding test in a com= plete theoretical equation. A com= mon way to come to such a theoreti= cal equation which depends only from the dimensions of the different sounding tools is the working equa= tion.It has been investigated if this simple equation is besides for non-cohesive soils also relevant for the Rhineland silt. This working equation is only valid for the dynamic pene= trometers. An other possibility is the plotting of the penetration re= sults as summarized works over the depth. With this presentation a much smaller scattering of the measured values can be observed at the compar= ison of different penetrometers.

The compared values were compiled from all available research works and also from soil mechanical reports. Therefore it is not surprising that the scattering is somewhat higher as only for a limited single investiga= tion area.

2 COMPARISON BETWEEN THE RESULTS OF THE STANDARD PENETRATION TEST AND THE HEAVY DYNAMIC PENETROMETER

For the Rhineland silt for 207 com= parable measured values of the blow-count n30 (SPT) and n20 (HDP) the re= gression analysis shows the highest correlation coefficient at a double-logarithmic transformation (Fig.1). According to the T-values of the student distribution for a fractile of 5% the equations are statisti= cally significant. At the normal distributions of the two parameters n30 and n20 the variation coeffi= cients are situated between 0,54 and 0,55 (Fig.1).

The common term of the working equation is

$$A^{x} = \frac{Q \cdot h \cdot n'_t}{F \cdot t'} \qquad (1)$$

with

Q = weight of the drop hammer
h = height of drop
n'_t = blow-count
F = square of the cone
t' = penetration depth

The theoretical equation between n20 and n30 is therefore with the different parameters for the two penetrometers (Biedermann 1979 and 1982):

$$n20 = 0{,}95 \cdot n30 \qquad (2)$$

which fits in a very good manner with the empirical equation (Fig.1).

With the mean values of the normal distributions of the blow-counts of the standard and heavy penetrometer a ratio of $\bar{n}20/\bar{n}30 = 0{,}839$ can be determined. In the contrary to the regression analysis is this ratio physically exact, because the straight line begins at the zero-point. But there is a greater dis= crepancy between the ratio of the mean-values and the result of the working equation (Fig.1). The curves according to the regression analysis on the other side do not begin at the zero-point (Fig.1).

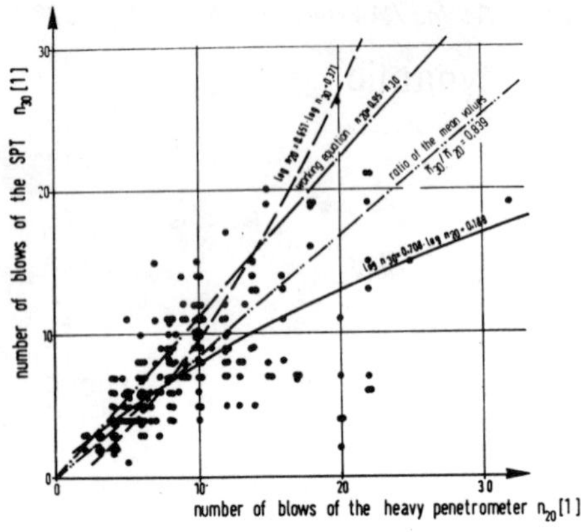

number of values : n = 207

type of soil : silt

regression equations	corr. coefficient	T	$T_{(n-2,95)}$
$n_{30} = 0.523 \cdot n_{20} + 3.002$	r = 0.630	11.611	1.972
$n_{20} = 0.759 \cdot n_{30} + 3.459$	r = 0.630	11.611	1.972
$n_{30} = 12.098 \log n_{20} - 3.114$	r = 0,656	12.436	1.972
$n_{20} = 12.539 \log n_{30} - 0.999$	r = 0,589	10.429	1.972
$\log n_{30} = 0.708 \log n_{20} + 0.189$	r = 0,679	13.241	1.972
$\log n_{20} = 0.651 \log n_{30} + 0.371$	r = 0,679	13.241	1.972

Fig.1. Results of the regression analysis between the blow-counts of the standard and the heavy dynamic penetrometer

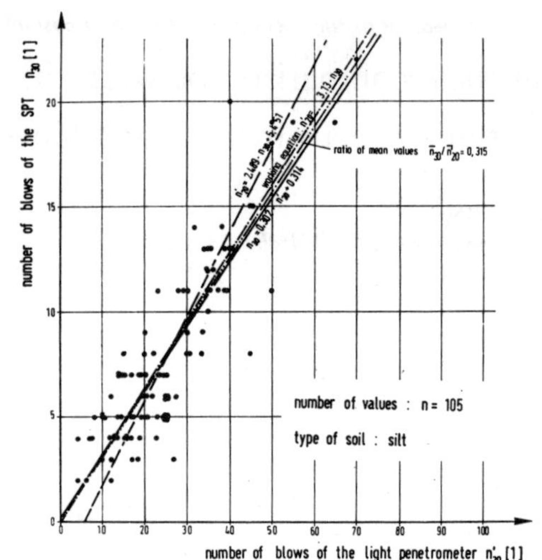

equations of regression analysis	correlation coefficient	T	$T_{(n-2,95)}$
$n_{30} = 0,302 \cdot n'_{20} + 0,314$	r = 0,867	17,681	1.983
$n'_{20} = 2,489 \cdot n_{30} + 5,451$	r = 0,867	17,681	1,983
$n_{30} = 14.373 \log n'_{20} - 11.395$	r = 0,773	12,354	1,983
$n'_{20} = 45.038 \log n_{30} - 12.570$	r = 0,830	15,084	1,983
$\log n_{30} = 0.792 \log n'_{20} - 0.225$	r = 0,905	13,776	1,983
$\log n'_{20} = 0.9[illegible]5 \log n_{30} + 0.658$	r = 0,905	13,775	1,993

Fig.2. Results of the regression analysis between the blow-counts of the standard and the light dynamic penetrometer

3 COMPARISON BETWEEN THE RESULTS OF SPT AND LIGHT DYNAMIC PENETROMETER

For 105 comparable values the regression analysis shows in a linear system the highest correlation coefficient with r = 0,867 (Fig.2). The equations are according to the statistical tests significant. The variation coefficients are lying between 0,50 and 0,55 for the two blow-counts n30 and n'20 (Fig.2). The ratio of the mean values determines $\bar{n}'20/\bar{n}30$ = 25,150/7,914 = 3,18, which fits extremely good with the results of the regression analysis and the working equation (Fig.2).

The common working equation shows according to the measurements of the two penetrometers the following ratio

$$n'20 = 3,13 \cdot n30 \qquad (3)$$

(s.a. Biedermann 1982).

The comparison between this two dynamic penetrometers shows very good agreement between the measured values and the theoretical ratio from the working equation.

4 COMPARISON BETWEEN THE BLOW-COUNTS OF HEAVY AND LIGHT DYNAMIC PENETROMETER

For 223 comparable values between the blow-counts n20 and n'20 statistically significant regression equations were calculated with a highest correlation coefficient of r = 0,574 (Biedermann 1982). In this case the values are coming only from one investigation area. The ratio of the mean values $\bar{n}'20/\bar{n}20$ = 2,998 differs only little of the working equation, which shows a value of 3,33 for the blow-counts of the light and heavy dynamic penetrometer. It is also possible in this case to calculate the values from one type of penetrometer into another with the working equation at a sufficient exactness for Rhineland silt.

5 COMPARISON BETWEEN THE RESULTS OF SPT AND THE CONE RESISTANCE OF THE STATIC PENETROMETER

Between the blow-count n30 of the SPT and the cone resistance of the static penetrometer 243 values could be compared from different investigations for Rhineland silt. The highest correlation coefficient with r = 0,779 was gathered at the regression analysis for a relationship in linear coordinates between the two parameters. The equations of the regression analysis are statistically significant at the usual statistical tests (Biedermann 1982).

Comparing the regression analysis with different equations from literature for silt shows a wide scattering of these ratios (Biedermann 1982). The main problem of these relationships is, the lack of informations how they were calculated. Generally no information about the number of comparable values and about the scattering of the values is present. Commonly only ratios between the two parameters are shown in the literature.

6 EVALUATION OF WORKING EQUATIONS FOR SOUNDING TESTS

Besides the direct comparison between the results of penetration tests for each depth, it is nowadays preferable to determine the results according to the equations for the work of the penetrometer summarized over the depth. The equations for each type of penetrometer was detailed given by Rollberg (1976).

For an example of a SPT in silt the interpretation of this method will be explained (Fig.3). With the given blow-count n30 and the measurement of the standard penetrometer (DIN 4094) the working equation is then

$$A(SPT) = 48{,}3 \Sigma n30 \cdot t \cdot (1/30) \,[kNm] \quad (4)$$

(s.a.Rollberg 1976,p.92). The results according to this working equation must be summarized over the depth from 0 to t, which will be drawn as a depth profile according to Fig.3.

It is possible to evaluate for each type of penetrometer the summarized work, which has the following condition for the heavy dynamic penetrometer (Rollberg 1976)

$$A(HDP) = 5{,}0 \,\Sigma\, n20 \cdot (1/20) \,[kNm] \quad (5)$$

and for the light dynamic penetrometer

$$A(LDP) = 1{,}0 \,\Sigma\, n'20 (1/20) \quad [kNm] \quad (6)$$

As theoretical values between the different dynamic penetrometers can be calculated according to Rollberg (1976)

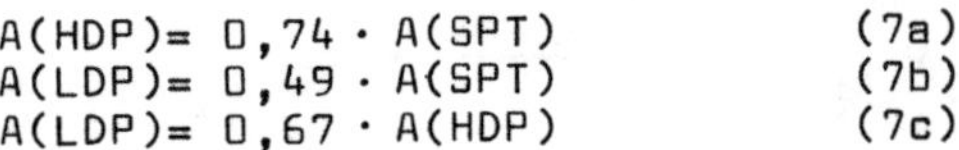

$$A(HDP) = 0{,}74 \cdot A(SPT) \quad (7a)$$
$$A(LDP) = 0{,}49 \cdot A(SPT) \quad (7b)$$
$$A(LDP) = 0{,}67 \cdot A(HDP) \quad (7c)$$

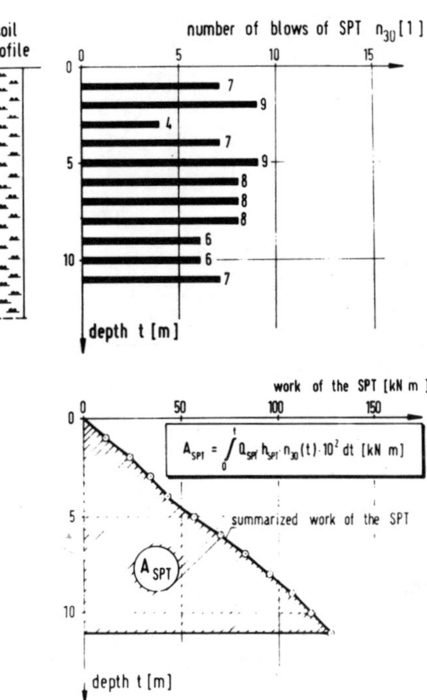

Fig.3. Result of a SPT diagram with calculation of the summarized work over the depth

7 STATISTICAL COMPARISON OF PENETRATION RESULTS AND SUMMARIZED WORKS

For a penetration profile between comparable values of SPT and HDP for the ratio between the blow-counts n20/n30, a mean value of 1,176 was calculated at 11 measured values (Fig.4a). The variation coefficient is 0,306 (Fig.4a) and represents one statistical characteristic value to evaluate the scattering of the measured values. Another possibility is the comparison of the mean value and the theoretical value according to equation (2) and the calculation of the probability of failure PF (s.a. Schultze 1977,1979a and b). At the present investigation the value PF can be adopted as a"reliability" between empirical and theoretical values (Biedermann 1982).

For the normal distribution of the ratio of summarized work between HDP and SPT the variation coefficient is

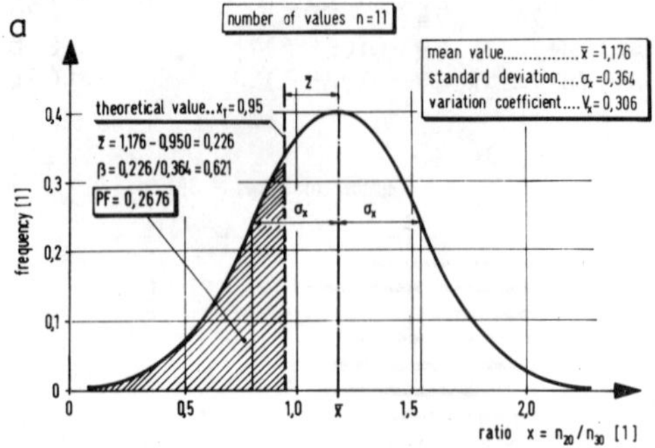

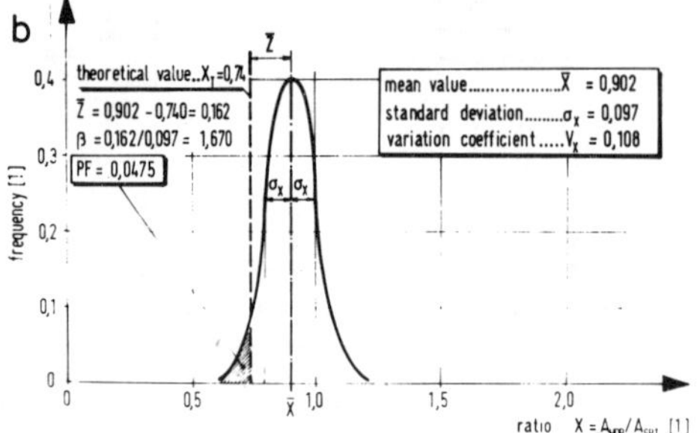

Fig.4. Statistical evaluation as standard normal distributions:
a) ratio of the blow-counts n20,n30
b) ratio of the summarized work of the heavy and standard penetrometer

only 0,108 and therefore much smaller than for the ratio of the blow-counts (Fig.4b). For the probability of failure PF, which represents the area of the normal distribution left of the cut-off by the theoretical value, the calculated value 0,0475 is also much smaller in this case, than for the ratio n20/n30 with 0,2676 (Fig.4).

An evaluation of 18 different penetration profiles between HDP and SPT in silt for summarized work shows for the mean values a range of 0,63 to 1,22 . The variation coefficients are lying between 0,04 and 0,21 , which are much lower than at the evaluation of the ratio n20/n30.

Comparing all the 127 values for the summarized work for each depth of 1 m, shows a mean value of 0,857. The standard deviation is 0,204 and the variation coefficient 0,238, which means a little scattering taking into consideration the different investigation areas and the different geological history of the silt. For the probability of failure the difference between the empirical and theoretical value means $\bar{z}$ = 0,857 - 0,74 = 0,117. For the probability coefficient follows $\beta = \bar{z}/\sigma z$ = 0,117/ 0,204 = 0,574. For the probability of failure can therefore be written PF= 0,2830, which means a value for the reliability between theoretical and measured values.

More detailed explanation of the theoretical background and statistical results between the heavy and light dynamic penetrometer will be presented by Biedermann (1982).

8 CONCLUSIONS

As a result of the present investigation can be ascertained:

1. The simple working equation is appropriate to calculate the results from one type of penetrometer into another with sufficient exactness also for silt,
2. The summarized work of penetration results fits better to the theoretical values because of its lower scattering and should therefore applied in the future for the conversion of penetration results.

9 REFERENCES

Biedermann,B. 1979, Dynamische und statische Sonden und ihre praktische Bedeutung in der Bodenmechanik. State-of-the-art-report,Symposium: Sondierungen und in-situ-Messungen,BVFA,Geotechnisches Institut, Wien.

Biedermann,B. 1982, Vergleichende Untersuchungen mit Sonden im Schluff,Doctor thesis,TH Aachen.

DIN 4094, Ramm- und Drucksondiergeräte, Part 1 and 2.

Rollberg,D. 1976, Bestimmung des Verhaltens von Pfählen aus Sondier- und Rammergebnissen, Forschungsberichte aus Bodenmechanik und Grundbau (FBG),Heft 4,Aachen.

Schultze, E. 1977, The probabilistic approach to soil mechanics design, Proc.9. ICSMFE,Tokyo,Vol.3: 5o1-5o3.

Schultze, E. 1979a,Slope Stability and the Bearing Capacity of shallow Foundations on Slopes,3.Int. Conf.Applic.Statistics and Probability,Sydney,Vol.3: 198-228.

Schultze, E. 1979b, Etat de l'art, stabilité des pentes et force portante de foundations superficielles, Journées: Approche Probabiliste de la Sécurité pour les Etudes de Mécanique des "Sols-Structures", Ecole Centrale des Arts et Manufactures, Paris.

Proceedings of the Second European Symposium on Penetration Testing / Amsterdam / 24-27 May 1982

Correlations between the results of sounding and laboratory tests for Rhineland silt and their use for calculations for foundation design

B.BIEDERMANN
Consulting Office for Soil Mechanics & Foundation Engineering, Würzburg, Germany

1 INTRODUCTION

In the existing investigation the results of the Standard Penetration test (n30), the heavy dynamic pene= trometer (n20), the light dynamic penetrometer (n'20) and the static penetrometer (σs) were compared with the results of laboratory tests like water content (wn), liquid limit (wL), plastic limit (wP), plasticity index (IP), consistency index (IC), degree of saturation (Sr), void ratio (ea) and the indices of compressibili= ty v and w from oedometer tests for the Rhineland silt. A comprehensive presentation of the soil mechanical properties of Rhineland silt was gi= ven by Kotzias (1963), Horn (1964), Krause (1966), Odendahl (1969), Schultze (1972) and Kahl (1972).

It is attempted to get from empi= rical relationships between the re= sults of sounding and laboratory tests with the blow-count or the cone resistance as well the shearing coefficients as the coefficients of compressibility. With these coeffi= cients it is possible to calculate the safety of ground failure, the slope stability and the settlements for the foundation design.

2 STATISTICAL RELATIONSHIPS WITH THE RESULTS OF THE STANDARD PENETRA= TION TEST

Between the results of the SPT and the results of the laboratory tests for Rhineland silt 101 till 164 measured values collected from re= search works and foundation reports were disposal. At the statistical evaluation for the single parameters the mean value, standard deviation and variation coefficient were cal= culated as significant indication for the scattering of the measured values. For the results of the labo= ratory tests the lowest variation coefficient with 0,20 was calculated for the plastic limit and the highest with 0,55 for the plasticity index. For the blow count n30 of the SPT, the variation coefficient reaches from 0,46 to 0,49 and the mean value from 9,5 to 10,7 which means an ac= ceptable order of magnitude for this type of soil.

Besides the statistical evaluation, linear regression were also executed to get relationships between the dif= ferent variables. For the signifi= cance of the equations of regression analysis the test statistic coeffi= cient and the confidence coefficient were examined (Fig. 1).

According to this criterion the empirical relationships between the blow count n30 and the liquid limit, the plastic limit, the plasticity index and the coefficient of com= pressibility v are significant (Fig. 1). The highest correlation coefficient was reached with the co= efficient of compressibility v with $r = 0,697$ (Fig. 1). In this case it is obvious, that for the equation of regression analysis the section of the axis is too high, which means at a blow count $n30 = 0$, already a value $v > 0$.

In a further step only those val= ues which belong, according to the diagram of Casagrande to the soil group CL, were compared. Then the variation coefficients of the results of the laboratory tests decrease partially considerable to values be= tween 0,08 and 0,44, whereas the variation coefficients of the blow count n30 are lying between 0,45 and 0,50. They have the same magni= tude like at the comparison with the investigation of all values. For this special soil group of Rhineland silt only between n30 and the water con= tent, the consistency index, the co= efficient of compression v and the void ratio statistical significant equations from the regression analy= sis could be calculated. The corre= lation coefficient between n30 and v increases to 0,801 which means a suf= ficient relationship for 74 compa= rable values.

General equation: $y = ax + b$

number	number of values	dependent variable	independent variable	regression coefficients		correlation coefficient	test statistic coefficient	confidence coefficient
	n	y	x	a	b	r	T	$T_{(n-2;95)}$
	1	1	1	1	1	1	1	1
1	2	3	4	5	6	7	8	9
1	164	n_{30}	W_n	-6,86	11,40	-0,077	0,986	1,975
		W_n	n_{30}	-0,001	0,21			
2	101	n_{30}	W_L	15,92	4,37	0,283	2,931	1,984
		W_L	n_{30}	0,005	0,27			
3	101	n_{30}	W_p	30,13	3,45	0,250	2,572	1,984
		W_p	n_{30}	0,002	0,18			
4	101	n_{30}	I_p	-15,13	11,31	-0,213	2,173	1,984
		I_p	n_{30}	-0,003	0,15			
5	101	n_{30}	I_c	0,87	8,70	0,080	0,792	1,984
		I_c	n_{30}	0,007	0,06			
6	150	n_{30}	S_r	-3,40	12,80	-0,136	1,666	1,976
		S_r	n_{30}	-0,005	0,849			
7	163	n_{30}	v/p_a	0,09	1,46	0,697	12,343	1,975
		v/p_a	n_{30}	5,45	42,59			
8	163	n_{30}	w	-1,28	10,84	-0,039	0,491	1,975
		w	n_{30}	-0,001	0,481			
9	119	n_{30}	e_a	0,01	10,65	0,000	0,002	1,980
		e_a	n_{30}	0,000	0,644			

Fig.1. Results of the linear regres= sion analysis between the number of blow-counts n30 of the SPT and the results of laboratory tests

Selecting only those values which belong to the soil group CM according to the Casagrande diagram, the varia= tion coefficients of the results of the laboratory tests decrease once more to a range of 0,07 and 0,29, whereas those of the blow-counts n30 decrease only to 0,40. The empirical equations between the results of the laboratory tests and the blow-counts show only with the coefficient of compressibility v a significant result with a correlation coefficient of 0,864, for 18 comparable values.

3 RELATIONSHIPS BETWEEN HEAVY DYNAMIC PENETROMETER AND LABORA= TORY TESTS

For the comparison with the blow-count n20 of the heavy dynamic pene= trometer only 16 to 33 values were available. At the statistical inves= tigation the variation coefficients of the results of the laboratory tests are situated between 0,09 and 0,42 and for the blow-count n20 be= tween 0,42 and 0,51. Only with the index of consistency and the void ratio significant linear equations from the regression could be calcu= lated, with a highest correlation coefficient of 0,595 between n20 and IC (see Fig.2).

General equation: $y = ax + b$

number	number of values	dependent variable	independent variable	regression coefficients		correlation coefficient	test statistic coefficient	confidence coefficient
	n	y	x	a	b	r	T	$T_{(n-2;95)}$
	1	1	1	1	1	1	1	1
1	2	3	4	5	6	7	8	9
1	33	n_{20}	w_n	-17,05	10,80	-0,266	1,536	2,040
		w_n	n_{20}	-0,004	0,22			
2	23	n_{20}	w_L	-2,405	14,668	-0,020	0,092	2,080
		w_L	n_{20}	-0,0002	0,316			
3	23	n_{20}	w_p	20,278	9,946	0,074	0,340	2,080
		w_p	n_{20}	0,0003	0,192			
4	23	n_{20}	I_p	-10,132	15,111	-0,067	0,308	2,080
		I_p	n_{20}	-0,0004	0,124			
5	23	n_{20}	I_c	9,393	4,055	0,595	3,392	2,080
		I_c	n_{20}	0,0377	0,525			
6	32	n_{20}	S_r	-1,16	8,37	-0,077	0,422	2,042
		S_r	n_{20}	-0,005	0,759			
7	33	n_{20}	v/p_a	0,021	5,36	0,239	1,371	2,040
		v/p_a	n_{20}	2,68	84,77			
8	33	n_{20}	w	2,47	6,55	0,098	0,551	2,040
		w	n_{20}	0,004	0,402			
9	27	n_{20}	e_a	-16,02	18,64	-0,383	2,073	2,060
		e_a	n_{20}	-0,009	0,77			

Fig.2. Results of the linear regres= sion analysis between the number of blow-counts n20 of the heavy dynamic penetrometer and the results of labo= ratory tests

4 RELATIONSHIPS BETWEEN THE LIGHT DYNAMIC PENETROMETER AND LABORA= TORY TESTS

The statistical results of the 18 to 34 comparable values determine for the laboratory tests variation coef= ficients between 0,09 and 0,66 and for the number of blows n'20 of the light dynamic penetrometer between 0,40 and 0,61 . At the regression analysis between the blow-count n'20 and the water content, the liquid limit, the plasticity index, the in= dex of consistency, the degree of saturation and the index of com= pressibility v could be calculated statistical significant equations (Fig.3).

General equation : $y = ax + b$

number	number of values	dependent variable	independent variable	regression coefficients		correlation coefficient	test statistic coefficient	confidence coefficient
	n	y	x	a	b	r	T	$T_{(n-295)}$
	1	1	1	1	1	1	1	1
1	2	3	4	5	6	7	8	9
1	34	n'_{20}	w_n	-72,92	32,22	-0,453	2,877	2,037
		w_n	n'_{20}	-0,003	0,24			
2	19	n'_{20}	w_L	-202,906	82,497	- 0,574	2,890	2,110
		w_L	n'_{20}	- 0,016	0,334			
3	19	n'_{20}	w_p	- 3,902	22,815	- 0,006	0,025	2,110
		w_p	n'_{20}	- 0,000	0,195			
4	19	n'_{20}	I_p	- 207,328	42,233	- 0,607	3,149	2,110
		I_p	n'_{20}	- 0,002	0,138			
5	19	n'_{20}	I_c	3,441	18,930	0,474	2,220	2,110
		I_c	n'_{20}	0,0065	0,765			
6	33	n'_{20}	S_r	-16,11	30,68	-0,425	2,611	2,040
		S_r	n'_{20}	-0,01	0,94			
7	34	n'_{20}	v/p_a	0,14	5,12	0,685	5,314	2,037
		v/p_a	n'_{20}	3,29	33,58			
8	34	n'_{20}	w	- 3,43	20,29	-0,058	0,330	2,037
		w	n'_{20}	- 0,001	0,50			
9	26	n'_{20}	e_a	4,70	17,13	0,037	0,180	2,064
		e_a	n'_{20}	0,00	0,70			

Fig.3. Results of the linear regres= sion analysis between the number of blow-counts n'20 of the light dynamic penetrometer and the results of labo= ratory tests

5 RELATIONSHIPS BETWEEN THE STATIC PENETROMETER AND LABORATORY TESTS

The variation coefficients of the cone resistance have values between 0,54 and 0,76 which means a great scattering of the results, while those of the laboratory tests are much smaller and lie only between 0,09 and 0,40. The regression anal= ysis shows only between the cone resistance and water content, plas= tic limit and coefficient of com= pressibility v significant equations (Fig.4).

A detailed description and further= more interpretations of the results especially also comparisons with results of the literature are pres= ented by Biedermann (1982).

General equation : $y = ax + b$

number	number of values	dependent variable	independent variable	regression coefficients		correlation coefficient	test statistic coefficient	confidence coefficient
-	n	y	x	a	b	r	T	$T_{(n-2,95)}$
-	1	1	1	1	1	1	1	1
1	2	3	4	5	6	7	8	9
1	62	σ_s/p_a	W_n	- 202,16	66,39	-0,508	4,573	2,000
		W_n	σ_s/p_a	-0,001	0,21			
2	27	σ_s/p_a	W_L	-137,31	65,40	-0,219	1,122	2,060
		W_L	σ_s/p_a	-0,000	0,304			
3	27	σ_s/p_a	W_p	-414,60	105,03	-0,392	2,129	2,060
		W_p	σ_s/p_a	- 0,000	0,203			
4	27	σ_s/p_a	I_p	-7,82	25,58	-0,013	0,067	2,060
		I_p	σ_s/p_a	-0,000	0,104			
5	27	σ_s/p_a	I_c	15,67	8,53	0,339	1,801	2,060
		I_c	σ_s/p_a	0,007	0,86			
6	54	σ_s/p_a	S_r	-19,50	44,44	-0,215	1,591	2,007
		S_r	σ_s/p_a	- 0,002	0,75			
7	62	σ_s/p_a	v/p_a	0,19	11,61	0,474	4,166	2,000
		v/p_a	σ_s/p_a	1,21	71,94			
8	62	σ_s/p_a	w	18,83	24,08	0,156	1,220	2,000
		w	σ_s/p_a	0,001	0,39			
9	43	σ_s/p_a	e_a	11,29	26,67	0,071	0,454	2,020
		e_a	σ_s/p_a	0,000	0,51			

Fig.4. Results of the linear regres= sion analysis between the cone re= sistance σ_s of the static penetrome= ter and the results of laboratory tests

6 CALCULATION OF THE ANGLE OF INTER= NAL FRICTION

With well-known and appropriate em= pirical equations it is possible to estimate the angle of internal fric= tion indirectly from the results of penetration tests. For the Rhineland silt the relationship between the an= gle of friction and the index of plas= ticity (Odendahl 1969) will be deter= mined to

$\varphi = 45,38 - 5,53 \cdot \ln IP \pm 3,99$ (1)

with IP in % (s.a.Schormann 1979). The statistical significant equation between n30 and IP for 101 comparable values is according to Fig. 1:

$IP = - 0,003 \cdot n30 + 0,149$ (2)

With the given blow-count n30 it is easy to calculate immediately the an= gle of internal friction with the equations 1 and 2. For relevant blow-counts for the silt n30 = 5 to 15, the following angles of friction can be estimated to n30 = 5 → IP = 0,134: $\varphi = 31,03 \pm 3,99$; n30 = 10 → IP=0,119: $\varphi = 31,68 \pm 3,99$; n30 = 15 → IP=0,104: $\varphi = 32,43 \pm 3,99$.

According to the results of triaxial tests for silt, the calculated values represent in a good manner the rele= vant angles of friction for silt. But it must be recognized, that for a wide range of the blow-count n30, the angle of internal friction scatters only very little.

With the mean value of the 101 blow-counts $\bar{n}30 = 9{,}48$ and the equivalent value for the plasticity index IP = 0,12 = 12[%] according to equation (2), the angle of friction according to equation (1) is then $\varphi = 31{,}64^\circ \pm 3{,}99$. The value 3,99 must be considered, because it represents the scattering of the measured values. Therefore the angle of friction has a range for the mean value $\bar{n}30 = 9{,}48$ between $27{,}65^\circ$ and $35{,}63^\circ$.

One possibility to introduce the reliability of the number of measured values is to get the lower safety coefficient for a normal distribution with an estimated fractile. For a mean value $\bar{n}30 = 7$ at 6 measured values of the SPT at a fractile of 0,10 and a variation coefficient of 0,50 the lower safety factor can be calculated to $\eta = 1/(1 - V \cdot T) = 1/(1 - 0{,}5 \cdot 0{,}603) = 1{,}432$. The equivalent values of IP and φ are now 13,43[%] and $31{,}02^\circ \pm 3{,}99$. Because of the very small sensitivity of the equation between n30 and IP, the scattering of the angle of internal friction is also very little.

On the other hand the lower coefficient of safety is also available for the calculated index of plasticity IP, which means at n30 = 7 with equation (2) a value of 12,8%. The multiplication with the safety factor of 1,432 gives for the reduced IP=18,3%, and means introduced in equation (1) an angle of friction of $\varphi = 29{,}3^\circ \pm 3{,}99$. For the lower range of the scattering a value of $\varphi = 25{,}3^\circ$ can be testified.

Besides the application of the safety factor on the measured values of the blow-count n30 or the index of plasticity IP, it is also possible to subject the calculated angle of friction with this coefficient. But in this case a relevant variation coefficient for the angle of friction must be introduced which can adopted by experience with V = 0,20 which leads to $\eta = 1/(1 - 0{,}2 \cdot 0{,}603) = 1{,}137$. For n30 = 7 → IP = 12,8% the result for the angle of friction is $\varphi = 31{,}28^\circ \pm 3{,}99$. Divided by the safety coefficient the angle of friction is then $\varphi = 27{,}5^\circ \pm 3{,}99$. The value for the lower range can be calculated to $\varphi = 23{,}51^\circ$. Applying the safety factor on $\tan\varphi$ the result for the angle of friction is then $\varphi = 28{,}12^\circ \pm 3{,}99$, and for the lower range $24{,}1^\circ$.

With the results of the other types of penetrometers the same procedure can be performed for calculation of the angle of friction and then to get the safety of ground failure or slope stability.

7 CALCULATIONS FOR SETTLEMENT PREDICTION

For the settlement prediction several methods are existing for the silt. First of all it is possible to calculate the modulus of compressibility Es from well-known empirical equations with the blow-count n30 for this type of soil.

According to Menzenbach(1959) is

$$Es = 12{,}0 + 5{,}8 \cdot n30 \qquad (3)$$

which means at a mean value of $\bar{n}30 = 9{,}5$ for Es = 67,1 (kp/cm2)=6,71 (MN/m2). The equation of Schultze (1972)

$$Es = 27{,}0 + 5{,}0 \cdot n30 \qquad (4)$$

brings Es up to 74,5 (kp/cm2) = 7,45 (MN/m2). At both equations (3) and (4) the modulus of compressibility is not depending on the contact pressure (s.a.Schormann 1979).

In this case also several equations are existing to calculate the coefficients of compressibility v and w from results of the standard penetration test or from laboratory tests and then to calculate the modulus of compressibility according to the equation

$$Es = v\,(\sigma)\ \mathrm{EXP}w \qquad (5)$$

Kahl (1972) has suggested to apply empirical equations for v and w which depend on the degree of saturation, the void ratio and the index of consistency. With the mean values of this laboratory tests the results are: v = 70,8 (kp/cm2)= 7,08 (MN/m2) and w = 0,65. With a contact pressure of σ = 150 (kN/m2) the modulus of compressibility according to equation (5) is then Es = 92,2 (kp/cm2)= 9,22 (MN/m2).

With the equation from Fig.1 the coefficients of compressibility v and w can be calculated at a mean value of $\bar{n}30 = 9{,}5$ to v = 94,4 (kp/cm2) = 9,44 (MN/m2) and w = 0,47, which means at a contact pressure of σ = 150 (kN/m2) a modulus of compressibility Es = 114,2 (kp/cm2) = 11,4 (MN/m2).

It is therefore easy to calculate with the modulus of compressibility, the measures of the foundation and the settlement factor the predicted settlements.

For a given single footing with dimensions of a = b = 1,2 (m), foundation depth t = 1,0 (m), thickness of the compressible layer ds = 2,4 (m) and contact pressure σ = 150 (kN/m2) the settlement can be calculated to

s1 = 150•1,2•0,6381/9220 = 1,1 (cm) with the Es-value according to Kahl (1972)

or

s2 = 150•1,2•0,6381/11400 = 1,0 (cm) with the Es-value according to Fig.1. Another possibility to get directly from the blow-count n30 to the calculated settlement is the empirical equation of Sherif (1973). With this equation and a factor of 0,1163, depending on the type of soil, the settlement for the example of the single footing is then s3 = 1,3 (cm) which is in the same order of magnitude as the values for the settlements calculated before.

8 CONCLUSIONS

According to the present investigation it is possible to calculate the angle of internal friction for silt with sufficient exactness from the results of penetration tests supported by empirical equations between soil mechanical coefficients from laboratory tests and the angle of friction. Safety coefficients according to statistical methods can also be adopted.

The settlement prediction can also be done by empirical equations based on penetration results. It is possible to determine the modulus of compressibility from coefficients of compressibility in consideration of the contact pressure. With the dimensions of the foundation and the convenient settlement-coefficient it is therefore easy to estimate the probable settlements.

9 REFERENCES

Biedermann,B. 1982,Vergleichende Untersuchungen mit Sonden im Schluff, Doctor thesis, TH Aachen.

Horn,A. 1964,Die Scherfestigkeit von Schluff, Forschungsberichte des Landes Nordrhein-Westfalen Nr.1346, Westdeutscher Verlag,Köln/Opladen.

Kahl,W. 1972,Geologie und Bodenmechanik des Rheinischen Schluffs, Mitt.VGB 54, Aachen.

Kotzias,P.C.1963,Die Zusammendrückbarkeit von Schluff, Mitt.VGB 28, Aachen.

Krause,J.1966, Das rheologische Verhalten von Schluff beim Kompressionsversuch, Mitt.VGB 35,Aachen.

Menzenbach,E. 1959,Die Anwendbarkeit von Sonden zur Prüfung der Festigkeitseigenschaften des Baugrundes, Forschungsberichte des Landes Nordrhein-Westfalen,Westdeutscher Verlag, Köln/Opladen, Nr. 713.

Odendahl,R. 1969, Die Scherfestigkeit des ungestörten rheinischen Schluffs, Forschungsberichte des Landes Nordrhein-Westfalen Nr. 2039, Westdeutscher Verlag, Köln/Opladen.

Schormann,K. 1979,Baugrunduntersuchungen des rheinischen Schluffs durch leichte Rammsondierungen, Forschungsberichte des Landes Nordrhein-Westfalen Nr. 2825, Westdeutscher Verlag, Opladen.

Schultze,E.1972, Bodenmechanische Probleme bei Schluff, Mitt.VGB 55, Aachen.

Sherif,G.1973,Setzungsmessungen an Industrie- und Hochbauten und ihre Auswertung, Mitt.VGB 57, Aachen.

Proceedings of the Second European Symposium on Penetration Testing / Amsterdam / 24-27 May 1982

Pore water pressures generated during dynamic penetration testing

C.R.I.CLAYTON
University of Surrey, Guildford, UK

S.S.DIKRAN
National Center for Construction Laboratories, Baghdad, Iraq

SUMMARY

This paper presents a study of the excess pore water pressures generated at the tip of a cone penetrometer during dynamic penetration testing in a fine sand. For this purpose a large triaxial chamber was constructed in which a sand specimen 435mm in diameter and 630-690mm in height could be housed. A brief description of the testing chamber is given together with details of the penetrometer used. The results show that negative pore water pressures are set up in medium dense, dense and very dense specimens; positive pore pressures are set up in loose specimens which may cause liquefaction of specimens which are confined under low mean effective principal stresses.

INTRODUCTION

In 1948 Terzaghi and Peck proposed a correction for SPT results in very fine or fine silty sands below the water table. When 'N' values were greater than 15 it was proposed that they should be reduced. The assumption was that dense sands would dilate during driving, leading to overestimates of relative density. Thus loose sands might be expected to generate positive excess pore pressures, and reduce driving resistance. The measurement of excess pore pressures during driving is difficult, requiring a system with a very rapid response time. Thus the introduction of the pore pressure transducer to the testing of soil, in the late 1950's, (for example Whitman, Richardson and Healy (1961)) marked the earliest time for this work to be undertaken.

Attempts to measure the excess pore water pressure generated during advancement of a penetrometer are relatively recent. The importance of this problem was pointed out at the First European Symposium on Penetration Testing in Stockholm, (Senneset 1974, Janbu and Senneset 1974, Schmertmann 1974, Hansbo 1974, Bemben and Myers 1974, and Kok 1974) and indeed in the Group Discussions on future developments it was pointed out that it is especially important to study the pore water pressure developed around the tip of a penetrometer and the rate effect during quasi-static penetration. Most of the attention, however, was given to quasi-static penetration and no mention was given to the pore water pressures developed around the tip of a penetrometer during dynamic penetration. Subsequently, work on the magnitudes of pore pressures generated during quasi-static penetration testing has been reported by Wissa et al (1975), Torstenson (1975), Massarch et al (1975), Baligh et al (1977), and Rocha Filho (1979).

For the specific case of pore water pressure generated during dynamic penetration of a cone, the information from literature is rather scarce. Where available, it deals with the excess pore pressures built up during pile driving, and this is mainly in clay soils. Pore pressures in sand have been measured by Möller and Bergdahl (1981).

This paper reviews an experimental investigation carried out in order to study the excess pore water pressures set up around the tip of a dynamic penetrometer.

GENERAL DESCRIPTION OF THE TRIAXIAL CHAMBER

The test chamber can be used to control stresses at the specimen boundaries in both vertical and horizontal directions. A cross-section of the chamber is shown in Fig.1.

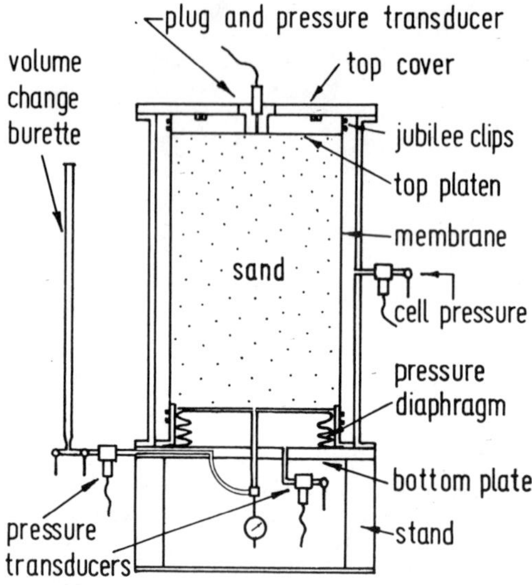

FIGURE 1. GENERAL LAYOUT OF THE TRIAXIAL CHAMBER

Vertical stresses are applied to the specimen via a pressure diaphragm in the base of the chamber. This diaphragm is filled with water and pressure is applied by means of an air-water pressure system. This compresses the sand specimen against the top platen. A stainless steel drainage tube is fixed to the centre of the diaphragm and emerges from the bottom plate through a brass bushing. Vertical movement of the specimen is measured by means of a dial gauge mounted on this tube underneath the chamber.

Lateral stresses are applied by increasing the pressure of the water surrounding the specimen, which is enclosed in a 1.2mm thick latex rubber membrane. The vertical and horizontal stresses are measured by means of two electrical pressure transducers which are fixed to the outer cylinder wall and the bottom plate. These pressures are read on digital readout units. A 50mm diameter hole is provided in the top platen. During consolidation of the specimen a plug seals this hole. A pressure transducer is mounted into this plug in order to measure pore water pressures provided by suction at the bottom during specimen preparation, and during consolidation. At the end of consolidation of the specimen along the desired stress paths, this plug is removed and replaced by another plug with a 22mm diameter hole through it. Through this hole the dynamic penetration testing is carried out.

SPECIMEN PREPARATION

The chamber was designed to contain saturated and dry sand specimens. To prepare saturated specimens, the rubber membrane is fixed with two Jubilee clips to the angle ring at the bottom of the chamber. The specimen former is then placed in position with the rubber membrane pulled up inside it and overlapped around the top of the former. The former with the membrane inside is then filled with water and the "sieve" lowered inside until it touches the base. This "sieve" has a diameter equal to that of the specimen and has equi-distant holes designed to achieve homogeneous specimens in suitable periods. Sand is then lowered onto the sieve by means of hand scoops. When the sieve is filled with sand, it is raised very slowly to a height equal to that of the sieve (150mm). This procedure is repeated until the desired height of the specimen is achieved. The sieve is then removed and the top platen lowered into position. The rubber membrane is fixed to it by means of two Jubilee clips. Before removing the specimen former, a suction of about 25kN/m^2 is applied to the bottom of the specimen by means of a vacuum pump. Once the pore pressure transducer in the plug indicates that the vacuum has extended to the top of the specimen, the former may be removed. The space around the specimen is now filled with water to the top of the specimen and the top cover is placed into position and bolted to the outer cylinder. The top platen of the specimen is sealed against the top cover. Vertical and horizontal stresses are then applied to the specimen in order to achieve a desired stress path. During application of these stresses readings are made of the volume change and the vertical movement of the specimen during consolidation. At the end of the consolidation of the specimen to the desired stress path, penetration testing can be carried out.

THE PENETROMETER

The penetrometer is shown in Fig.2. It consists of a 20mm diameter by 200mm long 60° cone ended stainless steel probe mounted on 15mm diameter stainless steel rods. The cone end of the penetrometer is formed of porous carborundum, and behind this tip is mounted a Druck (PDCR 81) 7 bar miniature pore pressure transducer. The penetrometer is driven into the sand using a free fall automatic trip hammer which allows a 10kg weight to fall freely through 431mm onto a striker plate screwed to the top of the

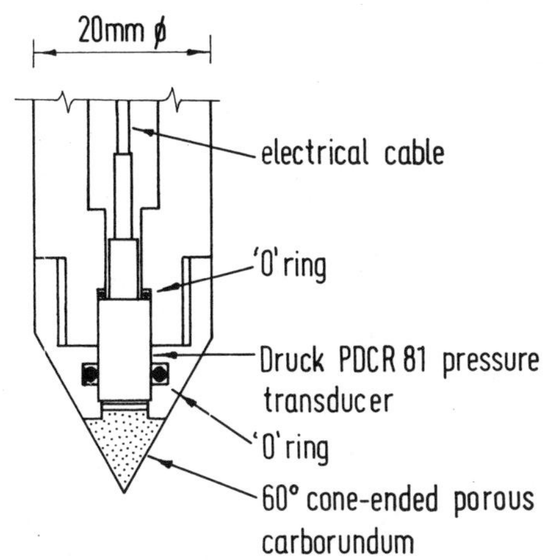

FIGURE 2. PENETROMETER TIP

penetrometer rods. Its design is a small scale version of the trip hammers commonly in use in site investigation in the U.K.

The weight and height of drop of the hammer were initially estimated on the basis of dimensional analysis to give a similar range of 'N' values as is achieved by the full scale SPT. Blow counts for the 164mm penetration distance given by this analysis have ranged between 2 and 27 for fine sand under low confining pressures.

The response of the tip to pore pressure changes in the surrounding soil is clearly of crucial importance. The PDCR 81 diaphragm has a volume take of 1.7×10^{-7} cm^3/bar, and the porous ceramic has an estimated conductance of 8.3×10^{-2} cm^3/bar/s. The carborundum tip presents a much lower resistance to flow, having a measured conductance of $4.8 cm^3$/bar/s. Assuming that all water flowing into the device, as a result of both the flexure of the transducer diaphragm and the compression of the water within the system, must flow through both the carborundum tip and the transducer ceramic, the time for 95% equalization of the device was very conservatively estimated as better than 0.2ms.

Three recording devices were used. The transducer was linked through an IEEE - 488 compatible strain gauge amplifier to a storage oscilloscope, a chart recorder, and a minicomputer. This system allowed the acquisition of both the rapid and slow transient pressures.

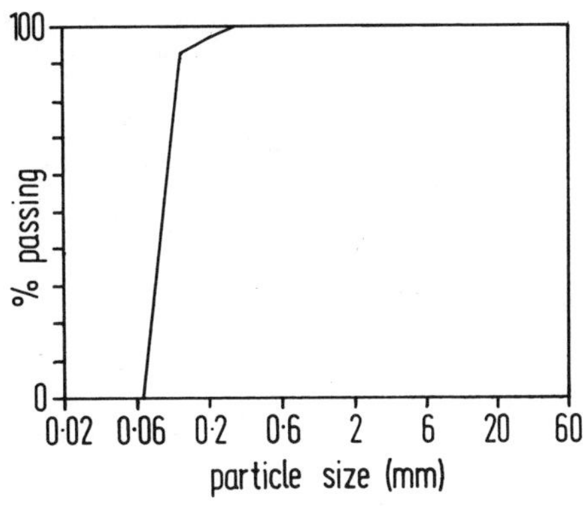

FIGURE 3. GRADING OF SAND

THE SAND SPECIMENS

The soil used in the experiments described below was subangular uniform fine quartz sand. (Leighton Buzzard, 100 - 200 mesh) (Figure 3). The sand was sedimented through water using three techniques in order to provide loose, medium dense, dense and very dense specimens, as follows:

Method 1 Height of fall of sand kept to a minimum by slowly raising the sieve. Typical relative density 25%

Method 2 Sieve raised in 15cm lifts, giving a maximum height of fall of 15cm. Typical relative density 40%

Method 3 Sand placed as in Method 2, and compacted in three layers using a poker vibrator. Typical relative density 75 - 95%

The minimum and maximum densities for the sand were determined according to the methods given by ASTM D2049 and Akroyd (1957) respectively. These were

$$\gamma d \text{ min} = 1.33 \text{ Mg/m}^3$$

$$\gamma d \text{ max} = 1.65 \text{ Mg/m}^3$$

After sedimentation, specimens were subjected to either isotropic or K_o stress increases. Some specimens were tested when normally consolidated, whilst others were over-consolidated. With the exception of one test, the final effective vertical stress level was between 50 and 65kN/m^2. One test was carried out with $\sigma_v' = 200$kN/m^2 and $\sigma_h' = 118$kN/m^2 to judge the effect of stress level on pore pressure response.

Table 1. Summary of pore pressure responses observed during dynamic penetration testing in a fine sand

Pore water pressure magnitude and elapsed time	Relative density before test (%)		
	0 - 35	35 - 85	85 - 100
Typical Values			
1st peak kN/m^2	+40 to +60	-40 to -80	-40 to -50
ms	(5), 20	3 to 10	10
2nd peak kN/m^2	-	+40 to +60	-
ms	-	20 to 30	-
Maximum Values			
1st peak kN/m^2	(-100), +64	-100	-82
2nd peak kN/m^2	-	+64	-

TEST RESULTS

A summary of the test results is given in Table 1. Typical shapes of pore pressure curves generated during single blows during various tests are shown in Fig.4. These figures were traced directly from photographs taken of the oscilloscope screen during each blow in the testing. The results were analysed by studying groups of photographs in each test. The results have shown that large pore pressures are generated, and that the general pattern and duration from moment of impact to significant dissipation differ considerably in the various tests. There seem no significant differences, however, in the magnitude of pore pressures in the various tests. The effects of the relative density, confining pressures and overconsolidation on the pore pressures are discussed below.

Relative density has a noticeable effect on the pore pressures generated during dynamic penetration tests, Table 1, as postulated by Terzaghi and Peck. However, in more than 300 observations of the pore pressures during driving it has been observed that

(a) the pore pressure response is broadly similar over the range of relative density between about 35% and 85%, and

(b) all tests show very rapid fluctuations in pore pressure in the initial part of the record, which are probably associated with rolling and sliding of soil particles during shear.

In the relative density range 35-85%, for the fine sand tested, pore pressures were initially negative for a period of about 6-15ms, after which they became positive. (Figure 4(b) and (c)). The peak negative pressures typically had a magnitude of 40-80kN/m^2 and occurred some 3-10ms after impact. Peak positive pressures typically had a magnitude of 40-60kN/m^2 and occurred some 20-30ms after impact.

In very dense specimens mainly negative pore pressures were generated by penetrometer driving. In any one specimen the blow to blow response was very much more uniform than in the case of medium dense and dense sand specimens. Typical peak pore pressures had a magnitude of 40-50kN/m^2 and occurred some 15ms after impact (Figure 4(e) and (f)).

The pore pressure changes for both medium dense to dense and for very dense sand are similar in trend to those observed at the side of a 20mm dia. model steel pile by Möller and Bergdahl (1981), who used a fine sand with a particle size "mainly between 0.125mm and 0.25mm".

The magnitude of the pressure changes in very dense sand reported here are, however, much greater than those of Möller and Bergdahl (cf. 40-50kN/m^2 and 10-20kN/m^2), whilst in medium dense to dense sand the peak pore pressures were generally slightly larger, and could be considerably larger (cf. -80kN/m^2 and -50kN/m^2, +60kN/m^2 and +20kN/m^2). Furthermore, the times to peak pressure in our experiments are longer than those reported by Möller and Bergdahl.

In loose sands (relative density 20-35%) pore pressures could show some rapid negative movement, but generally they became strongly positive immediately after impact, (Fig.4(a)). In both the specimens prepared loose, liquefaction occurred beneath the penetrometer during testing and was followed by a rapid unchecked downward penetrometer movement and the boiling of sand through the specimen top plate. In these cases, peak positive pore pressure changes of 60-65kN/m^2 were measured in a sand which initially had horizontal effective stresses of 30-32kN/m^2 and vertical effective stresses of 60-62kN/m^2. The pore pressures created by blows before

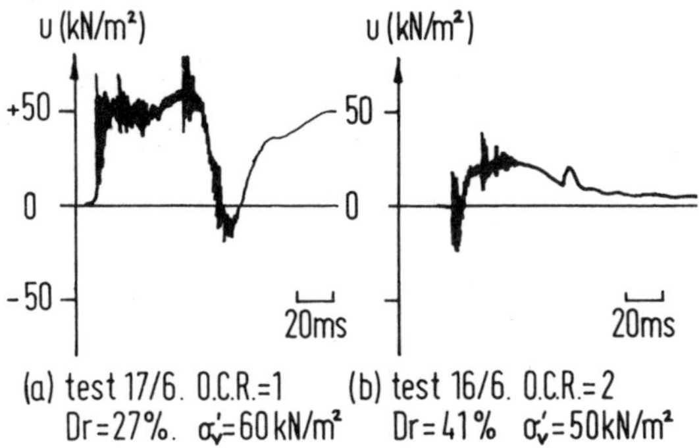

(a) test 17/6. O.C.R.=1 Dr=27%. σ_v'=60kN/m²
(b) test 16/6. O.C.R.=2 Dr=41% σ_v'=50kN/m²

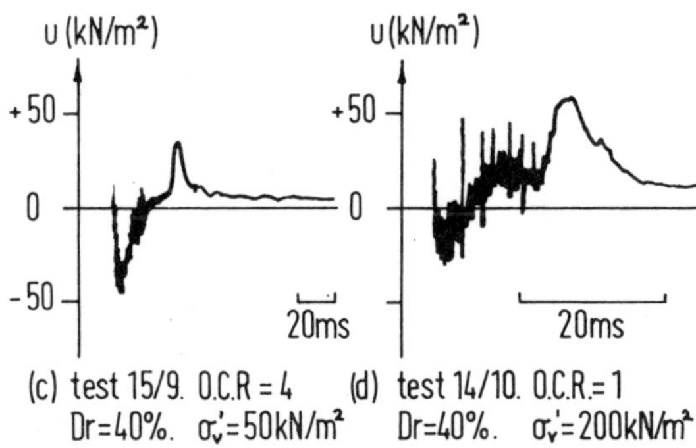

(c) test 15/9. O.C.R = 4 Dr=40%. σ_v'=50kN/m²
(d) test 14/10. O.C.R.=1 Dr=40%. σ_v'=200kN/m²

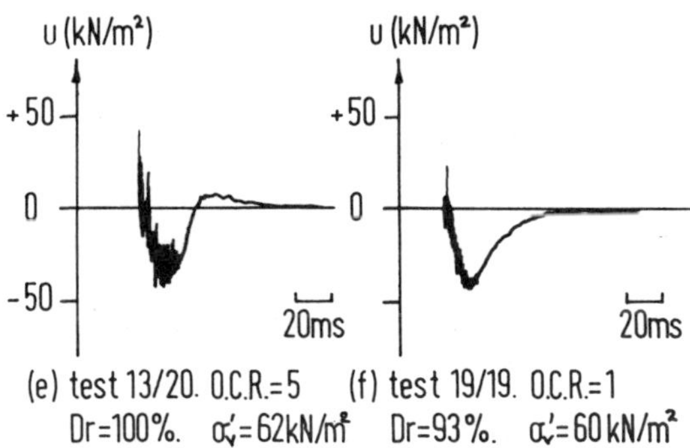

(e) test 13/20. O.C.R.=5 Dr=100%. σ_v'=62kN/m²
(f) test 19/19. O.C.R.=1 Dr=93%. σ_v'=60kN/m²

FIGURE 4. TYPICAL PORE PRESSURE CURVES

liquefaction occurred had values ranging from 20kN/m^2 to 56kN/m^2.

The decay of pore pressures as they dissipated through the sand was not as rapid as might be expected. Oscilloscope traces typically indicated that at least 90% dissipation would occur between 25ms and 80ms after impact but chart recorder output indicated that dissipation to within 1kN/m^2 of equilibrium pressure could take considerably longer. In test 17, for example, the pore pressures of the initial four blows dissipated to 1kN/m^2 within 2.0 to 2.5s. Pore pressures generated by blow 5 lasted for about 11s, while the liquefaction under blow 6 gave pore pressures which took some 50s to dissipate. Whilst overconsolidation (and horizontal effective stress) had no noticeable effect on either the magnitude or duration of pore pressures, it was observed that the duration of pore pressures was significantly shorter when specimens of similar relative density were tested under higher effective stress levels, compare Figs. 4(c) and 4(d).

CONCLUSIONS

(i) large pore water pressures have been observed during dynamic penetration testing

(ii) the pattern of pore pressure response is significantly affected by the relative density of the specimen

(iii) local liquefaction can be induced by dynamic penetration testing, if the relative density and effective confining pressures are low.

REFERENCES

Akroyd, T.N.W. 1957, Laboratory Testing in Soil Engineering, London Soil Mechanics Ltd.

A.S.T.M. D2049-69, Standard Test Method for Relative Density of Cohesionless Soils, American Society for Testing and Materials, Philadelphia.

Baligh, M. M., C. C. Ladd & V. Vivatrat 1977, Exploration and evaluation of engineering properties of marine soils for foundation design of offshore structures, Report of Massachussetts Institute of Technology, Report no. MIT SG77-18.

Bember, S. M. & D. A. Myers 1974, The Influence of rate of penetration on static cone resistance values in Connecticut River Valley varved clay, Proc. Europ. Symp. on Penetration Testing, Stockholm, 2.2 : 33-34.

Hansbo, S. 1974, Pore pressure sounding apparatus, General discussions on future developments, Proc. Europ. Symp. on Penetration Testing, Stockholm, 2.1 : 109.

Janbu, N. & K. Senneset 1974, Effective stress interpretation of in-situ static penetration tests, Proc. Europ. Symp. on Penetration Testing, Stockholm, 2.2 : 181-193.

Kok, L. 1974, The effect of the penetration speed and the cone shape on the Dutch static cone penetration test results, Proc. Europ. Symp. on Penetration Testing, Stockholm, 2.2 : 215-220.

Massarch, K. R., B. B. Broms, & O. Sundquist 1975, Pore pressure determination with multiple piezometer, ASCE Specialty Conf. on in-situ measurement of soil properties, North Carolina State Univ., Raleigh, 1 : 260-265.

Møller, B. & U. Bergdahl 1981, Dynamic pore pressure during pile driving in fine sand, Proc. Tenth Int. Conf. Soil Mech. & Found. Eng., 2 : 791-794.

Rocha Filho, P. 1979, Behaviour of cone penetration tests in saturated sands, Ph.D. thesis, Univ. of London, Imperial College of Science & Technology.

Schmertmann, J. 1974, Penetration pore pressure effects on quasi-static cone bearing, q_c, Proc. Europ. Symp. on Penetration Testing, Stockholm, 2.2 : 345-352.

Senneset, K. 1974, Penetration testing in Norway Proc. Europ. Symp. on Penetration Testing, Stockholm, 1 : 85-95.

Torstensson, B. A. 1975, Pore pressure sounding instrument, ASCE Specialty Conf. on in-situ measurement of soil properties, North Carolina State Univ., Raleigh, 2 : 48-54.

Whitman, R. V., A. M. Richardson & K. A. Healy 1961, Time-lag in pore pressure measurements, Proc. Fifth Int. Conf. Soil Mech. & Found. Eng., 1 : 407-411.

Wissa, A.E.Z., R. T. Martin & I. E. Garlanger 1975, The piezometer probe, ASCE Specialty Conf. on in-situ measurement of soil properties, North Carolina State Univ., Raleigh, 1 : 536-545.

Proceedings of the Second European Symposium on Penetration Testing / Amsterdam / 24-27 May 1982

Comparative study of DPB and IS dynamic sounding test for Tapi alluvium

M.D.DESAI & B.J.MEHTA
S.V.R. College of Engineering & Technology, Surat, India

1 INTRODUCTION

The different types of dynamic cones used in India have been reviewed. (Desai M.D., GRS Jain etc 1974) The work on uncased dynamic cone based on Indian Standard 4968 Part-I (1968) as modified by Desai M.D. and Singh has been published at ESOPT Stockholm 1974.

The economy, time constraint for investigation at remote sites as well as multinational involvements in many projects brought out a need to adopt a universal standard test with reasonably reliable prediction of soil characteristics for a given geological deposit. Later statistical compilation of such local correlations, can lead to a safe, though not always economical, predictions of soil properties for design of foundation. In developing countries it has an advantage of saving time and cost and can be used by any skilled untrained operator without involving personal factors as is the case with SP test.

Hence authors extended their work to change over to DPB test prescribed by a group of ISSMFE at Tokyo 1977. The IS and modified version used in practice as well as DPB specifications have been compared by Desai M.D. (1980). It has been observed that present practice with 50.8 mm dynamic cone with 60° apex angle can be brought within the tolerances of DPB without any major change. The results had to be reported in form of energy in Mega pascal MPa units. (1 MPa $=10^6 N/m^2$

The present work is an attempt to establish conservative predictions of engineering properties of Tapi alluvium around Surat using DPB test and modified IS test in use.

2 STANDARDIZATION

The dynamic cone test used for analysis has a cone 50.8 mm diameter with apex anle of 60°. The cone is fixed in a jacket of 41 mm OD, 41 mm long which is connected to 'A' size drill roads. The guide and anvil weight 32.30 kg. The hammer used is 65 kg and fall is adjusted 750 mm. The hammer used has additional excess mass of 1.0 kg in IS system. The rate of blow is less than 30 blows/min. It appears that difference in cone angle, energy and size of rods may not be large when resistance is recorded in form of energy required to drive this cone. Thus with minor changes, hardly influencing the resistances recorded in MPa, the modified IS cone by Desai can be converted into DPB specified international cone system.

On few sites therefore the modified IS cone (Desai, 74) and DPB tests have been conducted side by side in Tapi allùvium for comparision. The result of modified cone has been interpreted by correlations developed by Desai M.D. & Vyas M.J. (1976). These correlations have also to be modified in the light of correlation of driving energy for DPB and modified IS cone for a given subsoil at same depths.

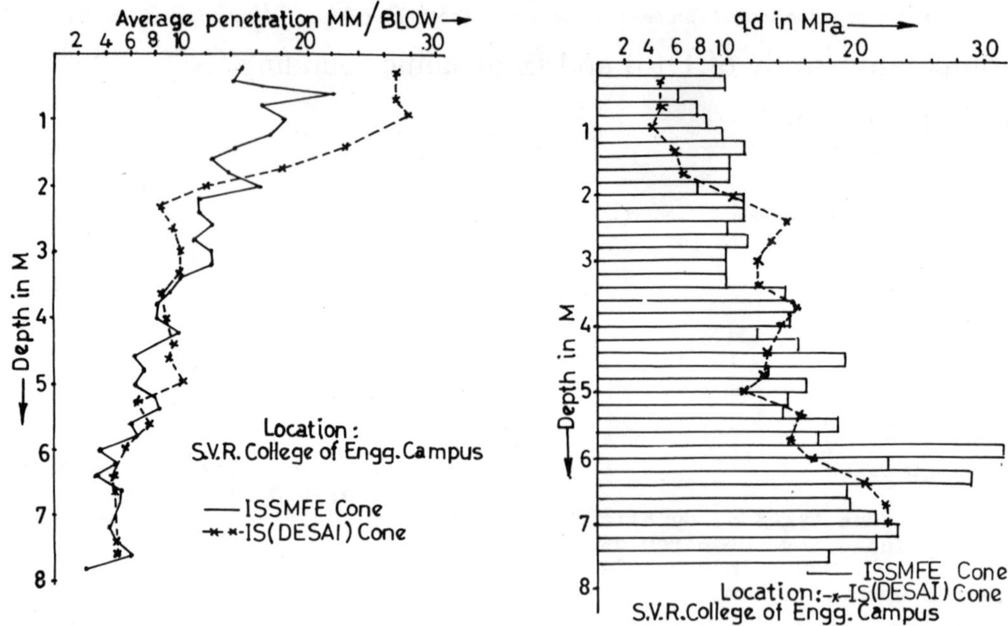

Fig 1 Depth Versus Average Penetration

Fig 2 Depth versus qd

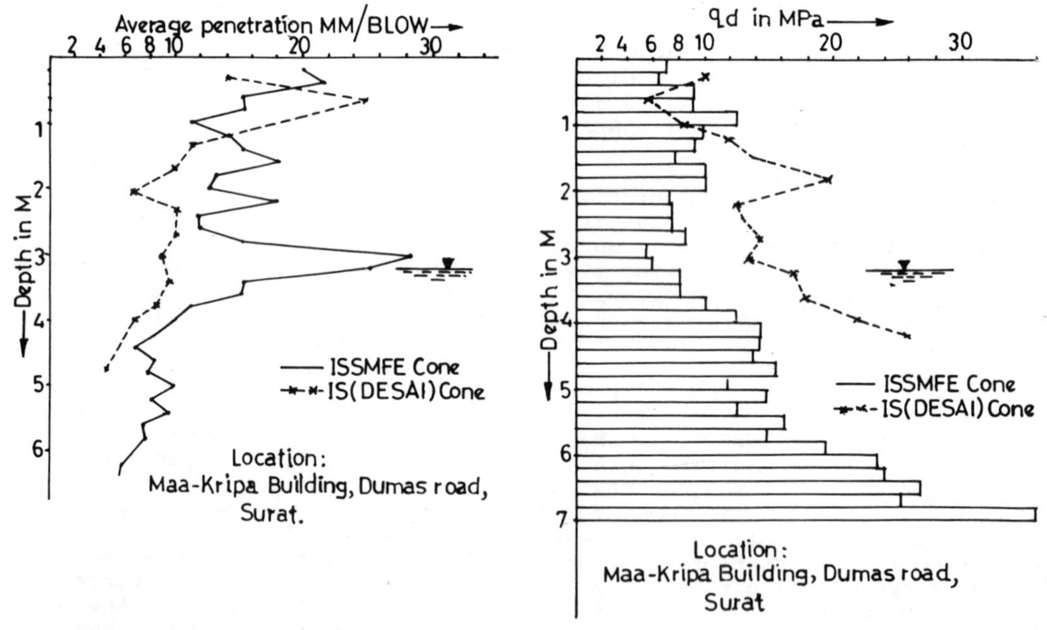

Fig 3 Depth Versus Average Penetration

Fig 4 Depth versus qd

3 INVESTIGATION AND RESULTS

The Tapi alluvium is fairly new deposit of silt fine sand and silty clay of variable thickness. There are two distinct zones having non-cohesive or very low plastic silty fine sand and medium cohesive clayey silt, deposits in top 8 to 10 m subsoil.

The investigations have been carried out at sites

1. Rachana Aptt. Navsari site-1
2. Maa-Krupa Building
3. Sir J.J. School site-1
4. S.T. Bus Station
5. 108 Jinalaya Dhaman
6. SVR College of Engg. campus
7. Rachana Aptt. Navsari site-2
8. Sir J.J. School site-2
9. Varachha Road site

The sites (1) to (5) falls in zone of non cohesive or low plastic alluvial deposits and rest are in zone of medium cohesive clayey silt deposits. All these sites have been explored for foundation of a multy storeyed structure. The drilling, sampling, SPT and other field observations have also been recorded.

The typical variation of average penetration in mm/blow for both the cones and set of sites are shown in Fig.1 and Fig.2.

The resistance in MPa by both the cones at typical sites in each zone has been plotted against depth in Fig.3 and Fig.4.

4 STATISTICAL CORRELATION

The comparison of variation of resistance to penetration at a given depth has been made for each site, using Karle pearsons method.

The bivariate universe can be correlated by method of Karl Pearson. Two variates can be correlated by correlation of x and y

$$r_{xy} = \frac{E\,(x-\bar{x})\,(y-\bar{y})}{\sqrt{E(x-\bar{x})}\,\sqrt{E(y-\bar{y})^2}}$$

$$= \frac{Cov\,(x,\,y)}{\sqrt{var\,(x)\times var\,(y)}}$$

$$= \frac{\mu 11}{\sigma x\,\sigma y}$$

r_{xy} is called products moment correlation co-efficient and σx and σy are called Standard deviation of x and y respectively. In our case, q_{dI} and q_{dD} can be taken as two variable, and correlated by line of regression as shown in Fig.5. The reliable correlations of q_{dI} international and q_{dD} modified IS cone have been superimposed in Fig.5.

These results have been used to project a conservative safer boundary for

a. noncohesive silty fine sand deposits zone I
b. medium cohesive clayey silt Zone II

5 INTERPRETATION OF DPB

The statistical correlations have been judged by empirically safer boundary curve to convert the q_{dI} into equivalent q_{dD}. The recommended conservative factor for alluvium in noncohesive deposits is

$$q_{dI} = \frac{q_{dD}}{1.8}\ \text{MPa}$$

For cohesive deposits $q_{dI} = q_{dD}$.

The q_{dD} is then converted into N_c resistance to penetration by modified IS cone test using the following conversion

$$q_d = \frac{M}{M + M'} \cdot \frac{MgH}{Ae}$$

M= mass of hammer = 65 kg

M'=mass of guide rod, Anvil and of drill rod = 32.30 kg + $4.3xD_f$

g =9.8/sec^2

H =fall of hammer = 0.75 m

A =cone base area = $20.19 \times 10^{-4} m^2$

e =average penetration mm/blow

D_f=depth at which cone test is performed in m

The value of e average penetration is converted into Nc blow/30 cm for which correlations have already been estaBlished. (M.J. Vyas 1976) Thus q_{dD} can be now tabulated against estimated relative density (I_D) angle (ϕ_d), the allowable bearing capacity for a permissible total settlement of 40 mm without ground water table correction. The results of the computations are drawn in a monograph Fig.6.

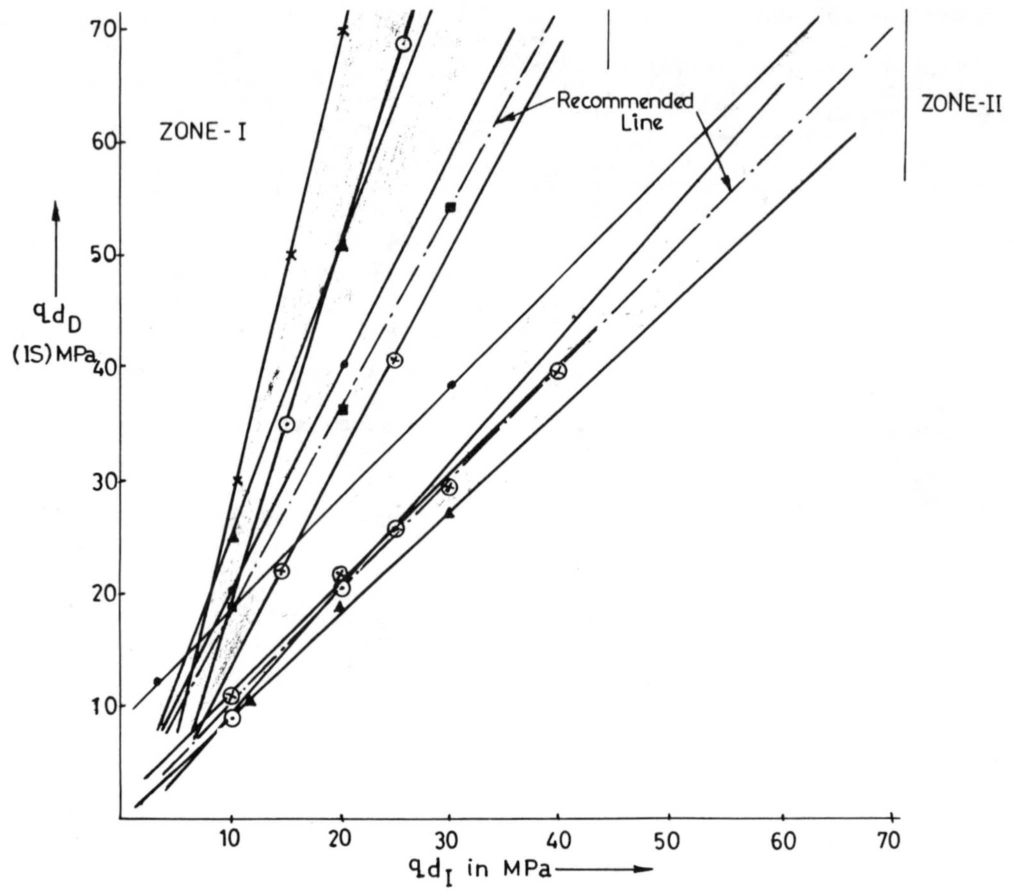

Fig 5 Statistical Correlation between qd_I and qd_D

Thus qd_I obtained at a site of non cohesive silty fine sand alluvium can be used to anticipate the probable engineering properties of the subsoil directly using Fig.6. For a given depth, approximately effective overburden pressure (P) is estimated in t/m^2. Then for given qd_I and p_o top left quadrant will read relative density on y axis. This horizontal line when extended will out the line in right quadrant to read allowable bearing capacity on the X axis in t/m^2. This value is for no ground water table within the bulb of pressure. It is constomary to reduce q_a by 30% for the foundations where in W.T. rise is expected. The bottom quadrant will read Ø' value for qd_I resistance directly.

This first hand information for very small structures can provide adequate initial data to know the expected soil profile ground water level which is shown by a quick increase following a slight decrease and the relative density. At many sites wherein designer has to proceed without any data, this data can be useful. The gradual refinements of the relations for wider range of geological formations may ultimately provide relatively much more information by a single test at low cost and in very short time.

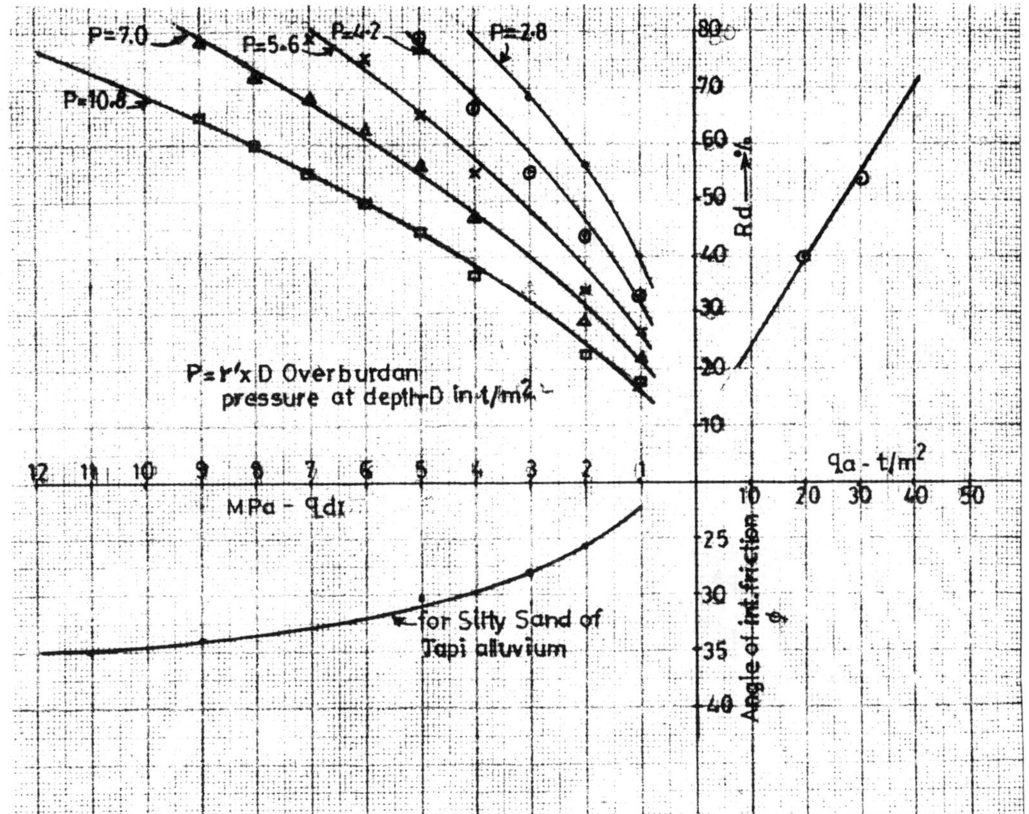

Fig 6 Interpretation of DPB test

6 CASE STUDY

To substantiate the approach taken by the authors the field data for a site is anticipated by Fig. 5 and verified by conventional other methods.

The site chosen is J.J. School at Shapore, city of Surat. The subsoil comprises of silty fine sand strata at 4 to 5 m depth. Clay content 8-10%, silt 25 - 30% and fine sand with 5% gravel, 67%, PI$<$10. The value of penetration resistance recorded by DPB cone test gives

$$qd_I = 14.54 \text{ MPa}$$

The overburden pressure P=4.8 t/m^2. Using Fig 6 we get

Rd$>$ 80%

$q_a >$ 50 t/m^2 (No water table)

$q_{awt} >$ 35 t/m^2 for Sp = 40 mm

ϕ' = 35°

The conventional practices would be analyse the data by insitu density test and standard penetration test.

The UD field density obtained from open pits indicate γ_d=1.44 t/m^3 (G = 2.72). This would, for γ_d max = 1.65 t/m^3 and γ_d min =1.35, t/m^3 mean Rd = 70%

For N by SPT = 30 blows/30 cm and using IS code, 1904;(1966)

R_d = 65%

ϕ = 36°

q_a = 40 t/m^2

7 ACKNOWLEDGEMENT

The work in the field was conducted with the assistance of Shri D.C. Brahambhatt, Chemist and drawings have been prepared by Shri Lad H.A., Draftsman of SVR College. The work would not have been feasible without assistance of M/s. S.B. Engg. Co., Surat. The permission to use the facilities of SVR College laboratories is hereby acknowledged.

8 REFERENCES

Desai M.D. and Jain (1974), Penetration Testing in India ESOPT - Stockholm (1974), Vol. 1.

Desai M.D. and A. Singh (1974), ESOPT Stockholm (1974) Vol. 2, and 3.

Desai M.D.(1980), Evaluation and standardization of penetrometers, Geotech. 1980, IIT, Bombay.

Vyas M.J. (1976), M.E. Thesis, Penetrometer, A review of practices for shallow foundation.

Mehta B.J. (1981) M.E. Thesis, Evaluation of Sounding Methods for Exploration in South Gujarat (under finalization).

Indian Standards No.4968, Part-I Method for Surface Soundings for soil.

IS 2131-1963, Method for Standard Penetration test for Soils.

IS 1904-1966, Code of Practices for Structural Safety of Building foundations.

Proceedings of the Second European Symposium on Penetration Testing / Amsterdam / 24-27 May 1982

A site investigation through penetration tests

M.U.ERGUN
Middle East Technical University, Ankara, Turkey

1 INTRODUCTION

Penetration data obtained in a dam site are presented. 38 m high Susurluk Dam is to be founded on 25 m thick alluvium. The site is approximately 400x500 m. Conventional laboratory and field tests have been carried out to estimate engineering and index properties of soils. Static cone penetration tests (CPT), standard penetration tests (SPT), dynamic penetration tests (DPT) and pressuremeter tests comprise the field tests. More than a single type of field test have been performed at many borehole locations. The following is a comparison of the penetration tests at several borehole locations. Among large number of penetration tests data are presented only at the locations where there are at least two types of penetration tests performed. Some observations about the potential capabilities of the tests in soil profiling and estimating the engineering properties will be summarised.

2 EQUIPMENT

Standard penetration equipment used has internationally accepted standards (50.8 mm diameter sampling spoon, 63.5 kg hammer, free fall from 76 cm etc.) Procedure of the test is quite standard, but there are many points that can be discussed as pointed out by a number of researchers (Ireland et.al. 1970, Fletcher 1965). Although it was stressed that the water level in the casing had better be kept roughly at the elevation of the water table especially in fine granular soils the Writer considers himself unsuccesful in making this practice applied by the few SPT crew as a routine at the site.

Static cone penetration apparatus used is a standard piece of equipment. It has 100 kN capacity, and its mechnical design employs a Begemann tip. Both end resistance and total resistance can be measured.

Dynamic penetration test uses a heavy hammer (159 kg), height of free fall is 61 cm and the cone has 60° apex angle, and has diameter of 63.5 mm. Number of blows for 15 cm penetration is recorded. Data may be plotted at 15 cm intervals continiously or number of blows for 30 cm may be plotted. The first procedure is adopted here.

3 SOIL AND PENETRATION DATA

Data are presented in Figure 1 (a) through (1). Detailed soil descriptions in the original borehole logs have been converted to letters using unifield soil classification system due to compact form of presentation and limitations in space. In doing so certain characteristics of the soil have lost the meaning but the basic nature of the soil can be identified, e.g. using letters SP or SM it is not clear whether it is a fine medium or a coarse sand. In preparing the soil profiles the records taken at the site during boring and observations on disturbed and undisturbed samples and the results of the index tests were studied. On several occasions static penetration data have been resorted to assist the profiling. It is not possible here to include the results of large number of tests on consistency limits and gradation of the soils, and they will be mentioned in relevant discussions. Undisturbed samples are shown by vertical rectangular boxes with crosses in them, and disturbed samples by small boxes, both on the right side of the cloumn for description of the soil.

Penetration profiles are self-explanatory. Points for standard penetration resistance are not joined together since 2 m interval has been considered large.

4 PENETRATION TESTS AND PROFILING

Having a close look at Fig.1 the following observations can be made :

1. Dynamic penetration test (DPT) penetration numbers excepting boulders and deep strata do not show significant changes with depth.
2. DPT is more reliable for shallow depths. Its use should be limited to shallow explorations. Although it is difficult to specify a depth, 10 m may be roughly taken as a limit. Examination of the data indicates that there is almost always an increasing pattern in penetration numbers in deeper strata irrespective of the type of soil. That must be due to accumulating friction on the rods, and penetration number can not respond to the changes in the soil character at greater depths, e.g. boreholes 125, 123, 132, 140.
3. Excepting very few strata static cone end resistance values are in accordance with the soil profiles obtained through borings before any modifications are made using CPT data. CPT cone is sensitive to variations in soil layers, and the Writer considers the static cone penetration test most reliable due to its virtues like frictionless end resistance measurements hence sensitivity to changes in soil layers irrespective of depth and independence to procedural and human factors together with its continious record facility to detect all bands of soil.
4. SPT numbers have been observed to correlate with CPT end resistances in some locations whereas there is complete deviation in others. This erratic relationship is most probably due to great number of factors influencing SPT numbers, compare for example, locations 133, 138, 137, with 130, 136, 126, 140.

Another problem with the standard penetration test is the interval between test levels. Intervals of 0.75 m to 2.50 m are generally used, but in extensive exploration programs like in this study an interval around 1.5-2 m is suitable. On the other hand this length is considerable, and there may be bands in between which may be critical in the analysis, e.g. a meter thick peat band. An inexperienced crew may miss it during boring operation, and there are many drilling teams in large site investigation programs, and one should not expect that they are all experts in this job.

The small amount of soil retained in the sampling spoon is an advantage over the other penetration tests. The soils between two consecutive samples are advanced by classical drilling methods and soil chips in the circulating water and the pit are not sufficient to identify the soil accurately. A SPT sample will disclose the true character of the soil. Since the most soils at this site have little cohesion it was possible to retain SPT samples where it was not possible to take an undisturbed sample using an ordinary open-ended sampling tube.

5. If, for a moment, one imagines that no CPT program were used in this investigation there would be hardly sufficient information about the compactness of the various strata. Soils on the right side of the valley are sands, and SPT numbers are very low, see locations 133, 135, 138. This is also true at many other locations near the right abutment where SPT was the only test performed. Notice how CPT end resistances are high. It is highly probable that the low N values are mainly caused by disturbance and boiling, and do not reflect the true relative densities.
6. When the cohesion of the layers gets higher invariance in the CPT end resistances is noticed. q_c values show increase for sandy and gravelly formations with little fines.

5 ENGINEERING PROPERTIES

The most abundant soils are gravelly and silty sands at the site. It was not possible to obtain undisturbed samples using ordinary sampling methods from such deposits. Split spoon samplers with inner thin aluminium tubes were used. Undisturbed samples were taken only in layers having some cohesion, and therefore the number of such samples was not many. The quality of undisturbed sampling may be considered poor due to thick cutter edge of the split spoon sampler tube.

The following summary laboratory data are about the cohesive soils at the site. They are silty, sandy clays, plastic silts and clayey sands which are located at the left hand side of the valley, i.e.locations 125, 140, 137, 136, 123, 132, 128 etc. No laboratory studies were planned using the artifically formed sand samples at various relative densities. Instead, field data on N and q_c were analysed to estimate the relative densities of strata to evaluate the liquefaction potential of the foundation soils and settlements under loading.

At this stage the main problem was the questions about the reliability of SPT results and lack of CPT data on similar alluvial soils for liquefaction evaluation. Discussion of relative densities and dynamic considerations will not be given here. Correlation between SPT and CPT results for the soils at the site are briefly reported in the next section.

A summary of results of laboratory tests are presented below :

1. CL, CL-ML, SC-CL, SM-ML groups of soils yield coefficients of consolidation, c_v, between $1\text{-}7x10^{-4}$ cm^2/sec in oedometer tests, but it should be admitted that these values have no meaning in the engineering sense for such variable alluvial soils.

2. Undisturbed samples from CL and SC-CL soils all between 9-15 m indicate compression indexes, C_c, around 0.14-0.16. Corresponding CPT end resistances and SPT values are 20-23 kg/cm^2 and 7-11 respectively. For increasing end resistance values in less cohesive layers C_c values remain approximately constant.

3. The soils at the left bank are so variable that lean clays and sandy silts give values of half the deviator stress in undrained shear from 10 to 120 kN/m^2 considering five unconfined and thirty unconsolidated undrained triaxial tests. The soils in the 10-25 kN/m^2 range are usually loose silts and lean clays. For relatively more cohesive layers like CL group it may be mentioned about undrained shear strengths and such results obtained on samples from locations 132, 140, 123, 125 between 11-21 m depth are presented in Table 1.

Table 1. Undrained shear strengths and penetration resistances for several cohesive layers.

$c_u(kN/m^2)$	$q_c(kg/cm^2)$	N	$I_p(\%)$	% 200
22.5	16	10	15	44
24.5	18	9	14	45
25	22	10	14	69
11	17	8-9	8	35
25	23	11	?	?
24*	19	7	10-12	53

*Unconsolidated triaxial test, others unconfined

4. Triaxial tests on undisturbed samples yield the failure envelope shown in Figure 2. Effective angle of shearing resistance varies between 24° and 36°, but $\phi'=32°$ is a good average fit. Effective cohesion intercept is almost zero.

6 CORRELATION BETWEEN CPT AND SPT

The soils at the site are so variable that it is difficult, if not impossible, to assign correlation factors between CPT end resistance q_c and SPT number N for various layers. A close look at the penetration data roughly reveals q_c/N ratios associated with the following soil groups : CL 2, SC 2-4, SM and SP 4-7, SW 9 and gravels mostly greater than 10. On the other hand there are few boreholes at the right bank, as mentioned before, in which SPT numbers are very low, and these result in very high q_c/N ratios. But it is interesting that standard penetration tests performed at other locations at the right bank by another drilling company yield higher N values.

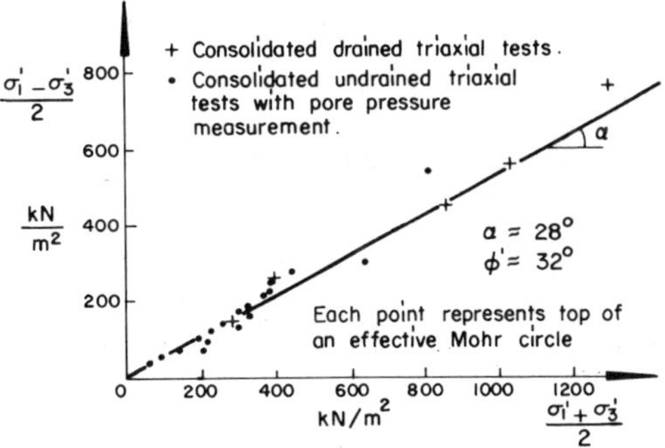

Figure 2. Failure envelope for plastic silts, silty-sandy clays and clayey sands at the site.

7 CONCLUSION

At the end of the site exploration by different penetration tests the following deserve to be emphasized :

1. Static cone penetration test (CPT) is superior to other penetration tests in soil profiling. Used together with careful drilling and sampling at a number of locations at a site where soils are not erratic it may be replaced by borings at many other locations hence cutting the time for site investigation and costs.

2. Standard penetration test is more liable to human factors and variations in testing procedure. It is slower and more expensive than the other tests, yet

there is not an accepted technical advantage over the other tests in making use of SPT data. Wide use of SPT in the past may be considered an advantage especially in liquefaction studies. Observing the soil in the sampling spoon is useful.

3. Dynamic penetration test may be useful in shallow explorations if equipment and procedure are standardized. Friction on the rods becomes a critical problem in deep strata. General use of DPT should be discouraged. A dynamic penetration test with frictionless tip design may be attempted.

4. Due to reasons given above estimation of engineering properties of soils like undrained shear strength, effective angle of internal friction and relative density and compressibility using CPT data is more reliable than those obtained through SPT numbers. In the Writer's opinion use of CPT should accompany every soils investigation.

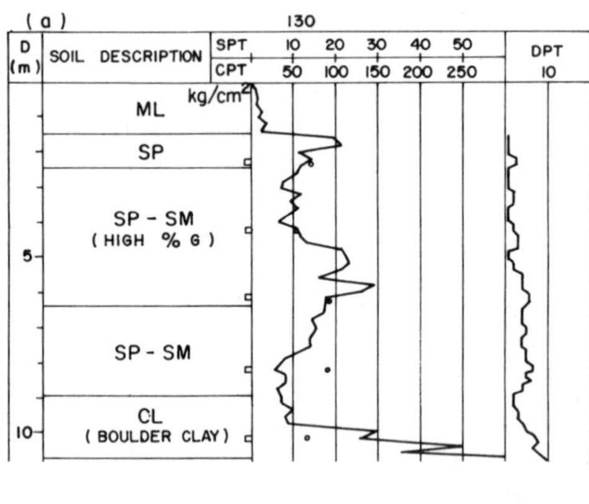

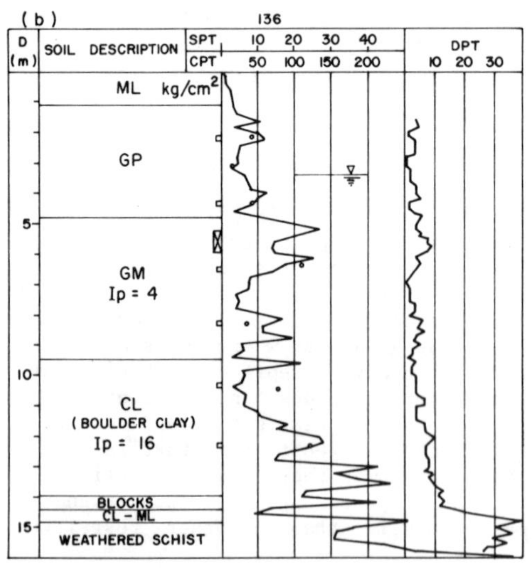

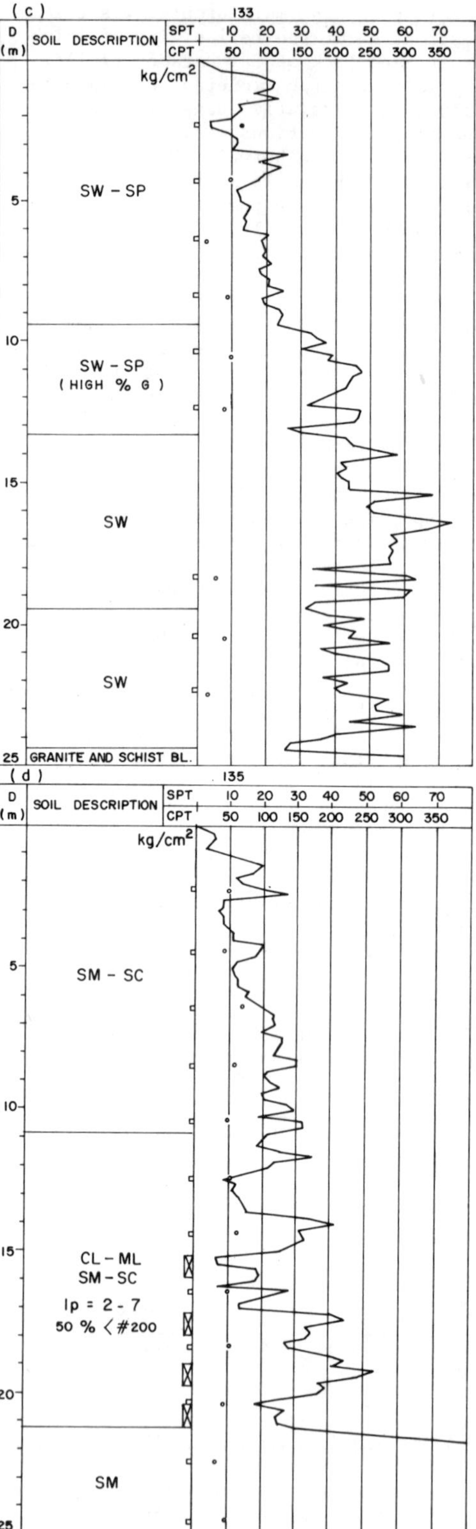

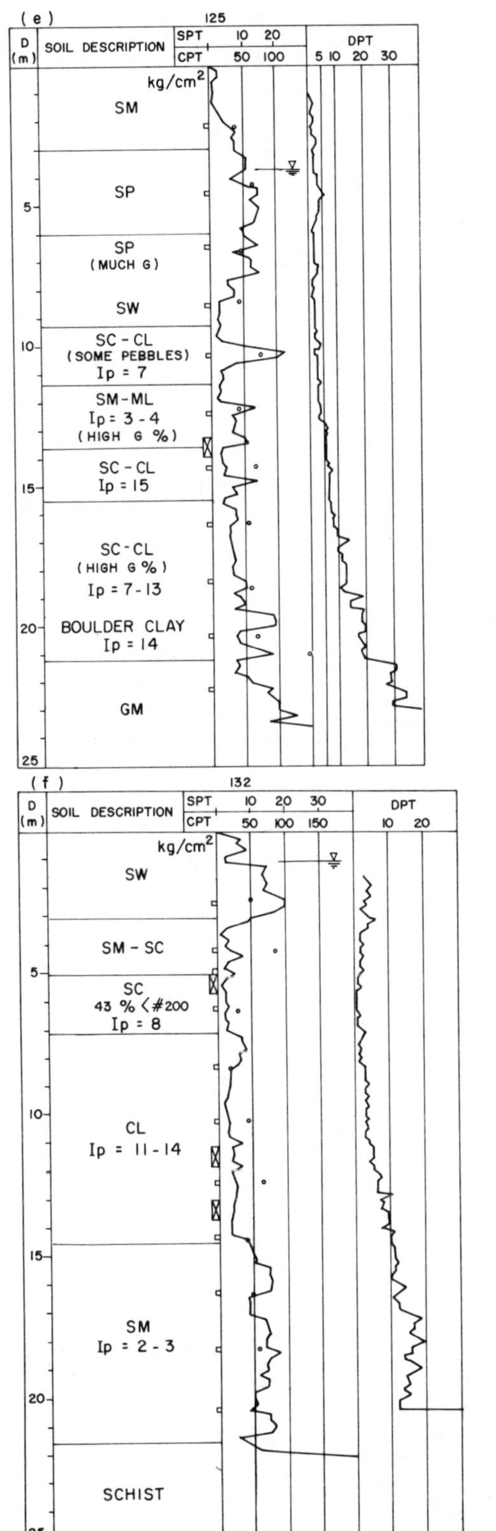
(e)
125
D (m)
SOIL DESCRIPTION
SPT 10 20
CPT 50 100
DPT
5 10 20 30
kg/cm2
SM
SP
SP (MUCH G)
SW
SC - CL (SOME PEBBLES) Ip = 7
SM-ML Ip = 3 - 4 (HIGH G %)
SC - CL Ip = 15
SC - CL (HIGH G %) Ip = 7 - 13
BOULDER CLAY Ip = 14
GM
5
10
15
20
25
(f)
132
D (m)
SOIL DESCRIPTION
SPT 10 20 30
CPT 50 100 150
DPT
10 20
kg/cm2
SW
SM - SC
SC 43 % < #200 Ip = 8
CL Ip = 11 - 14
SM Ip = 2 - 3
SCHIST
5
10
15
20
25

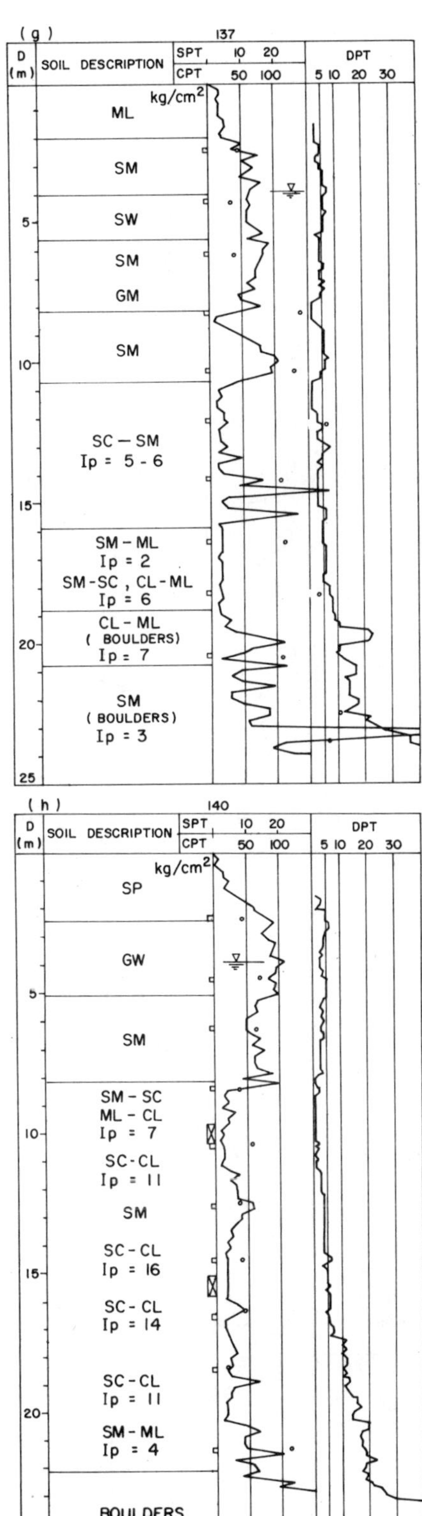
(g)
137
D (m)
SOIL DESCRIPTION
SPT 10 20
CPT 50 100
DPT
5 10 20 30
kg/cm2
ML
SM
SW
SM
GM
SM
SC — SM Ip = 5 - 6
SM - ML Ip = 2
SM-SC, CL-ML Ip = 6
CL - ML (BOULDERS) Ip = 7
SM (BOULDERS) Ip = 3
5
10
15
20
25
(h)
140
D (m)
SOIL DESCRIPTION
SPT 10 20
CPT 50 100
DPT
5 10 20 30
kg/cm2
SP
GW
SM
SM - SC ML - CL Ip = 7
SC-CL Ip = 11
SM
SC - CL Ip = 16
SC - CL Ip = 14
SC - CL Ip = 11
SM - ML Ip = 4
BOULDERS
5
10
15
20
25

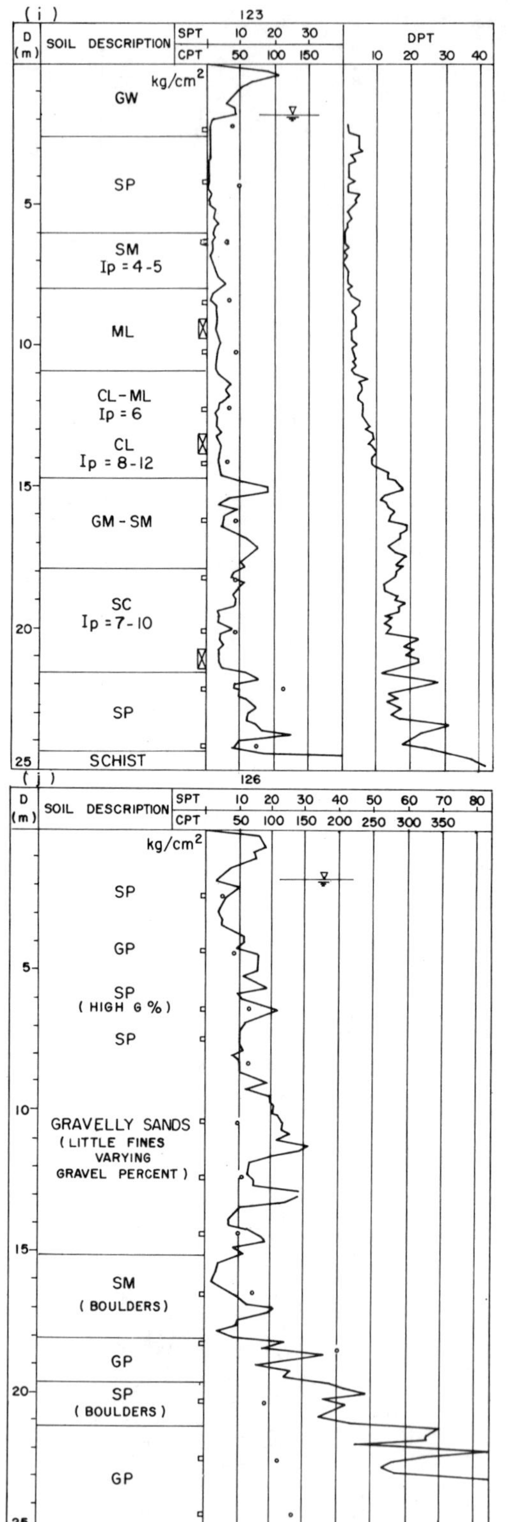

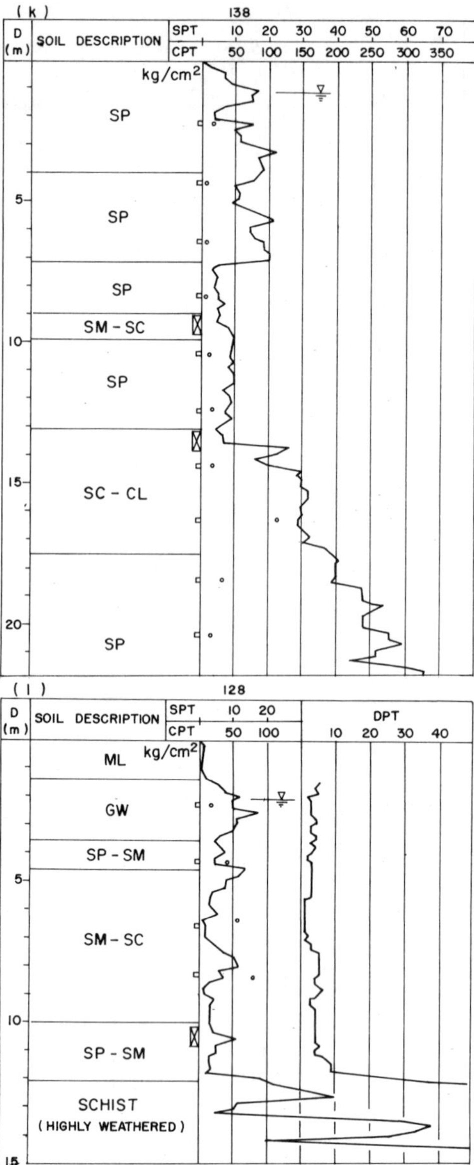

Figure 1. Penetration data.

8 REFERENCES

Fletcher, G.F.A. 1965, Standard Penetration Test: its uses and abuses, J. of Soil Mech. and Found. Eng. Div., ASCE, 91, SM4:67-75.

Ireland, H.O., et.al. 1970, The dynamic penetration test: a standard that is not standardized, Geotechnique 20, No.2:185.

Proceedings of the Second European Symposium on Penetration Testing / Amsterdam / 24-27 May 1982

Various types of penetrometers, their correlations and field applicability

M.C.GOEL
Water Resources Development Training Centre, University of Roorkee, India

1 INTRODUCTION

For exploration of sandy foundations various types of penetrometers are in use. A number of correlations of standard penetration test resistance with that for other tests have been developed from time to time but have been found to vary widely. The results obtained by various investigators for correlation in between static cone resistance (q_c) and standard penetration test (N_s) are summarised in Table 1 which would indicate that the ratio q_c/N_s varies from 2.5 to 18. The various sizes of cones, drill rods, energy used and correlation in between dynamic cone resistance (N_c) and standard penetration resistance (N_s), used by different researchers are given in Table 2. This table also reveals the diversity of opinion and practices adopted by various authors.

Field tests were carried out at a number of sites in U.P. State of India having varying foundation strata by different penetrometers such as standard penetration test (SPT), static cone, dynamic cone driven dry or with the circulation of bentonite slurry to develop correlation between various penetrometers to suit to local conditions. The scope of study was limited to sandy strata widely varying in gradation. The D_{50} size of the material met with varied from 0.01 mm to 0.2 mm.

2 EXPERIMENTAL SET UP

The split spoon sampler of outer dia 51 mm, inner dia 35 mm was driven into the ground by the impact of 63.5 kg drop weight falling freely under gravity through a vertical height of 76.2 cm imparting total energy of 48 kg.m. By the impact of drop weight, the spoon is initially driven 15 cm. Number of blows required to obtain a further penetration of 30 cm was taken as the resistance of soil penetration (N_s). The resistance measured is that due to the penetration of split spoon alone. No skin friction is mobilised along

Table - 1 Correlation of q_c versus N_s

	Author	Soil	q_c/N_s
1.	Meigh and Nixon (1961)	Fine silty sand	4
		Medium and coarse sand	8
		Gravelly sand	8 to 18
		Sandy gravel	12 to 16
2.	Meyerhof (1956)	Sandy silt	2.5 to 7 with avg. of 4
3.	Rodin(1961)	Fine to Medium sand	4.8
		Sand with some gravel	8
4.	Schultze & Knausenburger (1957)	Non-cohesive	3.6
5.	Sutherland (1963)	Sandy silt	2.5

Table - 2 Various Cones used and their correlation with SPT

Authority	Cone size mm	Apex Angles degrees	Outer dia. of drill rods mm	Driving Energy Drop weight kg	Height of fall cm	Casing used yes/no	Correlation N_o/N_s
Meyerhof (1956)	50.8	60	41.3	63.6	76.2	No	2
Palmer Stuart (1957)	50.8	60	50.8	63.6	76.2	Yes	1
Schultze and Kuausenberger (1957)	43.7	90	32.0	50.0	50.0	No	-
Rodin (i)	50.8	60	50.8	63.6	76.2	Yes	1
(1961) (ii)	63.5	60	48.0	159.0	60.0	No	-
Canadian Standards Association (1962)	57.2	60	44.5	-	-	No	1
Phalen (1964)	-	-	-	63.6	76.2	No	1
Schultze and Malzer (1965)	50.8	60	-	63.6	76.2	No	-
Desai et al (1968)	50.8	60	41.3	63.6	76.2	No	-
Mohan et al (i)	43.7	60	41.3	63.6	76.2	No	-
(1970) (ii)	50.8	60	41.3	63.6	76.2	No	-
(iii)	63.5	60	41.3	63.6	76.2	No	1.5 to 4
(iv)	63.5*	60	41.3	63.6	76.2	No	1
(v)	76.2	60	41.3	63.6	76.2	No	-

*Tests were performed with circulation of bentonite slurry.

the drill rods as they are driven freely in a cased bore hole.

Two types of dynamic cone tests were carried out viz (i) 51 mm cone driven dry and (ii) 63.5 mm cone driven with the circulation of bentonite slurry. The apex angle of cone in both cases with 60^0. The driving mechanism and energy supplied was the same as for SPT. During dry testing, the cone was loosely fitted at the lower end of standard A type drill rods and was driven continuously into the soil strata under the impact of drop weight falling freely under gravity. The number of blows N_c required to produce the penetration of the cone by 30 cm was recorded continuously. For test with the circulation of bentonite slurry, the slurry of 8-10% concentration was pumped by means of a reciprocating pump at a pressure of 10 kg/cm^2. The driving of cone and the pumping of the slurry was carried out simultaneously and the number of blows recorded per 30 cm penetration of cone was taken as the dynamic cone penetration.

The static cone test apparatus was standard

Table - 3 Details of Various sites explored

Name of site	Strata	Depth upto which foundation was to be probed	Depth probed (m)				
			Static cone	SPT	50 mm cone	63 mm dry	63.5 mm cone with slurry
Obra Dam	Coarse sand mixed with gravel	21	19	21	-	-	-
Girja Barrage	Fine and Medium Sand	24	24	24	-	-	-
Lower Sarda Barrage	Fine and medium sand	28.5	21	28.5	13	-	15
Sarda Sagar dam	Silty, fine to medium sand	8	8	8	-	8	-
Gomti Barrage	Sand, silt, clay mixed with gravel	28	10.5	28	6	-	9
Gomti Aqueduct	Sand, silt and clay mixed with gravel	45	7	45	8	-	8
Solani River bed	Fine sand and silt	-	14	13	10	-	13
Ranipur River bed	Fine sand and silt	-	14	12	12	-	12
Pathri River bed	Fine to medium sand	-	16	12	12	-	13
Dhanauri River bed	Fine sand	-	15	14	9	-	15

Dutch cone apparatus. The cone was of 37 mm diameter, apex angle 60° giving an end area of 10 cm^2. The cone was attached to the rod which was protected by a sleeve. Both the cone and sleeve were pushed separately and the penetration resistance required to push the cone through 7.5 cm, measured in kg/cm^2, was termed as standard cone penetration resistance (q_c).

3 DATA COLLECTION AND ANALYSIS

3.1 Strata

Penetration tests viz. static penetration, standard penetration, dynamic cone test using 51 mm and 63.5 mm. diameter cones driven dry and 63.5 mm cone also driven with circulation of bentonite slurry were carried out at a number of sites in U.P. having varying foundation strata. The depth to which each site could be explored by a penetration test along with the depth to be probed and the details of foundation strata are summarised in Table-3 indicate that the strata at the various sites where tests have been carried out are widely different from each other. The strata at Obra dam site comprised medium to coarse sand, the coarseness increasing with depth. The strata at Ghagra and Lower Sarda Barrage sites varied and comprised erratic deposit of fine, medium and coarse sand mixed with gravel. At Sarda Sagar dam the strata was silty fine sand while at Gomti Barrage and aqueduct sites the strata comprised sand mixed with silt and clay on the top

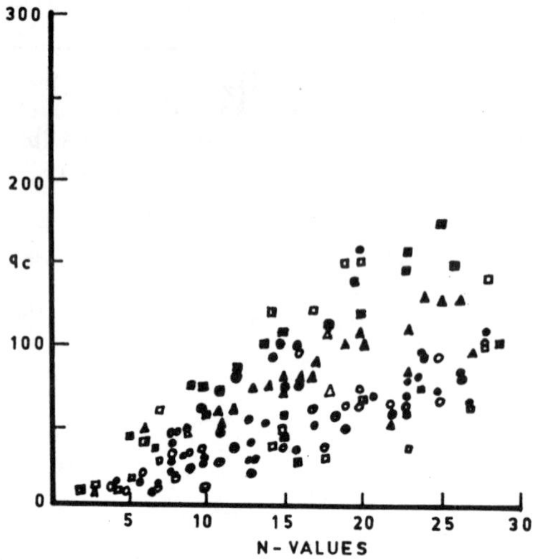

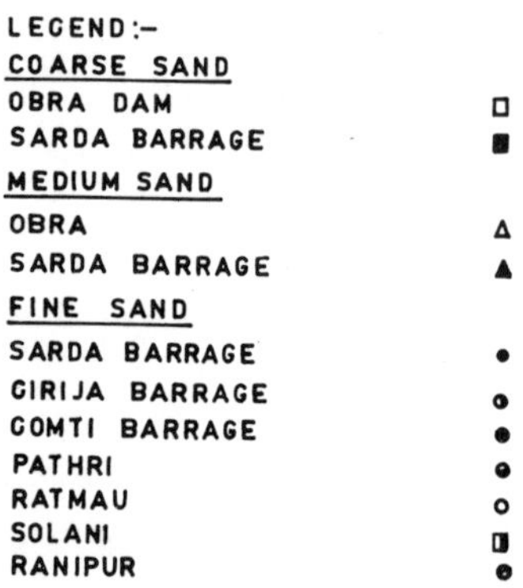

Fig. 1 Static Cone Resistance vs. SPT Blows

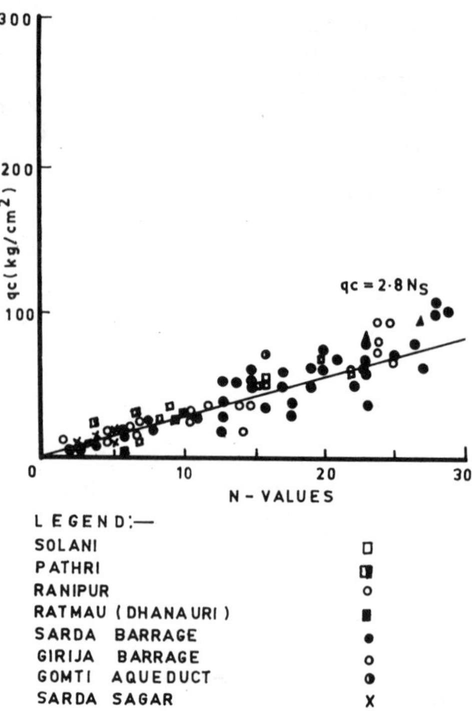

Fig. 2 Static Cone vs. SPT Resistance (Fine Sand)

followed by clay with kankar. The strata in Solani river and Ranipur comprised fine sand while at Pathri it is medium sand.

The data of penetration tests was analysed and the correlations obtained as a results of analysis of the data are given below.

3.2 Static cone vs Standard Penetration Resistance

The static cone resistance and corresponding standard penetration tests carried out at the various sites and plotted in Fig.1, indicate that no unique relationship exists between static cone resistance q_c and standard penetration blows N_s which may be due to the various factors affecting the test results of the two penetrometers. On grouping the data according to the grain size and plotting in Figs.2 to 4, the following average correlation between static cone resistance and standard penetration blows are indicated :-

Silty fine sand	$q_c = 2.8\ N_s$
Medium sand	$q_c = 5.0\ N_s$
Medium to coarse sand with gravel	$q_c = 7.0\ N_s$

The relation $q_c = 2.8\ N_s$ is the lower limit of the relation obtained by plots of Meyerhof (1956) which is for silty sand found at Scotia and Burlingtom and also obtained by Sutherland (1963) on the basis of semifield tests. The other relation of $q_c = 5\ N_s$ and $q_c = 7\ N_s$ compare well with the relation obtained by Rodin (1961) and Meigh and Nixon (1961) for similar sand gradation. Their results are also indicative of the limitations of Meyerhof's correlation $q_c = 4\ N_s$ (1956) being used generally which has also been supplemented by Narhari (1967).

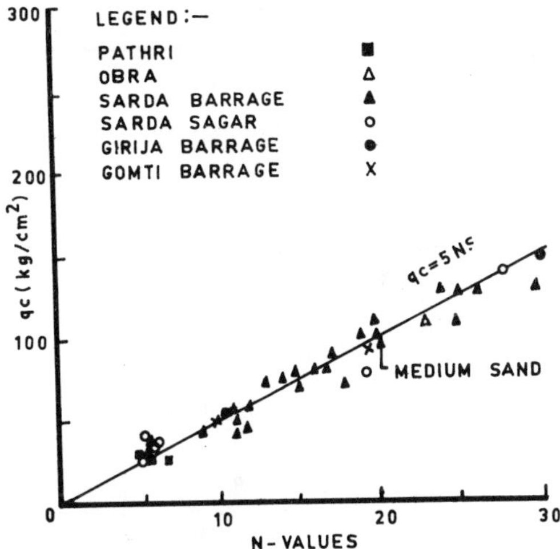

Fig. 3 Static Cone vs. SPT Resistance (Medium Sand)

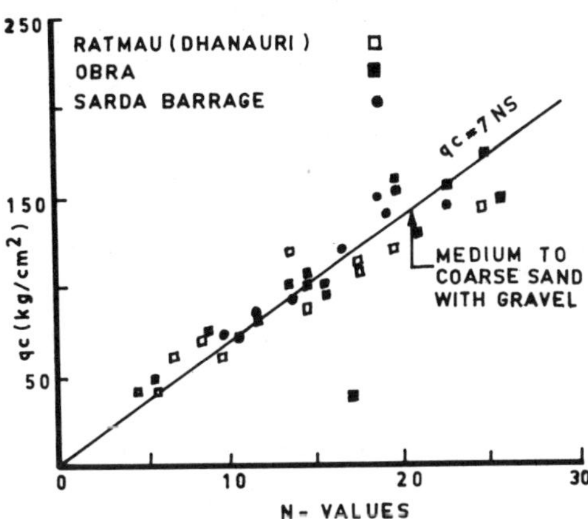

Fig. 4 Static Cone vs SPT Resistance (Medium to Coarse Sand)

A reference to Table 3 indicates that the depths to which the static cone test could be carried out varies from site to site depending upon the density of the strata and the capacity of the machine. In medium to coarse sandy strata with traces of fine gravels, the machine could penetrate upto a depth of 19 m, as at Obra, while it could penetrate upto 10 m only at Gomti Aqueduct and Barrage sites where pebbles were met and the total resistance of the strata was more than 500 kg/cm^2. Thus the static cone resistance could be used for the exploration of strata having fine to medium and coarse sand mixed with fine gravel, and upto a depth where the total resistance of the strata was less than 500 kg/cm^2.

3.3 Dry cone (51 mm) vs standard penetration resistance

The penetration tests carried out by 51 mm dry cone and standard penetration at various sites plotted in Fig.5(a) show that even for the same type of sand the data indicates large scatter varying between $N_c = N_s$ to $N_c = 5 N_s$ which may be attributed to the variable skin friction acting upon the drill rods. The data may be compared with the data plotted by Desai et al (1968) which also indicates similar scatter even within the narrow range of depth in which attempt was made to classify the data. In a comparative study of the various size cones with standard penetration test, Mohan el al (1970) have pointed out that 63 mm diameter cone when driven dry yielded minimum skin friction on the drill rods. As such the 63.5 mm cone was also driven dry at Sarda Sagar Dam site and the results of N_c vs N_s are plotted in Fig.5(b) and lesser scatter between $N_c = N_s$ and $N_c = 3.5 N_s$, has been indicated in this case. This difference in the behaviour of the two cones when pushed dry may be attributed to the fact that gap of only 5 mm between the drill rods and the hole made by the cone is left in the case of 51 mm cone as compared to the larger gap of 11 mm in the case of 63.5 mm cone resulting in greater tendency of the skin friction to be developed in the case of smaller cone than in the larger cones.

The depth of penetration by the 51 mm dynamic cone is restricted by the skin friction developed along the drill rods. At Gomti barrage site the number of blows by cone penetration was 50 at 5 m depth which rose to 97 at about 6 m depth corresponding to standard penetration test blows of 6 and 23 respectively. At Sarda Barrage it was, however, possible to penetrate upto 13 m depth. But no correlation was possible with standard penetration test. The ratio of N_c/N_s for 51 and 63.5 mm cone pushed dry have been plotted in Fig.6(a) and (b) respectively. For 51 mm cone the ratio for N_c/N_s indicates wide scatter. However, in top 3 m depth, the ratio N_c/N_s varies from 1 to 2, with average of 1.7 roughly. In the case of 63 mm cone, this ratio is a constant and is equal to

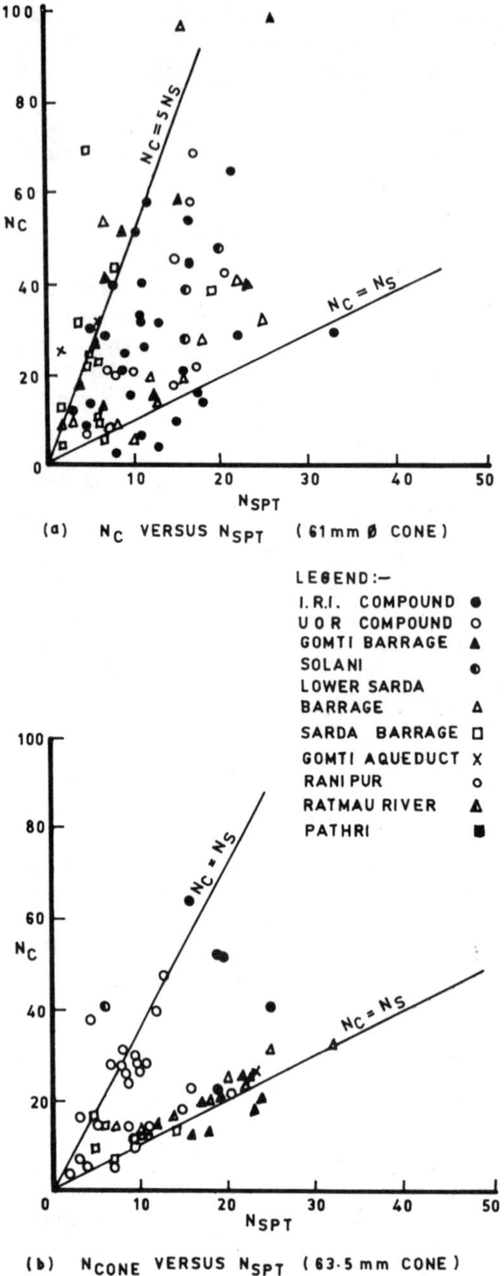

Fig. 5 Dynamic Cone vs. SPT Resistance

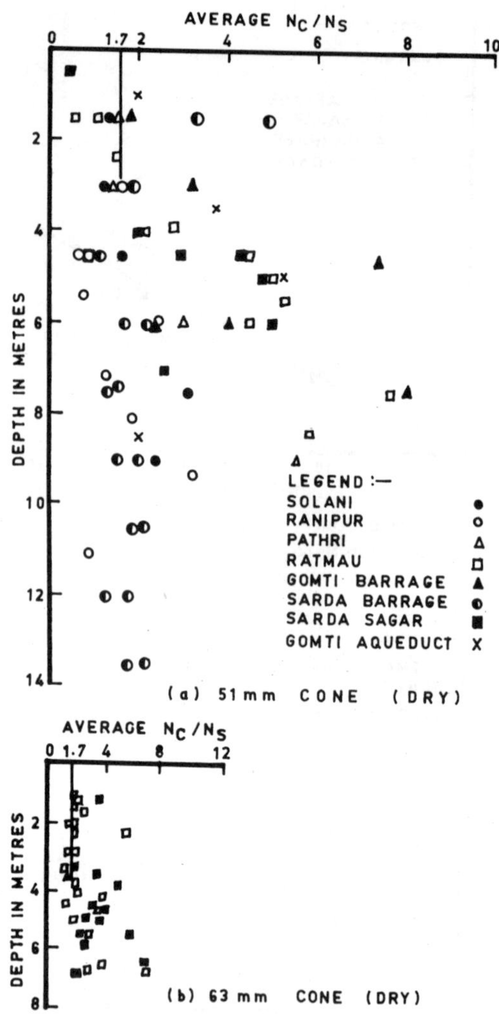

Fig. 6 Ratio of N_c/N_s vs. Depth

1.7 upto 4 m depth. Below 4.0 m depth there is lot of scatter and no relation holds good.

3.4 Dynamic cone (63.5 mm) driven with the help of slurry vs standard penetration resistance

The penetration tests by 63.5 mm dynamic cone driven after passing bentonite slurry and corresponding standard penetration test carried out at various sites are plotted in Fig.7. In this case most of the points lie between $N_c = 0.8\,N_s$ and $N_c = 1.5\,N_s$ with an average linear correlation of $N_s = N_c$. This finding confirm the earlier work carried out by Jain el al (1968).

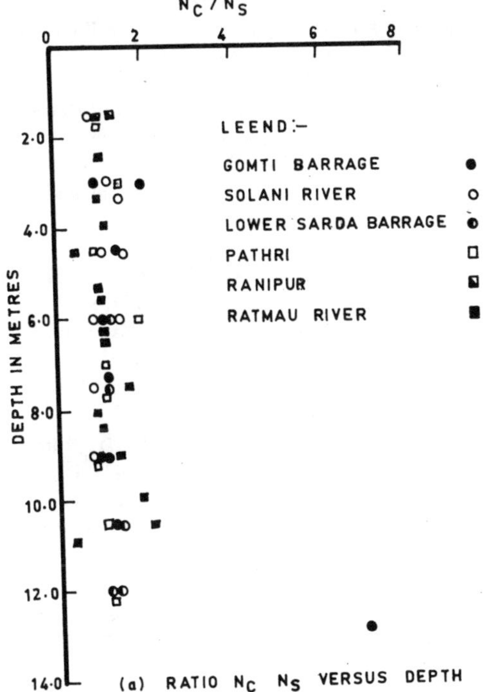

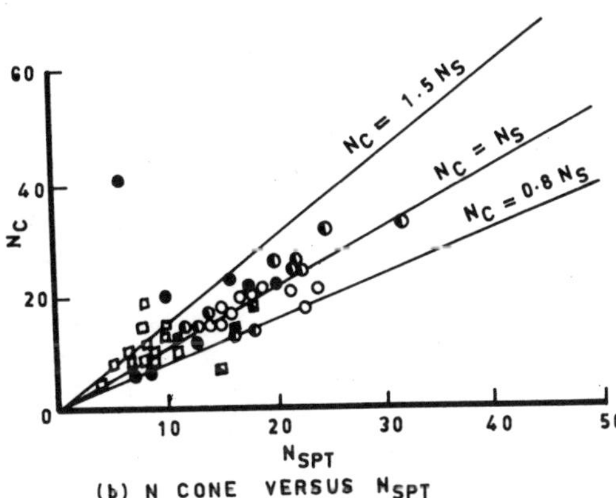

Fig. 7 Dynamic Cone vs. SPT Resistance

The 63 mm cone with circulation of bentonite slurry eliminate skin friction and hence the same can go deeper but its movement is restricted by the presence of gravel. In sandy strata at lower Sarda Barrage site, penetration test have been successfully performed upto 15 m. But at Gomti Barrage site where large gravel (greater than 50 mm size) was present, it could go only upto 9 m. Thus the method may be used for exploration of sandy foundations upto 15 m depth.

Dynamic cone tests may be considered superior to SPT because the former is devoid of bore hole effects, quick and inexpensive. But it has been experienced that in case of medium and dense strata the equipment viz. drill road etc. are damaged during the cone penetration test because the blows by 63.5 kg hammer are applied simultaneously all along the depth of exploration. The stresses caused by a simple blow of the hammer and the resulting deformation are added up throughout the test with the result that even in one bore hole the threads of the drill rods are damaged and the rods bend. In the case of tests with bentonite slurry, the correct consistency of the slurry is an important factor in the satisfactory performance of the tests. As such apart from its disadvantages, the standard penetration test still remains a versatile tool for insitu tests. As indicated in Table-3, the method has been used upto great depths for exploration of foundation having gravels of size approximately equal to the diameter of the penetrometer. The extra expense made in the performance of the tests due to the boring in SPT is partly compensated from the extra expenditure made in making a separate bore hole for logging and the damage caused to equipment during cone test. However, for obtaining correct results by means of standard penetration test great caution is necessary in the procedure of boring and driving. The use of high density drilling fluid with appropriate additives like bentonite or vel-clay in condition to severe boiling is also very necessary to obtain satisfactory results.

4. CONCLUSION

The study indicates that static cone resistance (q_c) has been found to equal to 2.5, 5 and 7 times the number of SPT blows respectively for fine, medium and coarse sands which confirms the earlier finding of Sutherland(1963) and Rodin (1961). Meyerhof's correlation $q_c = 4 N_s$ (1956) can be used only for medium sands and its use for fine and coarse sands is not likely to yield reliable results.

The 51 mm and 63.5 mm cone driven dry may be used to penetrate sandy strata upto shallow depths of 3 and 4 m only for which a correlation of $N_c = 1.7 N_s$ has been found to exist.

Below this depth, the test results show a wide scatter, 63 mm dynamic cone driven with the circulation of bentonite slurry has been found to penetrate sandy strata upto 15 m or so and the value of number of blows by dynamic cone and SPT have been to be almost similar. This test has not been found to be applicable in strata having gravel of size greater than 51 mm.

The recommended depth of probe for each penetrometer are given below:

Static cone	Upto the depth till total resistance is less than the capacity of machine.
Dynamic cone (51 mm driven dry)	Shallow depth upto 3.0 m
Dynamic cone (63.5 mm driven dry)	Shallow depth upto 4 m
Dynamic cone (driven with the circulation of bentonite)	Upto 15 m
Standard penetration test.	Upto any depth

5. ACKNOWLEDGEMENT

The matter for this paper is extracted from U.P.I.R.I.unpublished technical report No. 45 R.R.(S-143) which was prepared by the author under the guidance of Director,I.R.I. and with the assistance of staff of Soil Research Division, U.P.I.R.I. Roorkee.

6. REFERENCES

Canadian Standard Association(1962), Dynamic Cone soil penetration test, C.S.A.119.3.

Desai M.D. and Roy M.P.(1968), Correlation of dynamic cone and standard penetration test, Journal of Indian National Society of Soil Mechanics and Foundation Engineering, July, 1968, pp 311.

I.S.4968 - Part I & II,(1968), Method for subsurface soundings for soils, Part I-Dynamic Method using cone and Bentonite slurry.

Jain G.S.,Sen Gupta O., and Aggarwal V.S. (1968), Dynamic cone penetration test for site investigations, Symposium for Earth & Rockfill Dams, Talwara (1966).

Meigh A.C. and Nixon,I.K.(1961), Comparison of insitu test for granular soils, Proceedings Vth Conference, Soil Mechanics & Foundation Engg. Vol.I, pp 499.

Meyerhof G.G.(1956), Penetration test and bearing capacity of cohesionless soils, journal of Soil Mechanics and Foundation Engg. Proceedings ASCE,Paper No.866.

Mohan D.,Agarwal V.S. and Tolia D.S. (1970) The correlation of cone size in the dynamic cone penetration test with standard penetration test, Geotechnique, September 1970, pp 315.

Narhari D.R. and Agarwal V.S.(1967), Note on the correlation of standard penetration test and standard cone penetration test, Proceedings symposium on site investigations for foundation CBRI,Roorkee (March Vol.I, pp 153).

Palmer D.J. and Stuart, J.G.(1957), Some observations on the standard penetration and correlation of the tests with a new penetrometer, Proceedings IVth International Conference on Soil Mechanics & Foundation Engg. Vol.I, pp 231.

Rodin, Stanley (1961), Experiences with penetration testing for site investigations for pile foundations, Symposium on Bearing capacity of piles, CBRI, Roorkee.

Schultze E.and Knausenberger,H.(1957), Experiences with penetrometers,Proceedings IVth International Conference on Soil Mechanics and Foundation Engineering, Vol.I, pp 249.

Schultze, E. and Melzer, K.I.(1965), The determination of density and modulus of compressibility of non-cohesive soils by soundings, Proceedings VIth International Conference on Soil Mechanics and Foundation Engineering, Vol.I, pp 354.

Sutherland M.B.,(1963), The use of insitu test to estimate the allowable bearing pressure of cohesionless soils, Str.Engg. London, Vol.XXI, No.6, March.

Proceedings of the Second European Symposium on Penetration Testing / Amsterdam / 24-27 May 1982

The use of a light percussion sounding apparatus in road engineering

A.DE HENAU
Belgian Road Research Center, Brussel

INTRODUCTION

Since the ESOPT I Symposium at Stockholm in 1974 the Belgian Road Research Center (C.R.R.) has made a study of a light percussion sounding apparatus (Kindermans 1976 a) chiefly intended for making preliminary geotechnical surveys for road projects.

An operating procedure has been established (C.R.R. 1978) which standardizes the equipment and enables a bearing characteristic for the soil layers which the tip passes through to be deduced from the penetration (X) in mm per blow.

The tip consists of a cylindrical sleeve of equal height and diameter with a section of 5 cm^2 and a point which meets at an angle of 60° at the vertex. It is so constructed that it forms a whole with a 1m extension rod. This rod is screwed to a percussion rod which guides a 10 kg mass falling freely through a distance of 50 cm onto the striking plate.

Successive extension rods may be inserted each of which extend the sounding depth by one metre. The penetration (X) mm per blow of the tip can be correlated to the CBR up to depths of 2 m - 2 extension rods - and give a good approximation of the bearing capacity of each layer traversed (fine soils $D_{90} < 0,2$ mm).

In certain cases this relation can be used up to depths of 3 m (3 extension rods); beyond this depth the percussion curve gives information which is useful for the qualitative comparison of the strength of the traversed layers. Up to depths of five metres the sensitivity of the equipment is still very good (see below), but beyond that, though the apparatus may still be used with success, handling is no longer very practical and loses its simple and straightforward character. Other existing equipment can give higher yields in these cases.

The basic kit includes three extension rods (for soundings between 2 and 3 metres) (fig. 1). In road construction, geotechnical investigations seeking information on the subgrade to a depth of 2 to 3 m, though only concerned with a limited width, extend over great lengths crossing the most diverse geological formations in areas which are often not easily accessible to vehicles.

The light percussion sounding apparatus makes it possible to interpolate more accurately between more extensive, and more expensive, soundings (for example, static soundings for engineering structures), which may be spaced more widely.

Fig. 1 CRR 1611/6

Team at work.

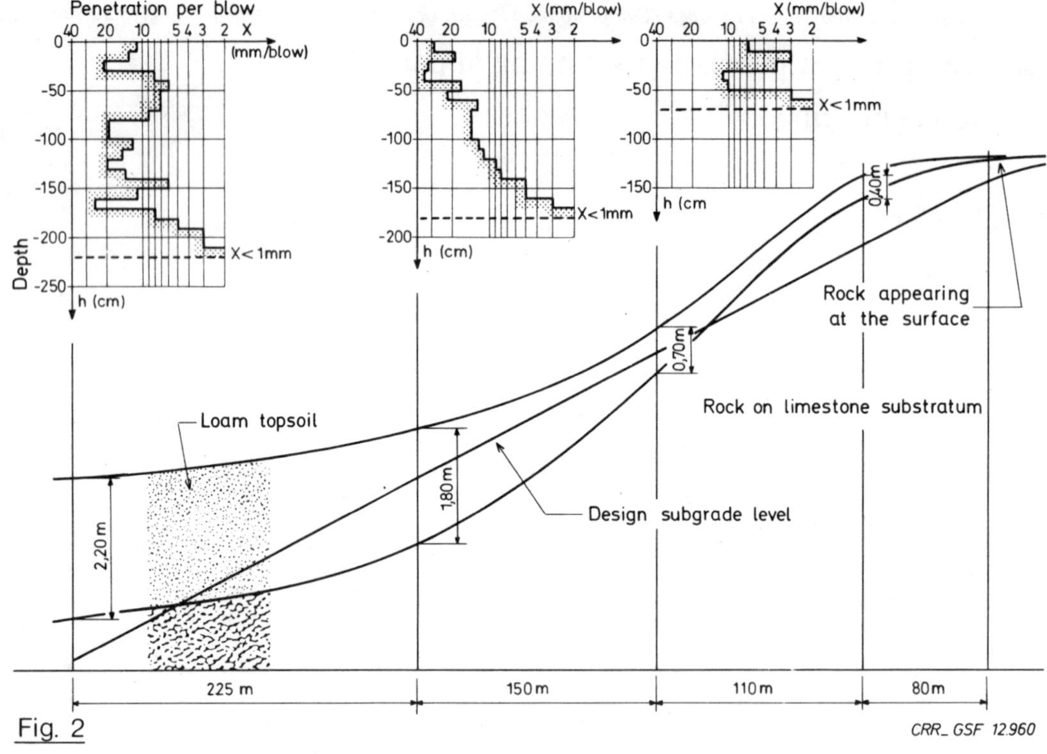

Fig. 2

Determination of the thickness of a loam topsoil.

EXAMPLES OF QUALITATIVE INTERPRETATION

a) First example (Kindermans 1976 b)

A road of minor importance is to be constructed by excavating in the side of a valley worn in carboniferous limestone, covered by a variable depth of loam.

The thickness of this loam layer, the gradient of the slope being known, determines the quantity of loose soil and rock which it is planned to excavate.

A series of percussion soundings (fig. 2) will make it possible to obtain extremely satisfactory estimates of the thickness of the loam.

b) Second example of qualitative use

A sounding made with the percussion sounding apparatus through a 5 m embankment made of a non-conventional material (Gorlé) is shown in the diagram of figure 3.

This diagram reflects the two construction methods used in detail. The first layers (fill base) were compacted every 40 cm while the upper layers were compacted every 20 cm.

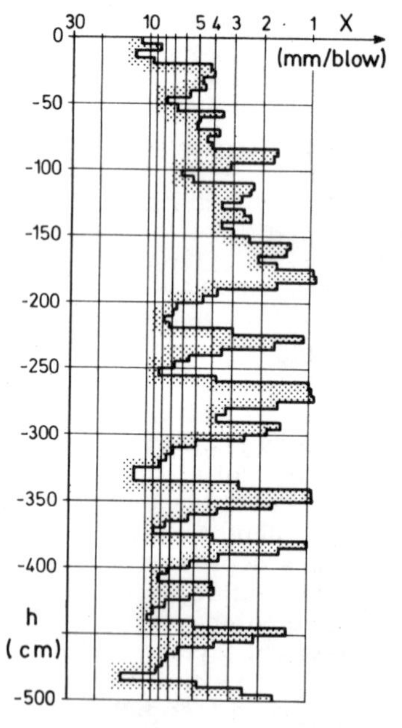

Fig. 3

Sounding diagram in a 5 m high embankment.

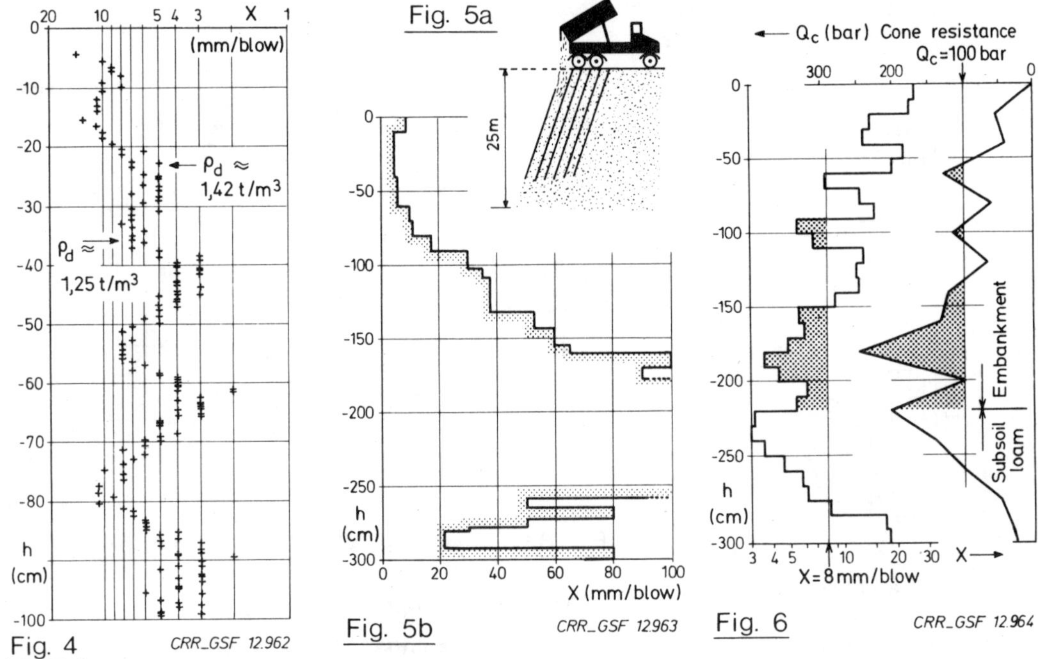

Fig. 4 Diagram blow per blow gradient in the layers of 20 cm.

Fig. 5b Penetration diagram at the top of a fly-ash dumpingground.

Fig. 6 Percussion sounding (X) versus static penetration (Qc).

A closer blow-by-blow analysis of the penetration diagram shows a compactness gradient in each individual 20 cm layer ($\rho_d \approx 1,42$ t/m^3 above and 1,25 t/m^3 below).

Figure 4 shows how the percussion sounding apparatus can be used for checking the homogeneity of work during construction.

However, even for qualitative interpretation, soundings must be standardized.

c) A further qualitative application

A stock of non-conventional material, dumped in an abandoned quarry of about 20 m depth, is being considered for the construction of an embankment.

The hole was filled by successive dumpings from the side (see fig. 5a), by lorries carrying a 25 ton load. The only compaction was due to the lorry traffic.

The surface of this dumping ground is encouraging and leads one to suppose that material could be removed with a loader. Nevertheless the penetration diagram obtained from the percussion sounding apparatus indicates (fig. 5 b) that the material is only properly compact and bearing for a thickness of 0.5 to 1.5 metres and consequently reveals that it would be impossible to use this stock as desired. There is even a real danger involved in removing the surface layer.

EXAMPLES OF QUANTITATIVE INTERPRETATION

a) Compaction control

The construction of an embankment from a non-cohesive fine material may lead to a danger of liquefaction in saturated areas when compaction is inadequate.

The acceptance criterion of such a structure can be based on the lower limit of the resistance to the static penetration of a test cone, for example Qc<100 bar.

This test cannot be easily used as a control method leading to an immediate reaction on the construction site.

An experimental 2 m high embankment was constructed and on this the light percussion sounding apparatus was used to make a series of dynamic soundings parallel to a series of static soundings.

Figure 6 shows, without attempting to establish a direct relationship, that a maximum limit for the penetration per blow (X) mm which should not be exceeded in order to guarantee a satisfactory value for Qc can be reasonably determined.

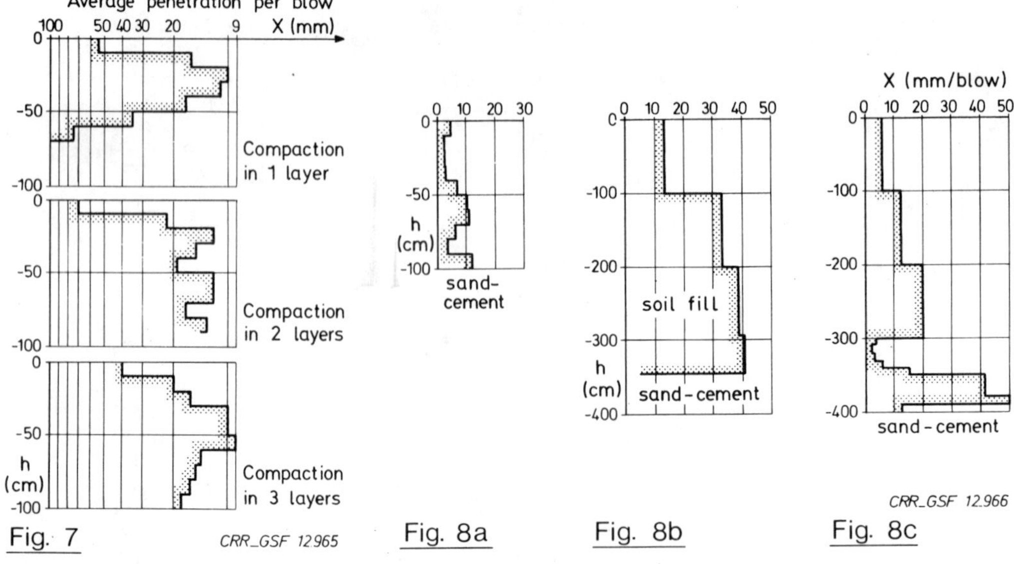

Fig. 7

Control in a narrow trench (effect of compaction).

Fig. 8a Fig. 8b Fig. 8c

Penetration diagram of the fill on a deep sewer.

In the quoted example we are tempted to say that, if during construction the percussion sounding apparatus gives a penetration per blow of X mm ≤ 8 mm for each layer, the static penetration test will probably give a satisfactory result.

b) Checking the filling of a trench

Numerous subsidiary works in road engineering involve the use of the subsoil and consequently involve excavations or trenches which must be filled so that their stability is the same as that of the neighbouring undisturbed soil.

This applies especially to the construction of water clearance systems (sewers, drains, etc.) or the installation of vertical capillary screens in order to protect the subgrade against edge effects (C.R.R.1981).

In the latter case the trenches are deep (150 cm) and narrow (15 cm).

For determining if the filling and compaction method proposed by the contractor is satisfactory, a practical control method uses a percussion sounding apparatus.

Let us consider three filling techniques : compaction in a single 90 cm layer, compaction in two successive layers or in three successive layers corresponding to the three penetration diagrams given in figure 7. The object was to obtain a penetration per blow (X) mm of less than 20 mm. It can be seen that a 2 layer method is adequate for the depth.

In another case a deep trench under a road (installation of a 200 cm diameter collector at a depth of 4 metres in sand-cement to half the height of the pipe) must be backfilled so that the original bearing capacity of the soil is restored.

Figure 8 a shows a penetration diagram in the sand-cement layer just after it has been installed along the length of the pipe.

Figure 8 b gives the diagram for the filled layers above the stabilized sand.

It can be seen that only the upper metre has been compacted and the layer adjoining the pipe has been compacted the least.

Figure 8 indicates that the trench has been satisfactorily compacted, but reveals an anomaly at the level of the cement stabilized layer around the duct.

The percussion sounder had not yet been introduced as a test method when this work was carried out. The road authority had specified that plate loading tests had to be carried out in the trench. The practical difficulty of performing this test under these conditions (2 men team in the trench - works halted during the test, etc.) means that this control technique was only infrequently employed.

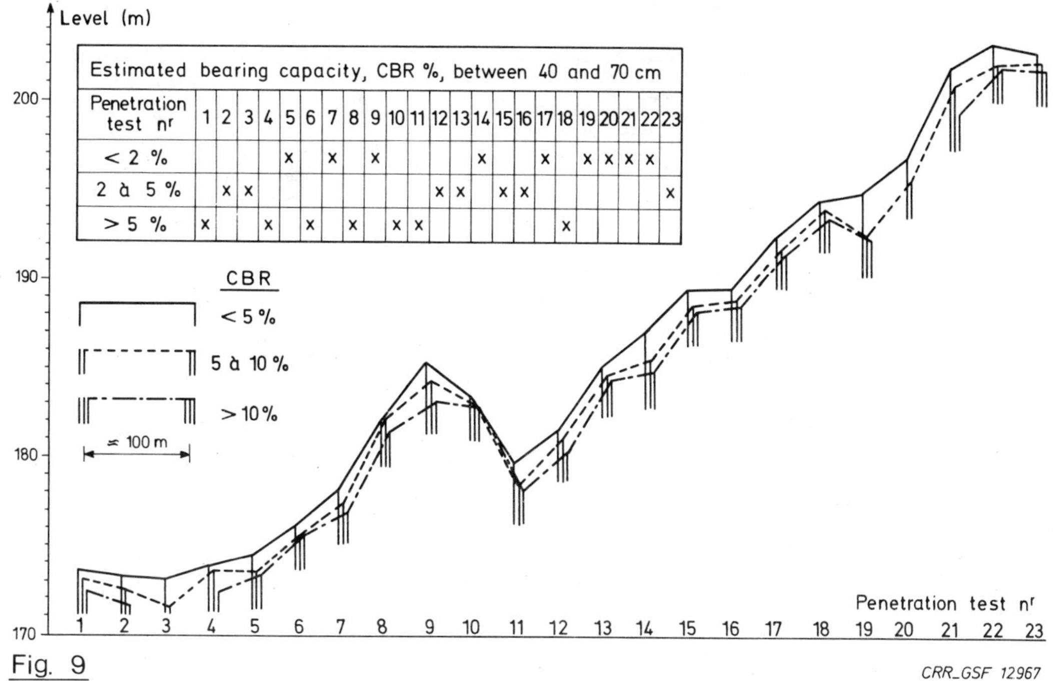

Estimated bearing capacity, CBR %, between 40 and 70 cm																							
Penetration test nr	1	2	3	4	5	6	7	8	9	10	11	12	13	14	15	16	17	18	19	20	21	22	23
< 2 %					x		x		x					x			x		x	x	x	x	
2 à 5 %		x	x									x	x		x	x							x
> 5 %	x			x		x		x		x	x							x					

Fig. 9

CRR_GSF 12967

Estimated bearing compacity, CBR %, to 2 metre .

c) Use of a percussion sounding apparatus when choosing road subgrade levels

The design of the thickness of the layers is affected by economic considerations; the bearing characteristics of the subgrade must be taken into account when determining the thickness of the layers of the structure.

Here it is vital that the designer determines the nature of the subgrade soil in order to avoid as far as is possible any surprises during execution, which almost always lead to the replacement of "bad" soil - an item on the bill of quantities which is always speculative - as the levels of the carriageway have already been designed. Here the light percussion sounding apparatus can render invaluable service during the geotechnic exploratory survey of the alignment of the road without increasing costs to unthinkable levels.

In the example quoted here the designer proposed constructing a semi-rigid pavement, the base of which consists of a cement-stabilized mixture (lean concrete).

However, as he knew that certain works in the area had encountered difficulties (soils with a low bearing capacity) he surveyed the route for 2000 m making soundings with the light percussion sounder at approximately every 100m in accordance with the operating procedure (C.R.R. 1978). By transferring the simplified penetration diagrams to an anamorphosed longitudinal profile as shown in figure 9, it can be seen that the soil in its natural state contains in its depth (1 to 2 m) an excellent bearing layer, a layer of variable thickness with a less good bearing capacity (CBR between 5 and 10 %) and a layer which has a generally poor bearing capacity (CBR of less than 5 %).

It can be seen at two places in the longitudinal profile the strong course is lacking. The records of the municipality revealed that these were due to two old quarries which had long been filled in and forgotten.

By regrouping the soundings in relation to the bearing capacity found between 40 and 70 cm under the natural ground level, which was the level originally projected for the subgrade by the designer, the results given in the table shown in figure 9 were obtained.

This table makes it immediately evident that it is not possible to cut the route into homogenous sections using the same construction technique of an adequate length.

The project must be designed on the basis of the poorly bearing soil. Figures 10a, 10b, and 10c give a detail of three characteristic penetration diagrams on which one has marked the position of the

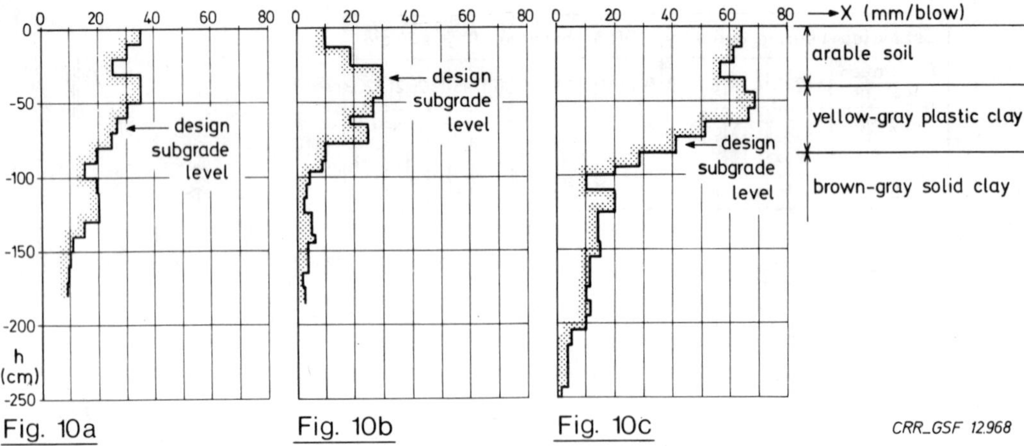

Three types of penetration diagrams.

subgrade as determined by the designer before starting the sounding programme.

It can be seen, from figure 10 b for example, that the specified earthworks would eliminate the upper bearing layer which is precisely the strongest layer or, from figure 10 c, that the earthworks would have stopped too early to gain from an excellent bearing course.

A core, which has also been schematically reproduced in figure 10 c, shows that the clayey soil is drier deep down but is in plastic state in the upper layer.

The designer was advised not to use a rigid layer in the structure if the subgrade could not rest on a homogeneously bearing soil.

Figure 10 allows the designer to select a reasonable soil bearing capacity which will not lead to excessive overdesigning or underdesigning of the future road, with optimal earthwork volumes.

CONCLUSIONS

The light percussion sounding apparatus developed by the C.R.R. is an apparatus which is inexpensive and easy to use.

It will rapidly supply a mass of qualitative and quantitative data which are extremely useful in road geotechnology.

It is suitable for use on all construction sites, from the very largest to the smallest (public and private roads, car parks), not only in the surveying stage, but also as a useful tool for checking the quality of the work as it progresses.

The standard method of use as set out in the operating procedure (C.R.R., 1978) is extremely simple and does not require highly qualified personnel.

It can thus be used both in industrialized and developing countries, and is of invaluable service to designers, builders employers, supervisory officials and contractors.

REFERENCES

Centre de Recherches Routières 1978, Mode opératoire-Estimation rapide de la portance des sols à l'aide d'une sonde de battage légère type C.R.R. MF 39/78 Brussels, C.R.R.

Centre de Recherches Routières 1981, Code de bonne pratique pour la réalisation d'écrans capillaires verticaux contre l'effet de bord sous les chaussées R 48/81, Brussels, C.R.R.

Gorlé, D. to be published, Etude spécifique du remblai expérimental en gypse résiduaire à Zelzate. Compte rendu de recherche, Brussels, C.R.R.

Kindermans, J.-M. 1976a, La sonde de battage légère en construction routière CR 5/76, Brussels, C.R.R.

Kindermans, J.-M. 1976b, Utilisation en construction routière de la sonde de battage C.R.R., La Technique Routière, XXI/3 : 1 - 10.

Proceedings of the Second European Symposium on Penetration Testing / Amsterdam / 24-27 May 1982

The application of a portable pavement dynamic cone penetrometer to determine in situ bearing properties of road pavement layers and subgrades in South Africa

E.G.KLEYN
Transvaal Provincial Administration Roads Department, Pretoria, South Africa

J.H.MAREE
National Institute for Transport & Road Research, CSIR, Pretoria, South Africa

P.F.SAVAGE
University of Pretoria, South Africa

1 INTRODUCTION

Direct evaluation of the in situ strength properties of road pavement layers has always been desirable but often too cumbersome. Thus pavement engineers have mostly resorted to indirect methods such as the laboratory CBR (California Bearing Ratio) which is basically a penetration-strength test on a soil sample.

A portable Dynamic Cone Penetrometer (DCP) (Van Vuuren 1969) was used during an extensive investigation of road pavement performance by the Roads Department of the Transvaal Provincial Administration (TPA) during 1973 in an endeavour to obtain some estimate of in situ strengths (Burrow 1975).

The DCP has been correlated with standard bearing and performance parameters (Kleyn 1975). At present it is being used to augment the Heavy Vehicle Simulator (HVS) programme (Freeme et al. 1981 and Appendix A).

The purpose of this paper is to introduce the portable DCP and to describe some elementary applications to road pavements.

2 THE INSTRUMENT

The pavement DCP basically consists of a 16 mm diameter steel rod with a 20 mm diameter, 60° cone of tempered steel at one end. (Figure 1). Impact is provided by means of an 8 kg sliding hammer, falling 575 mm. The total weight of the instrument is about 12 kg.

Operation of the instrument is best carried out by three persons, although two will suffice if speed of operation is not essential. One person holds the instrument upright whilst the second person operates the hammer, allowing the third person to record the readings.

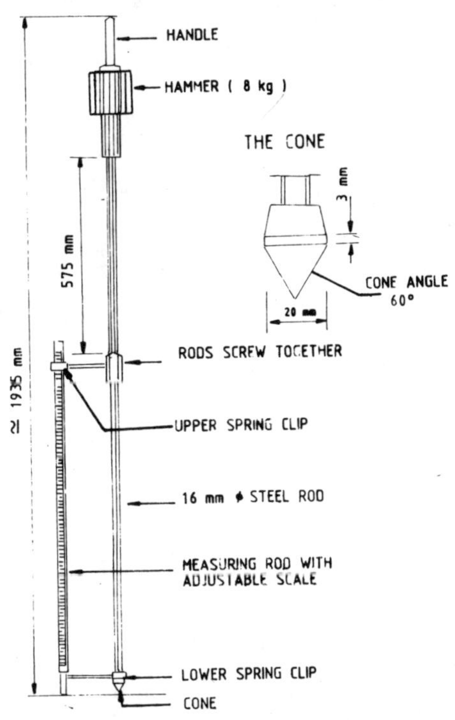

FIGURE 1

THE PORTABLE PAVEMENT DYNAMIC CONE PENETROMETER (DCP)

DCP soundings in road pavements are usually done to a depth of 800 mm below the surface. This is the depth beyond which the material normally has minimal effect on traffic-associated pavement performance.

3 REPRESENTATION OF READINGS

3.1 DCP Curve and DCP Number

The DCP Curve describes the number of blows to reach a given depth. Readings are normally recorded graphically on the field form as illustrated in Figure 2, affording an instantaneous visual illustration of the in situ material strength. Note that the slope of the curve at any point represents the resistance offered by the material - the flatter the slope, the higher the resistance. This slope, expressed in terms of mm/blow, is called the DCP Number (circled numerals in Figure 2).

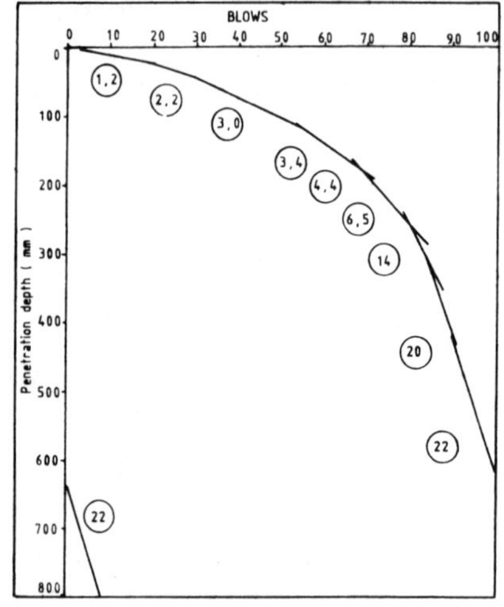

FIGURE 2

EXAMPLE OF FIELD FORM SHOWING TYPICAL DCP CURVE

DCP Numbers normally encountered in road pavements consisting of natural gravels and soils, fall in the range of 2-25 mm/blow - although the instrument is increasingly being applied to cemented soil and gravel layers, resulting in readings as low as 0,5 mm/blow.

3.2 Layer-Strength Diagram

A derivation of the DCP Curve is normally drawn, showing the DCP Number against depth, as illustrated in Figure 3 - called the Layer-Strength Diagram. This diagram makes the evaluation of soundings easier, since the relative position of a line is interpreted and not a gradient.

The CBR and UCS (Unconfined Compressive Strength) scales are also given to facilitate easy transformation from DCP Number to these well-known parameters (this correlation from section 4).

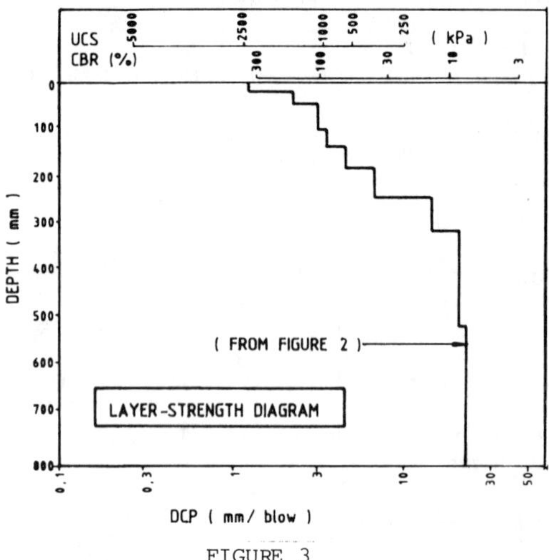

FIGURE 3

EXAMPLE OF LAYER-STRENGTH DIAGRAM

4 CALIBRATION OF THE DCP

Correlation of DCP with standard CBR was obtained (Figure 4) through penetration of some 2000 samples of various pavement materials in standard moulds directly following CBR determination (Kleyn 1975). It should be understood that the DCP essentially measures shear strength.

Initial calibration was done with a 30° cone and later converted to the 60° cone through parallel tests. The repeatability of DCP measurements is at least equal to that of other similar test equipment (Kleyn 1975).

The effect of various phenomena on the DCP calibration, such as mould confinement and density gradient, was evaluated as described by Kleyn (1975).

Similarly the DCP was also correlated with UCS, resulting in Figure 5 (Bester and Hallat 1977 and De Villiers 1980).

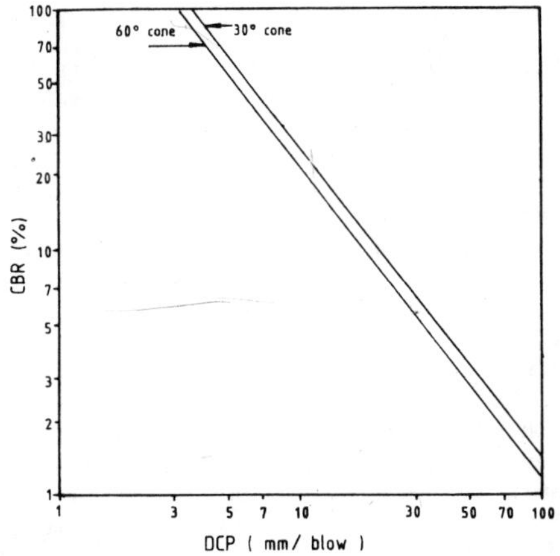

FIGURE 4

RELATIONSHIP BETWEEN CBR AND DCP FOR TWO DIFFERENT CONE ANGLES

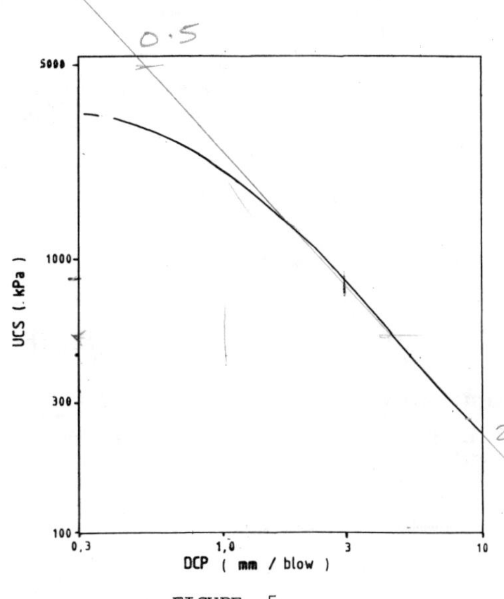

FIGURE 5

RELATIONSHIP BETWEEN UCS AND DCP

5 APPLICATION OF THE DCP TO ROAD CONSTRUCTION

5.1 Potentially collapsible soils

A DCP sounding may be done in the in situ material, after which the immediate area (≃300 mm radius) is inundated with water. Another DCP sounding is then done close to the previous sounding and within the now saturated area. If the readings from the two sets of soundings differ markedly, a potentially collapsible or sensitive soil is indicated - warranting a more sophisticated evaluation.

5.2 Construction control

The DCP affords a method whereby in situ shear strength may be monitored directly - even if only to ascertain whether the bearing capacity over an area is uniform. Typically, this applies when large fills are constructed and high lift rates render normal control techniques too time-consuming to be effective.

Through continuous supplementary DCP monitoring, non-uniform compaction can be spotted on site and remedied immediately by directing more compactive effort to the particular area.

5.3 Efficiency of compaction equipment

When investigating the effectiveness of different rollers or rolling techniques, DCP soundings present a quick method of evaluating the in situ effect at depth (Wolmarans 1977). DCP soundings may be done initially on the material to be compacted and the soundings repeated after a certain number of passes. The soundings may be represented graphically on the Layer-Strength Diagram.

5.4 Stabilized layers

Since the strength of a stabilized layer of material should increase with time and proper curing, various anomalies pertaining to badly stabilized layers may be detected through use of the DCP. These anomalies may include insufficient depth of stabilization, uneven mixing of the stabilizing agent ("lenses"), or improper curing of the upper part of the layer. The increase of strength with time may also be monitored to verify whether design strength objectives have been met (Marais 1981).

5.5 Centreline sampling

Centreline sampling may be augmented by less time-consuming DCP soundings. Normally the in situ bearing capacity of material is higher than that found during laboratory testing (std. CBR, saturated sample). If it is the other way round, very loose material and/or excess moisture is suspected and should be investigated.

6 PAVEMENT EVALUATION AND MONITORING

6.1 Structural evaluation of pavements

Standard practice in evaluating a section of pavement would be to obtain a set of resilient deflection measurements from which locations of interest may be identified. DCP soundings may be regarded as second-stage evaluation after deflection monitoring, giving the in situ strength of the pavement at any desired location.

Similar sections may be grouped together when drawing the Layer-Strength Diagram (section 3), thus obtaining a visual mean. The mean may also be computed from a number of such soundings and drawn on a separate Layer-Strength Diagram together with the minimum and maximum strength envelope. Additionally, the desired strength at the various depths may be added in the form of a directrix to facilitate easy evaluation.

Using the Layer-Strength Diagram, appropriate representative sites may be selected for trialpits, thus affording third-stage evaluation and verification of phenomena illustrated on the Layer-Strength Diagram.

6.2 In situ strength vs laboratory CBR

It is current practice in South Africa to design pavement structures using the laboratory soaked CBR as the strength parameter for the subgrade. The basic pavement thickness is determined from this soaked CBR. This approach may be overconservative as was shown in a special HVS (Appendix A) test. The laboratory soaked CBR of the clayey subgrade of the pavement was equal to 2%. For such a poor subgrade, a total cover of about 850 mm would be required for a design traffic of one million standard axle loads. The particular pavement, having a structure thickness of only 310 mm, was therefore underdesigned and should have failed rapidly. During the HVS test however, two million standard axles were applied, without a failure occurring. The test was done under relatively dry conditions, and the use of the laboratory soaked CBR-value was questioned. Subsequent DCP soundings showed that the subgrade had an in situ CBR of 12-15%. The pavement structure for such a subgrade strength is therefore of adequate thickness to carry about two million standard axles. The test very practically demonstrated that the DCP measures in situ CBR and that in situ strength is an important-factor when evaluating pavements.

6.3 Structural monitoring of pavements

Using the Layer-strength Diagram, the DCP may be used to monitor pavement structures. In other words one could evaluate what is happening and at what depth.
Recent tests with the HVS have shown that the moisture content in the pavement layers may have an overriding effect on the behaviour of pavement structures. The DCP could be used to quantify this sensitivity to moisture,as was illustrated in the HVS test on road P6/1. This road has a light pavement structure, and was built about 25 years ago. Figure 6 shows the structure and Figure 7 shows its behaviour during the HVS test.

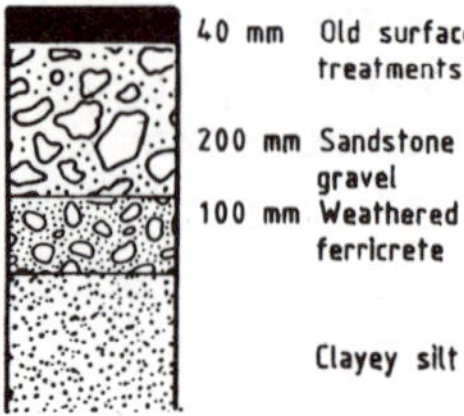

FIGURE 6

THE PAVEMENT STRUCTURE OF ROAD P6/1

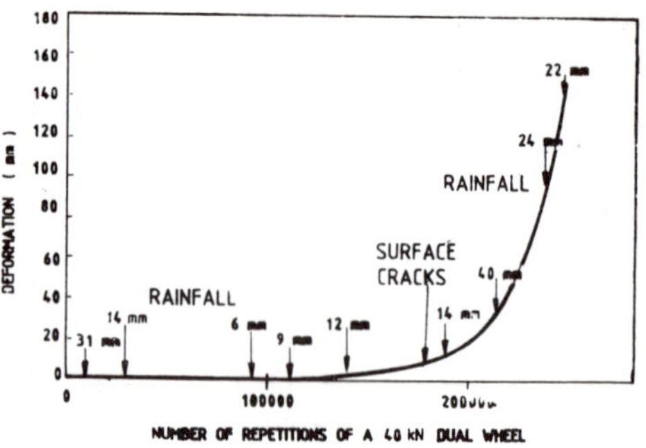

FIGURE 7

HVS TEST 45A4 ON ROAD P6/1 : PERMANENT DEFORMATION OF THE PAVEMENT STRUCTURE AGAINST NUMBER OF REPETITIONS OF A 40 kN DUAL WHEEL LOAD

The structure showed little deformation until water entered through the cracked surface. At that stage, the behaviour changed markedly and the rate of deformation increased rapidly. DCP soundings in the 'dry' and 'wet' phases of the test showed (Figure 8) that the upper layers lost their strength when the moisture content increased and this explained the

sudden increase in deformation. At the end of the test, trenches were dug across the section to establish the origin of the deformation. Most of the deformation (>80%) originated in the top 300 mm of the structure, confirming the DCP soundings (Figure 8).
The DCP could therefore quantify the effect of moisture.

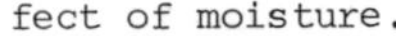

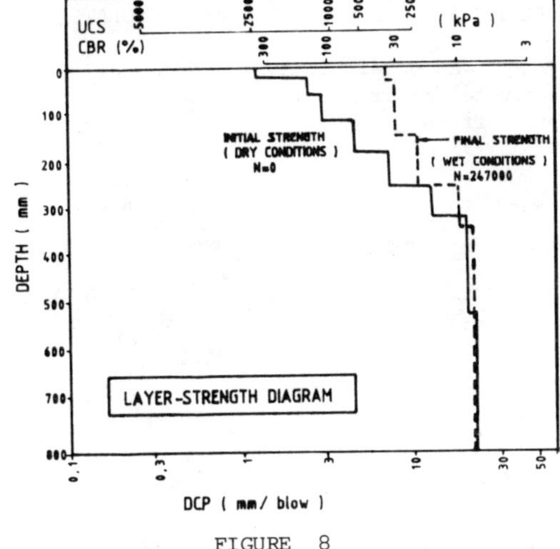

FIGURE 8

CHANGES IN THE LAYER STRENGTHS, AS MONITORED DURING THE HVS TEST ON ROAD P6/1

7 PAVEMENT DESIGN

The use of the DCP in pavement design is currently being investigated. As an interim measure the following may be done:

In section 6 it was mentioned that the desired strength at various depths may be drawn on the Layer-strength Diagram in the form of a directrix. Any DCP sounding in in situ material may be compared with this directrix to gain an estimate of the depth at which the in situ material may be utilized in the proposed structure. The directrix thus used, constitutes an elementary cover curve aimed at optimal utilization of in situ materials. The normal material standards for pavement layers will however still apply.

Figure 9 gives three directrices indicating the pavement bearing capacity compositions found to perform well in the Transvaal for light, medium and heavy traffic loading situations (TPA, 1978).

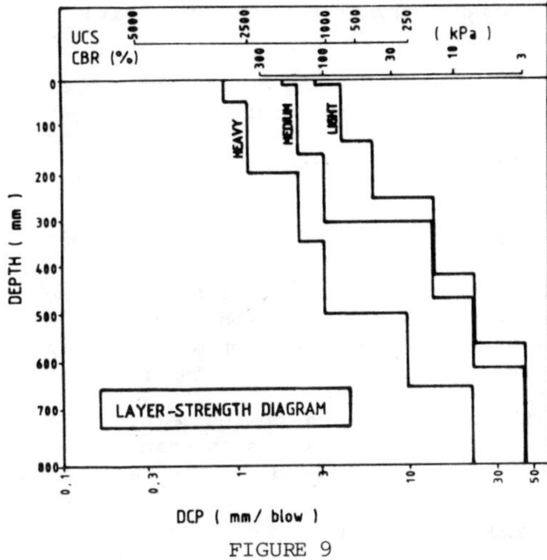

FIGURE 9

EXAMPLES OF BASIC PAVEMENT DESIGNS USED BY THE TRANSVAAL ROADS DEPARTMENT

8 FUTURE DEVELOPMENT

Current research will extend the applicability of the DCP to pavement design and evaluation to include the balance of pavement structures, a computerized DCP strutural number approach and correllations with resilient moduli and deflections as well as riding quality.

9 CONCLUSIONS

a) Methods have been developed to represent pavement DCP soundings i.e. the DCP Curve, DCP Number, and the Layer Strength Diagram.

b) The DCP was calibrated against commonly used strength parameters such as CBR and UCS and the repeatability of the instrument has been found to be equal to that of other similar devices.

c) Application of the DCP to road construction for identifying potentially collapsible soil, for construction control, evaluation of the effectiveness of compaction, monitoring stabilized layers and augmenting centreline sampling, has been illustrated.

d) The DCP may be used as a second-stage evaluation after deflection measurements.

e) Special Heavy Vehicle Simulator tests have demonstrated that the in situ shear strength derived from DCP sounding correlates better with the pavement's field performance than the laboratory soaked CBR values.

f) The DCP may be used to monitor the effect of moisture and traffic on road pavements. Use of the DCP in conjunction with the Heavy Vehicle Simulator can quantify this effect.

g) Directrices on the Layer-Strength Diagram indicating the pavement bearing capacity compositions for light, medium and heavy traffic conditions in the Transvaal have been developed. These may be used in pavement evaluation or design.

h) The evaluation and design methods presented may be regarded as interim as current research should lead to refinement.

10 ACKNOWLEDGEMENTS

The Director of Roads, Transvaal Provincial Administration and the Director of the National Institute for Transport and Road Research are thanked for their permission to publish this paper.

11 REFERENCES

Bester, M.D., and L. Hallat 1977, Dinamiese Kegelpeilstaaf (DKP). University of Pretoria, Pretoria. (Dynamic Cone Penetrometer (DCP)).

Burrow, J.C. 1975, The investigation of existing roads. Transvaal Roads Department, Report L1/75, Pretoria.

De Villiers, P.J. 1980, Dinamiese Kegelpeilstaaf korrelasie met Eenassigge druksterkte. University of Pretoria, Pretoria. (Dynamic Cone Penetrometer correlation with Unconfined Compressive Strength).

Freeme, C.R., R.G. Meyer and B. Shackel 1981, A method for assessing the adequacy of road pavement structures using a Heavy Vehicle Simulator. Proc. IRF World Meeting, Stockholm.

Kleyn, E.G. 1975, The use of the Dynamic Cone Penetrometer (DCP). Transvaal Roads Department, Report L2/74, Pretoria.

Marais, F.J. 1981, Sterktetoename van gestabiliseerde lae. Transvaal Roads Department, Report S2/81, Pretoria. (Strength increase of stabilized layers).

Transvaal Provincial Administration 1978. Pavement and materials design manual, Transvaal Roads Department, L1/78, Pretoria.

Van Vuuren, D.J. 1969, Rapid determination of CBR with the portable Dynamic Cone Penetrometer. The Rhodesian Engineer.

Wolmarans, C.H. 1977, The construction of roads and airfields on low-density windblown sands. ABG Annual Conference, Münich.

APPENDIX A : THE HEAVY VEHICLE SIMULATOR.

The HVS was developed by the National Institute for Transport and Road Research, South Africa.

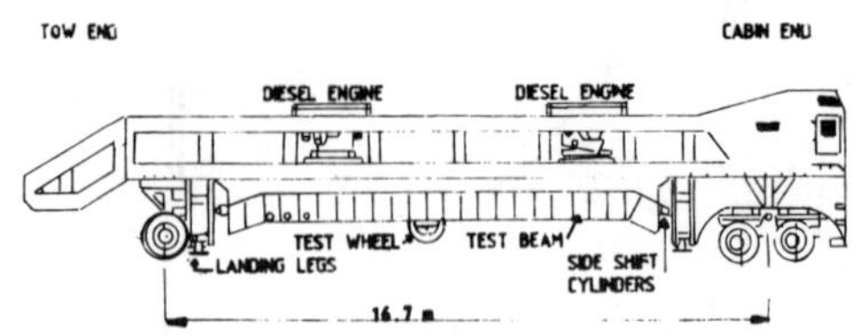

FIGURE A.

THE HEAVY VEHICLE SIMULATOR

Table A.Characteristics of the HVS

Overall length	22,56 m
Overall width	3,73 m
Overall height	4,20 m
Total Mass	57 000 kg
Trafficking speed	14 km/h
Trafficking length	8 m
Trafficking width	0,6 - 1,5 m
Repetitions/hour	1 200
Load - static	0 - 150 kN
- dynamic	0 - 100 kN
Inflation pressure	560 - 690 kPa

Proceedings of the Second European Symposium on Penetration Testing / Amsterdam / 24-27 May 1982

Investigation of possibilities of sounding technique for division of soil massive layers

B.I.KULACHKIN, V.P.OTREPIEV, M.I.SMORODINOV & A.Z.GHISTER
Research Institute for Bases & Underground Structures, Moscow, USSR

One of the main problems of sounding technique is the estimation of soil massive properties. Possibility of solution of such problem is the advantage of sounding technique comparing other methods of engineering geology, for the sounding technique allows to get the spatial picture of layers and their properties at any desirable details.

The basic problem of soil massive investigation is the determination of charater of laying and classify their types.

The criterium "t"(1) allows to classify soils to sands and clayish by means of relation of surface friction resistance P_δ to the cone resistance P_{SK} by results of sounding. Experimental investigations discovered this criterium was not always correct, especially at ultimate meanings, so it may be applied only as approximated.

As for dynamic sounding (DS), then some precautions are necessary because of effect of various factors at DS results.

Energetical treatment (3) for estimation of dynamical sounding indexes expansed the possibilities of this method and allowed to carry out investigations at modern level.

The method of static sounding with use of radioisotopes, specifically of soil natural radioactivity measurements, developed by Ferronsky (4), was the essential advance in this field.

Measurements of soil natural radioactivity may be used for classification of soil layers in combination with P_{SK} and P_δ data, but the sphere of application of this method is regional (4).

Closer definition of this method is the use of criterium "t" and soil natural radioactivities data.

Criterium "t" is assigned to approximate estimation of soil types, and soil natural radioactivity measuring - to more detail estimation.

This method and apparatus (PIKA-IO) are now developed in Research Institute for Bases and Underground Structures.

The apparatus PIKA-IO for measuring of P_{SK} , P_δ , and natural radioactivity, may be applied with any sounding device.

The following stage of estimation of soil massive properties is the investigation of every soil type.

Interpretation of many diagrams of sounding tests discovered that thickness of any soil layer was between 2 and IO metres.

The determination of minimum thickness of layer with homogenious properties by sounding tests is of much interest.

The cut off value of sounding parameters P_{SK} , P_δ is accepted to be the 5% of measuring value. The thickness of elementar layer with homogenious properties by sounding tests determined to be 20 sm.

This value was used for designing of apparatus and methodical manuals for static sounding tests.

Some investigations in various regions were performed for estimation of spatial variability of soil massive properties.

The dimensions of test sites in plane were 30x30 m.

The in-situ tests were performed under the following conditions:
-the same type of soil,
-the same depth of sounding,
- variations of sounding parameters are identified with non-uniformity of soil.
The later condition means that variations of sounding parameters includes the soil non-uniformity and measuring errors.
The errors of P_{SK} and P_{δ} measurings were no more, than 2,5%.
Each test consisted of two series. The results of statistical analysis of test data showed the following: the distribution of sounding parameters were not contradict to the Gauss law, $\bar{P}_{SK}$ were 3,53 МПа, $\bar{P}_{\delta}$ = 2,20 кПа, P_{SK} were 0,35 МПа, $\sigma_{P_{\delta}}$ = 18 кПа.
Similar results were got in various regions.
The effect of measuring errors at variations of P_{SK} and P_{δ} are neglectable. Thus the variations of P_{SK} and P_{δ} are caused by non-uniformity of soils.
Even in homogenious soil the variations of P_{SK} and P_{δ} are great enough and may reach 30% of average meaning of measuring properties. Consequently, the variations of mechanical properties connected with P_{SK} and P_{δ} will be in the same range of variation.
For getting of reliable values of soil characteristics it is necessary to get information about the character of lyaer, then perform the estimation of soil types, and, at last, determine the variations of every soil type properties.

R E F E R E N C E S

1. Instructions on soils penetration testings in construction. SN 448-72, Moscow, Stroiizdat, 1971.

2. Kulachkin, B.I., On the problem of estimation of soil type from the results of statical penetration. N.T.L. Construction and Architecture, Section B, Issue 6, Moscow, TsiNIS, 1978.

3. Kulachkin, B.I., Estimation of dynamic penetration showings from the energetical point of view. N.T.L. Construction and Architecture, Section B, Issue I, Moscow, TsiNIS, 1978.

4. Ferronsky, N.I., Penetration and carotazh methods. Moscow, Nedra, 1969.

5. Kulachkin, B.I. , Estimation of soils heterogeneity from the results of statical penetration testings. N.T.L. Construction and Architecture, Section B, Issue 4, TsiNIS, 1977.

6. Dunin-Barkovsky, I.V., Smirnov, N.V., Theory of probabilities and mathematical statistics in techniques, Moscow, Gostechizdat, 1955.

Proceedings of the Second European Symposium on Penetration Testing / Amsterdam / 24-27 May 1982

Dynamic probing research for pile driving predictions in the Netherlands

W.VAN LEIJDEN
Geotechnical Consultant, Zeist, Netherlands

H.M.A.PACHEN & J.V.VAN DEN BERG
Institute for Geotechnics & Foundation Engineering, Utrecht, Netherlands

SUMMARY :

The procedure of the dynamic cone penetration test (DPB) has a great similarity with the driving process of a foundation pile. Helas a simple correlation between the results of such a field test and the blow count experienced during pile driving, does not exist. The authors made efforts to establish which correlating factors exist in the two dynamic processes.

The object of the study was a prestressed concrete pile installed by means of a diesel hammer type Delmag D 36-02 in September 1978 at the site of the RAI-buildings in Amsterdam.

Favoring the research was the availability of :

- the complete soils information and pile driving data;
- a reliable wave equation computer program which fits the performance of a diesel hammer very accurately;
- the results of an extensive research program performed in the U.S.A. on standard penetration testing; and
- a computer program developed by the IGF based on extended energy-balance formulas.

1 INTRODUCTION

Penetration tests can be either of the (quasi) static type (e.g.CPT) or of the dynamic type (SPT, DPB). Both types of tests are used to determine the static soil parameters required to estimate the static bearing capacity of foundations. Similar to the application of pile driving formulas for the prediction of static bearing capacities of piles, use of SPT and DPB to such purpose requires experience and a general knowledge of the local foundation soils.

The dynamic penetration test should in principle, correlate better with dynamic soil behavior during for example the driving of a foundation pile, than a CPT.

To investigate which factors are involved in the driving of a pile and what influences the performance of a DPB, wave equation and energy-balance analyses were performed on both a pile driving case and on the outcome of a DPB-test executed at the site op the pile.

To that purpose first the driving resistance had to be estimated from known soil properties at the site. Second, the penetration speed and pile stresses experienced by the subject hammer-pile-soil respectively drop weight-DPB rods-soil were computed. And third, the obtained calculation results were compared with the actual driving record respectively DPB-blow counts and an assessment made of the accuracy of the estimated driving resistance.

2 PILE, HAMMER AND SOILS DATA

Fig.1 shows the subsoil profil as found when drilling a borehole at the pile driving site, together with graphs of the results of the dynamic cone penetration test (the latter

of the heavy type DPB according DIN 4094).
The driving record of the prestressed concrete pile is shown in the same figure. Relevant data on hammer, pile and soil is :

<u>Pile</u> :	
length	22.0 m
cross-section	0,4 x0,4sq.m
unit weight	25kN/cu.m
modulus of elasticity	50kN/sq.mm

<u>Hammer/DPB resp.</u>:	
Number of blows p.min.	37 - 53/30
weight of ram (kN)	36/0.5
drop height (m)	2.17/0.50
energy (kN.m)	78/0.25
weight anvil (kN)	9.1/0.18
weight pilecap (kN)	5.4/-
cushion material	azobe/-

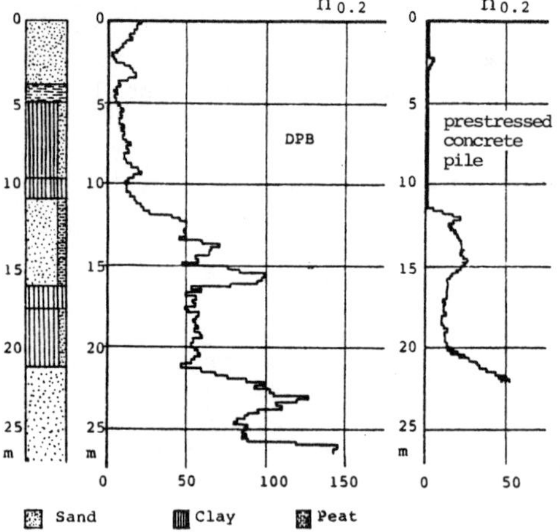

Fig. 1. Results of the DPB-test, the boring and blow counts during driving the concrete pile.

3 SIMULATION OF DRIVING PROCESSES

3.1 Wave Equation Approach

3.1.1 Driveability of a Pile

For the analysis a computer program called WEAP was applied which is considered to best represent in its mathematical model the real driving system of a diesel hammer because in it :

1.The ram is divided into segments to account for its flexibility : each segment is modelled as a mass and a spring;

2.Thermodynamic data (combustion pressure, ignition delay, expansion coefficient etc.) is part of the program input to simulate the particular features of the diesel hammer performance;

3.The spring stiffness above the anvil is combined with the stiffness of the lowest ram segment and since energy losses are associated with the impact of the ram, unloading spring stiffness incorporates a coefficient of restitution;

4.Similarly spring stiffnesses of capblock and cushion reflect for the unloading phase the influence of restitution coefficients thereby introducing the energy losses occuring within these materials.

The pile is divided into equally long segments of which the masses are connected by springs exhibiting the same phenomenon as those of capblock and cushion. The following restitution coefficients were applied:
hammer-anvil: 0.85 cushion: 0.40
capblock : 0.77 pile : 0.97
The main unknown factor in the wave equation analysis is the soil resistance : at one site soil conditions can already vary considerably. Other variables that change sometimes during driving concern capblock and cushion material properties. Driveability can also be influenced by set-up of the soil when driving is interrupted.

The soil resistance acting alongside each embedded pile segment and at the pile tip is considered in the analysis to consist during driving of a static component of resistance represented for each segment by an elastic spring. A friction block is used to model the ultimate of this static resistance. The dynamic resistance is introduced by a dashpot. For damping coefficients (dashpot property) values were used as applied in general practice in case of the subject soil types :
Damping coefficients in sec/m

Soil	Shaft	Tip
sands	0.16	0.49
clays	0.66	0.03

As maximum elastic ground compression (quake) was assumed 2.5 mm.
The static soil resistances were estimated from the available soils into following the method described in American Petroleum Institute RP2A.
This manual applies for :
shaft friction :
cohesionless soil : $f = k.\ p_v$.

cohesive soil : $f = k' \; c_u$.

tip resistance :

cohesionless soil: $q = p_v . N_q$.

cohesive soil : $q = 9 \; c_u$.

For all soil types was applied :
density (kN/cu.m) : 10.0
Other values used are :

$k = 1.0$ $\quad$ $= 30°$

$k' = 0.625$ $\quad$ $c_u = 60$ kN/sq.m.

$N_q = 40$

with :

f = friction

p_v = effective vertical soil stress

= friction angle soil/shaft

c_u = undrained cohesive strength

k = coefficient reflecting influence of pile shape

k' = ratio soil cohesion/shaft friction

N_q = bearing capacity coefficient

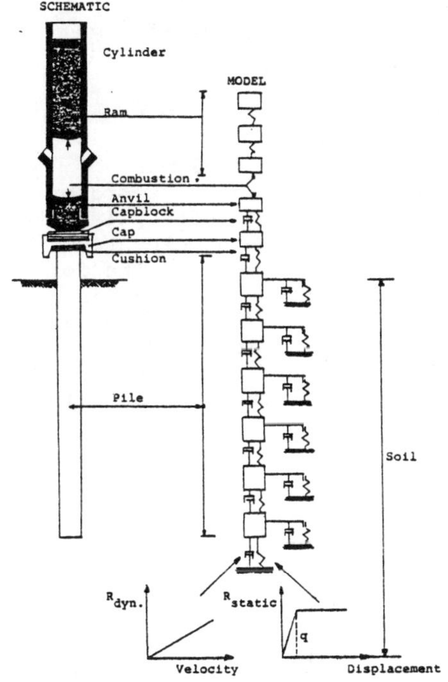

Fig. 2. Schematic and Model in Analysis.

3.1.2 Wave Equation Analysis of DPB-test

The application of the wave equation analysis technique to the dynamic cone penetration test with its well-controlled energy supply (as compared with the SPT) seems at first sight much simpler than to pile driving by means of a diesel hammer. Cushion and capblock lack and also no explosion takes place between drop weight and anvil.

But the direct impact between drop weight and anvil results in a very short compression wave which necessitates in the mathematical model dividing of ram and anvil into short segments. As a consequence calculation work increases considerably which means that a faster computer is required or allowance should be given for more computing time.

Another complication is that in the rather short drop weight many more wave traverses take place than in the DPB-rods in a certain period of time. As a result energy transfer is complicated.

Although DPB-soil systems were found of which wave-equation calculations resulted in approximately the same blow counts as actually measured with reasonable soil resistance estimates, it is the opinion of the authors that the DPB-procedure should incorporate recording of the transmitted energy at a point at the top of the rod string. A research program has been initiated and results should be expected in the near future.

3.2 Energy Balance Approach.

For this purpose the most extended formula used to predict driveability was chosen :

$$S_p \; . \; RU + 0{,}5 \; S_e \; . \; RU + E_t = = a.R.h. \frac{R+e^2}{R+W} - \frac{RU^2.L}{2.E.A} - V \qquad Eq(1)$$

or

$$E_p + E_e + E_t = E_r - E_e - V \qquad Eq(2)$$

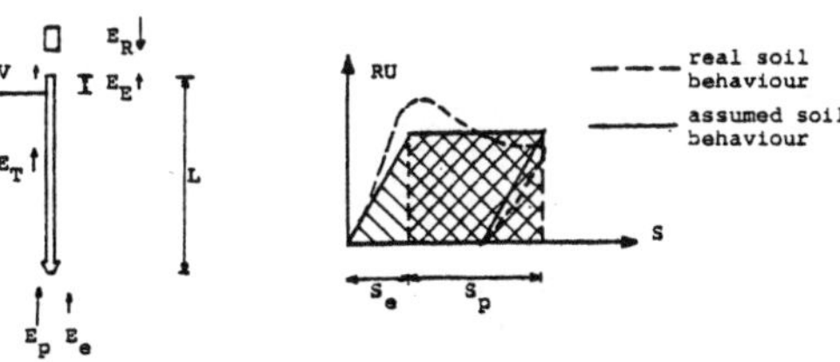

fig. 3. Energy balance pile or sounding rods.

fig. 4. Soil characteristics Dynamic resistance versus deformation.

E_p = energy due to plastic soil deformation.

E_e = energy due to elastic soil deformation.

E_R = energy of drop of ram.

E_E = absorped energy in the pile or sounding rod due to elastic deformation.

V = loss of energy due to kneading and heating of cushion material.

E_T = energy absorbed by shaft friction or along penetration rods.

S_p = plastic permanent deformation per blow.

S_e = elastic deformation per blow.

RU = dynamic soil resistance of pile or DPB.

R = weight of ram.

W = weight of pile or DPB-rods plus cushion weights and possibly anvil.

L = pile length or variable rod length

A_r = cross-section of pile shaft or DPB rods.

h = drop height.

a = efficiency of ram.

e = impact coefficient.

E = modulus of elasticity of pile material or sounding rods.

From the DPB-record S_p as a function of the depth is known. The elastic soil deformation of the cone is not measured during the test. Therefore this important part of Eq (1) is determined on basis on the Mindlin (4) solution of a point-load at an infinite halfspace and by relating the number of blows with the soil stiffness, as follows :

$$S_e = 0.66 \cdot \frac{RU \cdot D_t}{A_t \cdot \beta \cdot n_{0,1} \cdot p_a} \qquad \text{Eq(3)}$$

with :

D_t = diameter of pile or of penetration cone.

A_t = cross-section pile-tip or penetration cone.

p_a = reference pressure = 1 kN/m2.

$n_{0.1}$ =blows for 0.1 permanent penetration.

β =coefficient representing soil stiffness (e.q.clay 800, sand 4000).

When S_p and S_e of the D.P.B. with the other known data and of the DPB equipment are introduced into Eq(1) it gives the dynamic soil resistance RU for the cone RU_{cone}.

The prediction for the pile behaviour can than be made by introducing the pile data, driving hammer values and the calculated RU_{pile} into Eq (1).

The relationship between the dynamic resistance of the pile and the DPB can be written such as Eq (4).

$$\frac{RU}{A_t}\text{ (pile)} = \frac{RU}{A_t}\text{ (cone)} \qquad \text{Eq.(4).}$$

To perform these calculations a computer program, called "heislag" was developed. This program determines for every 0.2 m depth the number of blows necessary to lower the pile by means of the selected hammer.

An option of the program is the presentation of a corrected DPB diagram.
In this graph the influence of variation in length of the rod-string is eliminated as well as the shaftfiction along the rods.

4 COMPARISON OF PREDICTED AND ACTUAL DRIVEABILITY

4.1 Wave Equation Analysis.

Fig.1 shows the driving record of the prestressed concredte pile with in it plotted at three elevations the computed blow counts. The following tavle shows stresses (in N/sq.mm) and sets (in mm /blow) as calculated and as measured.
Fig. 5 and 6 show the force/time relationship for the pile in simple graph-form and in a three-dimensional presentation.

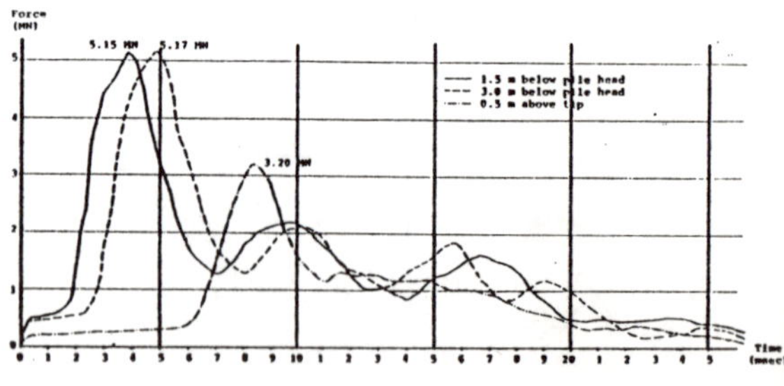

Fig.5 FORCE/TIME DIAGRAM.

Level in Pile		Pile Tip at Elevation 13.5 m - Calc.	Meas.	18.0 m - Calc.	Meas.	22.0 m - Calc.	Meas.
1,5 m below head	max	29.8	24.3	28.1		32.5	40.1
	min	0	-2.5	-1.2		0	-5.6
16,5m below head	max	25.4	21.9	19.9		24.9	33.7
	min	-1.6	-3.8	-9.9		0	-7.2
0.5 m above tip	max	16.9	14.5	7.2		20.0	26.1
	min	0	-1.0	-3.2		0	-1.2
Ultimate values	max	30.1	24.3	28.6	27.2	32.7	40.1
	min	-2.3	-4.2	-11.7	-6.3	-0.7	-7.6
Set		11	10	17	22	4	4.5

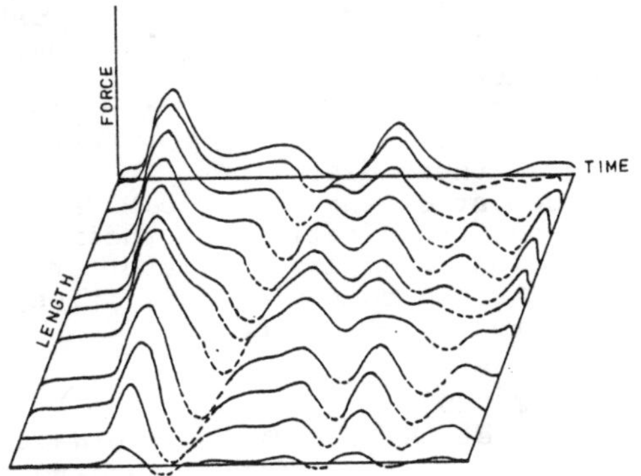

Fig. 6. Force/Time/Pile Length/Isometric.

(7.1) recorded number of blows

(7.2) corrected DPB-record. The DPB shown actually corresponds with a fixed rod length of 1 m. Shaftfriction is eliminated.

(7.3) represents the calculated dynamic bearing capacity of the cone in kN/sq.m. The influence of the quake of the cone has been taken into account.

4.2 Energy Balance Analysis

Fig. 7 to 8 are the computer prints of calculations of the program "Heislag".

In all figures values correspond with a standard 20 cm-penetration.

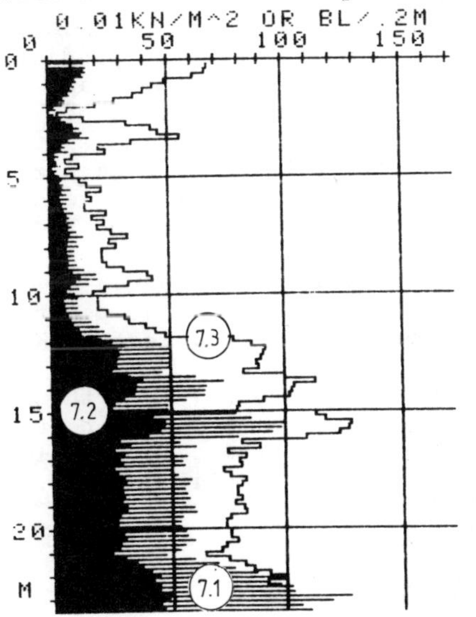

fig.7 DPB analysis: correlated blow counts and dynamic bearing capacity.

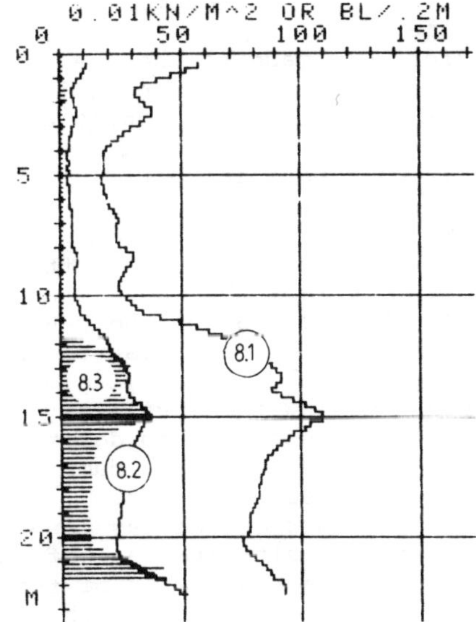

Fig.8 Prediction of blow counts.

(8.1.) again the dynamic bearing capacity but this time for the much larger pile diameter (cone 0.16 sq.m) by averaging the 7.3. - values over depth intervals of 3 x pile diameter.

(8.2.) computer prediction taking into account the quake at

the pile tip as well as the friction along the pile shaft. In the program is introduced a smaller modulus of elasticity and a larger relative friction where the pile perforates the clay layer.
(8.3.) for reason of comparison also the driving record is shown

5 CONCLUSIONS

The wave-equation analysis performed with the WEAP computer program supplied driveability figures that in the subject case of a 22.0 m long, prestressed concrete pile driven by a Delmag D 36-02 hammer were very close to measured values for set, within 25 percent for compression stresses and less reliable for tension.
It is emphasized that the results, as standard in the Smith-wave equations analysis, were obtained using a soil resistance estimated from static soil properties converted to dynamic by the introduction of damping coefficients.
An effort was made to use the wave-equation technique for an analysis of the DPB-test. Helas impact between drop weight and anvil, absence of softening material between anvil and rods result in such complicated wave pattern that direct measurement data of the driving energy in the rods is required. The mechanical, standard hammer drop of the DPB together with accurate measurements of energy by means of a dynamic load cell placed below the anvil, may create the possibility to determine the magnitude of the soil damping coefficients (see also ref.6).
Research to that extent has been initiated. The ultimate goal of the research is the replacment of the present mechanical model of the dynamic soil resistance by the results of in-situ measurements with the DPB-equipment.
From the calculations performed according the energybalance method it may be concluded that a pile driveabily prediction through D.P.B. measurements is feasible.
The comparison of the calculated values and the in-situ measurements indicates that the accuracy of dynamic elastic stiffness parameter and of friction (shaft-) values of the soil are essential for a good agreement between prediction and actual driveability. Helas these properties cannot be measured with the standard DPB-equipment.
Additional instrumentation of the DPB-test equipment as mentioned above in the form of a dynamic load cell placed below the anvil would make such research possible.

6 Acknowledgments

The authors are grateful to the CIAD-organisation for inclusing a dynamic cone penetration investigation in their research project and for placing the driving data of the test-piles at their disposal.
The suppost of the Frederic R.Harris (Holland) organisation and their permission to utilize computer program WEAP and computing facilities are gratefully acknowledged.

7 REFERENCES :

1. Gallet, A.J. : Use of the Wave Equation to Investigate Standard Penetration Test Fields Measurements; 1976. University of Florida USA.
2. Globe, G.G. and Rausche, Frank : "Wave Equation Analysis of Pile Driving"; U.S.Department of Transportation; Federal Highway Administration.
3. Jansen G.P.J., Pachen H.M.A., Execution of a D.P.B. to predict the blow-count of a foundation pile. I.G.F.-report 78.866 dd.14.8. 1980. CIAD-projektgroep heien.
4. Mindlin, R.D.: "A Preliminar Statement by the Galerkin Vectors, Force at a Point in the Interior of a Semi-Infinite Solid"; Physics 7, no.5, p. 195-202, 1936.
5. Pachen H.M.A., Dynamic penetration tests and S.P.T. lecture KIVI Delft, 1977.
6. Schmertmann, J.H.: "Use the SPT to Measure Dynamic Soil Properties? - Yes, But!", Dynamic Geotechnical Testing; ASTM STP 654, 1978, p. 341-355.

Proceedings of the Second European Symposium on Penetration Testing / Amsterdam / 24-27 May 1982

Estimation of the sensitivity of soft clays from static and weight sounding tests

B.MÖLLER & U.BERGDAHL
Swedish Geotechnical Institute, Linköping, Sweden

1. INTRODUCTION

An extensive investigation into localizing potential land slide areas is being carried out at the Swedish Geotechnical Institute. It is especially important to find areas where the soil consists of sensitive or quick, soft clays, since the extent of land slides in these areas can be enormous. In this investigation, previous site investigations are also used and these investigations often contain a number of soundings, especially weight soundings and mechanical static soundings. It is well known among soil mechanics eningeers that the penetration resistance in high-sensitive or quick clays is less than in medium-sensitive or low-sensitive clays. In Sweden clay sensitivity is classified as follows (determined from fall-cone tests on undisturbed and remoulded samples):

Low-sensitive	<8
Medium-sensitive	8-30
High-sensitive	30-50
Quick clays	>50

Some investigations are reported in the literature, e.g. Ladanyi (1967, 1969, 1973) found that the bearing capacity factor N_c is lower in sensitive or quick clay than in insensitive clays. Rygg (1978) has reported that for the Norwegian type of mechanical static sounding test the penetration resistance increases more with depth in insensitive clays than in very sensitive or quick clays (Fig. 1).

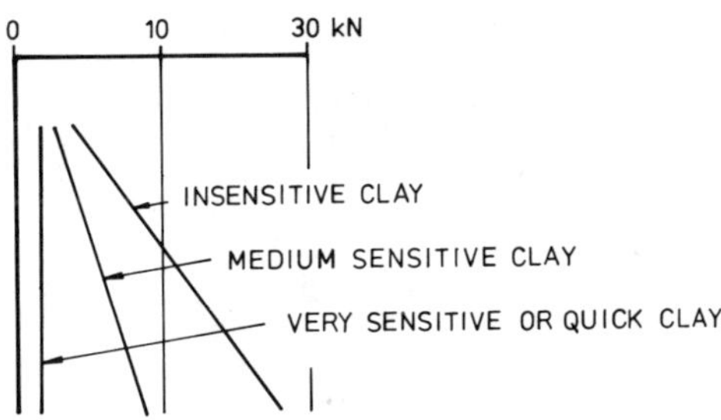

Fig. 1. The total static penetration resistance in clays of different sensitivities, according to Rygg (1978).

Fig. 2 shows the results of two weight sounding tests at two different places in the Göta River Valley, in the south-western part of Sweden. The shear strength at these

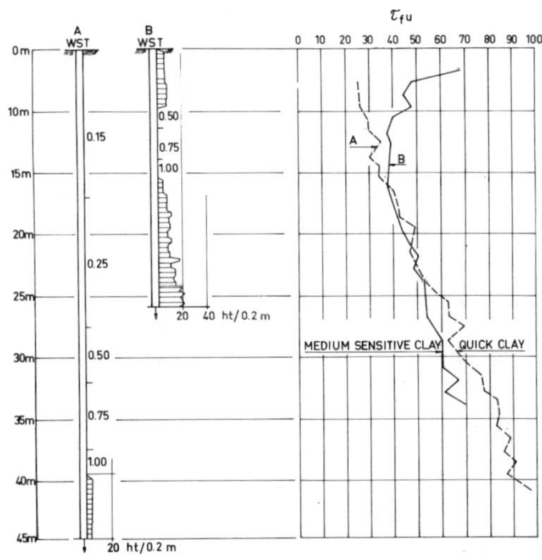

Fig. 2. Results from weight sounding tests in the Göta River Valley in a quick and a medium-sensitive clay.

sites has been measured by field vane tests as shown in the diagram. From this test, it is evident that relatively high penetration resistance may be obtained in clays of medium-sensitivity while the penetration resistance may be very low in quick clays.

Similar test results have been shown by Hansbo (1973) for both weight sounding tests and mechanical static soundings. The main reason for these differences is the reduction of the skin friction resistance along the rods.

The investigation reported below has been performed to find out whether it is possible to estimate the sensitivity of soft clays from weight soundings and mechanical static soundings.

2. INVESTIGATIONS

In these investigations, only sounding, laboratory and field vane test results from previous investigations have been used. These earlier investigations have been performed for buildings, bridges and road embankments. About 60 weight sounding tests and 20 mechanical static soundings were selected. A total of about 400 observations from weight soundings and 170 from mechanical static soundings have been used. Each observation contains a number of variables, such as depth, density, natural water content, liquid limit, sensitivity, shear strength according to the fall cone method and, of course, the weight sounding resistance or the total penetration resistance from the mechanical static soundings. The gradient of the resistance curve has also been evaluated from these latter soundings.

All observations have been divided into sensitivity ranges and different statistical analysis have been performed, such as linear-multiple regression analysis. All the different correlations obtained have been plotted in diagrams and some of them are shown below.

3. TEST RESULTS

3.1 Weight sounding

All weight sounding tests have been performed according to the recommended European standard on weight sounding.

The correlations of the weight sounding test results and the different variables mentioned above indicate that there is no direct relationship between the sensitivity of clay and the penetration resistance measured by weight sounding tests. However, further investigations were made to find correlations between the weight sounding resistance, the depth and different ranges of sensitivity. For this purpose the mean values of all observations in the different sensitivity ranges were calculated. Since the weight sounding test is performed in two different ways (Dahlberg 1974) the observations have been divided into two groups, one for resistances less than 1 kN and one for resistances above 1 kN and

Table 1. Mean values of some variables in different sensitivity ranges for weight sounding tests.

Sensitivity	Number of observations	Depth (m)	Shear strength (kPa)	Sensitivity	Weight sounding resistance (kN)	Weight sounding resistance (ht/0,2 m)
0-10	12	4.8	23	8	0.74	
	22	6.9	68	4		11
10-30	106	5.0	23	20	0.68	
	58	13.8	35.5	19		10
30-50	72	7.5	22	40	0.75	
	14	12.0	29	37		7
>50	102	7.9	21	100	0.63	
	9	10.6	35[1]	239		5[1]
All observations	292	6.6	22	52	0.69	
	103	11.8	41	37		10

[1] Only 9 observations

turning of the rods in ht/0.2 m = halfturns per 0.2 m of penetration. The results of these calculations are summarized in Table 1.

The values in Table 1 indicate that the weight sounding resistance decreases with increasing sensitivity and that the average shear strength is almost the same in all sensitivity ranges when the weight sounding resistance is less than or equal to 1 kN. On the other hand, for the weight sounding resistance above 1 kN, together with turning of the rods, the mean shear strength decreases with increasing sensitivity. The average depth for weight sounding resistance less than 1 kN is less than for weight sounding resistances above 1 kN together with turning of the rods. In Fig. 3 the limits of observations from the weight sounding tests are shown against depth and it is clear that the penetration resistances for the different sensitivity ranges fall in different areas. However, there is some overlapping in the diagram.

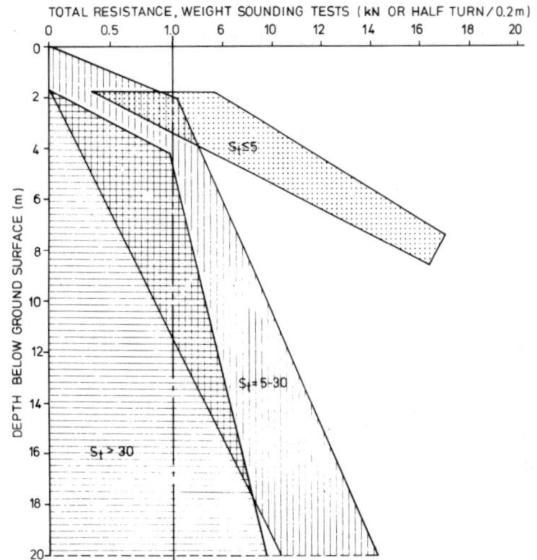

Fig. 3. Weight sounding resistance versus depth for different sensitivity ranges.

3.2 Mechanical static sounding

In this investigation, only test results from Geotech type mechanical static soundings, with a fixed weight sounding point have been used, Dahlberg (1974). This permits no separation between the point and skin friction resistances. This type of sounding is frequently used in the southwestern part of Sweden where there are very deep layers of soft clays. The analysis of the observations was performed in a similar manner to that used on the weight sounding results. The gradient of the penetration resistance curve is defined as the average slope (in degrees from the vertical) of the curve at different depths. The values are valid for the scales 1 kN = 10 mm and 1 m = 10 mm. Table 2 shows a summary of the mean values of some of the variables in the different sensitivity ranges.

The mean values summarized in Table 2 indicate that all the observations are relatively homogeneous, which means that the mean values for depth and shear strength are almost the same in all sensitivity ranges. The mean values also indicate that the penetration resistance and the gradient of the penetration

Table 2. Mean values of the depth, shear strength, the penetration resistance and the gradient of the penetration resistance in different sensitivity ranges.

Sensitivity	Number of observations	Depth (m)	Shear strength (kPa)	Total sounding resistance (kN)	Gradient of the penetration resistance, degrees
0-20	51	6.8	22	2.2	10.1
20-40	39	6.7	22	1.2	5.2
40-60	24	8.8	23	1.3	4.9
60-100	22	8.8	21	0.95	2.3
>100	34	7.5	20	0.75	0.51
All observations	170	7.6	22	1.4	5.3

Table 3. Equations for the penetration resistance of mechanical static soundings, shear strength and depth for different sensitivity ranges.

Sensitivity	Equations	Correlation coefficient
0-20	$q_{tot} = -1.84 + 0.0877\ \tau_{fu} + 0.311\ z$	0.94
20-40	$q_{tot} = -0.35 + 0.0485\ \tau_{fu} + 0.068\ z$	0.68
40-60	$q_{tot} = -0.94 + 0.0774\ \tau_{fu} + 0.060\ z$	0.87
60-100	$q_{tot} = -0.38 + 0.0348\ \tau_{fu} + 0.065\ z$	0.83
>100	$q_{tot} = 0.71 + 0.0336\ \tau_{fu} - 0.081\ z$	0.71
All observations	$q_{tot} = -1.09 + 0.0890\ \tau_{fu} + 0.072\ z$	0.63

resistance decreases with increasing sensitivity. Therefore the following assumption was tested by multiple regression analysis on the equation below:

$$q_{tot} = a + b\ \tau_{fu} + c\ z \qquad (1)$$

where

q_{tot} = total penetration resistance (kN)

τ_{fu} = undrained shear strength (kPa)

z = depth (m)

a,b,c = constants

The results of these analyses are shown in Table 3.

The results of the analyses indicate that acceptable correlations have been obtained in the different sensitivity ranges. The results also indicate that the penetration resistance in two sensitivity ranges, namely 0-20 and above 100, have a different dependence on depth than in the other ranges. The plotting of the analyses are shown in Fig. 4, where the areas of the different sensitivity ranges are indicated. It is evident from this diagram that there is some overlapping at lower depths. From a depth of about 4 m, there is a divergence of the different ranges and they are completely separated from a depth of about 8 m. From Fig. 4 it can also be seen that the sensitivity range 0-20 has been divided into two parts, namely 0-10 and 10-20 as it was evident from the analysis that it was possible to separate them definitely.

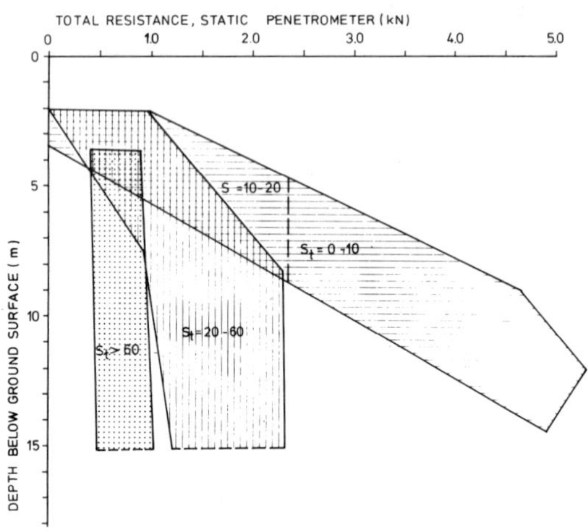

Fig. 4. Total penetration resistance versus depth for different sensitivity ranges.

Fig. 5 is a plot of the gradient of the sounding resistance diagram versus depth. The average of the gradient in the respective sensitivity ranges is lower for the highly sensitive clays than for clays of low sensitivity but the scatter is relatively large. From the diagram, it is possible to see that, for example, when the average gradient is less than 4 degrees the sensitivity is probably higher than 60.

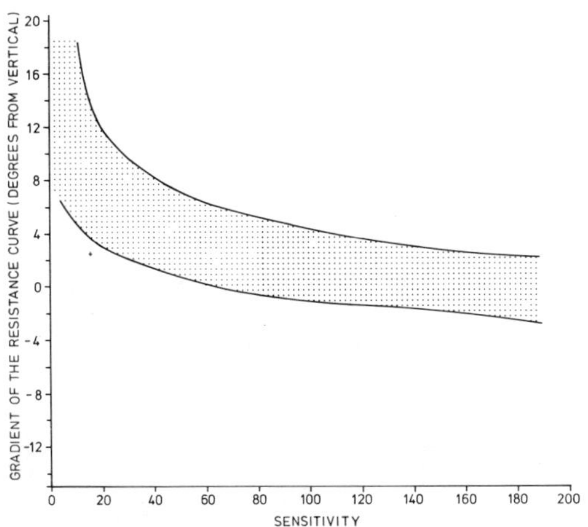

Fig. 5. The static sounding resistance gradient versus sensitivity of the clay for mechanical static sounding.

4. CONCLUSIONS

The above shown investigations indicate that there are differences between the weight sounding resistance with depth and different sensitivity ranges. However, the scatter is high. It is more accurate to separate the clays in low than in high sensitivity ranges. It could be concluded that a low sounding resistance at great depth can indicate a high sensitivity and, secondly, high weight sounding resistances in soft clays normally indicate clays of low sensitivity.

The investigations also show that the static sounding diagrams can be used for a rough estimate of sensitivity at different depths in a clay layer. Thus it is possible with good accuracy to separate the clay layers in the three sensitivity ranges 0-20, 20-60 and >60. The accuracy is higher the lower the sensitivity. This investigation thus confirms the principle of the penetration resistance diagram according to Rygg (1978), Fig. 1.

The investigation reviewed above indicate that it is possible to estimate the range of sensitivity of soft clays from sounding tests. However, it must be pointed out that the classification model shown, based on a limited number of areas, only is valid for the clay areas investigated. Similar relationships can probably be obtained for other regions. As the clays investigated are normally or slightly overconsolidated, the classification model should not be used for overconsolidated clays without further investigations.

5. REFERENCES

Dahlberg, R. 1974, Penetration testing in Sweden. Proc. ESOPT 1974, Vol. 1, Stockholm, pp. 115-131.

Hansbo, S. 1973, Use of Nilcon type static sounding equipment for field investigations. Symposium on cone penetration test. SGI Preprint and preliminary reports No. 58, Stockholm. (In Swedish)

Ladanyi, B. 1967, Deep punching of sensitive clays. Proc. Caracas, Vol. 1, 1967.

Ladanyi, B. 1969, Use of the deep penetration test in sensitive clays. Proc. Mexico City 1969.

Ladanyi, B. 1973, Bearing capacity of deep foundations in sensitive clays. Proc. Moscow 1973.

Möller, B. 1980, Assessment of the sensitivity of clays from weight and static sounding test results. SGI Varia, Linköping. (In Swedish)

Rygg, N. 1978, DREI Static sounding: interpretation of sounding test results. Vegdirektoratet, Veglaboratoriet, Internrapport nr 816, Oslo 1978. (In Norwegian)

Proceedings of the Second European Symposium on Penetration Testing / Amsterdam / 24-27 May 1982

Comparative study of static and dynamic penetration tests currently in use in Japan

T.MUROMACHI & S.KOBAYASHI
Kiso-Jiban Consultants Co., Ltd., Tokyo, Japan

1 INTRODUCTION

A number of different sounding techniques are presently in use in Japan. Among these, the standard penetration test (SPT) figures prominently accounting for a share in excess of 95 percent of all ordinary soil investigations. The second most common sounding method is the Dutch cone penetration test (DCT), which is, however, seldom used as a primary technique for surveys. In third order of popularity is the Swedish weight sounding method (SWS) which provides a simple but effective means of investigation.

These three sounding methods serve as important components of the Japanese Industrial Standard (JIS). Of the three, the SPT was the first to be accorded recognition in 1961 as JIS A 1219. The most significant result of this approval was the inclusion of the N value factor in governmental codes of practice as one of the most effective soil indices showing the in-situ relative strength of ground. Clear guidelines for test procedures and instrument dimensions are established in this standard. The DCT and the SWS methods also contributed significantly to this process and were accordingly recognized in 1967 as JIS A 1220 and JIS A 1221, respectively.

In addition to the above-mentioned methods, the portable cone penetration test (PCT) and dynamic cone penetration test (DPT) are widely used in investigations. The PCT of both single-rod and double-tube type are primarily employed to survey soft ground during reconnaissance. DPT's are often used as a simple and easy means of supplementing SPT's or as a substitute for the same. Among the various DPT's, the use of the Railway Technical Research Institute (RTRI) type and the Swedish automatic ram sounding (SRS) type is prevalent.

This paper summarizes the observed field relations of the above-mentioned six methods of penetration testing in Japan. Those related to vane tests and pull sounding are not addressed herein.

2 DESCRIPTION OF PENETROMETERS USED

Essential data regarding the three static cone penetrometers and the three dynamic cone penetrometers are given below in Tables 1 and 2, respectively.

Table 1. Essential data regarding three static cone penetrometers

Penetrometer Type	Point				Rod		Penetration Process	Max. Capacity (kN)	Penetr. Rate (cm/s)	Measuring Device	Recording	Index Value	No. of JIS
	Type	Angle (deg)	O.D. (mm)	Area (cm^2)	Type	O.D. (mm)							
Dutch Cone (Gouda)	Mantle Cone	60	35.7	10.0	D	T.28 R.15	Gear (human)	20	1	Oil Gauge Prov. Ring	Reading (every 0.25m)	q_c (bar)	A1220 (1976)
					D	T.36 R.15	Oil Pressure	100	1	Oil Gauge			
Portable Cone (RTRI, JNR)	Simple Cone	30	28.6	6.45	S	16	Hand	1	1	Proving Ring	Reading (every 0.1m)	q_c (bar)	—
					D	T.22 R.13	Hand	1	1				
Swedish Weight Sounding (S.G.I.)	Screw Point	Right Screw ℓ=200mm	33.3 (max)	—	S	19	Weight Loading and Half Turns	1	-	Weight 5, 10, 10 25, 25, 25kg	Sinking Weight	Wsw (kg)	A1221 (1976)
											Half Turns per Meter	Nsw	

Table 2. Essential data regarding three dynamic cone penetrometer

Penetrometer Type	Point: Type	Point: Angle/Length	Point: O.D/I.D (mm)	Point: Area (cm²)	Rod: Type	Rod: O.D. (mm)	Hammer Weight (kg)	Fall Height (cm)	Driving Method	Recording	Index Value	No. of JIS
SPT Device	Split Sampler	810mm	51/35	20.4 – 9.6 (10.8)	S	{40.5, 42.0	63.5	75	Rope and Drum (90%), Auto-hammer, Trigger	Counting of Blows (every 1 m)	$N(\frac{blow}{0.3m})$	A1219 (1961)
Dynamic Cone (RTRI, JNR)	Inverse Taper Cone (With Collar)	60°	50.8	20.3	S	40.5	63.5	75	Rope and Drum	Counting of Blows (every 0.1 m)	Nd63.5/10 $(\frac{blow}{0.1m})$	—
Swedish Ram Sounding (Borros)	Cylindr Cone	90°/90mm (Sylinder)	45	15.9	S	32	63.5	50	Auto-hammer	Auto-Counter Torque Measuring (every 1 m)	$Nd(\frac{blow}{0.2m})$	—

3 RELATION BETWEEN DYNAMIC AND STATIC PENETRATION TESTS

3.1 Relation between N values of SPT and q_c values of DCT

Based on numerous SPT and DCT penetration records, N values and q_c values for the same soils at the same depths are plotted in Fig. 1. This figure contains values for different soil types such as alluvial sand, clayey soil, diluvial gravel, sand, clay and volcanic ash. The relation between N values and q_c values on the whole show large scattering. However, the individual correlation for each soil type appears to be relatively good with a proportional ratio varying between 2 and 8 according to grading.

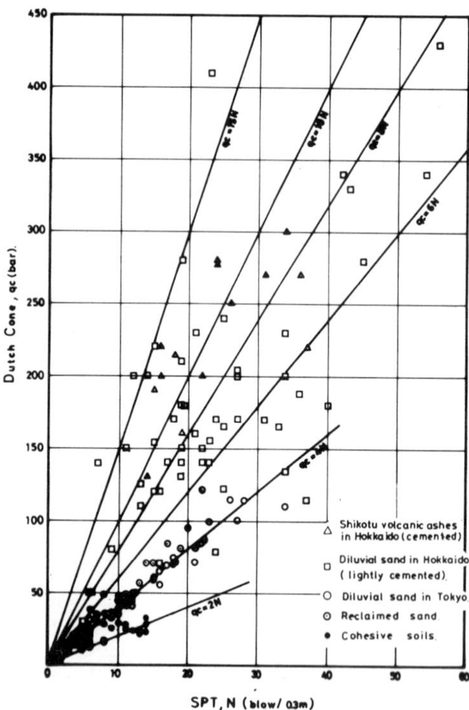

Fig. 1. Relation between q_c values of Dutch cone and N values of SPT.

To better ascertain this tendency as pointed out by S. Thornburn, data sets having a grading curve were extracted and the numerical ratio of q_c values in bar to N values, i.e. the static dynamic penetration resistance ratio, $\eta_{S/D}$, in expression (1) was plotted against mean grain size, D_{50}, as shown in Fig. 2 (Muromachi & Kobayashi, 1980).

$$q_c = \eta_{S/D} \cdot N \qquad \text{.....................(1)}$$

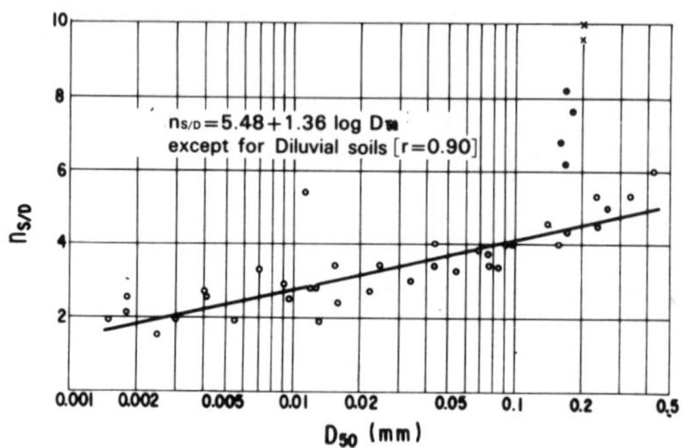

Fig. 2. Relation between static-dynamic penetration resistance ratio, $\eta_{S/D}$, and mean grain size, D_{50}.

It is clear from this figure that the ratio $\eta_{S/D}$ in expression (1) varies in accordance with mean grain size, D_{50}, except in the case of Shikotsu volcanic ash and lightly cemented diluvial sand in Hokkaido. This relation is expressed as follows:

$$\eta_{S/D} = 5.48 + 1.36 \log_{10} D_{50} \qquad \text{.......(2)}$$

Expression (2) is valid for $0.0015 < D_{50} < 0.5$ mm in ordinary alluvial soils and provides very good correlation as r = 0.90. As almost no data regarding grain size distribution for dense diluvial sand and volcanic ash exist, the tendency of $\eta_{S/D}$ in the range $q_c > 120$ bar is not clear. As for those soils excepted from expression (2), lightly cemented diluvial sand in Hokkaido in the range $q_c \geq 120$ bar show $\eta_{S/D} = 6 \sim 8$ for $D_{50} = 0.17$ mm and cemented Shikotsu volcanic ash in the same range show $\eta_{S/D} = 10$ for $D_{50} = 0.2$ mm. The tendency of these soils differs greatly from that of alluvial soils. It can be therefore observed that in these cases the numerical value of $\eta_{S/D}$ is not always dependent on the grading curve.

3.2 Relation between N values of SPT and N_{sw} values of SWS

The SWS has been widely used in investigation projects conducted by the Ministry of Construction and the Japan Highway Public Corporation.

The most practical relation between the N values of the SPT and the two index values of W_{sw} (kg) and N_{sw} (half turn/m) measured by the SWS is shown in Fig. 3. The values contained therein are the result of a comparative study conducted by the Japan Highway Public Corporation during construction of the Meishin Expressway between Kobe and Nagoya (Inada, 1960).

In Fig. 3, the general tendencies of two major soil groups, i.e. gravelly or sandy soils and clayey or silty soils, are distinguished and the practical relation between them is expressed individually as follows:

Gravelly or sandy soils:

$$\left.\begin{aligned} N &= 0.02\ W_{sw}\ \text{(kg)} \\ N &= 2 + 0.067\ N_{sw}\ \text{(half turn/m)} \end{aligned}\right\} \quad \cdots\cdots (3)$$

Clayey or silty soils:

$$\left.\begin{aligned} N &= 0.03\ W_{sw}\ \text{(kg)} \\ N &= 3 + 0.050\ N_{sw}\ \text{(half turn/m)} \end{aligned}\right\} \quad \cdots\cdots (4)$$

Since the SWS belongs to the single-rod type penetration test, the influence of skin friction increases with depth of penetration. Moreover, data measured at a certain depth are also influenced by the type and condition of soil layers already penetrated. It is due to this reason that the plots in Fig. 3 show relatively large scattering. Nevertheless, the relation obtained is very useful as it provides a certain amount of information on the strength of a soil layer rather than simply a record of a reconnaissance.

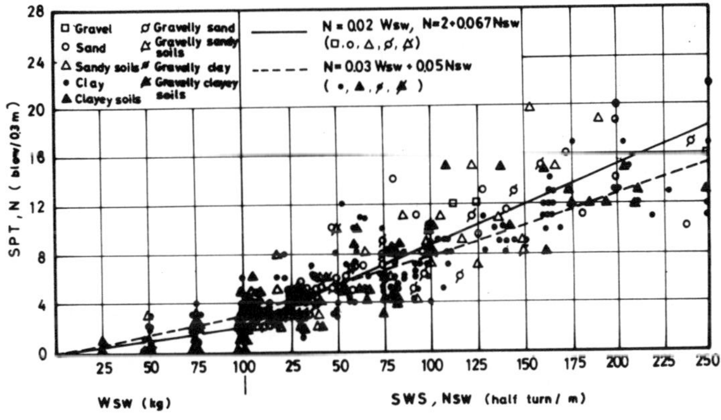

Fig. 3. Relation between N values of SPT and W_{sw}, N_{sw} values of Swedish weight sounding.

4 RELATION BETWEEN DYNAMIC CONE PENETRATION TESTS

4.1 N values of SPT and N_d values of RTRI-DPT

A DPT designed by the Railway Technical Research Institute (RTRI) of the Japanese National Railway (JNR) has been used mainly for JNR route surveys as a substitute for the SPT.

The typical relation between the N values of the SPT and the N_d values (blow/0.3 m) of the RTRI-DPT is shown in Fig. 4 as based on a comparative study conducted by the JNR during construction of the Hakata Station (Muromachi, 1964).

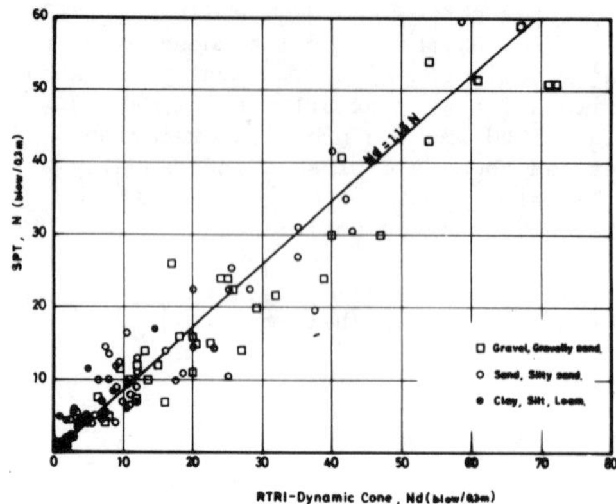

Fig. 4. Relation between N values of SPT and N_d values of RTRI-dynamic cone.

As can be seen in this figure, N_d values of the RTRI-DPT are somewhat larger than the N values of the SPT as a result of skin friction against the single-rod penetration tested. The actual relation between them is expressed as follows:

N_d(blow/0.3 m) = 1.15N ....(5)

It appears that both tests produce similar values if the effect of skin friction is negligible.

4.2 N values of SPT and N_d values of SRS

SRS, the most popular dynamic cone penetration test in Sweden, was recently introduced to Japan and has found wide acceptance.

In accordance with the SGS Standard, allowance for the effects of skin friction is made through measurement of the torque with a torque wrench at one meter intervals of penetration or less.

The relation observed between the corrected values of N_d (blow/0.2 m) of the SRS and the N values of the SPT is shown in Fig. 5. The relation between the two is expressed as follows (Satoh & Iwasaki, 1980).

N_d(blow/0.2 m) = N .... (6)

A method to allow for the effects of skin friction against a rod is given by R. Dahlberg and U. Bergdahl (1974) as follows:

$N_d = N_{dm} - 0.040\ M_v$ .... (7)

N_d : corrected value (blow/0.2 m)
N_{dm} : measured value (blow/0.2 m)
M_v : measured torque (N·m)

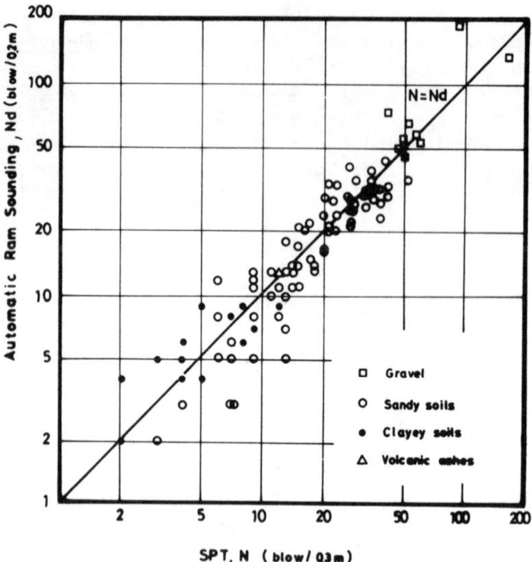

Fig. 5. Relation between N_d values of Swedish ram sounding and N values of SPT.

5 RELATION BETWEEN STATIC CONE PENETRATION TESTS

5.1 q_c values of DCT and N_{sw} values of SWS

The typical relation between q_c values of the DCT and N_{sw} values of the SWS is shown in Fig. 6, based on comparative studies carried out in the Tokyo area by the RTRI (Muromachi, 1971 & 1974). Though the scattering of measured data is large, the relation can be expressed as follows:

$q_c = 6.7 + N_{sw}$ (half turn/m) ....(8)

An example will serve to illustrate this expression. In the event that $N_{sw} = 0$ and q_c stands for $W_{sw} = 100$ (kg), then the critical q_c value at which the sounding rod sinks under the 100 kg weight is less than 7 bar.

5.2 q_c values of DCT and q_c values of PCT

The PCT as developed by the RTRI employs two methods of penetration; a single-rod type and a double-tube type. In order to compare q_c values of the DCT, a q_c value measured with the double-tube type penetrometer ($q_{cp\text{II}}$) is adopted. However as the PCT is used only for soft soil deposits, the range of comparison is limited to $q_{cp\text{II}} < 10$ bar.

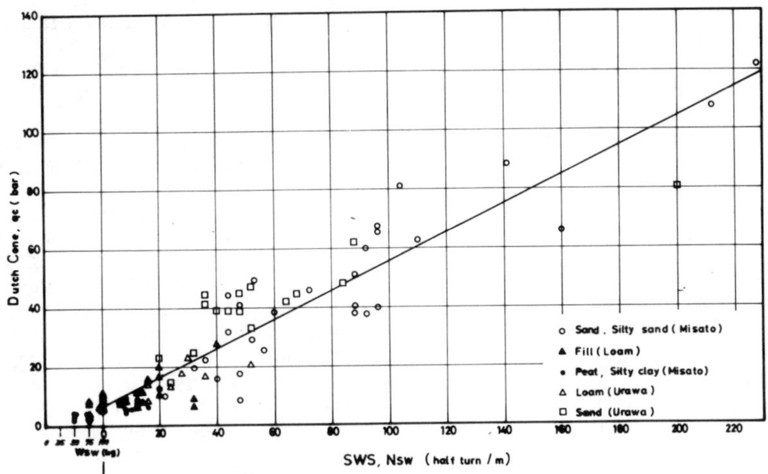

Fig.6. Relation between q_c values of Dutch cone and N_{SW} values of Swedish weight sounding.

Fig. 7 shows the relation between the q_c value of the DCT and the $q_{cp\mathrm{II}}$ value of the RTRI-PCT based on comparative studies of soft ground areas along JNR lines (Muromachi, 1971). The practical relation between them is expressed as follows:

$$q_{cp\mathrm{II}} = 0.741\ q_c \qquad (9)$$

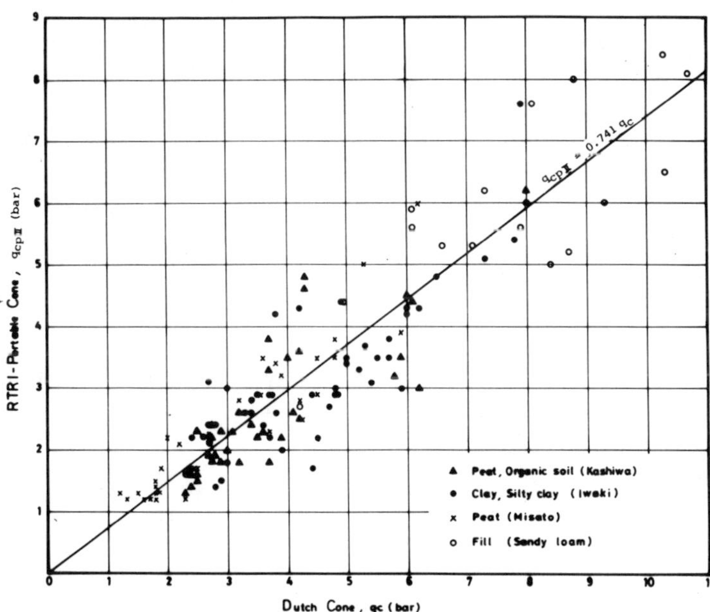

Fig. 7. Relation between q_c values of RTRI-portable cone and q_c values of Dutch cone.

6 CLOSING REMARKS

The primary importance of obtaining field relations between the results of penetration tests is the practical means they can provide to convert data from one sounding method to another for use in design.

Among the relations outlined in the foregoing pages, the one of most importance in Japan is that between the N value of the SPT and the q_c value of the DCT as the former is the foremost penetration test and the latter the second most popular.

Through the study of the relation between the N and q_c values of these methods, it can be seen that the ratio $\eta_{S/D}$, in the expression $q_c = \eta_{S/D} \cdot N$ varies depending on mean grain size, D_{50}, in ordinary alluvial deposits. As an illustration, it varies from 2 in pure clay to 6 or more in coarse sand.

The relation observed above is important because it gives factual evidence for rational conversion between N values and q_c values in alluvial soils and permits the modification of $\eta_{S/D}$ according to the grain-size distribution of a soil.

The substantial difference between static and dynamic penetration may be due to the different ratios of compression to shearing produced by each method of penetration; both compression and shearing are also influenced by the grain-size distribution of soils.

In-situ relations appropriately obtained between different penetration tests are invaluable in engineering practice. The more important the original relation, the more useful the transformed expression.

ACKNOWLEDGEMENT

The authors wish to express their thanks to Mr. D. Flory, Miss N. Ogino and Miss K. Yamakoshi for their editing and typing works.

REFERENCES

Dahlberg, R. & U. Bergdahl 1974, Investigations on the Swedish ram-sounding method, European Symposium on Penetration Testing, Vol.2.2, pp.93-102.

*Inada, M. 1960, On the use of test results of Swedish weight sounding, Tsuchi to Kiso, Vol.8, No.1, JSSMFE, pp.13-18.

*JIS Committee 1962, Method of standard penetration test for soils, JIS A 1219-1961, Japanese Standard Association.

*JIS Committee 1977a, Method of Dutch double-tube cone penetration test, JIS A 1220-1976, Japanese Standard Association.

*JIS Committee 1977b, Method of Swedish weight sounding, JIS A 1221-1976, Japanese Standard Association.

*Muromachi, T. 1960, Sounding: Soil Investigation Manual, Tokyo, JSSMFE, pp.67-132

*Muromachi, T. 1971, Experimental study on application of static cone penetrometer to subsurface investigation of soft subsoils, Railway Technical Research Report, No.757, RTRI, JNR.

Muromachi, T., J. Oguro & T. Miyashita 1974, Penetration testing in Japan, SOA Report, European Symposium on Penetration Testing, Vol.1, pp.193-200.

Muromachi, T. 1974, Experimental study on application of static cone penetrometer to subsurface investigation of weak cohesive soils, European Symposium on Penetration Testing, Vol.2.2, pp.285-291.

*Muromachi, T. & S. Kobayashi 1980, On the observed variation of q_c/N values due to grain size, Proc. of Sounding Symposium (Tokyo), JSSMFE, pp.151-154.

Muromachi, T. 1981, Cone penetration testing in Japan, Cone penetration testing and experience, Proc. of GT-Session, ASCE National Convention, St. Louis, pp.49-75.

Rodin, S., B.O. Corbett, D.E. Sherwood & S. Thorburn 1974, Penetration testing in United Kingdom, SOA Report, European Symposium on Penetration Testing, Vol.1, pp.139-146.

*Sato, K. & T. Iwasaki 1980, An experimental study of the Swedish automatic ram sounding for several Japanese soils, Proc. of Sounding Symposium (Tokyo), JSSMFE, pp.213-222

Swedish Geotechnical Society 1974, SGS standard for Swedish ram sounding, Appendices to European Symposium on Penetration Testing, Vol.2.1, pp.247-251

* Originally published in Japanese.

Proceedings of the Second European Symposium on Penetration Testing / Amsterdam / 24-27 May 1982

Theory and practice of a soil hardness tester YH-62

YASUSHI NAKAYAMA
National Research Institute of Agricultural Engineering, Tsukuba, Japan

1 PREFACE

The character of earth is identified by grain size distribution, consistency, the SPT, etc. In Japan, a soil hardness tester YH-62 was invented and has been used for planning and site selection in land-use. This paper deals with the theory and the measurement in pyroclastic materials. They are so complex in character that grain size distribution and consistency vary according to preparatory treatment, too.

2 SOIL HARDNESS TESTER YH-62

This instrument was invented by Dr. Kinjiro Yamanaka, an adviser of Pacific Consultants Co.,Ltd. It is a sort of static cone penetration tester, which was brought into earlier use of soil classification under cultivation. The purposes were an objective description of pedology in a test pit and classification of soil and producing capacity first and foremost. Gradually, it proved that the instrument was useful for a reasonable selection of automotive cultivators and combines on paddy field.

General use has been stretched into the evaluation of earth bearing capacity on civil engineering construction such as a fill-type dam, a high way and housing development. It can't be performed in a bore hole but on the wall of a test pit or an outcrop. The tester must be held horizontally. It is very simple with a cone, a coil spring and two tubes. A metal cone is 18 millimeters in bottom diameter, 40 millimeters in height with a head angle of 25°20'. A coil spring is 120 millimeters in length and 8 kgf/40mm in stiffness.

The method is to push a cone by a coil spring till a brim is in touch with the surface of soil and to read the shrinkage of a coil spring in millimeters. It is easy to get figures, x, if one sets an index rider on zero before measurement. Here, x means the shrinkage of a coil spring and (40-x) means the height of the inserted portion of a cone.

Bearing capacity is calculated by the formula as follows;

$$P = \frac{100\ x}{0.7952\ (40-x)^2}$$

where P : Bearing capacity, kgf/cm^2
x : Shrinkage of a coil spring or hardness index, mm

It is shown in Fig. 2. Bearing capacity is infinitesimal, if x is

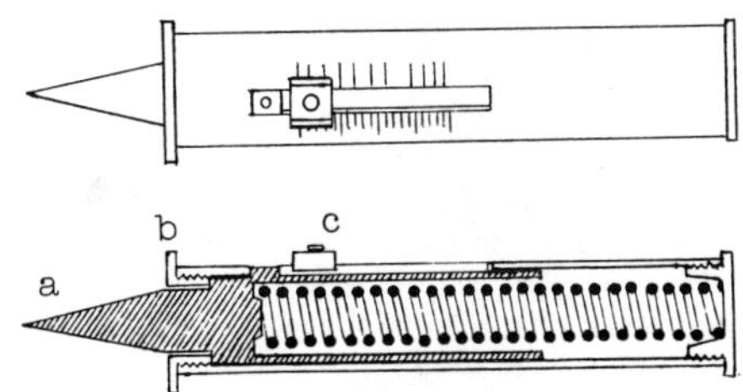

Fig. 1 Soil hardness tester YH-62
a: cone b: brim c: index rider

equal to zero; at infinity, if equal to 40. But, no such soil exists in fact.

The following is practical examples of measurement and classification in South Kyushu and Kanto regions.

2 CASE OF "SHIRASU" IN SOUTH KYUSHU REGION

"Shirasu" is non- or weakly-welded pumice flow and its secondary deposit. It means white sand in Japanese. The secondary one is perfectly a sort of soil; the primary pumice flow is an extraordinary layer of which the behavior is like earth or weak rock. There is "shirasu" in South Kyushu, Tohoku and Hokkaido regions. South Kyushu is solely dealt with in this paper.

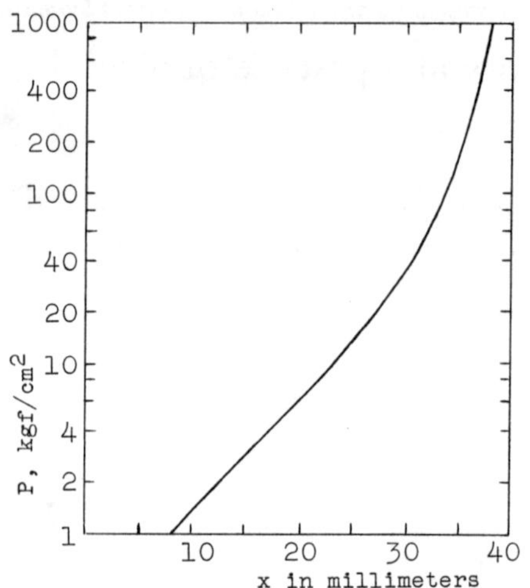

Fig. 2 Relation between hardness index, x, and bearing capacity, P

2.1 Geological description

There are Ata, Aira and Kakuto calderas in South Kyushu and Kikai in the sea. Aso (Central Kyushu) has one of the largest calderas in the world. It is 17 to 25 kilometers in diameter. Aira is famous for a vast mass of pumice flow ejection. Ata is the next.

Geology of Aira ejection is hypersthene-hornblende dacite flow, over 100 meters in thickness, bearing a dacitic pumice fall deposit, under 10 meters in thickness. Geological age is probably 21,000 to 23,000 years before present according to C^{14} dating.

A type of dacitic or andesitic volcanism takes a process of stratovolcano—glowing cloud—caldera—central or somma cones.

2.1.1 Stratovolcano

Repeated flows of lava and pyroclastic ejecta cover each other. A stratovolcano grows more and more. It looks usually like a tiny cone. A crater is very small in comparison with the mountain. The higher a level is, the steeper a slope is. It gets to about 40 degrees. The diameter of a base is about 30 kilometers; the thickness of the layers is about 1,000 meters maximally.

2.1.2 Pumice fall deposit

Magma becomes salic by degrees. Suddenly, porous grains of pumice and salic ash are ejected in the sky. They are carried by wind and classified. Coarse grains draw near thickly; fine ones wander far thinly. The thickness is homogeneous in the same distance whether on a hilltop or in a valley. Because of stability or erosion, they are absent on a steep slope.

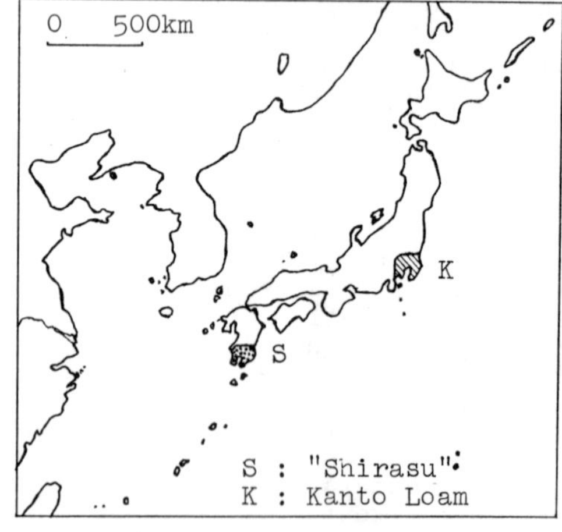

Fig. 3 Index map

Because of small unifirmity coefficient and looseness, pumice fall deposit, lying on a gentle slope, flows suddenly under the vibration of an earthquake.

2.1.3 Pumice flow (main member of "shirasu")

Glowing cloud or nuée ardente succeeds to pumice fall. Heated magma is ejected and runs down on the slope at a high speed being a sort of emulsion. Running over a hilltop, it stays thickly in a valley. The cloud is about 1,000 meters in height. An original surface of heated materials was flat. But, after cooling, some settlement appears in proportion as their thickness. The geological section of glowing cloud materials is very interesting to us. They will be soft "shirasu", if they are thin; they will be soft "shirasu"—hard one or welded tuff—soft one, if they are well thick. A grade of welding will be decided under the conditions below;

a. Small earth pressure and rapid cooling make soft "shirasu".

b. Large earth pressure and slow cooling make hard "shirasu" or welded tuff.

c. Large earth pressure and rapid cooling make soft "shirasu".

Such an eruption makes usually some caldera. Aira has a caldera of Glen Coe type, which bears a ring fissure.

2.1.4 Central and somma cones

There are several cones at the center and/or on the somma of a caldera. Mt. Sakura-jima is an example of the latter.

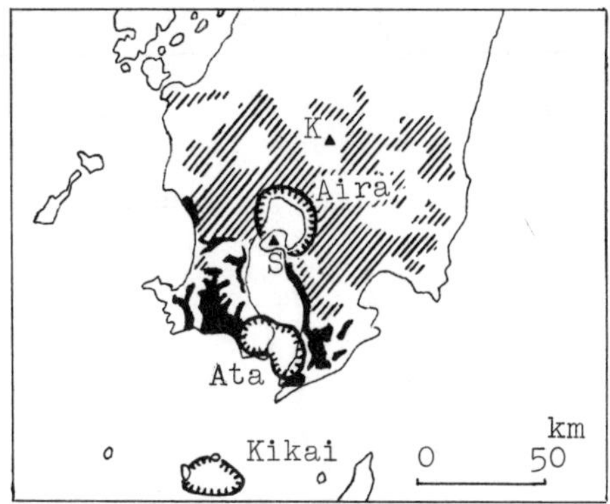

Fig. 4 Locality map of "shirasu" and calderas (After T. Matsumoto) K: Mt. Kirishima, S: Sakura-jima

2.2 JSF standard : M2-81

In April 1981, a standard on "shirasu" hardness was enacted by the Japanese Society of Soil Mechanics and Foundation Engineering. It took four and a half years to deliberate the draft in the Research Subcommittee on Standardization of "Shirasu" Classification. This classification demands two aspects of geology and soil hardness.

1. It is important to decide whether it is the primary "shirasu" or not. The primary one looks massive and partly welded. It holds often some degree of hardness which is explanatorily an effect of interlocking. Hardness will decrease, if the original structure is destroyed.

On the other hand, the secondary one is a layer which is sedimentary and shows bedding plane bearing often rounded gravels. Liquefaction will probably occur under saturation with an earthquake shock.

Both of the primary soft and the

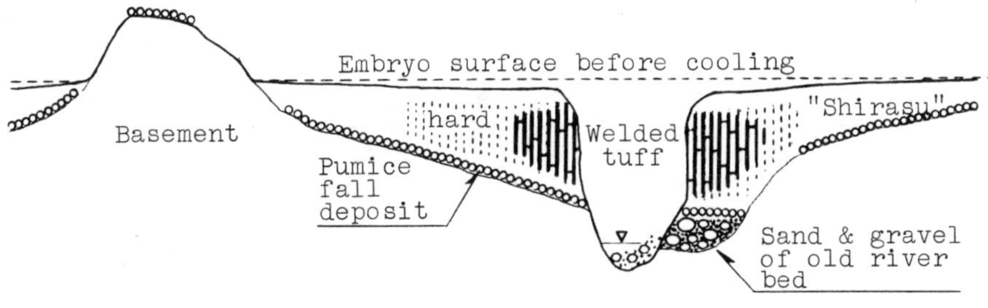

Fig. 5 Schematic section of "shirasu"

Table 1. Classification of primary "shirasu" by hardness

Classification	"Shirasu"				Welded tuff
	Very soft	Soft	Moderate	Hard	
Hardness index, mm	Under 20	20-25	25-30	30-33	Over 33

Established in 1981

secondary "shirasu" can be eroded by rushing water.

2. Neither grain size distribution nor consistency can describe the character of "shirasu". Hence, a soil hardness tester is adopted. The subcommittee certified that the values could be divided into four groups (cf. Table 1) and that each group was endowed normal distribution by means of χ^2-, F- and t-testings.

The instrument is the soil hardness tester YH-62. The measurements must be over five times on a vertical cutting plane which is more than 30 centimeters high.

The subcommittee also proposed additionally a tentative plan for working guide (cf. Table 2).

3 CASE OF KANTO LOAM

3.1 Geological description

Kanto Loam lies terrestrial (aquatic in some area) on the diluvial terraces. They consist of four members ——Tama, Shimosueyoshi, Musashino and Tachikawa loam formations. Geologic time covers from 400,000 to 20,000 years before present.

Their superposition will be shown in Fig. 6.

3.1.1 Tama loam formation

This is the oldest in Kanto Loam. It is rich in pumice fall deposit. Its total thickness is 20 to 30 meters. Its locality is limited in

Table 2. Control and counterplot of classified "shirasu" cutting works

Classification		"Shirasu"					Welded tuff	Pumice	Loam
		Very soft	Soft	Moderate		Hard			
Hardness index, mm		Under 20	20-25	25-30 Plantation easy	25-30 Plantation difficult	30-33	Over 33	-	-
No spring	Gradient	1/1.0 1.5	1/0.8 1.2	1/0.8 1.0	1/0.5 0.8	1/0.5 0.8	Over 1/0.5	1/1.0 1.5	1/1.0 1.5
No spring	Slope controls	1°2° 3°5°	1°6° 7°8°	1°6° 7°8°	9°	9°	0°	2°3° 5°	6°8° 10°
Spring found	Gradient	1/1.0 1.5	1/1.0 1.2	1/1.0 1.2	1/1.0 ±	1/1.0 ±	Over 1/0.5	1/1.0 1.5	1/1.0 1.5
Spring found	Slope controls	2°4° 5°	2°4° 5°	1°6° 7°8°	2°4° 5°	2°4° 5°	0°	2°4° 5°	6°8° 10°

A tentative plan by the Research Subcommittee on Standardization of "Shirasu" Classification, 1980

0° No treatment, 1° Framework & grass, 2° Framework & gravels, 3° Framework & blocks, 4° Affix of blocks, 5° Concrete, 6° Turf, 7° Plantation (deep rooted), 8° Mat of grass, 9° "Shirasu"-cement mortar, 10° Seed sowing

the western part of Kanto Plain. Geography is not generally even.

3.1.2 Shimosueyoshi loam formation

This formation bears several sheets of pumice fall deposit. They are "three-colored ice", Pm-1, etc. The superposition is "three-colored" pumice tuff, reddish loam, "oyako" fall, Pm-1 fall, reddish loam and chocolate loam in ascending order. They are generally aqueous in the east and terrestrial in the west. Chocolate loam and "three-colored ice" pumice tuff are sedimented in a lake or a lagoon regionally and often bear carbonized wood. Both of them are called Joso clay.

At the top, there is a layer of hard sandstone in the neighborhood of Lakes Kasumigaura and Ushikunuma.

3.1.3 Musashino loam formation

This formation is pale-brown-colored and rich in vertical cracks. It is 5 meters in thickness at the type locality. In the eastern and middle parts of Kanto Plain, it is about 2 meters thick.

Its sedimentary environment is generally terrestrial, making an exception of Kakioka basin. It seems that the basin was a lake in which this loam formation was settled.

Tokyo pumice fall deposit or TP is near to the bottom. It was ejected out of Hakone volcano.

3.1.4 Tachikawa loam formation

This is the uppermost formation of Kanto Loam. It is reddish loam with kuroboku or black soil band from Mt. Older Fuji, which has a body of basaltic mud flow. Mt. Younger Fuji is a stratovolcano with basaltic lava and lapilli tuff.

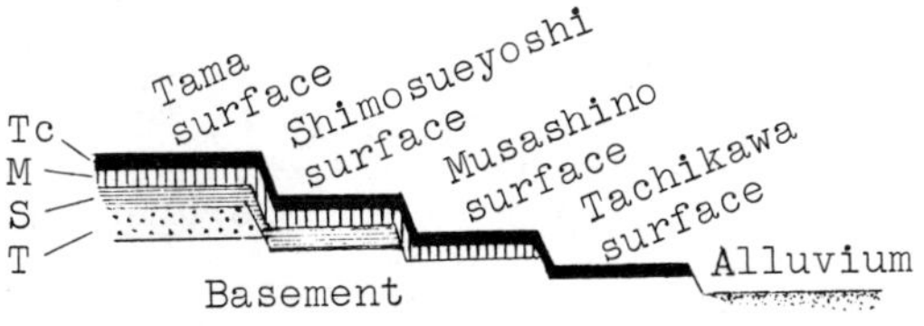

Fig. 6 Schematic section of Kanto Loam superposition

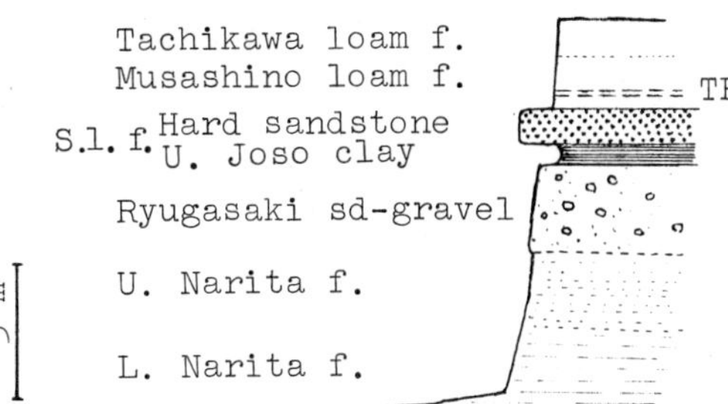

Fig. 7 A typical outcrop of Kanto Loam at Kidamari, Tsuchiura City

At the bottom, there is a layer of pumice fall deposit, which came out of Mt. Akagi. It is named Kanuma pumice fall or KP. Its volume is estimated 6.6 km^3, the greatest mass of pumice fall deposit in Kanto Loam.

Kanuma pumice fall deposit is used for a bonsai, which is an artificial landscape in a shallow bowl with distorted mini-plants, moss and cobbles. Usable soil needs holding of moisture and drainage.

3.2 Characters of soil mechanics

Moisture of Kanto Loam has several strength of combination with earth grains.

Movements of water are not generally reversible. Test piece weathered dry shows low liquid and plastic limits (cf. Fig. 8).

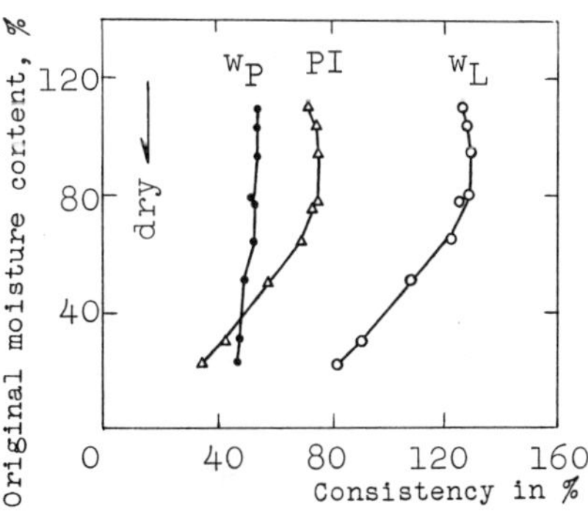

Fig. 8 Characteristic consistency of Kanto Loam (After Public Corporation of Road Construction)

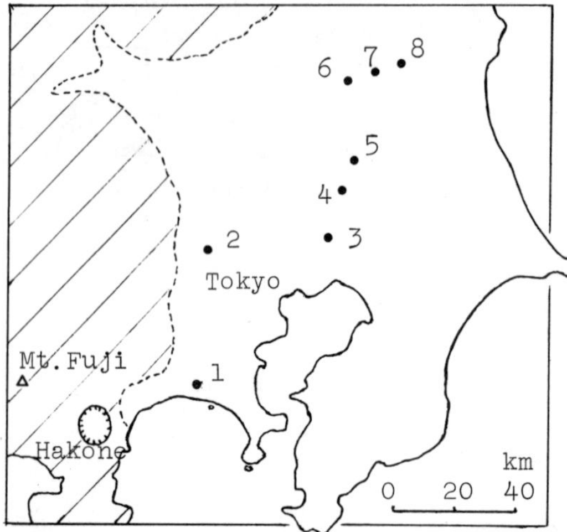

Fig. 9 Locality map of measurement

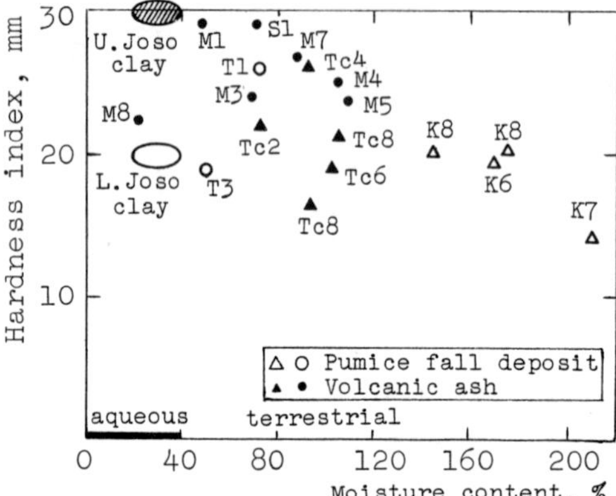

Fig.10 Mois.con.-hard.index graph

3.3 Soil hardness and moisture content

An engineering description on Kanto Loam needs hardness index and moisture content unlike "shirasu".

3.3.1 Volcanic ash

Hardness index is about 20 to 30 millimeters. It decreases in the east or in proportion to a distance from the source. Moisture content increases in the east and decreases under 40% in an aqueous layer (Tc: Tachikawa, M: Musashino loam f.).

3.3.2 Pumice fall deposit

This hardness is generally smaller than that of volcanic ash (K: KP, T: TP).

3.3.3 A layer of hard sandstone

It is about 33 millimeters in hardness index. This is a maximum value in Kanto Loam (abbreviated in Fig. 10).

3.3.4 Joso clay

Joso clay, belonging to Shimosueyoshi loam formation, consists of two members——the upper and the lower. The upper is 30 millimeters in hardness index and about 40% in moisture content; the lower 20 millimeters and 40%.

Upper Joso clay is easily eroded with slaking (cf. Fig. 7).

4 CONCLUSION

1. Soil hardness is useful for pyroclastic material classification. An example of measurer is the soil hardness tester YH-62.
2. The threshold of "shirasu" hardness index is 20, 25, 30 and 33 millimeters. They are useful only to South Kyushu region.
3. Kanto Loam needs a combination of hardness index and moisture content. Hardness index of volcanic ash is larger than that of pumice fall deposit. The nearer to the source, the larger the index is. Moisture content of terrestrial sediments is larger than that of aqueous ones. The nearer to the source, the smaller it is.

5 REFERENCES

Kanto Loam Research Group 1965, Kanto Loam. Tokyo, Tsukiji-shokan.

Kuno, G. et al. 1974, Tuffaceous clay. In the Japanese Society of Soil Mechanics and Foundation Engineering (ed.), Specific soil of Japan, p. 21-83. Library of JSSMFE 10.

Machida, H. 1977, Volcanic ash speaks all. Tokyo, Soju-shobo.

Yamanouchi, T. et al. 1974, "Shirasu". In JSSMFE (ed.), Specific soil of Japan, p. 203-361. Library of JSSMFE 10.

Yamanouchi, T. et al. 1981, On the classification of "shirasu" by hardness, Jour. of JSSMFE 279: 45-48.

Proceedings of the Second European Symposium on Penetration Testing / Amsterdam / 24-27 May 1982

Classification of submerged sediments by dynamic penetrometers

BEHNAM NIKAKHTAR, JOSEPH N.SUHAYDA & MEHMET T.TUMAY
Louisiana State University, Baton Rouge, USA

1 INTRODUCTION

As offshore work moves into deeper water, conventional methods of determining in-situ soil type and strength are becoming more difficult, time consuming, less reliable, and less efficient.

Soil strengths are presently measured to a great extent by boring, sampling and testing of undisturbed samples. This process of boring, sampling and testing has several inherent deficiencies such as high cost, and undesirable effects of mechanical sample disturbance especially if gassy sediments are present in the soil.

It is therefore generally agreed among practitioners and researchers that there is a clear need to develop new and improved in-situ methods for determining soil type, strength and other necessary soil properties that would reduce the cost and the effect of disturbances.

In this study (Nikakhtar, 1981) attention was directed toward dynamic penetrometers. The primary objective was to classify ocean bottom sediments by dynamic penetrometers. The secondary objective was to estimate the shear strengths of soil by dynamic penetrometers.

Dimensional analysis was used to combine the effect of mass, diameter, and the impact velocity of the penetrometer, together with the total depth of penetration and the shear strength of the soil, into dimensionless ratios. The equation of motion was also solved by considering the forces involved in dynamic penetration. The solutions of the equation were expressed in three dimensionless ratios (h/D, V^2/gh, SD^2/Mg).

The data used in the analysis were collected in the laboratory and field. The penetration data of other investigators were also used to confirm the results of this study.

Historically, the impact phenomena of projectiles have been man's concern mostly in the field of armor penetration and passive protection against bombing. The major objective of investigators such as Wang (1971) has been been to find the depth of penetration of a given projectile under given conditions. Because of the complexity of the nature of penetration, this objective has been met by using empirical equations in predicting depth of penetration.

2 EXPERIMENTAL PROCEDURE

The experiments were done by two non-instrumented penetrometers of 1.6 and 2.54 cm diameters. The penetrometers were fabricated of stainless steel and were designed to provide convenient attachment of additional weights. The overall length of the penetrometers were 1 meter for the 1.60 cm diameter and 2 meters for the 2.54 cm diameter penetrometer. Two cones of blunt and 60° tip were used in the tests.

In the laboratory, dry powdered bentonite clay was placed in a large concrete mixer and mixed with a measured amount of water to obtain a homogenous mixture of 5 kPa shear strength, simulating the ocean bottom conditions.

The target was constructed in a cylinderical steel mold of 40 cm diameter and 81 cm height.

The penetrations were performed by the 1.60 cm diameter penetrometer and by adding to the penetrometer the desired weights for different weight to diameter ratios and by adjusting the height of fall for different velocities at impact with the surface of the soil. The penetrations were performed as many times as possible without adversely influencing the results.

The field experiments were performed on two sites. The sites were located at St. Andrews Bay, Florida and at Pass Christian, Mississippi. The soil at both sites was classified as poorly graded soil with 80-90 percent fine sand.

The tests were performed during low tide period of 20 cm maximum height of water above the sand during penetration. The two penetrometers of 1.60 and 2.54 cm diameter were used with different weights at different impact velocities.

A steel frame was constructed for support and for adjustment of the height of fall.

The penetrometers were dropped three times and the maximum penetration values were chosen for analysis. This was decided due to the fact that shells existed in the site and the possibility of collision of the penetrometer with the shells were great.

3 THEORETICAL CONSIDERATIONS

The relationship between the physical characteristics of the penetrometer and the geotechnical properties of the sediments were obtained by first a dimensional analysis, and second by solving the equation of motion of the penetrometer.

3.1 Dimensional analysis

The primary variables considered in this analysis were the air mass of the penetrometer (M), the diameter of the penetrometer (D), the impact velocity (V_o), the soil shear strength (S), the maximum depth of penetration (h), and the acceleration of gravity (g). The unit weight of the soil is neglected in this analysis. The units and the dimensions of these variables are listed in Table 1. Using M, D and g as the repeating variables, the dimensionless ratios are h/D, V_o^2/gh and SD^2/Mg.

The first term h/D represents the number of penetrometer diameters the penetrometer penetrates. The second term, V_o^2/gh represents the kinetic energy over the potential energy of the penetrometer. The third term, SD^2/Mg represents the inverse of weight per area of the penetrometer over the shear strength of the soil. This ratio is related to the bearing capacity of the soil.

Table 1. Units of Primary Variables.

Variable	Dimensions	Units
M	Mass	kg
D	Length	m
V_o	Length/Time	m/sec
S	Force/Area Length	kN/m²
h	Length	m
g	Length/Time²	m/sec²

3.2 Prediction and penetration

Most of the proposed solutions of projected penetration are based upon the following general relationship summarized by Fuchs (1963)

$$M \frac{dV}{dt} = R \tag{1}$$

where M = mass of projectile, V = velocity of projectile, t = time and R = sum of the forces on the projectile.

For this first order analysis R was assumed to have two components; the downward force of gravity, and the upward resistance of soil. The downward force of gravity was taken to be the immersed mass of the penetrometer (M') multiplied by acceleration of gravity (g). The upward soil resistance was taken to be related to the soil shear stress (τ). Therefore, the total force on the penetrometer was related to by the shear stress multiplied by the cylindrical area of the penetromter, πDZ.

The final form of soil resistance was given by

$$\tau \pi DZ = N_p DZS$$

where N_p is the penetrability factor, which may include the effect of friction angle, cohesion, velocity, etc.; S is the shear strength of the soil; and Z is the depth of penetration at any time.

The equation of motion for the penetrometer is as follows:

$$M \frac{dV}{dt} = M'g - N_p DZS \tag{2}$$

The initial condition is that at t = 0 and at Z = 0, V = V_o. The final condition is that at $t = T$ and $Z = h$, $V = 0$.

The particular solution for the depth and velocity are

$$Z = \frac{a}{b} = \frac{V_o}{b^{1/2}} \sin b^{1/2} t - \frac{a}{b} \cos b^{1/2} t \qquad (3)$$

$$V = V_o \cos b^{1/2} t + \frac{a}{b} \sin b^{1/2} t \qquad (4)$$

where

$a = (M'/M)g$

$b = N_p (D/M) S$

These solutions define a relationship between the penetrometer characteristics, the sediment properties and the maximum depth of penetration.

Solving equation (3) and (4) results in the following equation

$$h = V_o \left(\frac{M}{N_p DS}\right)^{1/2} \qquad (5)$$

Equation (5) can be rewritten as the dimensionless ratios obtained previously:

$$h/D = (V_o^2/gh)/(SD^2/Mg)N_p \qquad (6)$$

Therefore, the penetrability factor, N_p can be calculated by the following formula of dimensionless ratios:

$$N_p = (V_o^2/gh)/(SD^2/Mg)(h/D) \qquad (7)$$

4 DISCUSSION OF RESULTS

Figure 1 shows the results of penetrations into laboratory clay of 5 kPa with the 1.6 cm diameter penetrometer. Figure 2 and 3 show the result of penetration into sand at the two previously mentioned sites. These figures show the linear relationship between the impact velocity and the maximum depth of penetration to velocities up to 8 m/s. They also indicate that the maximum penetration increases as the impact velocity increases.

Figure 4 shows the plot of two dimensionless ratios, h/D and V_o^2/gh. The third dimensionless ratio SD^2/Mg combines the effect of shear strength, diameter and the mass of penetrometers into one constant that decreases from 0.42 to 0.05.

As shown, all the points with the same value of SD^2/Mg fall approximately on one line. The data points are from two

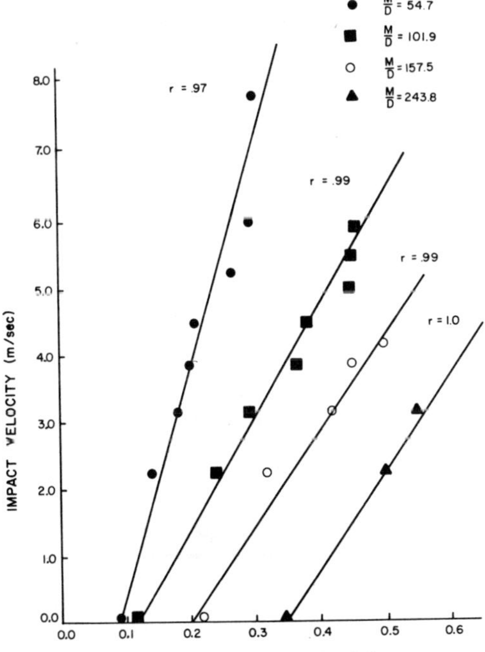

Figure 1. Impact velocity versus depth of penetration in clay.

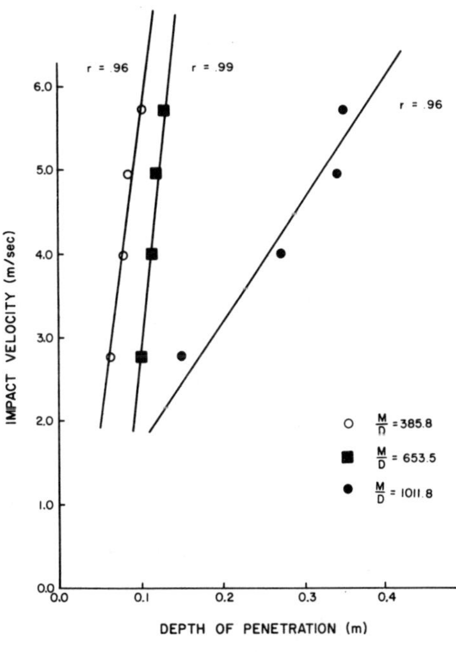

Figure 2. Impact velocity versus total depth of penetration in sand (Pass Christian).

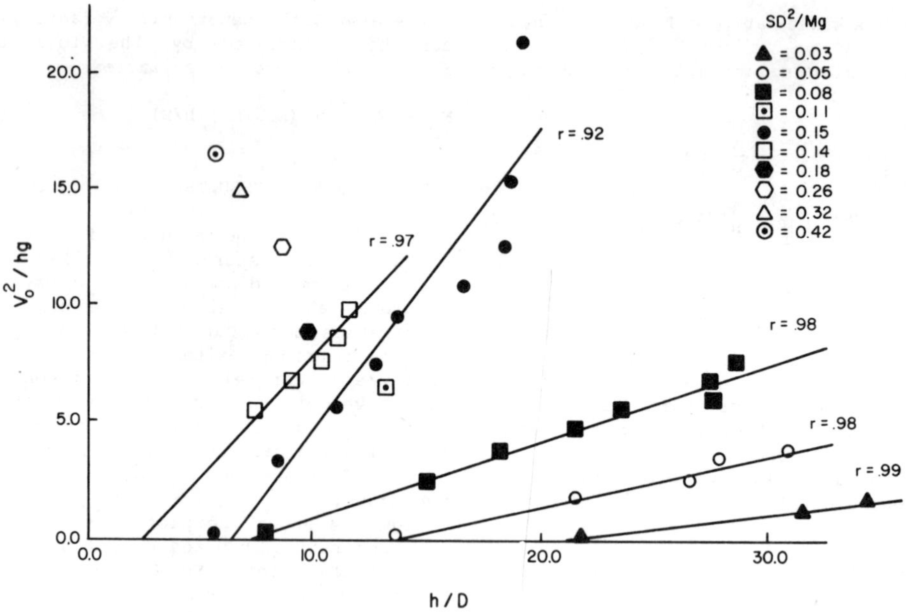

Figure 4. Plot of two dimensionless ratios for determination of clay shear strength (shear strength varies from 5-52 kPa)

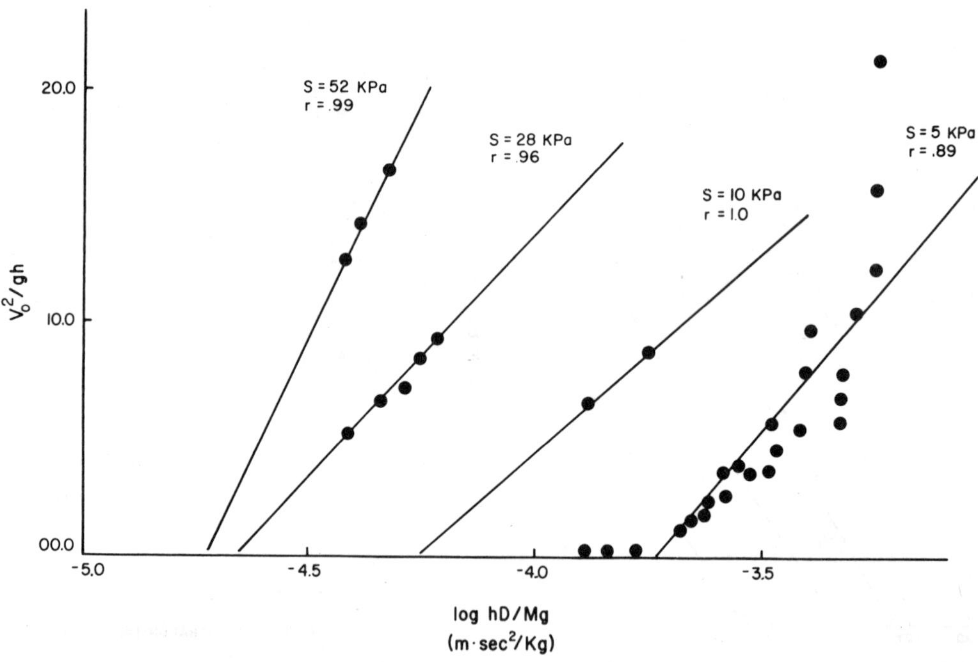

Figure 5. Determination of clay shear strength by normalization of all penetrometer characteristics.

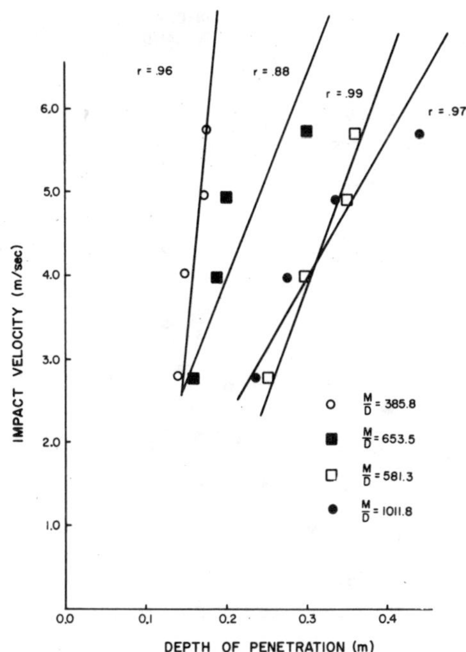

Figure 3. Impact velocity versus depth of penetration in saturated sand (St. Andrews).

sources. First from the results of this study with 5 kPa clay shear strength, and second from test results given in literature by Dayal and Allen (1980). This figure depicts valuable data for obtaining the shear strength of clay with dynamic penetrometers. A penetrometer with any mass, diameter and velocity limited to the range of the velocities in this study (< 8 m/s) can appreciably penetrate into clay, and by referring to this figure and determining the estimate of SD^2/Mg, the shear strength of the soil can be estimated.

In Figure 5, the effect of mass and diameter were normalized for classification purposes. The dimensionless ratio h/D was replaced by the dimensional ratio hD/Mg. This figure shows that all the clays with the same shear strength fall approximately on one line.

Figure 6 combines the data points of both sand and clay on one plot for classification purposes. The points for clay of 5 kPa and 10 kPa clearly fall on the right side. However, points with higher shear strength clay such as 28 kPa and 52 kPa mix with the saturated sand on the left. The sand in this figure has a saturated unit weight of 18.6 kN/m^3 and 20.2 kN/m^3 with the angle of internal friction of 35° and 40°, respectively. In the ocean bottom sediments, overconsolidated clay with shear strength of 52 kPa is very rarely encountered. Therefore, Figure 6 could be used as a method to classify the ocean bottom sediments.

The ratio Mg/hD with h at zero impact velocity gives the estimate of shear strength of clay. For example, a 1.6 cm penetrometer weighing 0.875 kg dropped into clay resulted in 9 cm of penetration when released with no initial velocity. Substitution of the above values into Mg/hD gives 5.9 kPa. The shear strength of the clay measured by fall cone was 5 kPa.

N_p, a dimensionless number calculated by equation (7) also indicates a clear distinction between sand and clay. Figure 7 shows that the log N_p is less than one for clay and greater than one for saturated sand.

The effect of pore pressures in sand are highly significant. The depth of penetration decreases greatly when pore pressures are present.

5 CONCLUSIONS

1. Ocean bottom sediment types can be classified by dynamic penetrometers, given the total depth of penetration, the mass of penetrometer, the diameter of the penetrometer and the velocity at impact.
2. Shear strength of submerged clay can be obtained by dynamic penetration, provided the penetrometer has adequate velocity and mass to sufficiently penetrate into the clay.
3. Logarithm of N_p, the penetrability factor, is less than one for clay and greater than one for sand.
4. Dynamic pore pressures in sand are significantly high to cause large reductions in penetration.
5. The need for further research with instrumented dynamic penetrometers which can measure tip resistance, sleeve friction and pore pressure is recommended.

6 ACKNOWLEDGEMENT

The financial assistance provided by the Office of Naval Research and Environmental Protection Agency is gratefully acknowledged. The contents of this paper reflect the views of the authors who are responsible for the facts and accuracy of the data presented herein. The contents

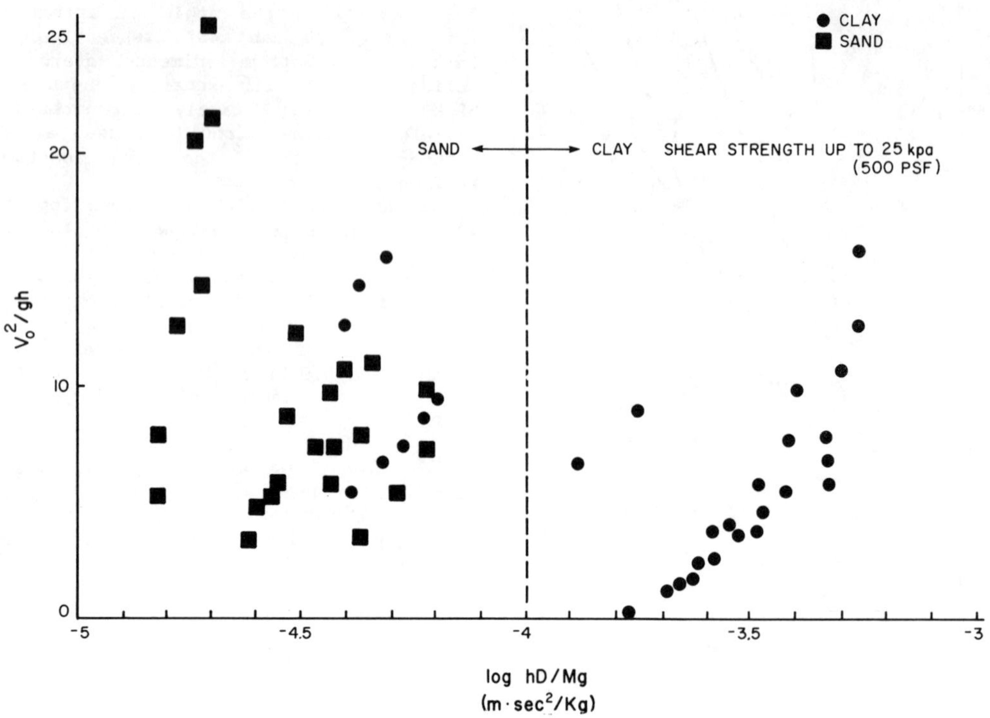

Figure 6. Classification of soil type by normalization of penetrometer characteristics

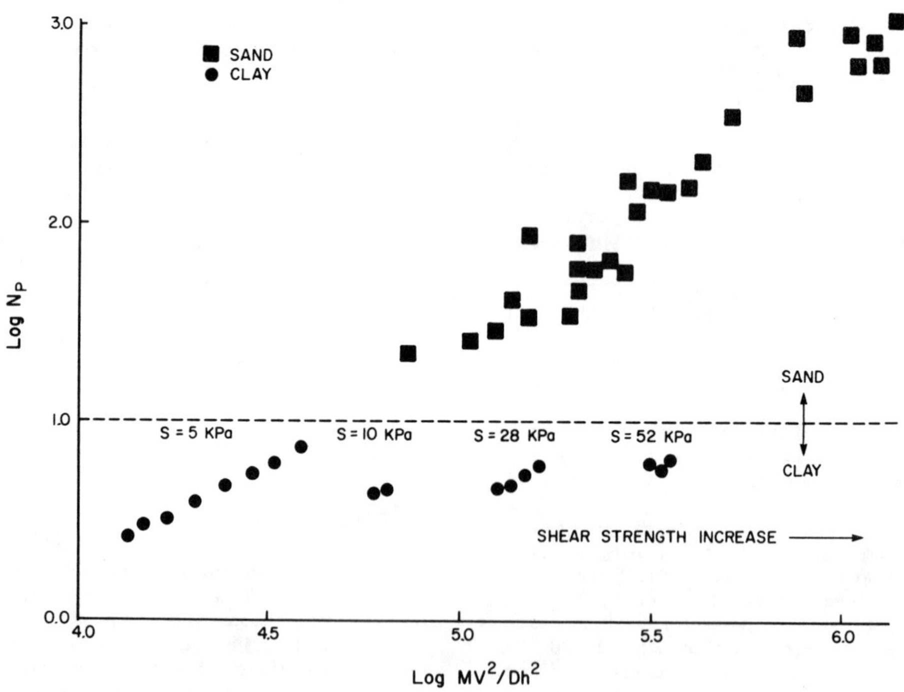

Figure 7. Classification of soil type by the penetrability constant

do not necessarily reflect official views or policies of the ONR and EPA. This paper does not constitute a standard, specification, or regulation.

7 REFERENCES

Dayal, V., Allen, J.H., & Reddy, D.V. 1980, Low velocity projectile penetration of clay, Journal of the Soil Mechanics and Foundation Division, ASCE 106:GT8 919-937.

Fuch, Otto P. 1963, Impact phenomena, AIAA Journal 1:9 2124-2176.

Nikakhtar, B. 1981, Classification of ocean bottom sediments by dynamic penetrometers, Masters Thesis, Louisiana State University, Baton Rouge, Louisiana.

Wang, W.L. 1971, Low velocity projectile penetration, Journal of Soil Mechanics and Foundation Division, ASCE 97:SM12.

Proceedings of the Second European Symposium on Penetration Testing / Amsterdam / 24-27 May 1982

Some aspects concerning the study of foundation soils, using cone penetration tests

M.PĂUNESCU & AGNETA GRUIA
Polytechnical Institute 'Traian Vuia' Timişoara, Romania

1 INTRODUCTION

In situ exploration of grounds by the penetration method has become widespread in our country, particularly in the last few years, along with the development of improvement technics for weak foundation soils and is used for quality testing of these operations (Gruia 1979, Păunescu 198o).

Penetration soundings permit continuos exploration of the ground (in the depth) in order to estimate quality of the improvement of weak foundation soils by supplying primary data at short intervals (usually lo cm), used to estimate the quality of the improvement of weak foundation soils, while cone penetration may also be used for natural foundation soils (Gruia 1979).

In the paper a description is given of the use of static and dynamic penetration soundings for the quality testing of grounds improved by means of ballast columns, vibrothrusting and plots, applied to increase bearing capacity of the foundation soil or reduce the liquefaction of sand soils located in seismic zones.

2 SOIL IMPROVEMENT TESTED BY PENETRATION SOUNDINGS

Civil engineering works performed in regions with weak (or difficult) foundation soils have raised a series of technical and economical problems. Indirect foundation technics (piles, caissons, columns etc) offer a general solution to foundation problems in difficult grounds; however, material consumption (concrete, steel etc.) and costs are great and not always warranted. These shortcomings have conducted to the development of improvement methods of weak grounds, i.e. increase of their mechanical strength, in order to make them appropiate for direct foundation (Păunescu 198o).

Depth improvement methods developed by the Chair of Roads and Foundations of the Polytechnical Institute "Traian Vuia" in Timişoara are based on the vibration technique (Păunescu 198o). These currently applied methods are as follows : depth improvement by vibropressed ballast columns and improvement by vibrothrusting methods officially acknowledged in our country by the standard C 29-77 (reinforcing of weak foundation grounds by mechanical procedures). Recently depth compaction by means of plots punched by vibropressing was also tested with great success.

2.1 Ballast columns or other granular materials

Soil improvement with ballast columns made by vibropressing consists in compaction of the natural ground by means of a tube provided at its lower end with a couple of valves, driven by vibropressing, and introduction from the tube - under continuous vibropressing - of an amount of ballast filling the space created by falling, simultaneously extracting the metal tube. In very weak soils the operation may be repeated once or several times (in the same location), thus obtaining a double-or multivibropressed column (Păunescu 198o). The columns are disposed in the points of an equilateral triangle, side length 2 - 3 d, d being the metal tube diameter. By the use of the vibropressing aggregate AVP1 (Păunescu 198o) developed in our country based on the results of the research work carried by the Chair mentioned, ballast columns up to 9 m in length can be obtained.

2.2 Vibrothrusting

Depth compaction by vibrothrusting is an appropiate improvement method for grounds composed of fine sands, and of loose sand fillings located below underground water level.

The technology of vibrothrusting consists in the introduction by vibration, of a metal housing to the depth of compaction, vibration of the same for two minutes without permitting further penetration into the ground, and extraction also by vibration (Păunescu 198o). The locations where vibrothrusting is performed are coincident with the points of an equilateral triangle with side length 2.0 - 3.0 m depending on the length of the metal housing (4 - 6 m) and/or thickness of the weak layer.

Compaction by vibrothrusting is based on a marked decrease of the frictional forces between the fine sand particles under the effect of vibration, conducting to their fluidization and rearrangement, respectively, thus allowing for a marked reduction of porosity, i.e. a high degree of compaction.

2.3 Plots punched by vibropressing

Depth improvement of the soils by plots punched by vibropressing was performed also at the Chair.

The method consists in a heavy rammer (weighing 1.5 - 3.0 tons) shaped a truncated pyramid, being driven by vibropressing. In the hole left after the rammer is extracted by vibration, ballast, sand, slags, etc. are introduced and compacted by the rammer being re-driven in the same manner.

The plots are arranged similarly with the ballast columns, the distance between the plots, however, is greater, i.e. 1.50 - 2.0 m, while in case of strip foundations, they are arranged in rows corresponding to the location of the foundantions.

3. PENETROMETERS USED FOR QUALITY TESTING OF THE GROUNDS

For the tests the 100 kN hydraulic static INCERC type penetrometer was used; cone penetration resistance (R_p)and skin friction (F_ℓ) were ascertained at 1o cm intervals.

Dynamic penetration soundings were performed using the hand-operated, light dynamic penetrometer PDU and the mechanically actuated, medium-heavy dynamic penetrometer PDM-G, fig. 1, built at the Chair. The particulars of the three penetrometers are given in Table 1.

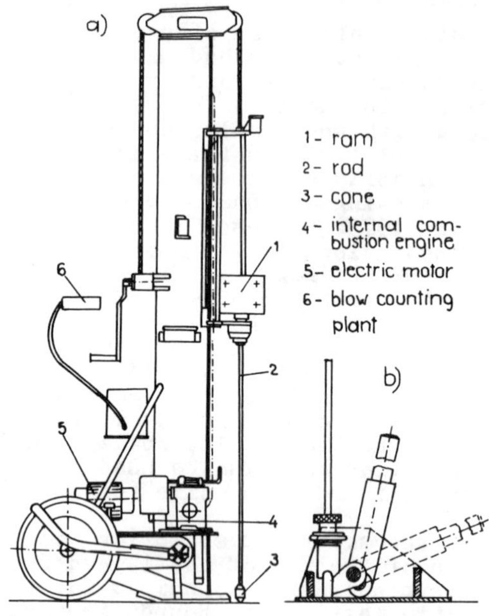

Fig. 1 The mechanically actuated, medium-heavy penetrometer.

Table 1. Technical particulars

Type of the penetrometer	Øcone mm	Ø rod mm	Rammer wgt. kg
Light - PDU	35.6	22	1o
Medium -PDM	43.7	32	35
Heavy - PDG	43.7	32	5o

The falling weight of the rammer amounted to 5o cm in all cases, the angle at the cone tip 90°. Extraction of the rod column is made by the cone-and-roller extractor (fig.1.b), that by its construction exerts only axial stress on the rods.

The medium and heavy dynamic penetrometer has been developed as a single apparatus provided with a 35 kg-rammer for the medium penetrometer; by adding two additional components a rammer weight of 5o kg is obtained, corresponding to the heavy dynamic penetrometer.

In order to ensure safe and accurate recording of the primary data the medium-heavy dynamic penetrometer was equipped with an electric rammer-blow counting device for a lo cm (N_{10}) penetration. An automatic recording system is now being developed.

4 QUALITY TESTING OF IMPROVED GROUNDS

Quality testing of improved grounds is made by two series of penetration soundings : control soundings in the natural ground, indexed "m" and survey soundings in the improved ground, indexed by "c".

Comparison of the survey penetration diagrams with the control diagrams for assessment of the depth down to which the effect of ground improvement, manifest by an increase of the static penetration resistance, R_p, or the number of blows, N_{10}, is obvious.

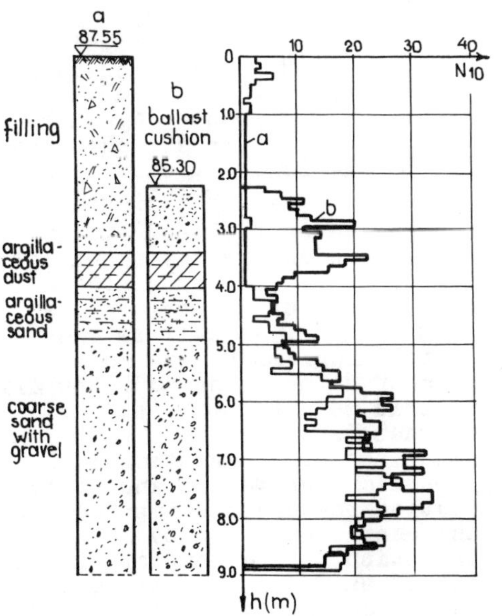

Fig.2 Control (a) and survey (b) penetration diagrams.

Fig. 2 shows the dynamic penetration diagram (a) in the natural ground on the side of a lo-storey apartment building where direct foundation on ballast cushion, 1.2o m in thickness, was adopted. Since the ballast hadn't been adequately compacted, subsequent compaction was made by ballast columns, penetration diagram (b) showing ground improvement to a depth of 6.o m, equalling column lengths.

The foundation ground made up of medium and fine sand strata with interbedded loose or medium-compacted dusty sand on the site of a combine group of enterprises required experimental use of depth compaction operations to enable selection of the optimum solution depending on the peculiarities of the individual enterprises in the combine. Quality of these operations was tested by static and dynamic penetration. Thus, for instance, survey penetrations performed in the experimental cell compacted by ballast columns have demonstrated net improvement of the foundation terrain, in that the sand had passed into the compacted domain, R_p ranging from 12o to 2oo daN/sq.cm

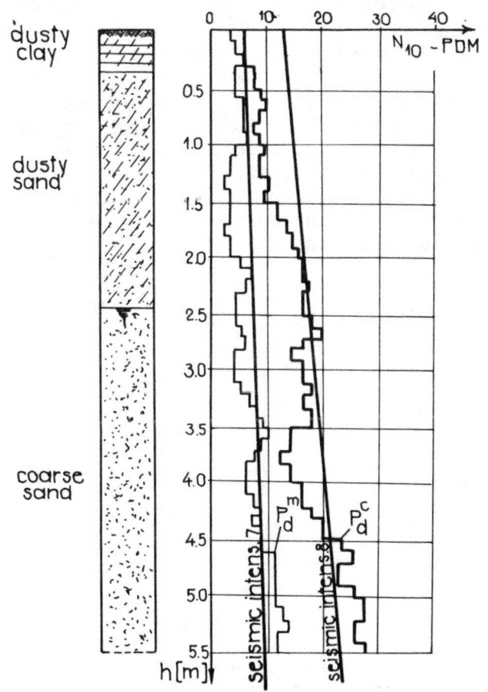

Fig.3. Estimation of the liquefaction potential.

The terrain of a district-heating electric power station located in a 7. degree seismic zone, composed of sand strata with some interbedded argillaceous layers, was proven to be able of liquefaction due to the reduced degree of compaction and the increased level of underground water. In order to diminish liquefaction potential, depth compactions were made with the methods shown in §.2. Fig.3 illustrates the average control and survey penetration diagrams carried out by the use of the medium dynamic penetrometer PDM, superimposed by the standard diagrams yielded by the studies of Ishihara with the standard SPT penetrometer and adapted to the PDM (Perlea 1979). From the control diagram it is seen that liquefaction of sand is possible down to a depth of 4.70 m with seisms of the 7. degree (MKS), subsequently liquefaction persisting with seims of greater intensity. After compaction by vibrothrusting, liquefaction potential decreases.

For depth compaction of the foundation ground for a 5-storey apartment building made of great panels and prefab foundations, vibropressed plots were provided according to the technology described in §. 2.3. Fig.4 illustrates the penetration diagrams; $P^c_{d_1}$ is the survey diagram made in the body of the plot, $P^c_{d_2}$ was made at half the distance between them. Very good compaction of the sand in the natural ground and the ballast in the body of the plot can be noticed.

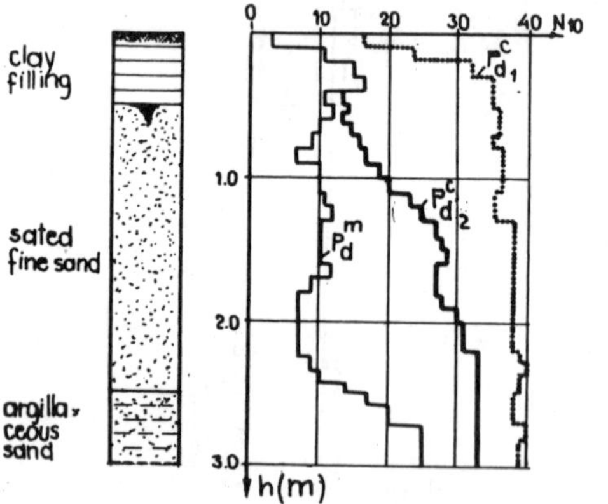

Fig.4 Penetration diagrams in a ground compacted by plots.

5 INTERPRETATION OF PENETRATION DATA

5.1 Static penetration

Primary data of the static penetrations were processed and interpreted according to the Prescriptions C 157-73, Construction Pamphlet No. 1, 1976. Dry volumetric weight, "γd", porosity "n" and pore index "e" were assessed with the graph in fig.5.

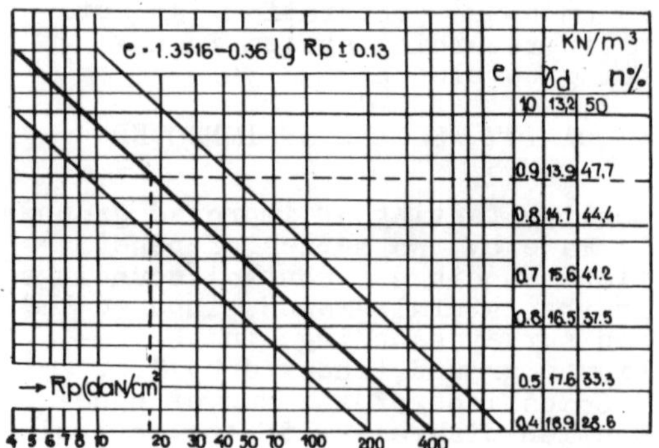

Fig. 5 Graph for the assessment of γd, n and e.

The modulus of deformation M was computed by equation :

$$M = \alpha R_p$$

where: α = 1.5 for sand with $R_p > 45$ daN/sq.cm
$2 < \alpha < 5$ for loam sand and compact clay with $15 < R_p < 30$ daN/sq.cm
$5 < \alpha < 10$ for soft clay with $R_p < 10$ daN/sq.cm

In fig. 6 the control and survey penetration diagrams are shown with the range of variations of the R^c_p values in case of improvement by ballast columns, as well as the variation diagrams of the edometric modulus of deformation "M", the dry volumetric weight "γd" and the pore index "e".

Comparison of the geotechnical soil particulars before and after compaction reveals the improvement of the characteristics by 10 - 20%.

Estimation of the compaction condition of sand was made based on the cone penetration resistance R_p according to the data in Tab.2 for penetration depths down to 5.o and 1o.o m.

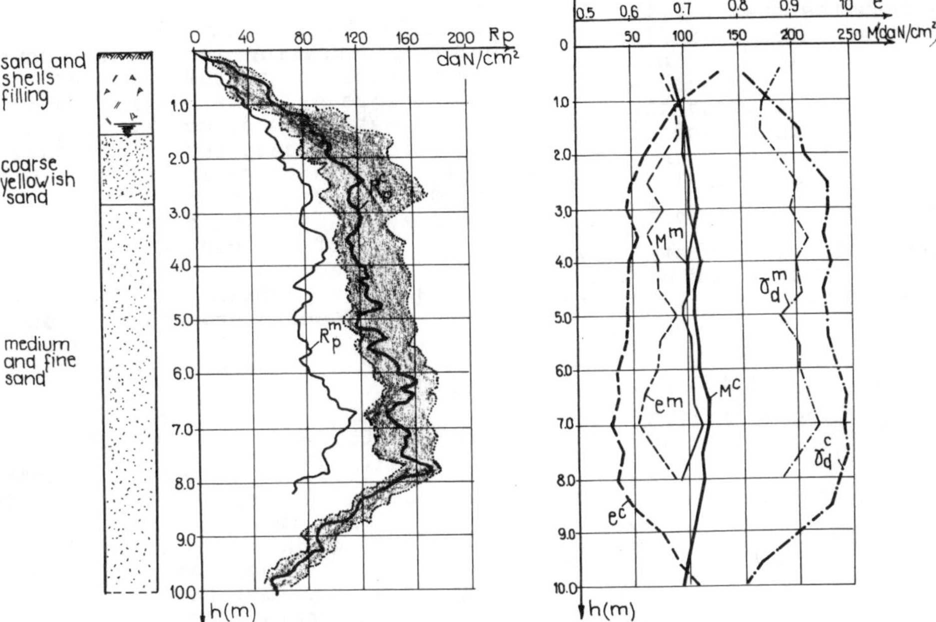

Fig. 6 Control and survey diagrams and the variation of some geotechnical soil particulars.

Table 2. Compaction vs. R_p (daN/cm^2)

Soil type	h m	Compacted	Medium compaction	Loose
Coarse	5	15o	15o–1oo	1oo
sand	1o	22o	22o–15o	15o
Medium	5	1oo	1oo– 6o	6o
sand	1o	15o	15o– 9o	9o
Fine	5	6o	6o– 3o	3o
sand	1o	9o	9o– 4o	4o

Quality of the soil improvement work is usually tested by effecting several static penetration soundings (about one penetration sounding per 75 sq.m of compacted ground), whence a great amount of primary data to be processed results. Since manual processing of these data is tedious the KARSOIL computer program was prepared by the authors, enabling computerized processing of the data (Gruia 1979); in addition to cone penetration resistance and skin friction, some geotechnical soil particulars are also obtained by this program.

5.2 Dynamic penetration

On several sites investigated medium and fine sand were encountered, because of which the interest of the authors was focussed primarily toward the use of dynamic cone penetration.

The experimental explorations of the authors using the light dynamic penetrometer PDU as well as model penetrometers in medium and fine sand, under controlled lab conditions, were used to set up a correlation between relative density Dr and the number of blows N_{10} – PDU. Statistical processing of the data has yielded equation (2), graphically illustrated in fig.7.

$$\log D_r = o.554 \log N_{1o} + o.98 \pm o.25 \quad (2)$$

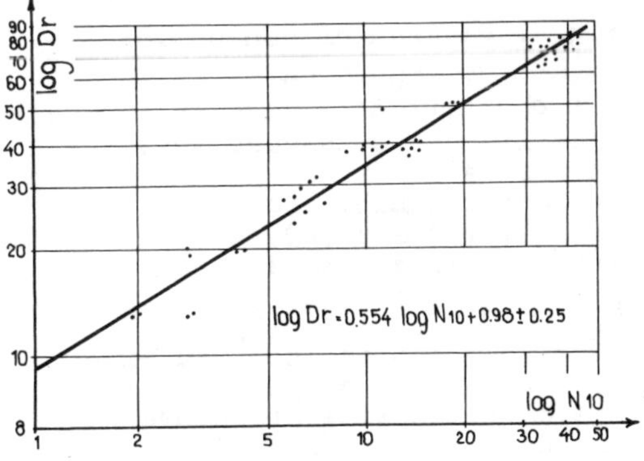

Fig. 7 D_r vs. N_{10} –PDU in medium-fine sand.

Accurate assessment of relative density requires that N_{10} in eq. (2) be not affected by influence factors (Melzer 1967, Păunescu 1966) f.i. depth, skin friction etc.

The great amount of static and dynamic penetration soundings made sinchronously in medium and fine sand has permitted establishment of the connection between R_p and N_{10} -PDU.

By correlation of the static with the dynamic, control and survey penetration data (Gruia 1979), fig. 8, the following equations were obtained :

$$R_p = 2.03N_{10} + 29.84 \pm 31.72 \quad (3)$$
$$R_p = 2.10N_{10} + 39.96 \pm 41.32 \quad (4)$$

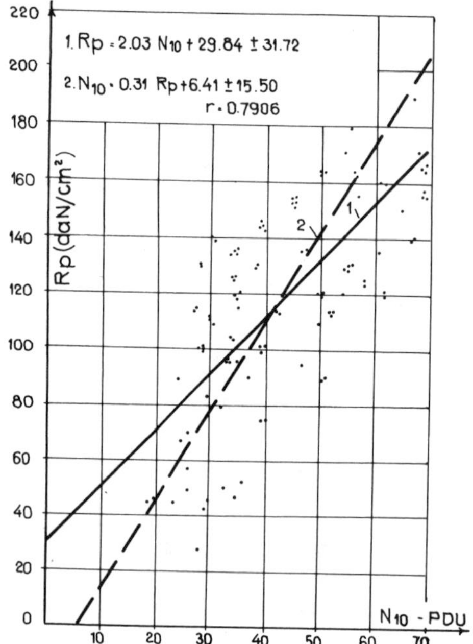

Fig.8. Diagram of connection $N_{10\text{-PDU}} - R_p$

In a computation made on the safe side it may be assumed that the negative value of the mean standard deviation annihilates the free term in eqs. (3) and (4), conducting to the simplified expression :

$$R_p = 2.03N_{10} \quad (3a) \qquad R_p = 2.10N_{10} \quad (4a)$$

Eq.(3a) can be accepted with a 3.5% error for compacted soils too, where the initial structure was disarranged during the compaction operations.

6 CONCLUSIONS

Depending on the control and the survey penetration resistance, some geotechnical characteristics of the soil before and after improvement can be established, characteristics that are necessary for an estimation of the bearing capacity of the soil and the deformations occuring under the load transmitted by the building. By the use of the penetration soundings the amount of the high-cost -price survey drillimgs is reduced, where the results are often inconclusive (f.i. in the case of sand located below underground water level).

7 REFERENCES

Gruia,A. 1979, Unele aspecte privind corelarea dintre penetrarea statică şi dinamică cu con (Some aspects concerning the correlation between static and dynamic cone penetration) Bul. ştiinţific şi tehnic al IP Timişoara, Tom 24(38).

Gruia,A. 1979, Contribuţii la cercetarea terenului "in situ" prin metoda penetrării dinamice cu con (Contributions to the "in situ" soil investigation by dynamic cone penetration) Teză de doctorat-conducător ştiinţific M.Păunescu, Timişoara.

Melzer,K. 1967, Sonderuntersuchugen in Sand (Special investigations in sand) Technische Hochschle Aachen.

Păunescu,M. 198o, Imbunătăţirea terenurilor slabe în vederea fundării directe (Improvement of weak soils for direct foundation) Editura tehnică, Bucureşti.

Păunescu,M., Gîdea,A.and Gruia,A. 1966, Despre factorii care influeţează rezultatele penetrării dinamice cu con (Contribution to the influence factors of the results of dynamic cone penetration).Rev. Construcţiilor şi a materialelor de construcţii, nr. 8.

Perlea,V.,M.Perlea 1978, Prognoza comportării construcţiilor fundate pe depozite lichefiabile(Prognosis of the behaviour of constructions founded on deposits with liquefaction potential)Schimb de experienţă.Comportarea"in situ" a construcţiilor. Călimăneşti.

Proceedings of the Second European Symposium on Penetration Testing / Amsterdam / 24-27 May 1982

Comparison of some dynamic, nondestructive test results for different geological conditions

JOANNA PINIŃSKA
Warsaw University, Poland

SUMMARY

The paper deals with the researches of the relation between dynamic field testing by the Schmidt -hammer method and geophisical field and laboratory tests for the rocks. As a geophisical method the acoustic waves propagation in the material is used.

The main purpose of these investigations is to define the connection between two quick and available, nondestractiv methods of determing of the geotecnical parameters in field and laboratory conditions.

As a field observations the Schmidt hammer reaction force and the acoustic waves velocity for different parts of geological profil are measured. In laboratory the acoustic testing of the chosen samples are condacted.
For both acoustic tests the high waves frequancy -range 1MHz - was applied. Such frequancy allows the similar tests conditions of unlimited medium for the small sample as for the field testing.

Finding the physical properties of rocks and rock massifs is a time-consuming task and asks generally for complex research instruments. The latter comprise mainly a defining of resistance and deformability of the rock media against a background of their differentiated lithologic composition and a comprehensive geologic evolution.

In the laboratory the rock properties are defined on the basis of some small fragments of a rock medium /samples/ whereas in the field - of large rock massif fragments /blocks/. Results of both types of analyses are badly comparable due to a so-called scale effect and such a small sample used in a laboratory as well as a block are not usually representative for all the features typical for a rock massif.

Both research methods possess advantages and disadvantages of their own. The laboratory analyses are considerably more detailed, more abundant and can comprise some modelled extra factors /humidity, strains, temperature, etc./. But they do not take into account the whole complexity of a rock medium /a collaboration of all the fragments, a ruggedness of a surface, cracks, etc./ that is in fact fulfilled by the studies in a site. So, the latter should be controlled by the laboratory analyses. At the same time such research methods should be sought that diminish the effects of a scale error.

A certain solution of this problem seems to be found by an introduction of geophysical methods to the rock investigations in a site as well as in a laboratory. Particularly the acoustical methods with an application of ultrasounds are useful. The ultrasound studies, using the artificially generated impulses of short wave lengths, enable to consider even a small sample, of several cubic centimetres, for an indefinite medium. By the ultrasound prospecting method several phenomena that define a material characteristics, can be noted at a wave route. Also, in agreement with the principles of an elasticity theory, an elasticity modulus or a medium density can be defined. This method is simple and enables the repeated studies, supplying with results of a great accuracy. For the laboratory investigations this way of finding the physical properties seems to be more and more reasonable [1 , 2].

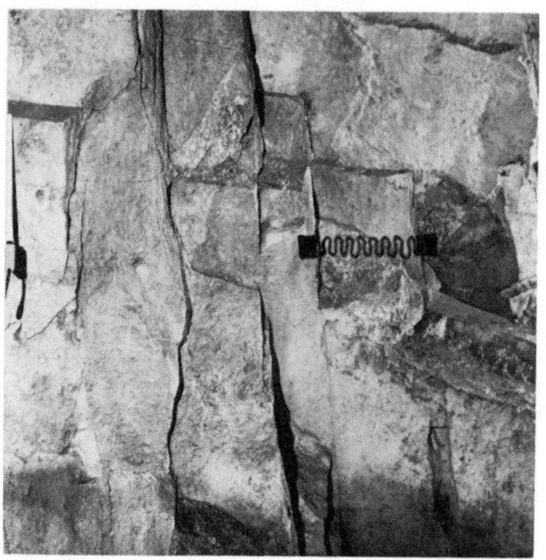

Fig. 1 An example of a bedding, favourable for ultrasound measurements "in situ"

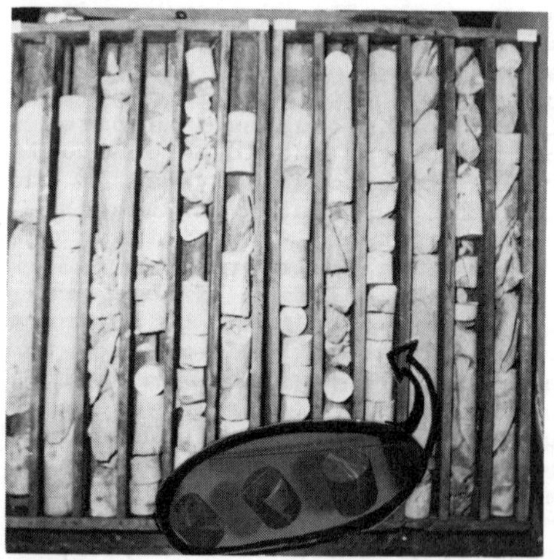

Fig. 2 Sections of a borehole core used for cutting the cylindrical samples

There are more difficulties if studying a massif as a prospection is either limited by a borehole direction [3] or by a bed arrangement in a rock massif /Fig. 1/. On the other hand it is well known that good results for a quick control of the rock massif properties in the field can be obtained if applying the Schmidt's hammer. The ultrasound investigations as well as kick studies with a use of the Schmidt's hammer possess some common features in spite of different principles of their measurement. They include the non-destructed, dynamic investigations and enable a repeated control of the parameters in any fragment of a rock massif or a sample, make it possible to consider the directed geologic factors. Among the advantages there is a simplicity of the observations. These methodical properties have been the basis to the undertaken, in this paper, comparative analysis of both the mentioned methods against a background of finding a resistance to a compression defined in a laboratory.

The acoustical measurements have been carried through by a detection method, with a use of the ultrasonic flow detector Unipan 541 and the replaceable generating and receiving sound boxes, with a frequency 40 kHz to 1 MHz as needed. At the same time, an absence of a dispersion was checked at small samples.

The measurements of a reaction index /τ / were done with a use of the Schmidt's hammer, Swiss made /Swiss patent 283099/; and for small samples a dispersion analysis of results was done, depending on their masses.

The investigations in a rock massif were done at the bed sites accessible to place there the head--boxes; the beds had quite smooth surfaces and an epidian was applied as a connecting medium.

In the laboratory the studies were carried through at cylindrical samples with polished surfaces; the samples came from a borehole core /Fig. 2/. They were then subjected to resistance analyses.

The experiments were done with a use of various sedimentary, volcanic and metamorphic rocks:

- gaizes and marls of Upper Creta-

ceous,
- limestones, sandstones and siltstones of Lower Jurassic,
- rhyolites, diabases, metadiabases, metamorphic slates - connected with Permian volcanism.

Marls and gaizes are quite homogeneous rocks; they compose of carbonates, silica and clay minerals, there are also abundant remains of sponges, colchiolites, pelecypods and echinoids.

The Jurassic limestones include mainly the rocky limestones with organic remains; they are macroporous with stylolite planes and cracks mineralized with a calcite.

The Jurassic sandstones compose mainly of medium-grained and fine-grained quartzic sandstones with a muscovite, poorly consolidated.

The siltstones contain but a muscovite admixture, the streaks of a carbonaceous matter. Among the rhyolites there are amygdaloidal and cast rhyolites. Diabases, metadiabases and slates are intensively tectonicly engaged, with a frequent mineralization with barite, quartz or pyrite.

The analyzed rock matter represented then quite a vast range of features resulting from geologic premises and so, enabled to formulate some methodical generalizations.

An application of ultrasound measurements is troublesome in the field With the Unipan instruments it is only possible if we have enough smooth and parallel surfaces. So, it is not possible to use it in the case of an intensively fissured massif /Fig. 3/. In this

Fig. 3 Fissured rock massif, unfavourable for measurements in the field

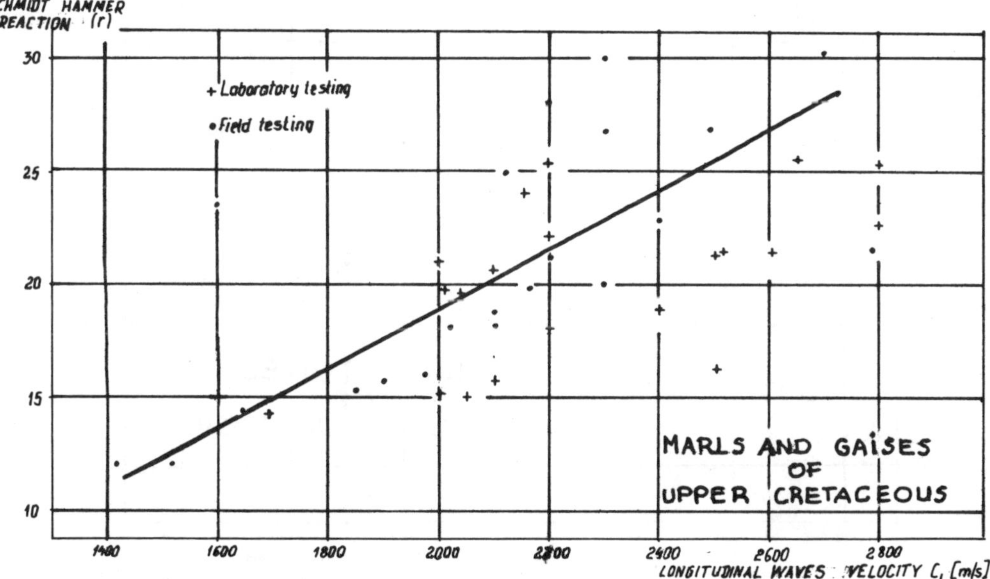

Fig. 4 A reaction index /r/ in a velocity function of longitudinal wave /C_L/

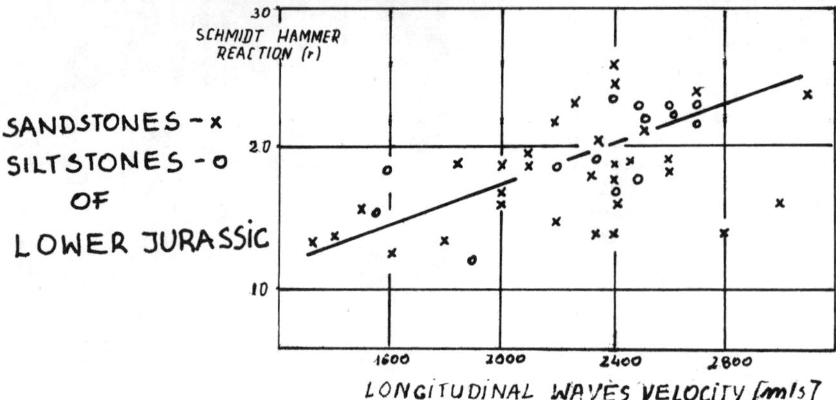

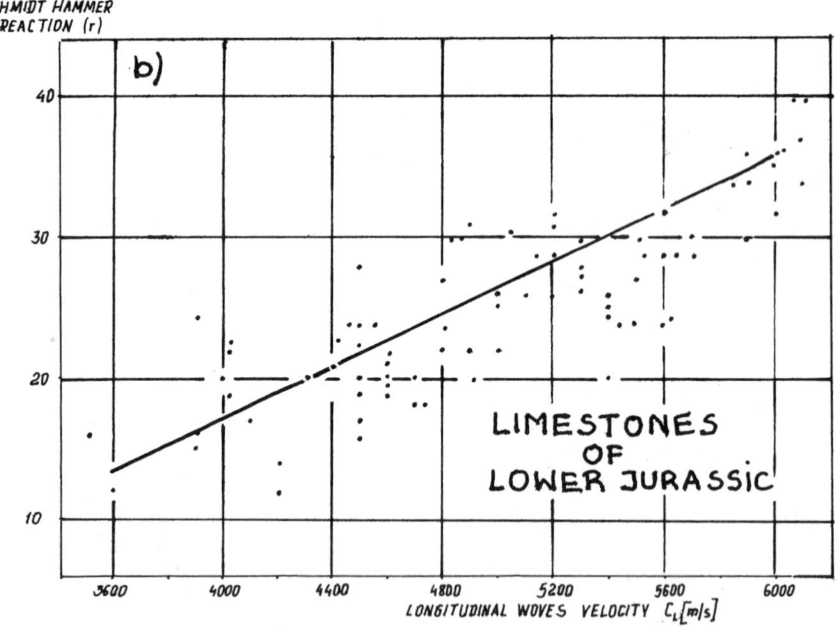

Fig. 5.A reaction index /r/ in a velocity function of longitudinal wave /C_L/.

case there are also some troubles with a use of the Schmidt's hammer.

If analyzing a relation of a propagation velocity /C_L/ and reaction indices /r/, in the case when such measurements are possible, we can note a great convergency of results of field and laboratory measurements /Fig. 4/.

For quite soft rocks /VI th category in the Protodiakonov's scale/, a considerable dispersion of results is possible, particularly there where the rocks contain an organic debris /gaizes, marls/; a distinguishment of Jurassic sandstones and hardpans on this ground is also difficult /Fig. 5a/, the rocks of the VIth category after Protodiakonov as well. A distinct convergency is on the other hand created for the rocks of the IIIrd category of density as it is exemplified by the Jurassic limestones /Fig. 5b/. For the very compact rocks /category II/ a dependency r=f/C_L/ for each rock can be interpreted with a great accuracy /Fig. 6/

tions similar values of a reaction index / / were received independently of a sample mass At the sa-

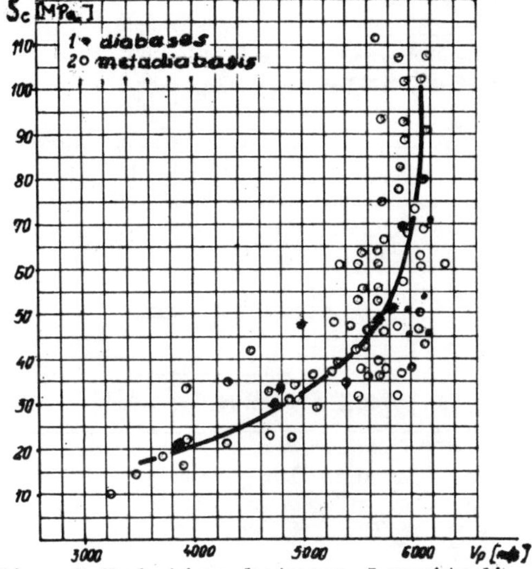

Fig 7 Relation between Longitudinal wave velocity /C_L/ and compression strength /S_c/

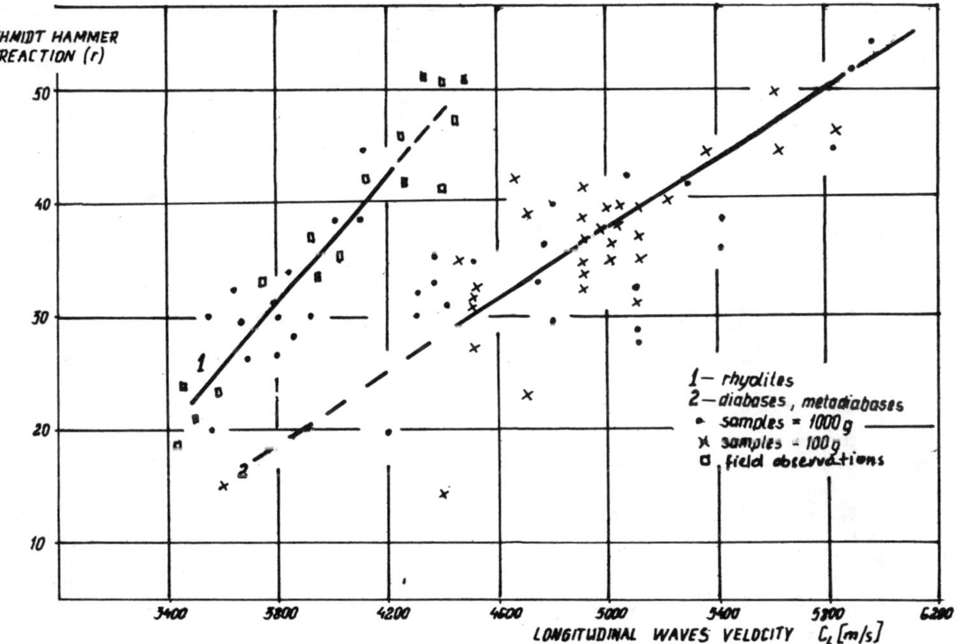

Fig.6. A reaction index /r/ in a velocity function of a longitudinal wave /C_L/.

It should be emphasized that in the carried laboratory investiga- me time the results of studies over hammer reaction / / for the compact rocks effected in lower values than the data for a massif.

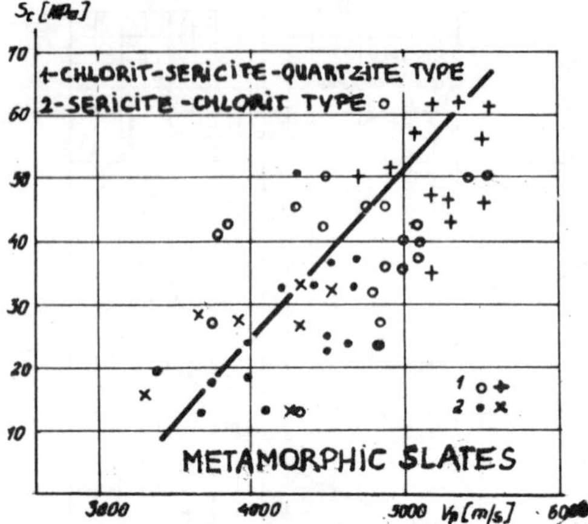

Fig. 8. Compression strength /S_c/ as a function of longitudinal wave velocity /C_L/

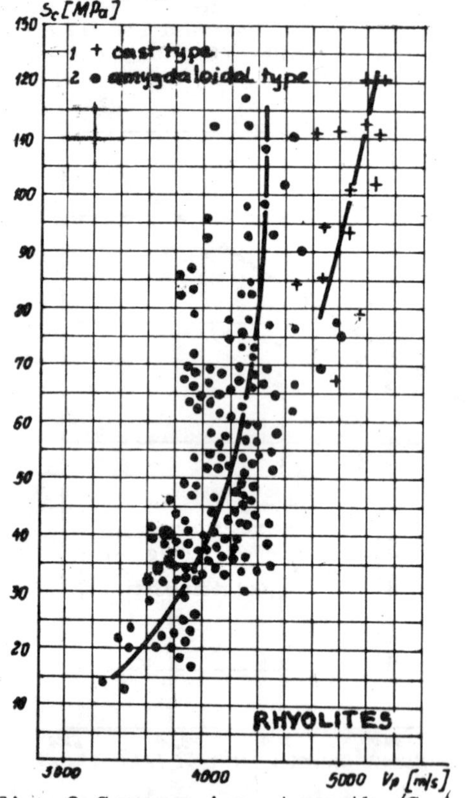

Fig. 9. Compression strength /S_c/ as a function of longitudinal wave velocity /C_L/

The samples testified by ultrasounds and the Schmidt's hammer were then subjected to resistance analyses, defining their compression strength S_c.

The considerations are supplemented with data on known exponential relations, connecting a compression strength and a propagation velocity of a longitudinal wave inside a medium /4/. These data /Fig. 7, 8, 9/ were used to prepare a diagram, showing the relations of a propagation velocity of a longitudinal wave /C_L/, compression strength /S_c/ and a reaction index /r/ /Fig. 10/.

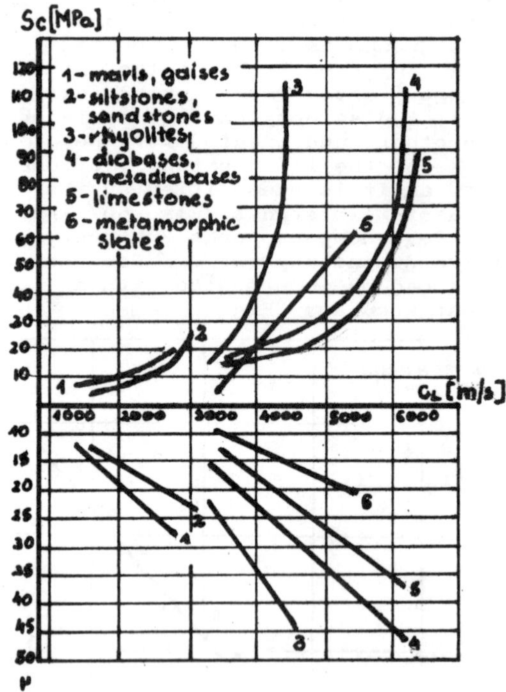

Fig. 10. Diagram of relations of longitudinal wave velocity /C_L/, compression strength /S_c/, reaction index /r/

It can be used to show the relations of individual dependences. If a medium density is known then a dynamic modulus of elasticity /E_d/ can be also defined, taking its relation to a reaction index.

The presented correlations are naturally approximate only as an effect of inaccurate measurements with a use of the Schmidt's hammer to small samples but also, due to a defining of the wave velocity at natural bedding planes in a rock massif. It is also influenced by a complex constitution of a rock medium, considerably different than in theoretically accepted assumptions for elastic wave propagation in an unlimited and homogeneous medium.

All the above considerations are to be used for a quick control of rock parameters in a massif, applying the Schmidt's hammer. Then, due to correlative connections, the compression and resistance parameters can be found. A preliminary estimation of the usability of a bedrock is then simplified as much as possible.

References

1. Pinińska J. 1977, Correlation between mechanical and ocoustic properties of the flyschys sandstones - Procc. Int. Symp. on the geotechnics of structurally complex formationes.Capri.
2. Michalopulus at all. 1979, Measurement selection and use of dynamic sail properties in design parameters in geotechnical engineering. Procc. of VII Conf. of E.C.S.M.F.E. Brighton.
3. Takimato K., Nakamura I, Fudo R. 1981, Application of ocoustic emision in situ test. Procc. of X Conf. I.C.S.F.M.F.E. Stocholm.
4. Pinińska J. 1978. Acoustic anisotropy of the flyschys sandstones. Procc. of the VIIth Int. Conf. of I.A.E.G. Madryt

Proceedings of the Second European Symposium on Penetration Testing / Amsterdam / 24-27 May 1982

Effect of velocity on penetrometer resistance

T.J.POSKITT & C.LEONARD
Queen Mary College, University of London, UK

1 INTRODUCTION

The use of cone resistance and sleeve friction measurements for designing piles is well established. The in-situ determination of these values can now be made routinely. For fine grained soils it is well known that the results depend upon the penetration velocity. This is due to viscosity.

Recent tests on soils in the laboratory Litkouhi (1980) have shown that the side friction on a penetrometer moving at 20 mm/s can typically be 25% greater than that observed when the penetrometer is moving at 0.2 mm/s. This is clearly of importance when interpreting penetrometer data. In the design of driven piles it is of particular significance since it implies the driving resistance will be greater due to viscous effects.

In order to study the influence of velocity more closely a hydraulically actuated penetrometer rig was constructed, Fig. 1(a). This drives the penetrometers, shown in Figs. 1(b) and (c), into soil samples at known constant speeds.

This equipment is now used on a routine basis for testing soil. At the slowest speed, 0.2 mm/s, the skin friction and cone resistance are obtained. By operating the penetrometer at a range of speeds between 0.2-2000 m/s the variation of resistance with velocity is obtained. This is illustrated for a till obtained from Cowden in North Humberside.The use of this data in driveability studies for two instrumented piles installed at the site is discussed below.

2 VELOCITY DEPENDENCE

Cowden till is fairly uniform over the top 6 m with a moisture content of 18%, PL 20% LL 40%, clay friction 35% and $c_u = 120$ kN/m². The soil has layers and lenses of fine sand within the sequence and the adhesion factor is about 0.5. Ten 250 mm diameter by 300 mm long undisturbed samples were obtained from a trench 2.5 m deep.

Because Cowden till is quite stoney a 10 mm diameter penetrometer was used. The penetrometer needs to be sufficiently strong to punch through small stones or push them to one side. For soils free of stones, a smaller penetrometer, 5 mm in diameter can be used. This gives results in agreement with the 10 mm instrument. The smaller size has considerable advantages since U4 samples can be tested in the tube.

For the cylindrical penetrometer it is assumed that the side and point resistances are given by

$$(R_d/R_s)_{side} = 1 + JV^N \quad (1)$$

$$(R_d/R_s)_{point} = 1 + J'V^{N'} \quad (2)$$

respectively, where:
R_d = dynamic resistance (side or point)
R_s = soil resistance (side or point) at V = 0.2 mm/s
J,N = viscous parameters
V = penetration velocity

The nonlinear form of the viscous resistance is now well established. The power laws adopted are common in rheological work. If N and N' are unity the laws become identical with those used by Smith (1960) in his famous wave equation analysis. A further advantage of these laws is they enable data to be analysed particularly easily.

The results of the tests on Cowden till are plotted in Fig. 2. This data is re-plotted on log-log scales in Fig. 3. From this the values of J and N can be readily obtained. These are given on Fig. 2.

3 APPLICATION TO PILE DRIVING CALCULATIONS

In pile driving calculations the viscosity is known to be very important. For wave equation studies this is incorporated using the Smith model which corresponds to N and N' being taken as unity in eqns (1) and (2). The parameters J and J' are then obtained by back analysis. The procedure gives an empirical value for the viscous terms.

Because of limitations in this procedure the Authors have in the past few years measured viscous resistance directly using their variable speed penetrometer. In all cases the results have shown a strong nonlinearity and the problem has arisen on how this data might be correlated with the extensive body of results which have been interpreted using Smiths law. For piles less than 30 m in length the following procedure has been evolved.

The relationship between the dynamic resistance R_d and the quasi-static resistance R_s is given by

$$R_d/R_s = 1 + JV^N \qquad (3)$$

Smith's law is

$$R_d/R_s = 1 + J_sV \qquad (4)$$

and the problem is to correlate J_s with J and N.

One method of correlation would be to fit the data of Fig. 2 to eqn (4) using a Least Squares technique. If this is done the line shown in Fig. 2 is obtained and a value of $J_s = 1.9$ s/m is obtained. This is over three times the value which would be obtained by back analysis pile driving data, hence this method is not acceptable.

For pile design a better method is to look at the energy components during driving. This can be done by modifying the Hiley formula to take account of soil viscosity. When this is done a good agreement can be obtained between the set per blow obtained using the wave equation and Hiley's formula modified to take account of viscosity.

Referring to Fig. 4(a) this shows a pile of weight P just prior to release of the hammer W. In this condition P is balanced by the soil resistance r_o.

On release of the hammer the kinetic energy at impact is

$$kWHg = \frac{Wv_H^2}{2}$$

where k is the hammer efficiency and v_H is the impact velocity.

The energy loss on impact according to Newton's Laws of impact is

$$U = kWH\frac{P}{W+P}(1 - e^2)g$$

where e is the coefficient of restitution.

The energy imparted to the pile is therefore

$$kWHg - U = kWHg\left(\frac{W + Pe^2}{W + P}\right)$$

and is stored in the form of Kinetic energy.

The velocity which this imparts to the centre of gravity of the pile (= V) is given by

$$\frac{PV^2}{2} = kWHg\left(\frac{W + Pe^2}{W + P}\right) \qquad (5)$$

As a result of impact the centre of gravity of the pile moves off with an initial velocity V. After some time t the equation of motion of the centre of gravity, see Fig. 4(b) is

$$P\frac{d^2u_c}{dt^2} = -\hat{r} \qquad (6)$$

It should be noticed that equation (6) is exact for the motion of the mass centre C of the pile and pilecap when acted upon by a resultant resisting force r. u_c is the total displacement of C, and consists of the rigid body motion of the pile together with the elastic compression of pile.

In Smith's treatment the resistance of soil elements is assumed to be velocity dependent thus

$$\hat{r} = \bar{r}(u_c)\left[1 + J_s\frac{du_c}{dt}\right] \qquad (7)$$

where $\bar{r}(u_c)$ is an idealised elastic-plastic law as shown in Fig. 4(d).

For the nonlinear treatment

$$\hat{r} = \bar{r}(u_c)\left[1 + J\left(\frac{du_c}{dt}\right)^N\right] \qquad (8)$$

Writing

$$\phi_N = 1 + J\dot{u}_c^N \qquad (9)$$

Equations (7) and (8) can be written

$$\hat{r} = \bar{r}(u_c)\,\phi_N \qquad (10)$$

Observing that $d^2u_c/dt^2 = \dot{u}_c\,d\dot{u}_c/du_c$

and substituting equation (10) into equation (6) gives

$$P\,\dot{u}_c\,d\dot{u}_c = -\bar{r}(u_c)\,\emptyset_N\,du_c \qquad (11)$$

Referring to Fig. 4(d), when

$$u_c = 0\;;\qquad \dot{u}_c = V$$

and

$$u_c = s + Q\;;\qquad \dot{u}_c = 0$$

Integrating equation (11) between these limits

$$P\int_V^o \frac{\dot{u}_c\,d\dot{u}_c}{\emptyset_N} = -\int_{oab} r(u_c)\,du_c \qquad (12)$$

Writing

$$F = \frac{2}{V^2}\int_o^V \frac{\dot{u}_c\,d\dot{u}_c}{\emptyset_N} \qquad (13)$$

Then equation (12) becomes

$$\frac{PV^2}{2}\,F = R(\bar{s} + Q/2) \qquad (14)$$

Values of F are given in Fig. 5 for N = 0 to 1.

Now $\bar{s}$ is the displacement of the centre of gravity of the pile C. What is required is the deflection of the pile head (= s). Just prior to rebound (b in Fig. 4) the pile is at rest. The pile and pile cap are assumed subject to uniform shortening RL/EA and R/k_1 respectively. k_1 is the stiffness of the pile cap. Rebound is considered to be a decaying oscillation about the centre of gravity C. After the motion has decayed the pile will have expanded equally on either side of C as shown in Fig. 4(c). Thus

$$\bar{s} = s + \frac{RL}{2EA} + \frac{R}{2k_1} \qquad (15)$$

Writing

$$c = Q + \frac{RL}{EA} + \frac{R}{k_1} \qquad (16)$$

which is called the temporary elastic compression of the pile

$$\frac{PV^2}{2}\,F = R\,(s + \frac{c}{2}) \qquad (17)$$

NOTE: in the absence of viscosity $\emptyset_N = 1$ and F = 1 and equation (12) becomes the well-known Hiley formula.

For the case of N = 1 (Smith's Law), Litkouhi (1979) has made an exhaustive comparison of equation (17) with wave equation calculations. This considered forty-three piles and showed that the average agreement was to within 5% (s.d. 30%). All the piles were load tested and for each the soil resistance at the time of driving was estimated. On average the test results agreed with those predicted by equation (17) to within 7% (s.d. 34%). The results given by the wave equation were within 4% (s.d. 25%) of the field values. The wave equation results are therefore only marginally better than those given by equation (17).

4 CORRELATION OF VISCOUS PARAMETERS

Referring to equation (17) it will be seen that the effect of soil viscosity is contained in the term F. If F is known for any values of N and J then the equivalent J_s can be obtained from Fig. 5. The procedure is illustrated on the open ended test pile driven into Cowden till, Rigden (1979). For this pile point resistance can be ignored.

Weight of hammer W	= 3.5 tonne
Velocity of hammer at impact v_H	= 4.14 m/s
Coefficient of restitution e	= 0.89
Length of pile	=12.6 m
Weight of pile and cap P	= 3.2 tonne

Writing $kWH = \frac{W\,v_H^2}{2g}$ in equation (5) gives

$$V = 4.11 \text{ m/s}$$

Taking N = .27 and J = 1.0 s/m Fig. 2 then r = $1.00 \times 4.11^{.27}$ = 1.46. From Fig. 5 F = .439. The value of r corresponding to N = 1 is r = 2.10 hence J_{smith} = .512 s/m. This should be compared with a nominal value of .656 s/m commonly used in offshore driveability calculations.

5 CONCLUSIONS

A variable speed penetrometer has been described which enables the relationship between soil resistance and penetrometer velocity to be determined. This is found to be highly nonlinear.

The viscous parameters obtained from the test are correlated with empirical parameters used in pile driveability calculations. The results confirm the magnitudes commonly used in practice and provide the means whereby measured values may be converted into equivalent empirical values (i.e. into Smith's damping constants).

6 REFERENCES

Litkouhi, S. and Poskitt, T.J. 1980, Damping constants for pile driveability calculations. Geotechnique 30, No. 1, 77-86.

Litkouhi, S. 1979. The behaviour of foundation piles during driving. Ph.D. thesis, University of London.

Rigden, W.J., Pettit, J.J., St.John, H.D. and Poskitt, T.J. 1979. Developments in piling for offshore structures. BOSS'79, London 28-31 Aug. 1979.

Smith, E.A.L. 1960. Pile driving analysis by the wave equation. Jnl. S.M.F. Div., A.S.C.E., Vol. 86, SM4, August 1960.

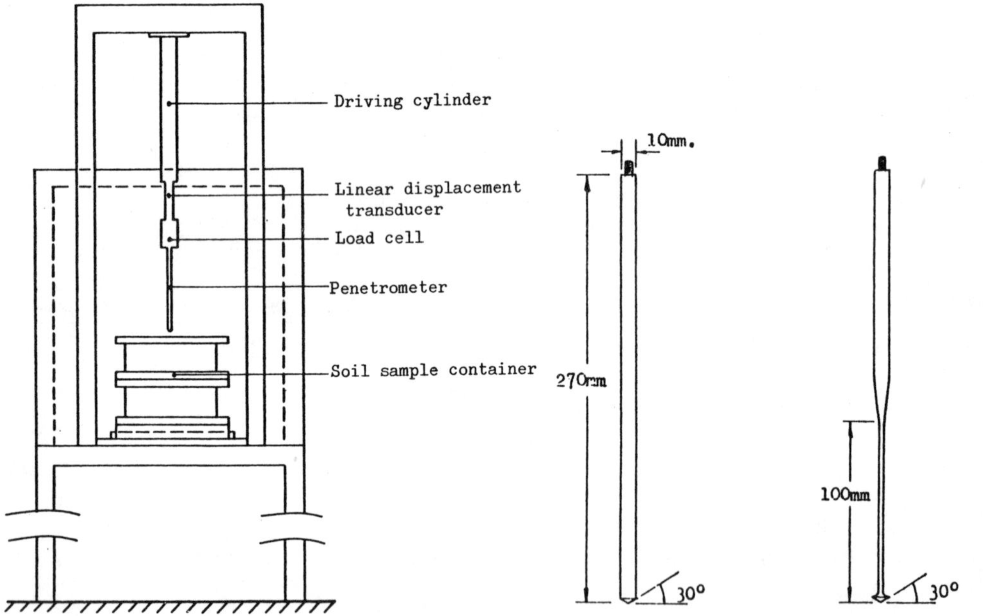

a) Constant velocity test rig

b) Penetrometer for side friction tests

c) Penetrometer for cone resistance tests

Fig. 1. Penetrometer Test Equipment

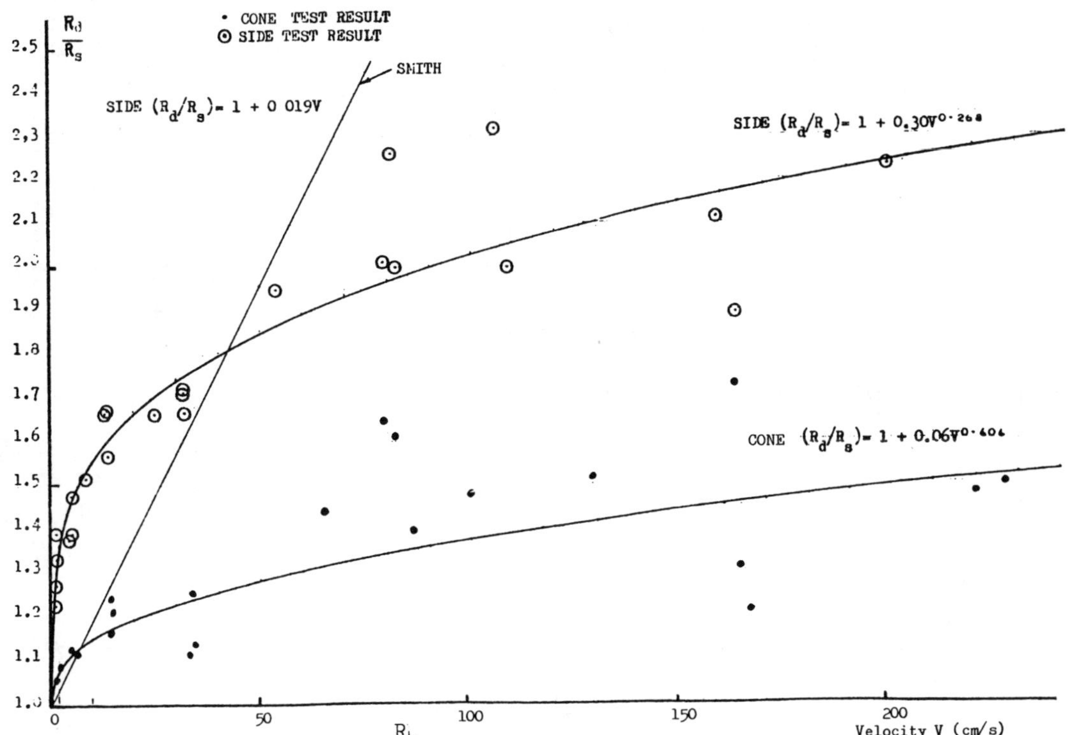

Fig. 2. Relationship between $(\frac{R_d}{R_s})$ and velocity for Cowden Clay

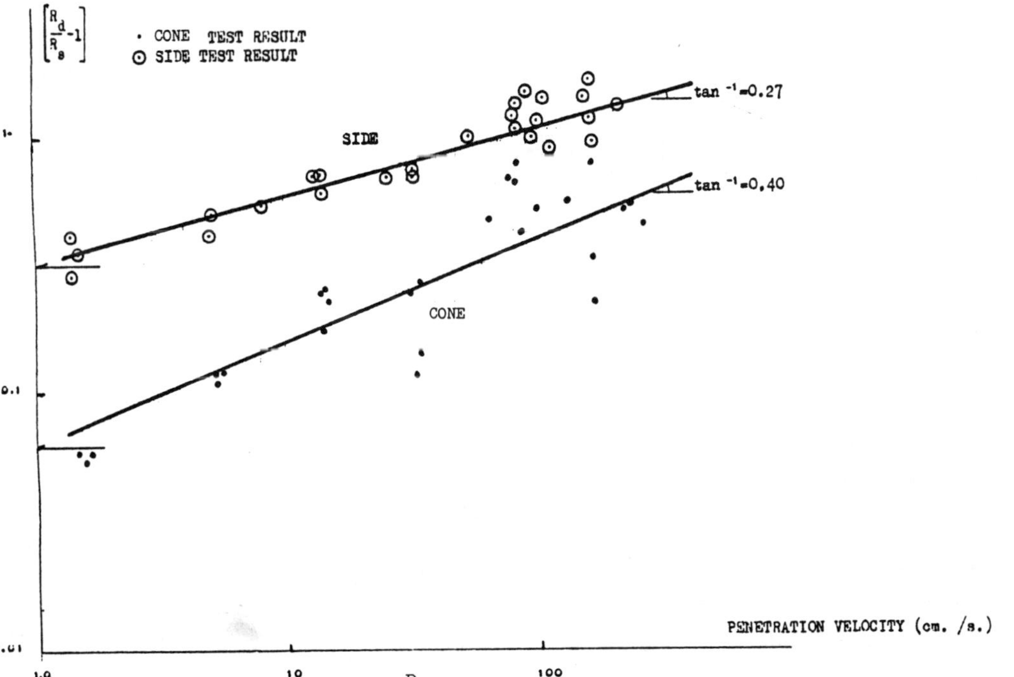

Fig. 3. Relationship between $\log[\frac{R_d}{R_s} - 1]$ and log velocity for Cowden Clay

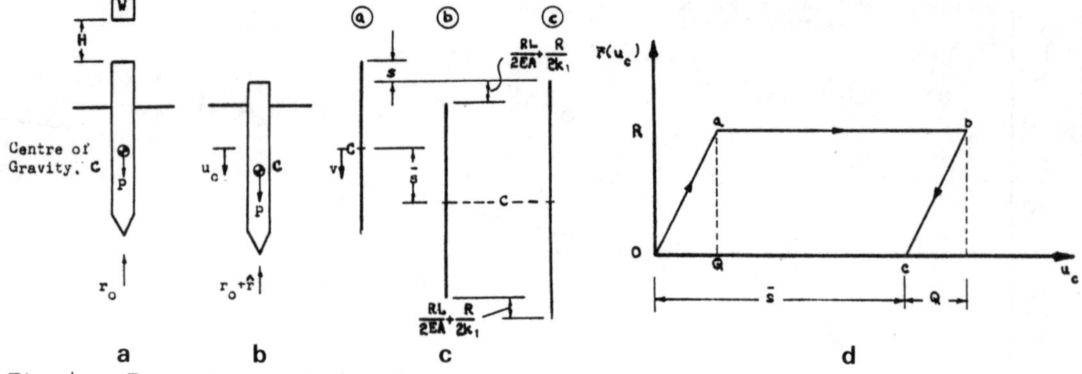

Fig. 4. Energy components in pile driving

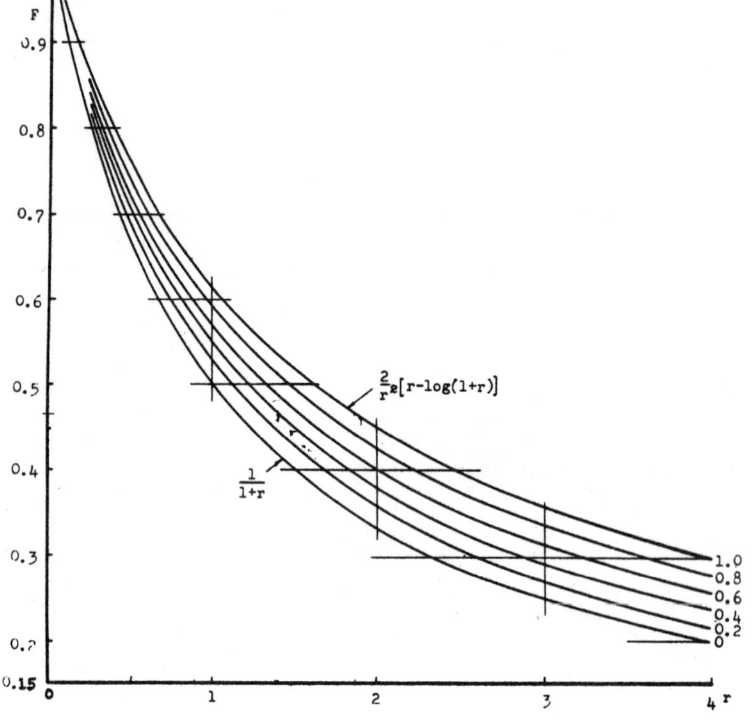

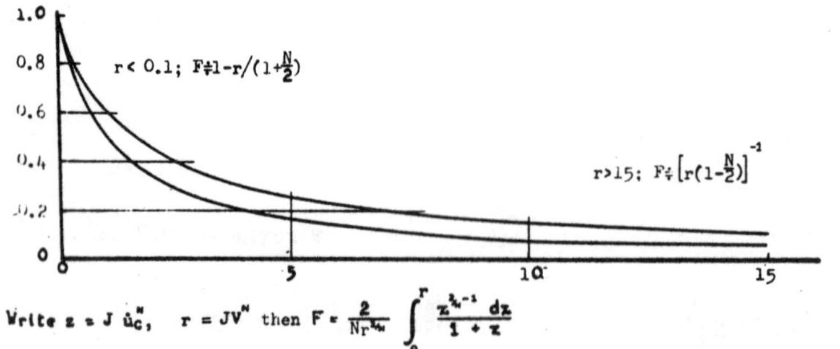

Write $z = J\dot{u}_c^N$, $r = JV^N$ then $F = \frac{2}{Nr^{2/N}} \int_0^r \frac{z^{2/N - 1}\,dz}{1+z}$

FIG. 5 - Evaluation of

$$F = \frac{2}{V^2} \int_0^V \frac{\dot{u}_c\, d\dot{u}_c}{1 + J\dot{u}_c^N}$$

Proceedings of the Second European Symposium on Penetration Testing / Amsterdam / 24-27 May 1982

Dynamic cone probing tests in gravelly soils

B.GOVIND RAO, D.R.NARAHARI & G.R.BALODHI
Central Building Research Institute, Roorkee, India

INTRODUCTION

Soil Stratum containing large percentage of gravels and cobbles Fig. 1 presents several problems during investigations. behaviour of the soil foundation system. Typical results showing stress-deformation behaviour at Antibiotics factory site at two locations are shown in Fig 2a, 2b and 2c. Consisting of boulder deposits with

Fig 1 GRAVELLY SOIL STRATUM

Laboratory testing for evaluating strength characteristics of such a soil deposit is generally not possible as the un-disturbed samples in the real sense can not be procured from the field. Also, native state of particle arrangements in virgin ground cannot be reproduced. Thus laboratory tests on remoulded samples are ruled out. The only alternative left to the investigation is to depend on well chosen field tests.

In-situ Load Tests

Among the field tests, in situ load test on cast in-situ concrete footings supplemented by a trial pit or a bore log (Narahari, Rao and Balodhi 1968) is the best as it provides the stress-deformation

filler material having predominantly sand and sandy soils (SM). Figs 3a and 3b show the test results at Ram Nagar site where predominantly gravelly soil deposit consisting of silt (SM) as filler material. It would be seen that Figs 2a, 2b and 3a clearly demonstrate the effect of footing size. While, Figs 2c and 3b show the effect of filler material on the stress-deformation behaviour.

In-situ Shear Test

In view of the cost and the time required for load tests, in situ shear tests consisting of either Boulder-Boulder Test (BBT) or Concrete Block Shear Test (CBT) may provide an alternative to the load test (Narahari, Rao and Jain 1968,

Mohan, Narahari and Rao, 1971). The Boulder Boulder Test consists of shearing a soil block, Fig 4 and in the concrete Boulder Test, a concrete block is sheared under a given normal load. Fig, 5 . Typical test results of BBT and CBT tests at Antibiotic Factory site is shown in Fig. 6 and 7.

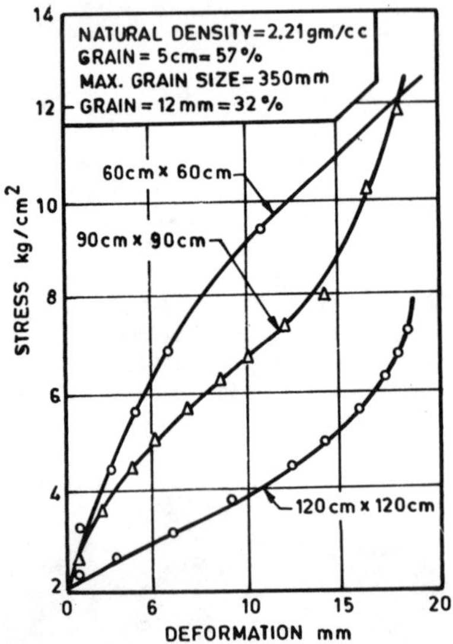

FIG 2(a) FIELD LOAD TEST RESULT ON CASTIN SITU FOOTING ON BOULDERS AT ANTIBIOTIC FACTORY SITE

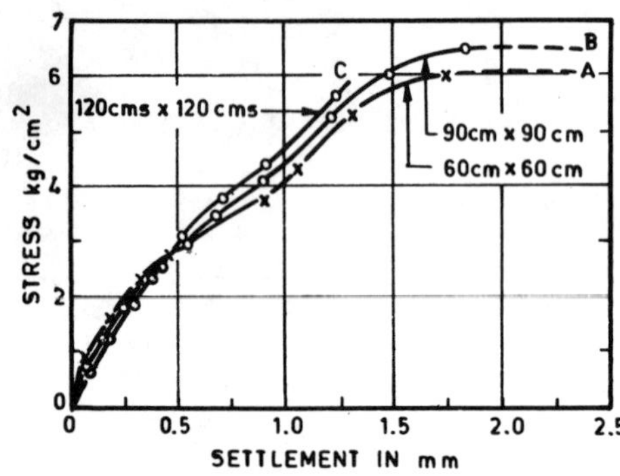

FIG.2 (b). STRESS DEFORMATION BEHAVIOUR OF BOULDERY SOIL AT ANTIBIOTIC FACTORY SITE

The standardisation of BBT and CBT, their relative merits and demerits have been discussed in detail elsewhere (Narahari, Rao, and Jain, 1976, Narahari and Rao 1978). The residual shear so obtained from the BBT and CBT has been related with allowable soil pressure in large size insitu load test for a number of sites and is expressed by eqn 1 (Mohan, Narahari and Rao, 1971)

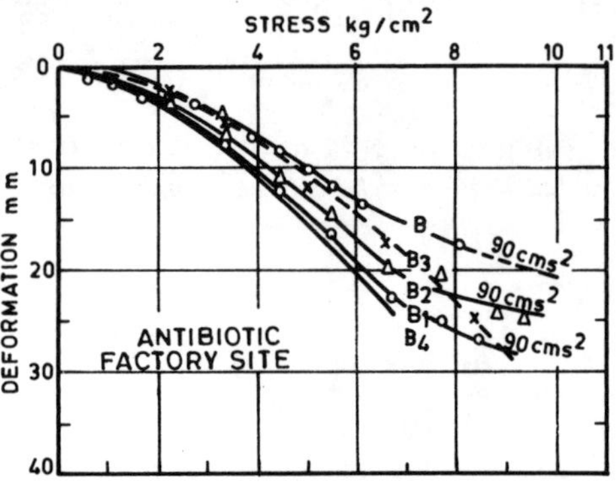

FIG.2(c). LOAD DEFORMATION CURVES SHOWING FFFECT OF COMPRESSIBILITY INDUCED BY INTERESTITIAL MATERIAL

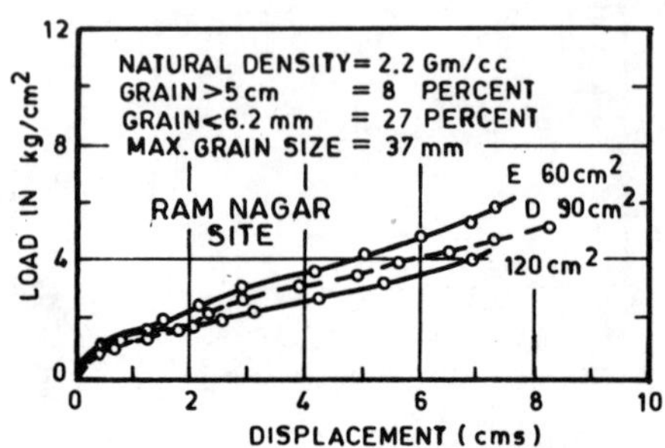

FIG.3(a). LOAD DEFORMATION CURVES SHOWING EFEECT OF FOOTING SIZE (SITE V)

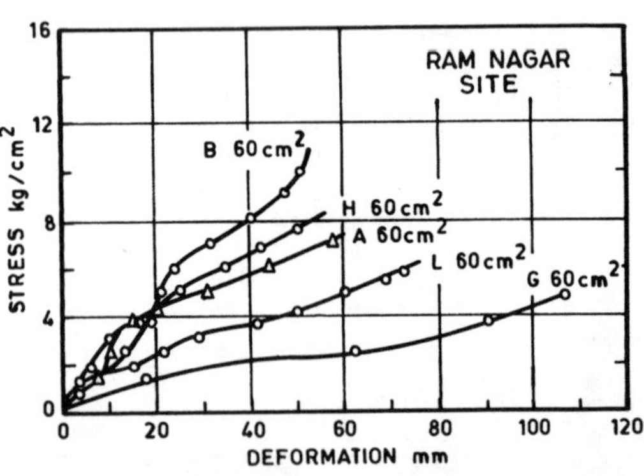

FIG.3(b). LOAD DEFORMATION CURVES SHOWING FEFECT OF COMPRESSIBILITY INDUCED BY INTERESTITIAL MATERIAL

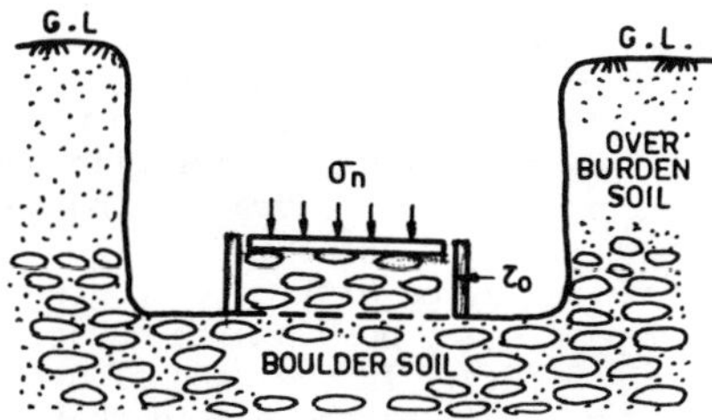

FIG.4-BOULDER BOULDER TEST (BBT)

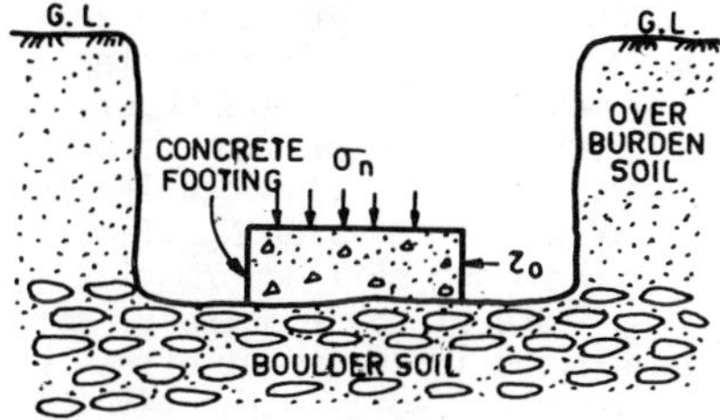

FIG.5-CONCRETE BOULDER TEST (CBT)

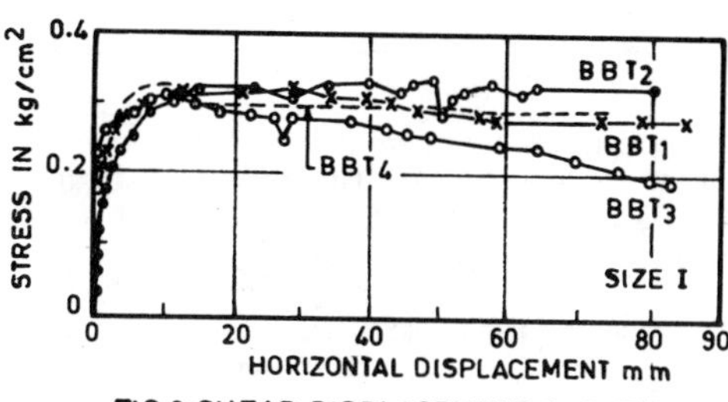

FIG.6. SHEAR DISPLACEMENT CURVE (B.B.T.)

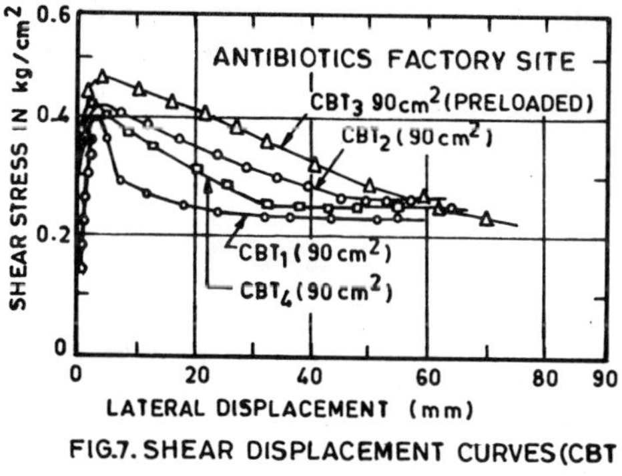

FIG.7. SHEAR DISPLACEMENT CURVES (CBT)

$$q_a = 6.25\tau_o\left[\frac{B+0.3}{B}\right]^2 \cdot \gamma \qquad \ldots 1$$

where q_a is the allowable load at 12mm deformation in (t/m^2) τ_o is the residual shear stress (t/m^2), B is the width of the footing (m) and γ is the unit bulk weight of the deposit (t/m^3). The co-efficient of co-rrelation in this case was found to be 0.8.

Utilising angle of shearing resistance obtained from in situ shear tests (BBT or CBT), the bearing capacity of gravelly soil deposit was computed from bearing capacity equation developed for rigid plastic solids and compressibility corrections were applied in accordance with Vesic (1975) approach using rigidity Index, I . The computed values were found to agree well with the load corresponding to 25mm deformation obtained from insitu load tests on footing sizes larger than 90cm, (Rao, Narahari and Balodhi, 1981).

Field Consolidation Tests

Besides BBT and CBT which help in computing strength of the gravelly deposits, field consolidation tests using mild steel rings (30cm-60cm dia and 30cm height) have recently been used to predict the settlement of such deposits under the design load (Rao, Narahari and Balodhi, 1981). A schematic diagram of the test set up and a typical test result is shown in Figs 8 and 9. The coefficient of volume compressibility m_v was found from the relationship expressed by equation 2

$$m_v = \frac{\delta h}{h} \cdot \delta p \qquad \ldots 2$$

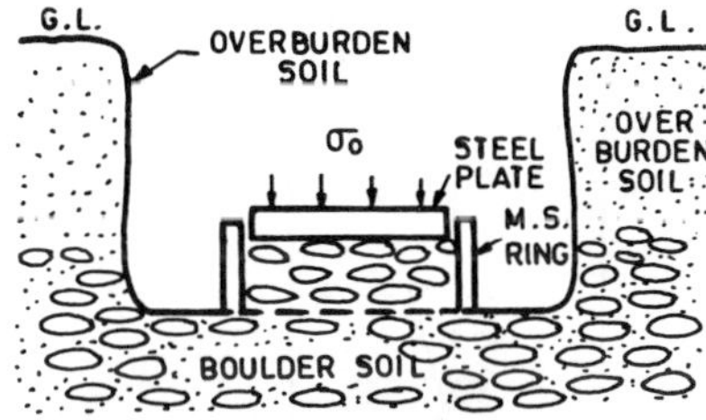

FIG.8. FIELD CONSOLIDATION TEST

(E_s) was found from Boussinesq equation, utilising insitu load test results at three different sites with varying type and amount of filler material. The relationship between (E_s) values and per cent filler material is shown in Fig 10 for two cases when the filler material is in the form of silt and clay and in the other case the filler material is coarse sand or crushed stone. The influence of amount and type of filler material on elastic modulus of deformation

E_s is clearly demonstrated. The investigation demonstrated that the settlement computed from m_v values (field compression tests) and those observed from the insitu load tests for the same intensity of stress were found to be in close agreement.

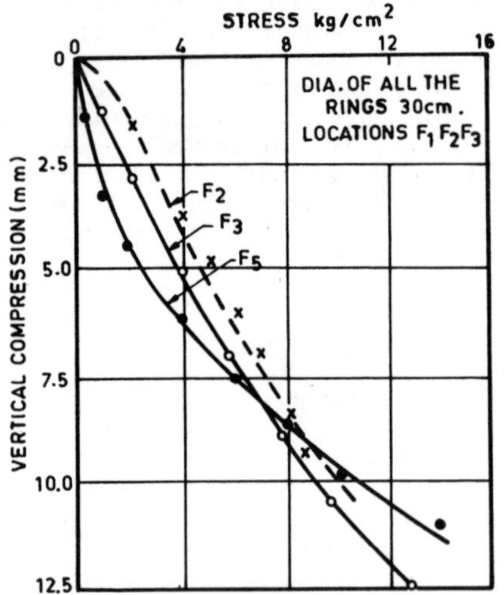

FIG.9. FIELD COMPRESSION TEST RESULTS

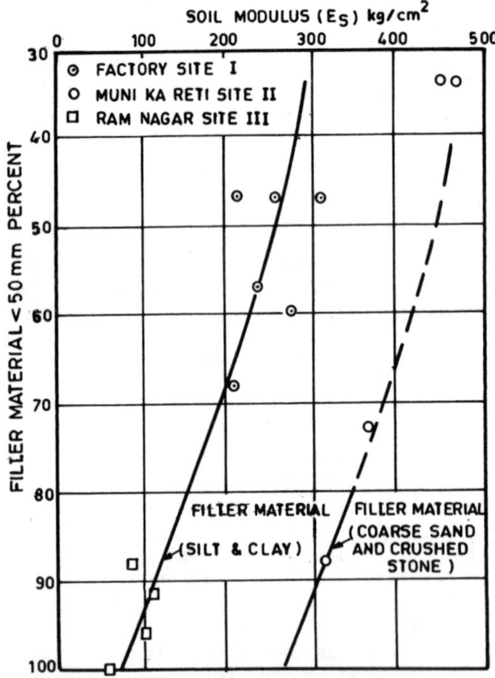

FIG.10. RELATION BETWEEN (E_s) VALUES AND AMOUNT OF FILLER MATERIAL (PERCENT)

Dynamic Probing Tests

Of the probing tests, static cone penetration tests (SCPT) is a versatile test and is a popular one. However, the test equipment requires anchoring to the ground which may be difficult in gravelly deposits and need a high capacity machine (15-20t). In the case of dynamic probing tests, it has been said that the test is not suitable for gravelly soils, particularly when the grain size approaches the size of the penetrating point (Meyerhof 1956). This observation has been made with respect to standard penetration test (SPT). Recently, the standard penetration value N_{SPT}(corrected) was related to the soil deformation S, (inches), width of the footing B, (ft) and the applied intensity of stress p (t/ft^2) for sandy soils with gravel. The relationship is expressed by equation 3 (Meyerhof (1974)

$$S = \frac{2\ p(B)^{0.5}}{N_{SPT}} \qquad \ldots 3$$

The computed settlement is to be doubled for silt deposit and further increased by 100 per cent for the presence of water table.

When the over-all size of the penetration tool is of the order of 50mm where a cast iron cone, at the end of a drill rod is driven with blows of standard drop weight 63.5 kg is possible even in coarse grained soils as there is not taking in of soil as in SPT sampler. The cumulative cone penetration value N_c has been related to allowable soil pressure for gravels and gravelly soils. (Mohan, Narahari and Rao, 1971) and is expressed by eqn 4.

$$q_a = \frac{1}{2.54}\ \frac{N_c\ S_a}{D_c\ B_f} \qquad \ldots 4$$

Where q_a is the allowable load (t/m^2) for a deformation S_a (cm), N_c is the cumulative number of blows corresponding to a depth diameter ratio (D_c/B_c, D_c in meters and B_c is the cone dia. in meter (0.0625m). B_f is the width of the footing. Comparision with observed values of allowable loads for footings upto 1.2m size showd good correlation coefficient equal to 0.9.

For fine and medium sand deposits using 62.5mm cone with 60° apex angle Fig. 11,

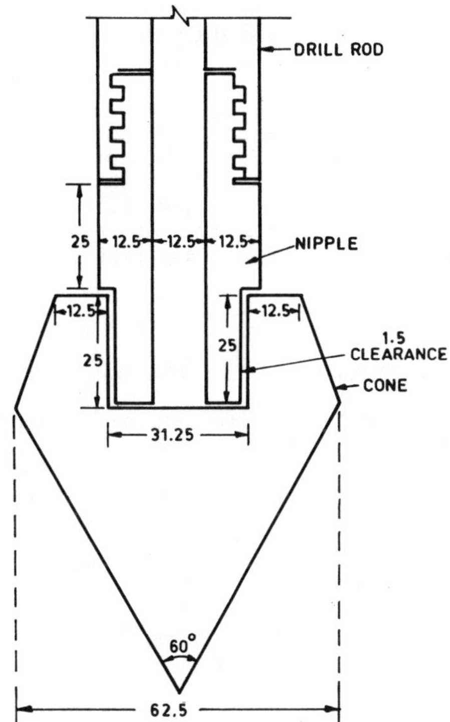

FIG.11. DIAGRAMMATIC SKETCH OF 625mm DIA. CONE

the dynamic cone penetration value N_c was related to N_{SPT} values (Mohan et.al 1970),(IS: 4968 - Pt II - 1976). The dynamic cone penetration test can be used upto 9m in silty sand (ML) and (SP) soil group. The relation with N_{SPT} is expressed by eqn 5 and 6

$N_c = 1.5\ N_{SPT}$ for 1-4m depth ...5

and $N_c = 1.75\ N_{SPT}$ for 4-9m depth...6

Typical test results at various sites in gravelly soil is shown in Fig 12a. and in pure gravel in Fig 12b (Narahari and Rao 1967).

Present Study

In the earlier study (Mohan, Narahari and Rao 1971) conducted dynamic cone penetration tests at several sites on gravelly soils and silty sand. The cumulative number of blows N_c with depth were plotted. The point B Fig 12a where the curve showed a distinct break was taken as D_c/B ratio and the corresponding cumulative number of blows N_c at point A was taken as penetration values.

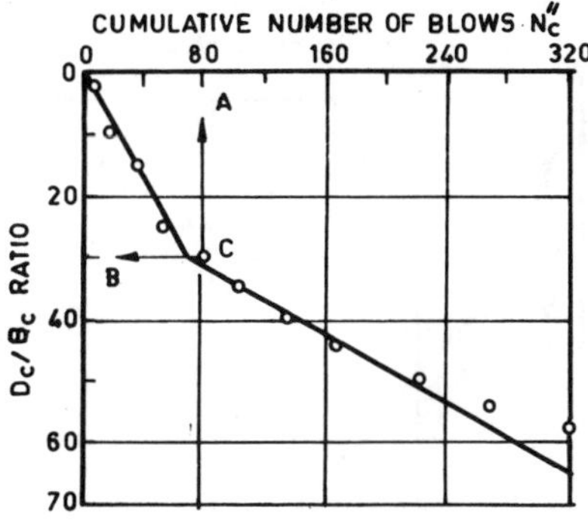

FIG.12 (a) TYPICAL PLOT OF D_c/B_c AND N_c''

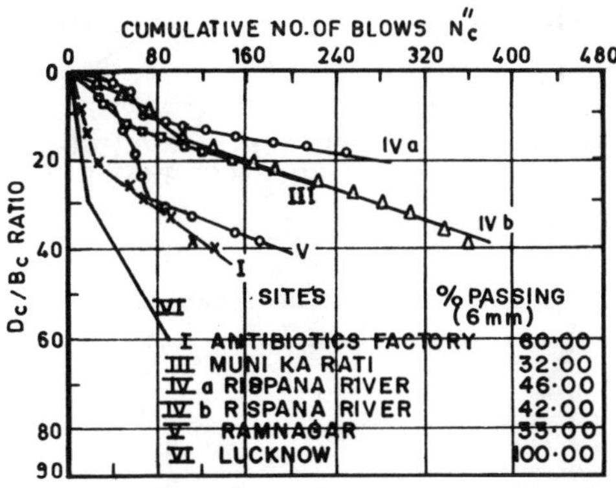

FIG.12 (b) RELATION BETWEEN NO.OF BLOWS AND DEPTH / WIDTH OF CONE RATIO AT VARIOUS SITES IN THE FIELD

These two parameters were used in equation 4 to predict the allowable loads for various sites, Fig 12b. The empirical equation 4 has served well during last decade during routine site investigations. However, during recent past, while dealing with gravelly soil deposit sites, in river valleys, the authors experienced that the relationship between cumulative number of blows N_c from the dynamic cone penetration tests and D_c/B ratio , does not give a unique break as in Fig 12a. point C, but in many cases often two or three breaks are observed point C_1 and C_2 in Figs 13a, 13b and also in several cases only a straight line is observed without any break point. This led to an ambiguity as to what value of

N_c is to be taken for the computation of allowable soil pressure in eqn 4.

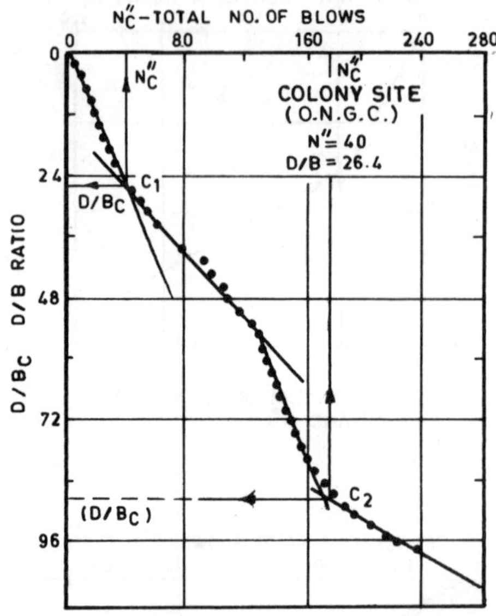

FIG.13(a).RELATION BETWEN N_c AND (D/B_c) RATIO SHOWING MORE THAN ONE BREAK POINT

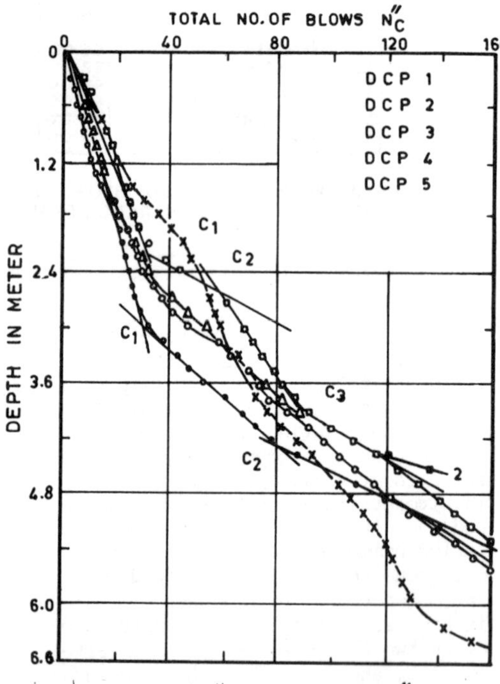

FIG.13(b).TOTAL NUMBER OF BLOWS N''_c AND DEPTH RELATIONSHIP

In view of this fact all the available data, was re-assessed by plotting dynamic cone penetration values N_c (per 30cm penetration) under the impact of 62.5 kg hammer falling through 75cms, with depth. The average N_c values below the level of the footings, upto a depth equal to 2B, was taken from several sites. These N_c values were used to predict the allowable soil pressure by a simple equation which is a modification of eqn 4 as expressed below:

$$q_a = K\,N_c \qquad \ldots 7$$

DISCUSSION OF THE TEST RESULTS

The available data from (Mohan, Narahari and Rao 1971) and the present series of tests consisting of dynamic cone penetration tests and insitu load test on footings and plate bearing tests were analysed in the light of eqn 7. The dynamic cone penetration test results on two sites are shown in Fig 14 for gravelly soils and Fig 15 for silty soil deposit. The corresponding grading curves for all the cases are shown in Figs 16 (a to d)

The type of soil whether gravelly or silty sand or soft clay, the type of the filler material and the uniformity coefficient (u) obtained from the grading curves from different sites demonstrate a definite influence on the value of constant of correlation k in the equation 7, Table 1.

The study of the table 1 indicate that the value of k for soft marine clay and saturated silt is 0.5 while those of silty sand (ML) and sandy silt (SM,SP) group the value of k is found to be 1. On the other hand in the case of gravelly soils with silty sand or silty clay as filler material the value of k may be taken as 1.5. However, when the filler material is coarse to medium sand or crushed stones which has low compressibility - the k value works out to be 3.

The computed values of allowable soil pressure (q_a) in t/m^2 from eqn 7 using suitable value of k for different sites, was compared with the stress- deformation behaviour of the soil deposit settlement corresponding to the computed allowable pressure from the insitu load test was observed to be well within 40mm.

Table 1 - Value of k for different filler material

Site	Type of soil	Filler material	Uniformity coefficient	Value of k	Remarks
Laboratory	Pure gravels	-	-	10	
Antibiotics factory	Gravelly soil	Silty sand	-	3	
Muni ka Reti	-do-	Coarse to medium sand	6	3	
Ram Nagar	-do-	Sandy silt	2.27	1.5	
Rispana	-do-	-do-	6	1.5	
IPE Dehradun	-do-	-do-	35-60	1.5	
Tel Bhawan	-do-	Sandy silt	20	1.5	
Tel Bhawan	Silty soil		23	1.5	
Sarsawa	Sandy silt			1.0	
Badaun	Sandy soil			1.0	
Haldia	Silty soil/ Clayey silt/ Saturated silt			1.5	

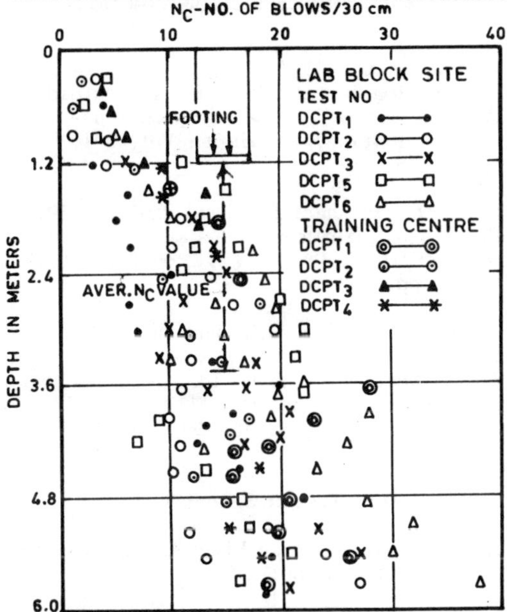

FIG.14. DYNAMIC CONE PENETRATION TESTS IN GRAVELLY STRATA

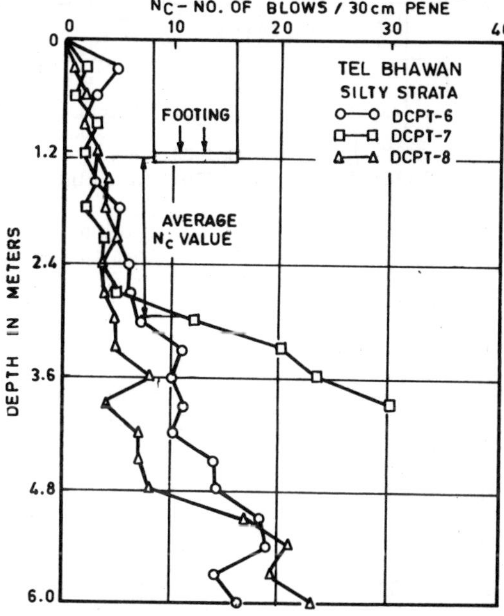

FIG.15. DYNAMIC CONE PENETRATION IN SILTY STRATA

CONCLUSIONS

The test results have demonstrated that :

1. Dynamic probing tests may be suitable and used in gravelly sub soil deposits when the grain size of the gravel does not exceed 100mm. It may also prove useful test in non-gravelly soils.

2. Analysis of the available data and the present series of tests on different sites have demonstrated that the average dynamic cone penetration value (N_c) measured within the depth of significant stress (equal to twice the footing width) may be related to the allowable soil pressure in the form of eqn 7. The value

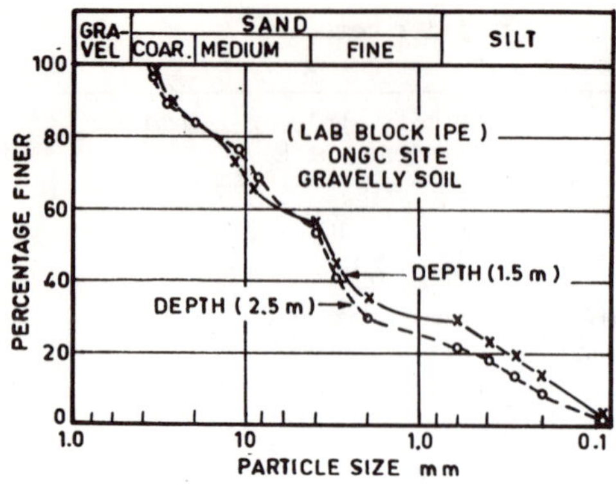

FIG.16(a).GRAIN SIZE DISTRIBUTION CURVE

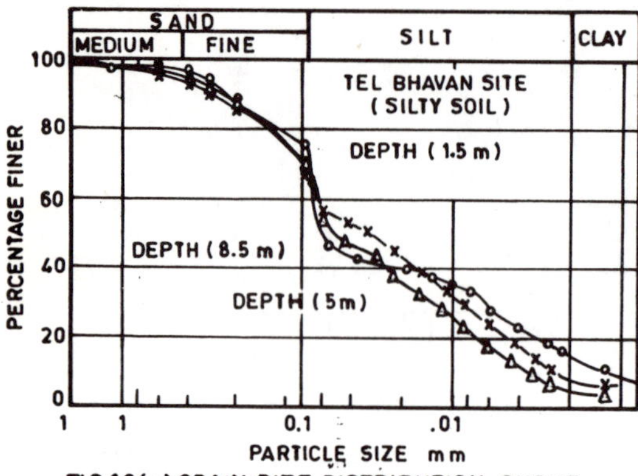

FIG.16(c).GRAIN SIZE DISTRIBUTION CURVE

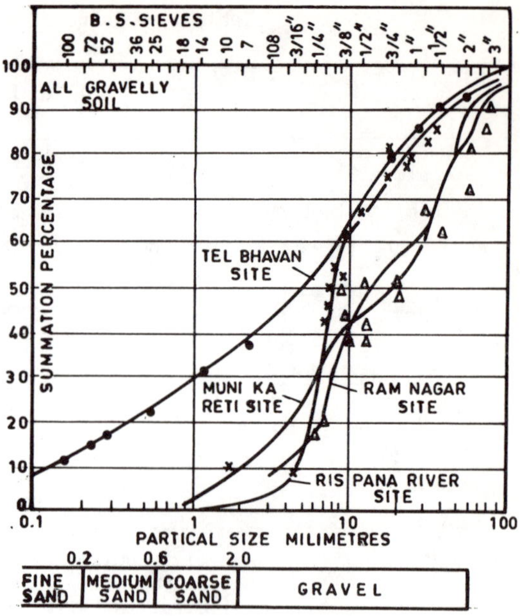

FIG.16(d).PARTICLE SIZE DISTRIBUTION

of empirical constant of correlation 'k' may suitably be selected from table 1 depending upon the type of deposit and the filler material.

ACKNOWLEDGEMENT

The work reported formed a part of normal research programme of the Central Building Research Institute and the paper is published with the permission of the Director.

REFERENCES

Deb A.K. Sharma, D. and Rao B.G. (1964)
Studies on Load Bearing Capacity of Gravelly Soil Stratum Including Boulders Ind. Jr. Tech Vol. II, No. 3, pp. 178-80

IS:4968 (Part-II) - 1976
Method for Sub-surface Sounding for Soils (First Revision)

Meyerhof G.G. (1958)
Penetration Tests and Bearing Capacity of Cohesionless soils, Jr. SM&FE Div, Proc ASCE, Vol 92, pp. 866

Meyerhof G.G. (1974)
Penetration Testing in Countries outside Europe. Proc. European Symp on Penetration Testing, Stockholm, Vol. 2.1, pp. 40-48.

Mohan D., Narahari D.R. and Rao B.G.(1971)
Field and Laboratory Tests on Gravel and Boulder Soils. Proc. IV ARC on SMFE Vol.1 No. 2, pp. 133-42.

Narahari D.R. Rao, B.G. and Jain R.C. (1968)
Field Shear Tests in Boulder Deposits Proc. of Symp on Earth and Rockfill Dams Talwara, Jnl. of INS of SMFE Vol. I, pp. 249.

Narahari D.R. Rao, B.G. and Jain R.C. (1976)
Insitu Shear Tests in River Gravel ISI Bulletin 28, pp. 170.

Rao B.G. Narahari D.R. and Balodhi G.R. (1981)
Footing Foundation in Bouldery Soil with Filler Materials. Proc. Symp on Engg Behaviour of Coarse Grained Soils Boulders and Rocks Hyderabad Vol. 1, pp. 93-98.

Vesić A.S. (1975)
Bearing Capacity of Shallow Foundations Foundation Engg. Hand Book Edited by H.F. Winterkorn and H.Y. Fang, pp. 121.

Proceedings of the Second European Symposium on Penetration Testing / Amsterdam / 24-27 May 1982

The use of dynamic soundings in evaluating settlements

L.H.SWANN
Dames & Moore, London, UK

1 INTRODUCTION

The light dynamic sounding machine is used extensively in Europe. Generally it is used qualitatively rather than quantitatively. However, correlations exist (DIN 4094, 1974) between the dynamic sounding and the standard penetration test (SPT) that suggest that, like the SPT, the dynamic sounding may be used in the evaluation of settlements.

This paper describes how the light dynamic sounding was used at a site near Jubail, Saudi Arabia, to determine the need for soil pretreatment and, subsequently to help control the soil pretreatment.

2 SITE CONDITIONS

The site covered an area approximately 1.5 x 1.5 kilometres adjacent to the Arabian Gulf. Prior to construction starting, the site consisted of an area of sand dunes along the coast extending 600 metres inland, with futher sand dunes along the southern border of the site extending 200 to 600 metres into the site. The sand dunes consisted typically of fine to medium, dense to very dense sand at elevations varying between +3.0 to +7.0 metres above sea level.

The remainder of the site, at an elevation of +1.1 to +2.0 metres, consisted of a sabkha (salt flat), a highly saline soil with properties governed by an evaporative environment (Akili & Fletcher, 1979). Typically the sabkha had a surface crust a few centimetres thick. Underlying the crust were up to 5 metres of very loose, fine sands, silty sands and very soft silts, and clays, the sands being predominant. There was considerable variation of soil type and density in lateral and vertical directions within the sabkha.

Underlying the sabkha and the sand dunes at an elevation of -2.0 to -4.0 metres were highly fractured, moderately strong calcarenites.

The first phase of site development involved regrading the sand areas and placing a mechanical and hydraulic fill over the sabkha areas to a minimum elevation of +3.5 metres. The maximum fill thickness was 4.0 metres. Investigations during construction demonstrated that the fill was very dense to dense.

The ground water was generally at an elevation of +0.45 metres, though in localised areas it was as high as +2.2 metres on completion of the hydraulic filling.

A typical log of boring along with a series of dynamic soundings from within 5 metres of the boring is presented in Figure 1. The boring and soundings demonstrate the variability described above.

2 SOILS INVESTIGATION PROCEDURES

2.1 Initial Investigations

Prior to construction starting, investigations were performed using borings. The preliminary investigation indicated that large settlements might be expected for shallow foundations. Thus more extensive investigations were performed by borings and Dutch Cone Penetration Tests. These second investigations confirmed that, for structures on shallow foundations within the origianl sabkha area, large settlements were likely. Accordingly, a further study was per-

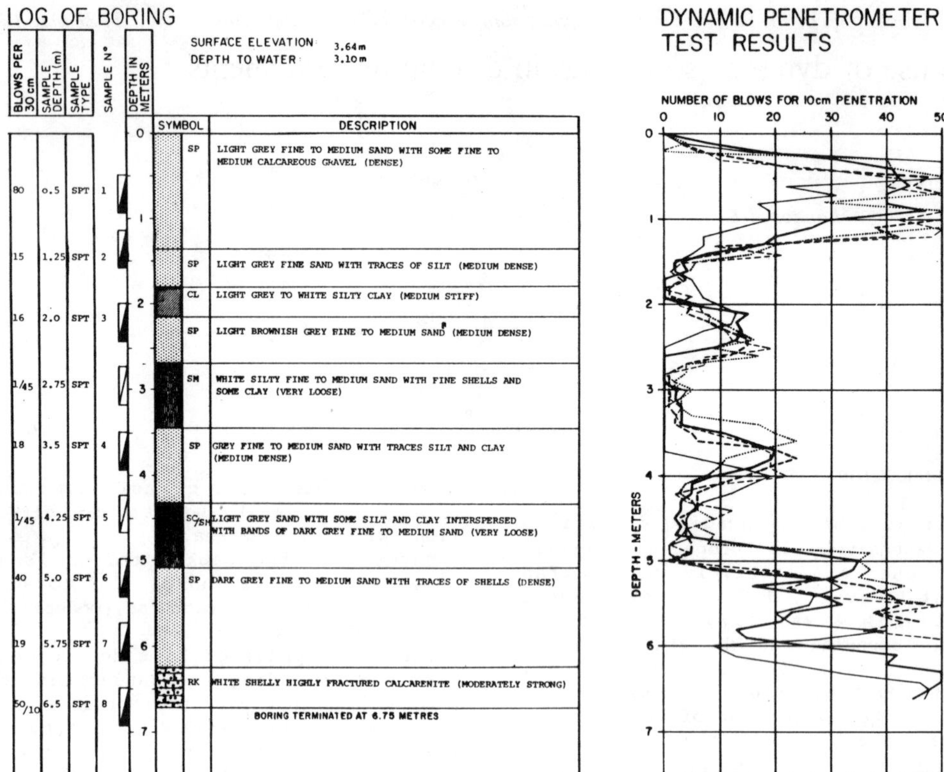

Figure 1. Typical log of boring and dynamic penetrometer test results.

formed using a trial embankment, trial footings and the dynamic sounding machine. A grid of dynamic soundings at 100-metre spacings was performed to define initially an area where soil improvement or deep foundations might be required.

2.3 Detailed Investigations during Construction

Having confirmed that soil treatment or deep foundations were required over parts of the site, trial embankments were constructed and trial areas of dynamic compaction were performed. A dynamic sounding was performed before construction and after removal of most trial embankment. Pressuremeter testing and soundings were performed before and after the dynamic compaction trials.

Subsequently, to define the amount of treatment required on each building area, dynamic soundings approximately 1 per $1000m^2$ were performed. Where areas were selected for treatment by dynamic compaction, pressuremeter tests were also performed prior to treatment. Dynamic soundings were also performed after the soil treatment was completed to confirm the success of the treatment. Approximately 700 dynamic soundings were performed over the site.

4 SETTLEMENT ANALYSIS USING DYNAMIC SOUNDINGS

Thirteen Dutch Cone Penetration Tests were located at or close to the original grid of dynamic soundings. A comparison of the two tests indicated that at this site, particularly within the sabkha areas, there was a good correlation between the two tests. As shown on Figure 2, an approximate relationship of:

$$N_{10} = 2C_r$$

where C_r = static cone resistance in MN/m^2

N_{10} = dynamic cone resistance in blows/10cm penetration

is valid for this site.

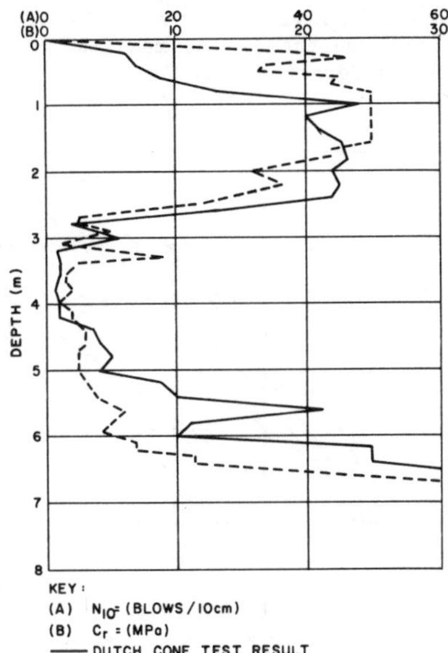

Figure 2. Relationship between Dutch Cone Penetration test and Dynamic sounding.

Accordingly it was considered that a method of analysis developmed for the static sounding might also be used for the dynamic sounding.

Since the site consisted generally of a thin, highly compressible layer (compared with loaded area) sandwiched between relatively incompressible layers, it was considered that the method of settlement analysis suggested by De Beer & Martens (1957) was the most appropriate Thus the settlement, δ , may be estimated using the relationship:

$$\delta = \frac{H}{C} \log \frac{p_o' + \delta p}{p_o'} \text{ mm}$$

and compressibility $C = \dfrac{K \times N_{10}}{p_o'}$

where H = layer thickness (metres)
p_o' = initial effective stress at centre of layer (kN/m^2)
N_{10} = average dynamic resistance (blows/10cm)
δp = change in stress at centre of layer due to loading
K = constant (relating compressibility to dynamic resistance)

The change in stress δp was estimated using the Westergaard (1938) stress analysis, which is considered to be most appropriate for layered or non-isotropic soils. At the site under discussion, this method also allowed a larger redistribution of stresses within the very dense upper fill.

Layer thicknesses were selected as thin as possible in order to increase the accuracy of the analysis. A minimum layer thickness of 0.2 metres was used.

In several soundings, N_{10} values of 0 or 1 were obtained. Nearby borings indicated that in these cases the material encountered was probably clay. Where these low values were obtained, a compressiblity value of 0.010 m^2/MN was used, based on results obtained in oedometer tests.

The soundings were generally stopped after three refusals within the calcarenite (N_{10} > 50). To allow for the compressibility of the calcarenite, an N_{10} values of 50 was assigned to a depth of 15 metres. Below 15 metres, the rock was assumed to be incompressible.

4.1 Determination of K

The compressibility constant K was determined by back analysis of the trial embankments. Settlements of the trial embankments varied between 3 and 191 millimetres. The settlements were rapid, with a maximum time for 90% consolidation of 36 days. The average time for 90% consolidtion was 13 days. The back analysis was based on trial embankment settlement with 100% consolidation completed.

The first six trial embankments when back analysed gave an average value of K of 0.66 with a range of 0.36 to 0.77. This value was used in all subsequent settlement analyses. However, later subsequent analysis of all trial embankments yielded an average K value of 0.77 with a range of 0.23 to 1.76.

The K value was based on the total settlement of trial embankments with the total load imposed. Many of the trial embankments were constructed in two stages. Where analysis was performed using the settlement after the first stage, higher K values were observed. These values were generally 10% higher than for the whole embankment though in one case the value was as much as 100%. The slighlty higher K values on the first stage loading are thought to be due to a small degree of overconsolidation of the site soils.

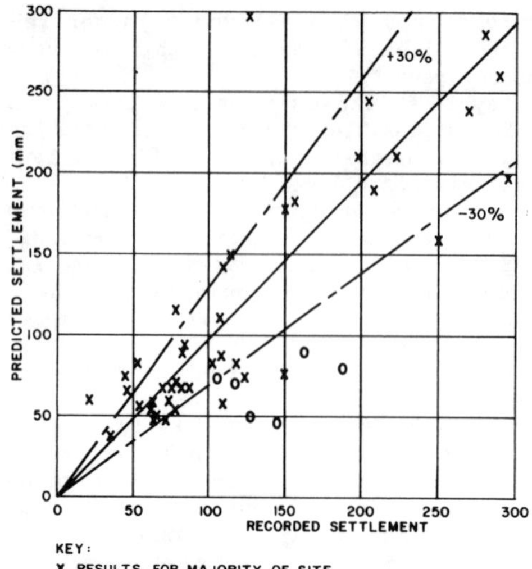

Figure 3. A comparison of predicted maximum preload settlements with recorded maximum preload settlements.

5 RESULTS OF ANALYSIS

The results of the settlement analyses performed using the dynamic sounding are presented on Figures 3,4 and 6.

Figure 3 presents a comparison of predicted with recorded maximum settlements for areas subjected to preloads. Although there is quite a large scatter of results, the mean ratio of predicted to recorded settlements is 0.98. Approximately 75% of all results where the settlements were underestimated by over 30% occur in one particular area of the site. In this area, three trial embankments indicated an average value of compressibility constant K of 0.3. The results for this area are indicated on figure 3.

Figure 4 presents a comparison of maximum recorded building settlements with settlements predicted using the dynamic sounding. The 'recorded building settlements' are based on observations during construction with 70 to 90% of the final load imposed, and are projected to give settlements on completion of construction. The results for buildings constructed following the performance of soil treatment are indicated separately.

For buildings constructed in areas where no soil improvements was considered necessary, the mean ratio of predicted to 'actual' settlements is 1.17 with 70% of all results within 30% of the predicted values. For 25% of the buildings monitored, the maximum settlement was underestimated.

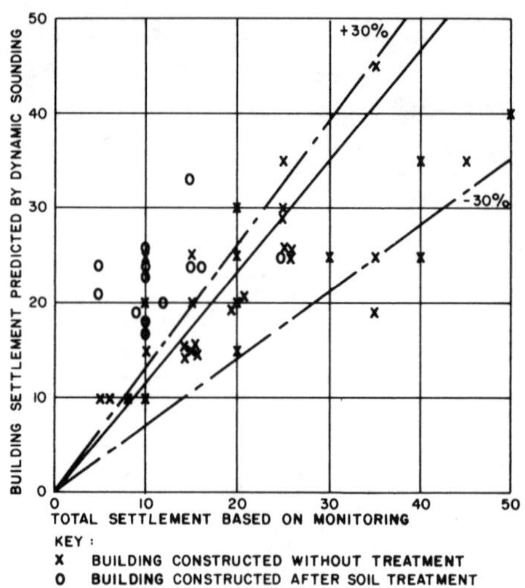

Figure 4. A comparison of maximum recorded building settlement based on observations with predicted maximum settlements.

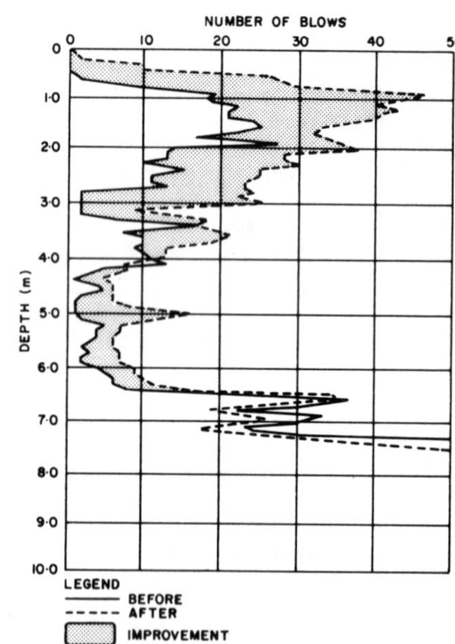

Figure 5. A comparison of dynamic sounding results before and after preloading.

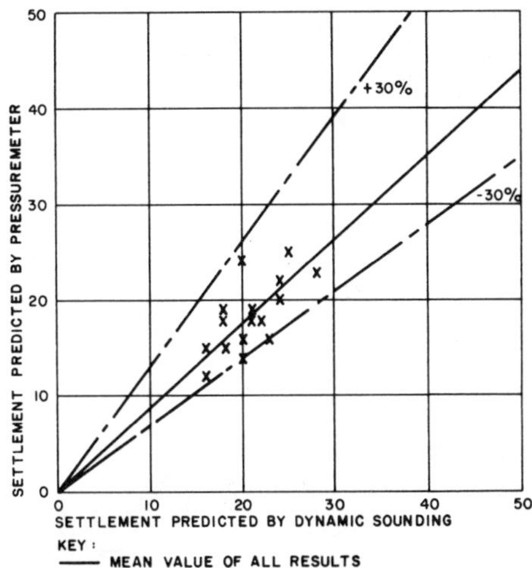

Figure 6. A comparison of settlement predictions using dynamic sounding machine with settlement predictions using a pressuremeter.

For buildings monitored after soil improvement was carried out, the method of analysis described above tended to overestimate settlement by at least 50% as shown in Figure 4. This demonstrates that the soundings do not fully reflect the degree of overconsolidation imposed by the soil treatment. However, as shown in Figure 5, the soundings did demonstrate that improvements of the soil had occurred. Out of 130 comparative soundings that were obtained, only four did not demonstrate an improvement in dynamic sounding characteristics after soil treatment.

Figure 6 presents a comparison of settlement prediction by two completely independent methods. Where dynamic compaction was performed, both the pressuremeter and the dynamic sounding were used as a form of quality control. The method of analysing the pressuremeter test results was that described by Menard (1975). Figure 6 shows that there is reasonable correlation between the two methods with the dynamic sounding predicting an average 11% less settlement than the pressuremeter.

6 CONCLUSION

The dynamic sounding machine is a light weight machine that can be operated by relatively unskilled labourers at sites with difficult access.

This paper shows that the dynamic sounding can be used with reasonable accuracy for predicting settlements of normally consolidated soils. However, as a result of the large variation at the site and the variability of the results, the methods of analysis described above should be used with extreme caution at other sites.

REFERENCES

Akili W. & Fletcher E.H. 1979, Ground conditions for housing foundations in the Dhahran Region, Eastern Province, Saudi Arabia, Proceedings IAHS international conference on housing problems in developing countries Dhahran, Saudi Arabia.

De Beere E. & Martens A 1957, Method of computation of an upper limit for the influence of heterogeneity of sand layers in the settlement of bridges, Proc. 4th International Conference Soil Mechanics and Foundation Engineering, London..

DIN 4084 Part 2 1974, German standard, dynamic and static penetrometers; application and evaluation of results.

Menard L. 1975, The interpretation of pressuremeter test results. Soils 26, Paris.

Westergaard H.M. 1938, A problem of elasticity suggested by a problem in Soil Mechanics: soft material reinforced by numerous strong horizontal sheets, Contributions to the Mechanics of Solids, Stephen Timoshenko, 60th Anniversary Volume, The Macmillan Company, New York.

Proceedings of the Second European Symposium on Penetration Testing / Amsterdam / 24-27 May 1982

Cone penetration tests (CPT) on clay and silt

M.TAMMIRINNE & V.LEINONEN
Technical Research Centre of Finland, Espoo

1 SCOPE OF THE TESTS

In Finland, the in situ properties of the fine grained soils are determined mainly with weight sounding and vane tests. On the basis of the weight sounding it is possible to estimate the boundaries between soil layers differing clearly from each other and the surface of the firm bottom layer. To some extent it is also possible to estimate the strength of the soil layers. However for design purposes the strength of the soil layers are usually determined with the vane tests, which again cannot indicate the boundaries between the soil layers.

The use of the cone penetration test (CPT) in practical soil investigations has been almost non-existent in Finland. About two years ago the Geotechnical Laboratory of the Technical Research Centre of Finland acquired equipment for CPT and this equipment has been used on fine grained soil. The aim of these penetration tests is to clarify the applicability of the CPT to determining of the properties of clay and silt layers. The hope is that the soils could be indentified and in situ strength values of the soil layers could be determined on the basis of the CPT in normal design cases.

This article presents results from some performed tests. As the material is very limited no generalizing or general conclusions have been made so far.

2 TEST METHODS

The CPT-equipment is shown in Fig. 1. Fig. 2a presents the in the soundings used tip (type Gouda) which is in agreement with the European Standard /2/. The sounding themselves have also been performed in agreement with the recommendation accepted by the ISSMFE.

Fig 1. CPT-equipment.

The weight soundings (WST) have performed in accordance the European Standard an equipment defined in the same Standard (Fig 2b).

The vane tests have been performed with an equipment presented in Fig.2c. The size of the vane was either 55 mm x 110 mm or 65 mm x 130 mm (H = 2D). The speed of rotation of the vane was about 0,1 °/s.

The soil samples for the laboratory tests were taken with a piston sampler (Swedish Standard Sampler ST II).

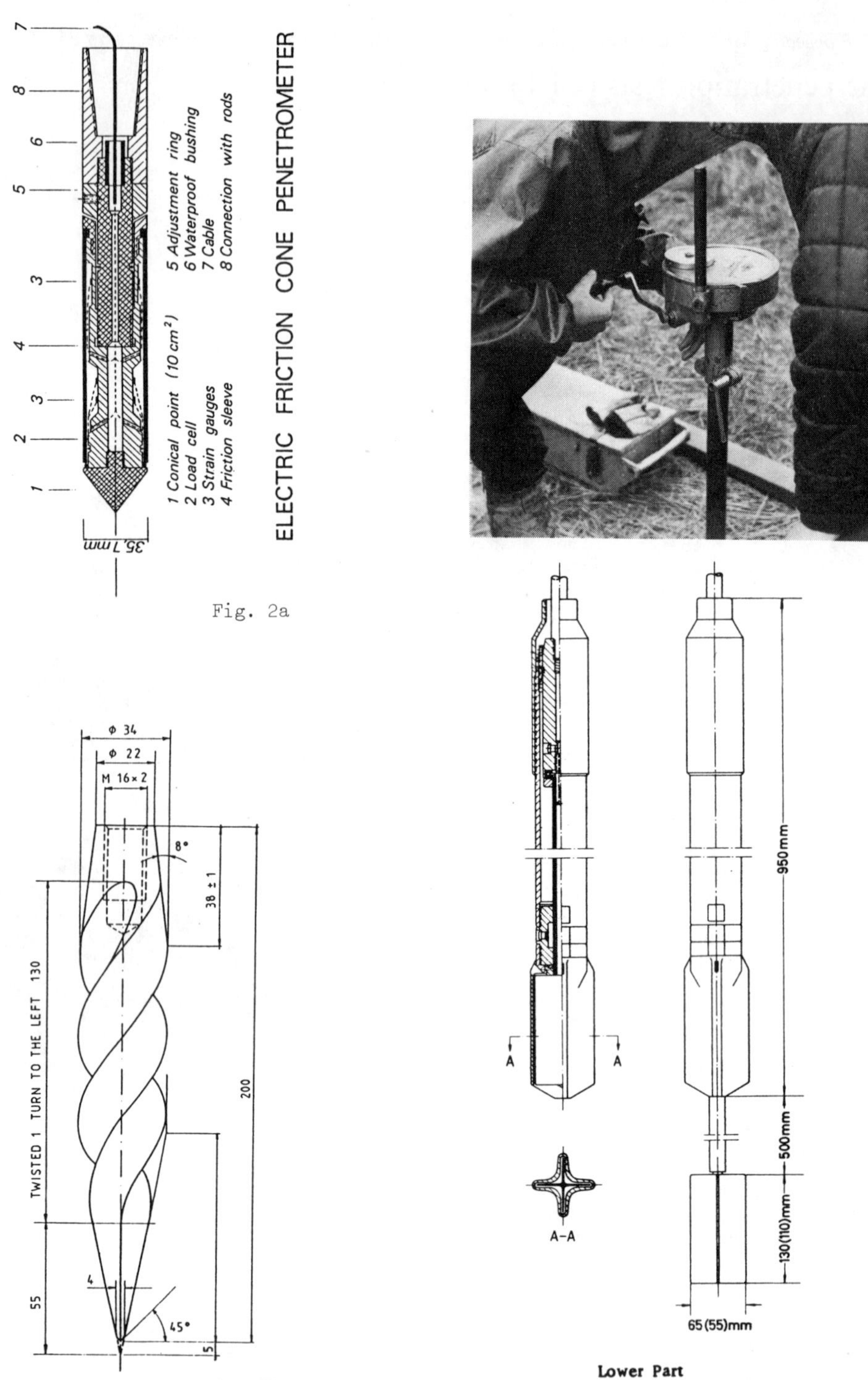

Fig. 2a

Fig. 2b

Fig. 2c

Fig 2. a) CPT-tip, b) WST-tip, c) Vane test equipment.

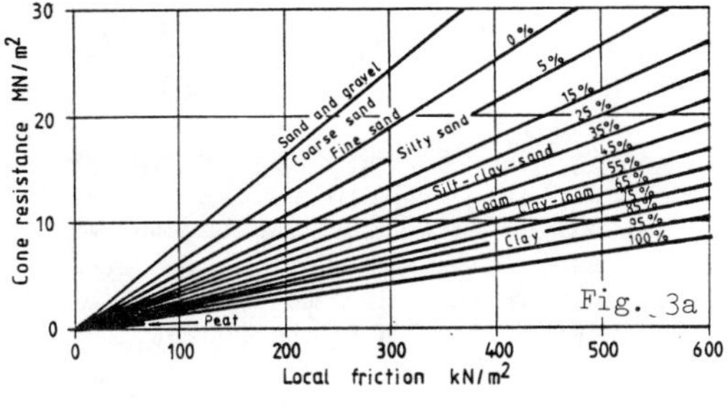

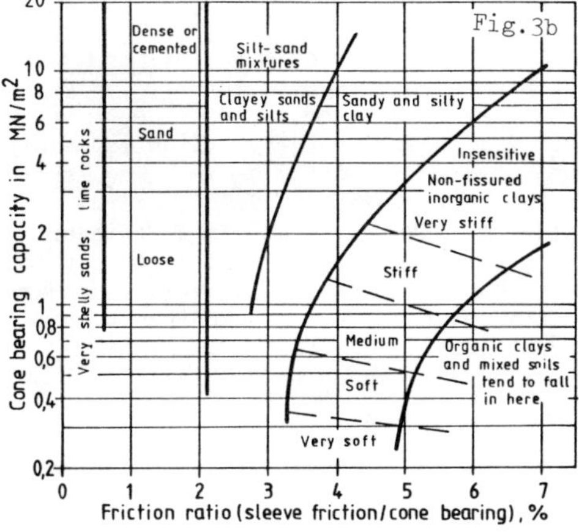

Fig 3. Soil interpretation from CPT-results
a) after Begemann /1/
b) after Schmertmann /4/

3 TEST SITES AND RESULTS

3.1 Salo

The total thickness of the clay layers on the test site is about 40 m. The upper part of the deposit is Littorina gyttja-bearing clay. Some geotechnical properties of the layer are presented in Fig 4a. Fig. 4b shows the results WST and CPT. Fig. 4b shows also the strength of the clay layer determined with vane tests and the shear strength ($q_c/24$) calculated on the basis of the cone resistance at the CPT.

Fig 4b shows that the skin friction at the cone penetration test in this type of clay layer is very small. For this reason the use of it would lead to entirely erroneous interpretations. The skin friction also shows the sticking of the sleeve to clay caused by the extension of the casing (interruption in the penetration).

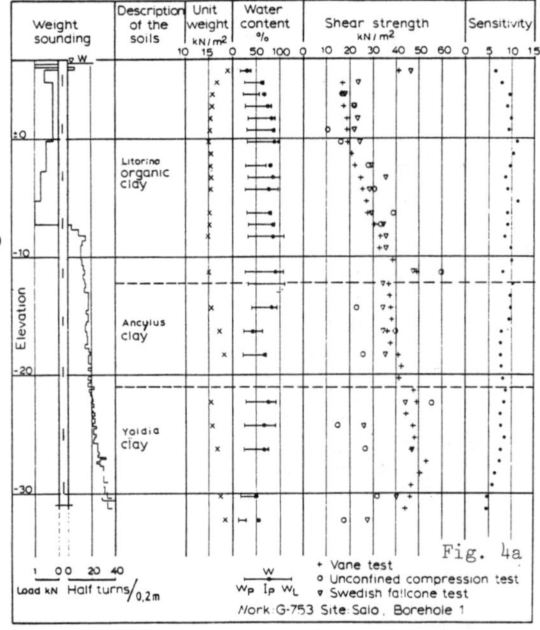

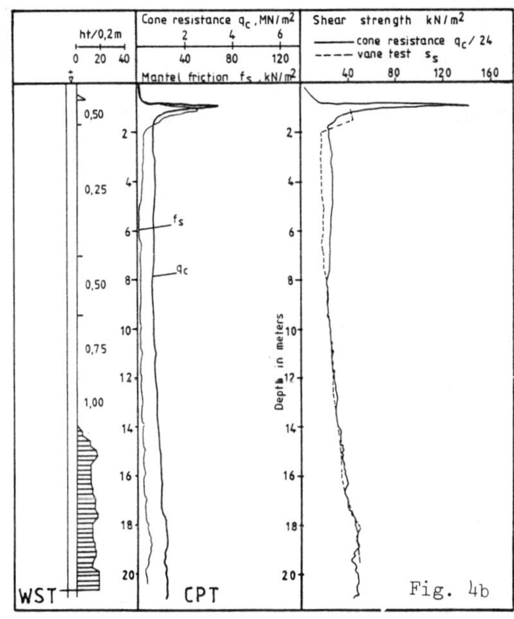

Fig 4. Test site Salo
a) Soil data
b) Testresults

The cone resistance at the CPT divided by the constant (= 24) corresponds well with the shear strength measured with vane tests. The difference between the shear strengths in the depth range of

2 - 8 m may be caused by the overconsolidation of the soil layers.

3.2 Vantaa

The soil layers to the sounding depth consist of clay. On the surface there is an about 3 m thick overconsolidated clay deposit the surface of which being dry crust. In greater depth the clay is normally consolidated partly varved Yoldian clay, the clay content being in places 60 - 90 %.

Some properties of the soil layers are presented in Fig. 5a. In Fig. 5b the results from WST and CPT are presented. At the CPT the skin friction is very small as in the above-mentioned case. In this case the data on the soil layers obtained from the point resistance at the CPT are not significantly more plentiful than the data obtained from the WST.

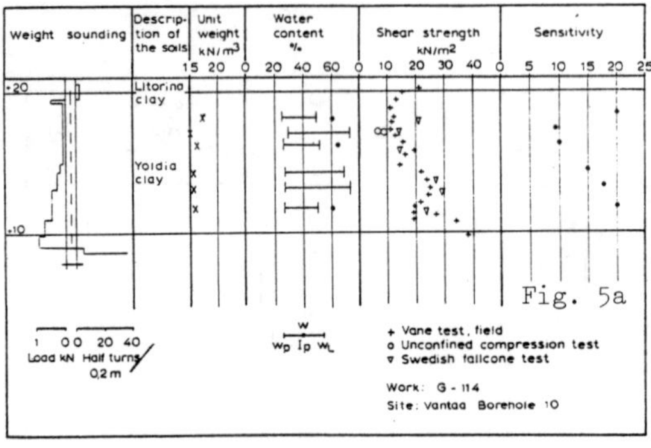

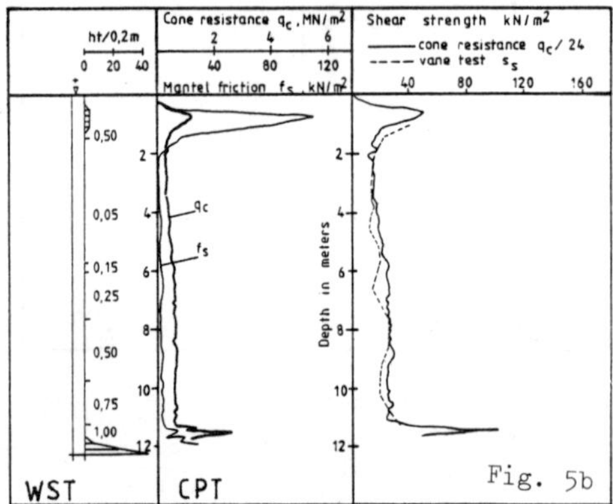

Fig 5. Test site Vantaa
a) Soil data
b) Testresults

The strength calculated on the basis of the cone resistance at cone penetration test ($q_c/24$) compared with the strength determined with the vane tests is presented in Fig. 5b. On the basis of Fig. 5b almost the same conclusions can be drawn as in the abovementioned case. However the strength obtained with the vane tests is lightly smaller than the strength obtained by dividing the cone resistance by the constant 24.

3.3 Lahti Site I

The test site is located in the marginal zone of the Salpausselkä esker. The surface layer of the deposit to a depth of about 3 m is Yoldian clay, which is mainly highly overconsolidated dry crust. The layers underneath represent sediments of the Baltic Ice Lake and consists of thinner layers with various clay and silt contents. The thickness of the soil layers is about 15 m to which depth the soundings were extended. Some properties of the soil layers are presented in Fig. 6a.

Fig. 6b presents, besides the results from WST and CPT, also the soil type identified on the basis of the relation between cone resistance and skin friction at CPT according to Begemann /1/ and Schmertmann /4/. Also soil types indentified on samples in laboratory are presented in Fig. 6b.

In this case the data on the varved structure of the soil layers obtained from the cone resistance at CPT are versatile compared with the data given by weight sounding resistance. The soil types identified on the basis of the cone penetration resistance are in good agreement with the soils indentified in the laboratory. Especially the q_c/f_s-curve gives a very good picture of the varved structure of the layer.

3.4 Lahti, Site II

This test site is also located in the marginal zone of the Salpausselkä esker. The layers under the firm dry crust layer are sediments of the Baltic Ice Lake, clay and silt. The properties of the soil layers are presented in Fig. 7a. The cone penetration tests have been extended to the depth of about 10 m. The results from the soundings are presented in Fig. 7b.

In general, the pictures of the varved structure of the ground given by the WST and CPT are similar. Naturally the cone resistance at the CPT reveals thinner

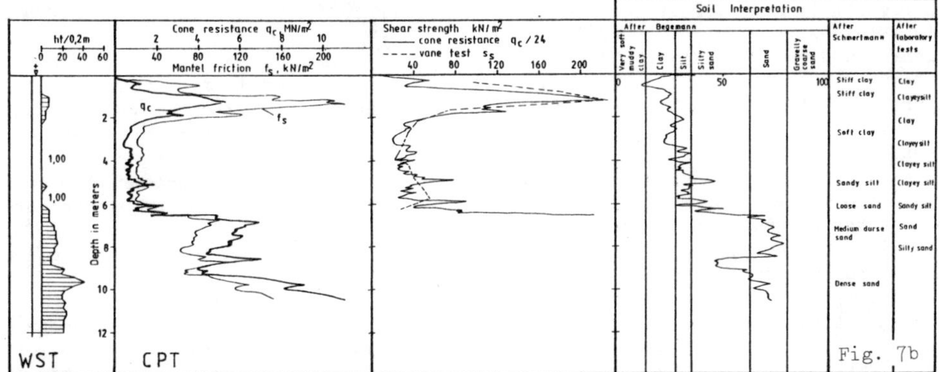

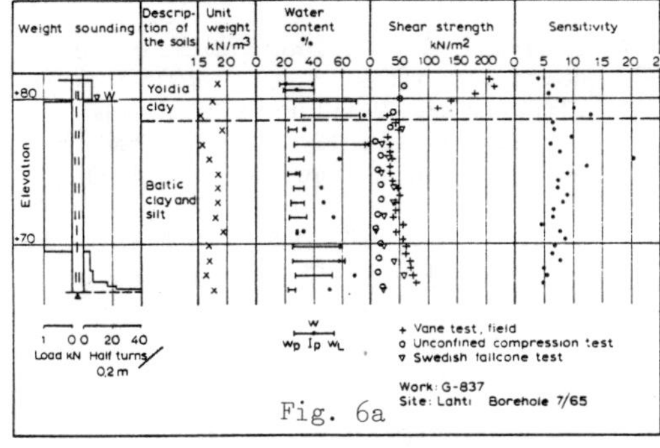

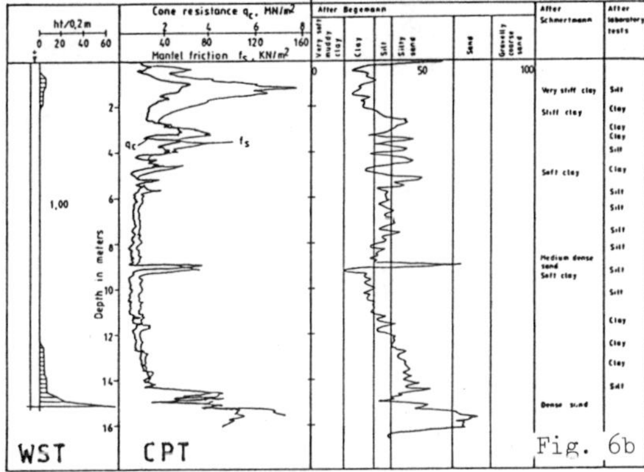

Fig 6. Test site Lahti I
a) Soil data
b) Testresults

layers than the weight sound.

Fig. 7b presents, besides the resistance at WST and CPT, the comparison between the shear strength determined on the basis of the cone resistance ($q_c/24$) and the strength determined with the vane tests, as well as the indentifications of the soiltypes according to Begemann and Schmertmann. Soil types in some depths identified on the samples are also indicated in Fig. 7b.

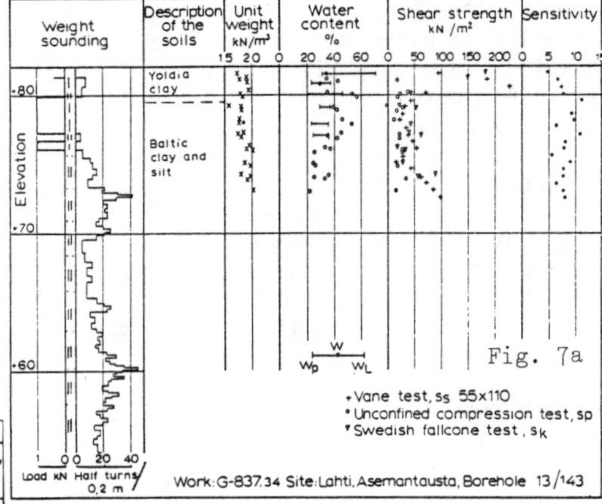

Fig 7. Test site Lahti II
a) Soil data
b) Testresults

The curves comparing the shear strengths show that the strength of the silty soil layers determined with the CPT corresponds fairly well with the strength determined with the vane tests. However, there is a much greater variation in the strength values of CPT depending on the "continuity" of the curve. The vane strength values are measured only at certain depths with depth intervals of 0,5 m.

The soil types identified with the CPT correspond well with the soil indentifications in the laboratory.

4 SUMMARY

The performed soundings indicate clearly that the information on the soil layers obtained from cone penetration tests (CPT) are remarkably more versatile than the data obtained from weight sounding (WST). Weight sounding has however the advantage of being able to penetrate into hard soil layers i.e. it has a wider range of use. The data on the geotechnical properties of the soil layers obtained from weight sounding are always very general and only indicating the order of size.

The shear strength of the soil layer, determined on the basis of the cone resistance of the CPT, cone resistance divided by the constant 24, seems to correspond well with the strength values determined with vane tests, both in clay and in clayey silt. Vivatrat /5/ has found that the value of the constant varies in the range 11 - 24. Sanglerat /3/ suggests for overconsolidated clays values in the range 22 - 26. In coarse silt the strength determined with the vane tests may often be erroneous because of disturbance occurring in connection with test. There is not enough experience of the applicability of the cone penetration test to estimation of the strength of coarse silts.

The studied cases have not given very encouraging evidence for determining the strength of the soil layers on the basis of the mantel skin firction.

In the identification of the soils the results from the cone penetration tests are, on the basis of the performed soundings, very useful and fairly reliable when the mantel skin friction is not very small. This, in spite of that the method of interpretation of Begemann has been made up for a mechanical tip with a form differing from the form of the tip used in the present study.

In very soft soils the indentification of the soils leads to incorrect result. On the other hand, when the mantel skin friction is found to be very small, this fact is enough for the soil being identified as very soft clay or gyttja.

All in all, the performed, so far very few, sounding tests have shown that the cone penetration test is a method to be considered and a very useful one both in determining the strength of fine soils and in identification of soils also in the Finnish geological conditions. To get more experience and to strengthen the grounds of interpretation the test soundings will be continued in fine grained formations with different properties.

REFERENCES

/1/ Begemann, H.K.S.Ph. 1969. The Dutch static penetration test with the adhesion jacked cone. LGM meddelingen. Deel XII - No 4 - April 1969:69-100.

/2/ Report of the sub-committee on the penetration test for use in Europe. 1977. International Society for Soil Mechanics and Foundation Engineering. Tokyo.

/3/ Sanglerat, G. 1972. The penetrometer and soil exploration. Elsevier Publishing Company. Amsterdam. London. New York. 464 p.

/4/ Schmertmann, J.H. 1969. "Discussion of Thomas (1968): Deepsounding Test Results and the Settlement of Spread Footings on Normally Consolidated Sands". Geotechnique 19(2):316-317.

/5/ Vivatrat, V. 1978. Cone Penetration in Clays. Massachusetts Institute of Technology. 427 p.

Proceedings of the Second European Symposium on Penetration Testing / Amsterdam / 24-27 May 1982

Dynamic probing and practice

E.WASCHKOWSKI
Laboratoire Régional des Ponts & Chaussées, Blois, France

1. INTRODUCTION

The dynamic probing is undoubtedly one of the oldest method of site investigation, which arises from wood pile driving method used several centuries ago.
After several years of research, we can put forward the factors which disturb the dynamic penetrometer tests, mainly in cohesive soils, and we hold up some simple means to neutralize them.
So, dynamic probing can become a full field test of soil mechanics, and has to find some applications within the categories compatible with its performances. Nevertheless, it is necessary to standardize the equipments and the operating procedure, if we want that the unit dynamic point resistance, qd, becomes a geotechnical parameter characterizing soils and may be indentical for all soil mechanics engineers.

2. USING CONDITIONS OF THE DYNAMIC PENETROMETER

If we want to consider dynamic probing as a full field test, it must be put under operating rules and,its possibilities and limits of use must be defined.

2.1 Operating rules

The dynamic penetrometer tests must be so conducted that they give the unit dynamic point resistance characterizing perfectly a slice of soil at any depth without any influence of the above soil layers, and of the equipment conception.

To achieve this aim, one must use an enlarged point with either casing pipes or bentonite injection. This latter, which consists of a simple filling of the annular void between the rod-side and the sounding wall, is easier (Fig.1) than operating casing pipes, which require particular elaborated equipment. Moreover, in cohesive soils, casing can be stopping by an excessive lateral friction resistance and it is not ever easy to pull it out.

Fig.1 Equipment for bentonite filling.

Some dynamic penetrometer test are conducted without casing pipes or bentonite filling. In this case, one recommands the rotation of the rods to measure the parasitic friction resistance. This operating process seems to us as ineffectual because it has many defects among others:

- the rod rotation is not uniform with the depth
- the torsional skin friction is nearer a static than a dynamic loading. We know that for these two loading conditions, the soil behaviour around the rods is very different
- the rod rotation does not take into account dynamic parasitic effects such as the increase of lateral stresses above the point and the falling of the soils in the sounding annular space.

Moreover, we recommend to adapt the impact energy to the soil resistance by modifying the hammer mass and keeping constant the height of fall.

2.2. Possibilities of the dynamic penetrometer

The dynamic penetrometer offers some interesting characteristics :
- the equipment can be dis mounting in independant elements easy for carrying.
- its setting up is possible in marshy or hilly sites
- it does not need a heavy anchor block as the static cone penetrometer
- it can be use in any soils without boulders
- it is impossible to remould the soils by a previous boring as for the standard penetration test

2.3 Limits of use

Now, with current equipments, one can compute the only one parameter, qd, which does not allow to characterize the soil nature. This limit can be removed with a more elaborated equipment.

A second limit concerns the probing depth according to the real result reporting. Indeed, we think that beyond a depth of 25 or 30 m, it is necessary to take into account the effect of the extension rods on the transmission of the impact energy and on the point behaviour.

3. SITE INVESTIGATIONS

The dynamic penetrometer is well adapted for preliminary investigations and can bring a certain contribution in :
- site zoning, in definition of soil aptitude to bear some types of buildings or in selection of foundation desings
- definition the position of bedrock surface to supplement some geophysical tests and/or some borings
- preliminary study of a landslide to set up geometry of soil layers and their density or consistency (Fig.2) and to place field measure apparatus such as piezometers, inclinometers ...
- detection of limestone cavity collapses (Fig.3) and areas of remoulded soils. Nevertheless, in this particular case, the lost point must be well fitting and possible bentonite sinks recorded. The dynamic probing can also be alternating with percussive sounding drilling with recording drilling parameters.
- preliminary investigation of marshy sites, where dynamic boring can complete tests conducted with a hand static penetrometer (Fig. 4).

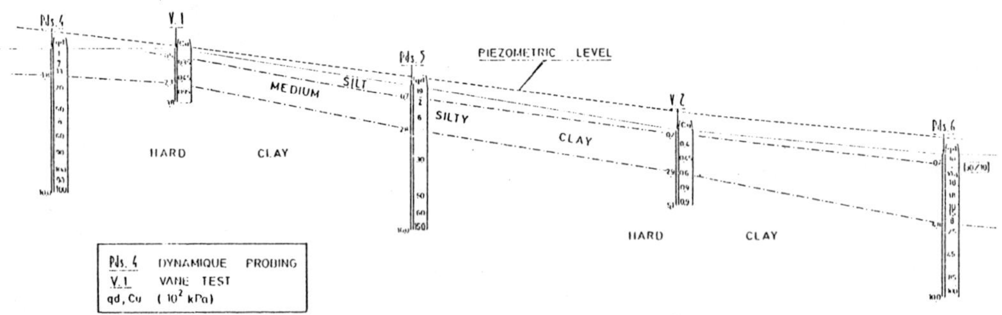

Fig.2 Investigation of a superficiel landslide

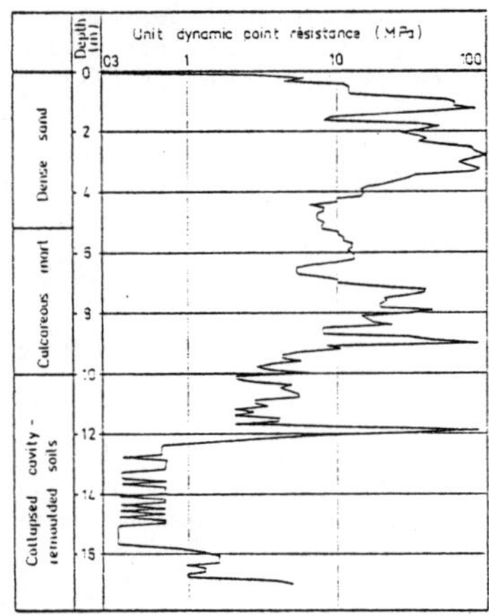

Fig.3 Detection of remouled soils in collapsed cavities of a karstic site.

4. DETERMINATION OF SOIL PROFILES

To determine soil profiles, it is necessary to can compute, qd, with a significative manner, in consideration of which one can :
- define the geometry of layers (Fig.4) and draw up horizontal correlations between soundings considering on the one hand the values of qd, and on the other hand , for exemple, the naturel radioactivity variation of soils.
- verify soil improvement or compacting effect
- measure characteristics of gravelly and stony soils for which the use of the static cone penetrometer is not recommended.

5. PREVISION OF STIFF SOIL AND WEAK OR FISSURED ROCK PENETRABILITY BY SHEETPILES AND DRIVEN PILES

For the prevision of soil penetrability by sheetpiles and driven piles, the dynamic penetrometer has a preferential position, because it acts upon soils in the same conditions that sheetpiles or driven piles, except the scale effect.

However, for such a study, we counsel two types of dynamic tests :

- a test with a point with the same diameter that rods. So it will be able to show up the magnitude of the frictional resistance, by the measurement of the total dynamic resistance

- a test with an enlarged point and a bentonite annular space filling, which characterizes the resistance of different soils layers and makes easier the estimation of the minimum length of driving elements, or the risks of an untimely isolated or systematic refusal.

Nevertheless, the tests must be conducted with an impact energy which is almost adapted to the soil resistance so as to achieve the prescribed depth.

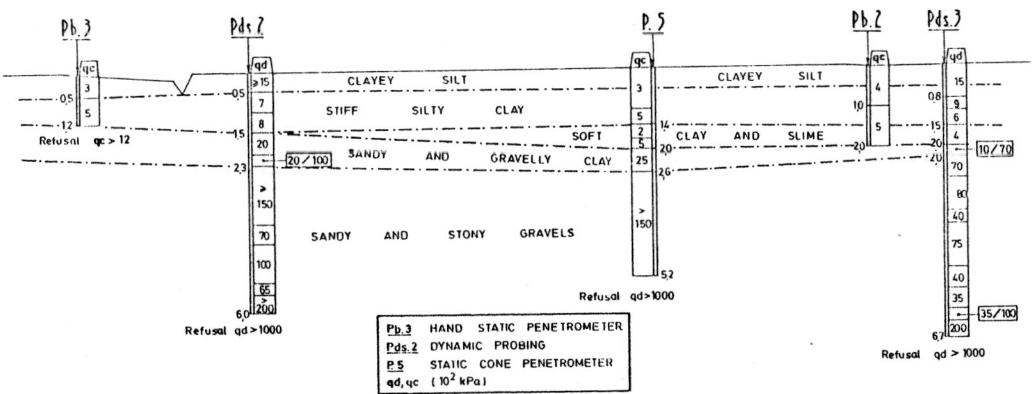

Fig. 4 Investigation in soft soils with hand static penetrometer, static cone penetrometer and dynamic probing

Thus, the dynamic penetrometer tests allow two approaches to the problem :

5.1 A total characterization

So, with a standard dynamic cone pentrometer one can compute :

- the unit dynamic point resistance which allows to define a depth which will be undervalue of 1 to 2 m for sheetpile and steel H piles,

- the total dynamic resistance which allows to value the depth of an displacement pile refusal. We have noticed that, a displacement pile and a dynamic penetrometer fitted with a point having the same diameter that rods, present generally a not very different refusal level,

- the unit dynamic point resistance which allows to estimate the soil heterogeneities and to select the setting up of one or several test driving areas.

5.2. A specific characterization

If one makes impact energy vary, one can value its effect on the soils resistance.

Nevertheless, in fissured rocks it may be useful to perform pressuremeter test which allowto specify the penetrability.

6. FOUNDATION BEARING CAPACITY

Presently, there are not satisfying methods to estimate foundation bearing capacity, directly from dynamic cone penetrometer data.

6.1 Possibilities from dynamic probing

We propound, according to the state of the art, to limit the dynamic probing use :

- to select foundation designs and to value summarely the bearing capacity in preliminary studies.

- to estimate foundation bearing capacity when the safety against failure is large, the sounding is deep enough and the settlement magnitude is considered as negligible. If one of these conditions is not satisfied, it will have to be done supplementary more elaborated tests

- to complete pressuremeter tests, so that their number may be restricted without trying and doing thoughtless extrapolations.

6.2. Correlations

From the great number of tests performed on the same sites with :

- a dynamic penetrometer Sermes, with an enlarged point and bentonite filling,

- a field vane test apparatus,

- a static penetrometer with a fixed or free point,

- a standard pressuremeter.

we can have established variational intervals of the ratio between the principal parameters, as summary in table 1 :

- qd : dynamic unit point resistance,

- qc : static unit point resistance,

- Pl : pressiometric limit pressure,

- E_M : pressiometric modulus,

- Cu : undrained cohesion measured by field vane test.

The aim of table 1 is to show that there are some relationships between the parameters measured by the different field tests.

Table 1 shows also that in comparison with static cone penetrometer, the dynamic probing gives smaller values in dense sands and gravels, equal values in normally consolidated clays, silts and slines, loose and medium sands and higher values in overconsolidated clays and silts. For this latter case, one must take into account the cylindroconical point of the dynamic penetrometer which mobilizes some frictional resistance.

Table 1 - Ratio between qd, qc, Pl et E_M according to the nature and the natural state of soils.

Overconsolidated clays and silts.	$1 < \frac{qd}{qc} < 2$ $3 < \frac{qd}{Pl} < 5$ $0.2 < \frac{qd}{E_M} < 0.4$
Normally consolidated clays, silts and slimes ; loose and medium dense sands.	$\frac{qd}{qc} \simeq 1$ $1.4 < \frac{qd}{Pl} < 2.5$ $0.4 < \frac{qd}{E_M} < 0.5$
Dense sands and gravels	$0.5 < \frac{qd}{qc} < 1$ $5 < \frac{qd}{Pl} < 10$ $0.4 < \frac{qd}{E_M} < 0.5$

The values presented in table 1 underline the importance of the nature and the natural conditions of soils.

Nevertheless, to buttress up our argument which consists to say that the dynamic penetrometer can be performed in any soils, we give on the Fig. 5 - 6 and 7, examples of correlations between Pl and qd or Cu and qd in clays and sands.

The correlations are hopeful because they allow to place the dynamic penetrometer in comparison with the othed field tests, with a satisfactory manner.

7. CONCLUSIONS

We think to have persuaded of the dynamic penetrometer utility, for solving the problems of soil mechanics.

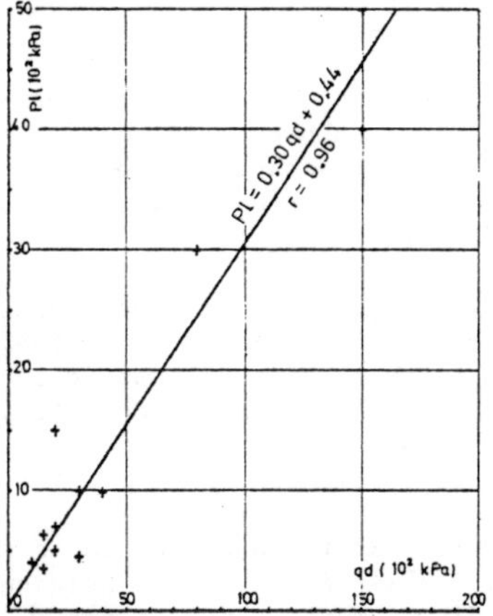

Fig. 5 Correlation Pl - qd for clays

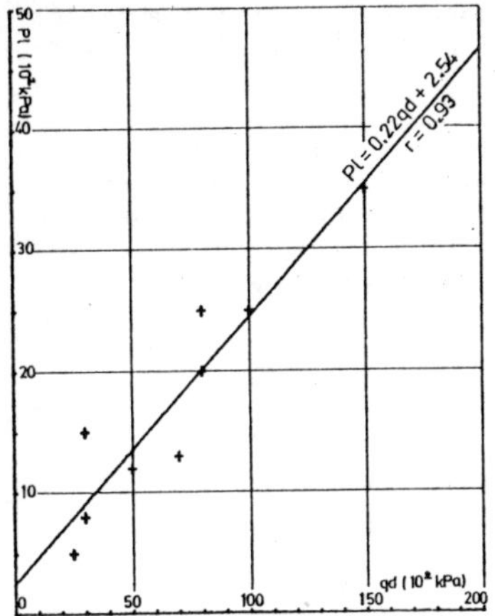

Fig. 6 Correlation Pl - qd for sands

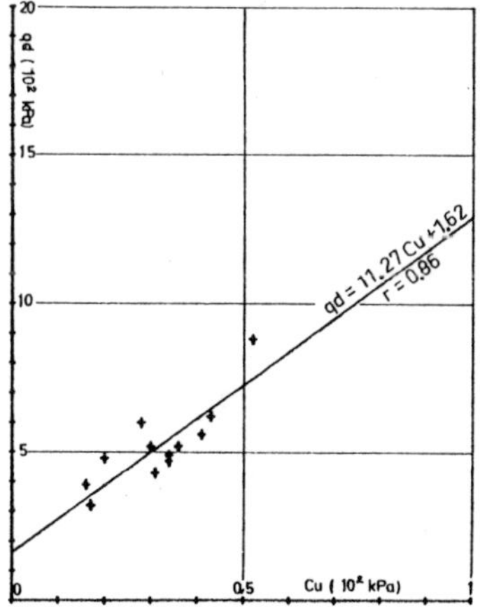

Fig. 7 Correlation qd - Cu for silty clays

7.1 Actual knowledge

The dynamic probing, correctly performed with a standard equipment and operating practice, allows to characterize perfectly soils according to their nature and natural state.

The unic dynamic point resistance presents some interesting correlations with the pressuremeter test, which allows to look to a new development of this apparatus.

7.2. Futur developments

We have shown that the dynamic penetrometer data are sensitive to the driving rate and to the impact speed of the hammer.There effects may be quantified.

Also, it is necessary that the impact energy can be calibrate and verify from time to time.

At last, for the foundation bearing capacity one will have to reconsider, according to the given data by a standard dynamic probing, all the "tricks" which have been established before.

Proceedings of the Second European Symposium on Penetration Testing / Amsterdam / 24-27 May 1982

Dynamic probing and site investigation

E.WASCHKOWSKI
Laboratoire Régional des Ponts & Chaussées, Blois, France

1 INTRODUCTION

The dynamic cone penetrometer has been too long considered as a rough and unsatisfactory investigation devise, because, though it is based on a simple principle, it includes many defects due to its conception and use.

1.1. Brief recall of kown conceptions

The dynamic cone penetrometer can be reduced to four essential elements (Fig.1) :

- the hammer, power supply, has generally a free fall with more or less friction and a more or less constant height of fall
- the anvil has sometimes a damping cushion and a fixed or friction connexion with the extensions rods
- the rods have very different weight and sizes
- the point is the most sophisticated element. Its characteristics concern the apex angle, the sizes, and the point diameter to rod diameter ratio.

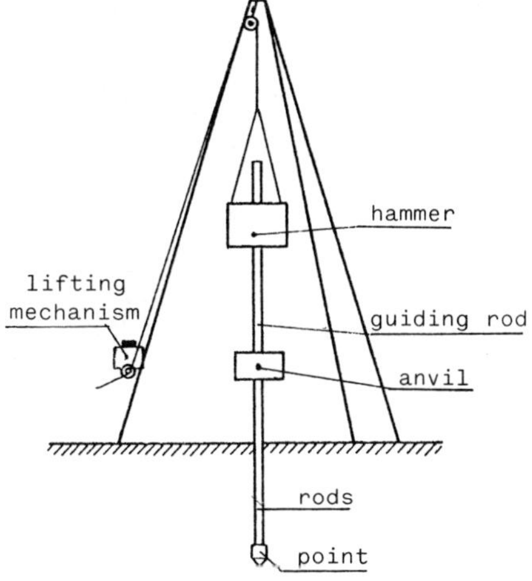

Fig.1 Schematic conception of a dynamic penetrometer.

1.2. Recall of the current practices

Generally, the dynamic penetrometers are fitted with an enlarged point. According to the different countries, one records the number of blows required to drive the penetrometer a distance of 10 - 20 or 30 cm, that we call respectively Nd_{10}-Nd_{20} or Nd_{30}.

Nevertheless, Nd cannot be a significant characteristic of dynamic probing, because it is influenced by many factors such as the impact energy, the driving rate, the diameter and the enlargement of the point, and the striking mass to struck mass ratio.

1.3. Reporting of test results

The dynamic probing results are generally presented in three manners, as a function of depth by :

- the number of blows, Nd, for a given driving
- the unit dynamic resistance, rd, calculated with one of the numerous driving formulae
- the unit dynamic point resistance, qd, estimated from the Dutch formula

$$qd = \frac{M.g.H}{A.e} \cdot \frac{M}{M+M'} , \qquad (1)$$

where :

M : mass of the hammer

M' : total mass of the extension rods, the anvil and the guiding rods

H : height of the fall

e : average penetration per blow

A : cross sectional area of the point

g : acceration of gravity

2 TYPES OF DYNAMIC CONE PENETROMETER TESTS

The four tests presented below, differ only in conception and sizes of the point and the extension rods, therefore in interaction mode between the dynamic penetrometer and soils.

The analysis submitted below, concerns comparative tests performed by us, on several sites, in homogeneous soils of different natures.

2.1. Test with a conical point with the same diameter that rods

This test corresponds to a driving model pile, which mobilizes simultaneously, but at different degrees according to the nature, density or consistency of surrounding soils, ultimate point and frictional resistances. One must notice that in this practice, the computed unit dynamic resistances refer to the top of the rods and square with a total value, which takes into account the unit dynamic resistance at the point level and the skin friction along the extension rods. This value cannot characterize the unit dynamic resistance of a slice of soil at any depth. Moreover, in such a practice, the results will have to report the variation of a total dynamic strength as a function of depth.

2.2. Test with a conical enlarged point

In spite of an annular space made by the point enlargement, the computed dynamic resistance corresponds to an unit dynamic point resistance, increased by a parasitic lateral friction mobilized either near the point or along the rod side, when the soil becomes closer or falls in. So, one deduces an over - estimate value of the unit dynamic point resistance. The influence of parasitic friction is more espacially important as the soil is finer and as the depth is greater.

2.3 Test with a conical enlarged point and with casing pipes

This test consists of driving alternately an enlarged point and the extension rods, then an outer casing.

The first application is due to Haefili (1944) then Cassan (1969), Gadsby and Meardi (1971)

In this test, the skin friction is eliminated, but the next disadvantages must be noticed :

- a double operating of rods and casing
- a longer duration of the test
- a possible parasitic friction between rods and casing, increased by an input of soil between the rods and casing near the point
- a possible stopping of the casing by an excessive lateral friction resistance.

2.4 Test with an enlarged point and a drilling mud filling

The annular space, made by the point enlargement, is filling with drilling mud, continually injected through hollow rods,just above the point.

Such applications were realised by Mohan (1970), Pfister (1973) and Baudrillard (1974), but it was a continuous forced circulation of bentonite slurry. As soon as 1973, we have observed that a simple filling of the annular space whas enough to prevent skin friction.

So, bentonite must stabilize the boring wall and reduce skin friction to a minimum. Now we use biodegradable drilling mud which allows, after a rest delay to take piezometric observations.

3. TYPE OF TESTS AND NATURE OF SOILS

In clean and dense sands or gravels, the unit dynamic point resistances are not influenced by the point conception, except for the heterogeneities (Fig.2)

This justifies why it is usually accepted that dynamic probing must be executed only in cohesionless soils.

This particular point is due by the mere fact that, during an impact, the relative displacement between rods and soil is very small, and the transmission of compression waves down the rods induces vibrations in sands or gravels surrounding the rods.

For all the other soils, such as slime silt, clay, soft chalk, marl, clayey sand, silty sand ..., the dynamic resistances are a function of the point conception and the skin friction mobilization, as much above the water level as under this one.

In these cases, the greatest values are measured with a point which has the same diameter that the rods (Fig.2)

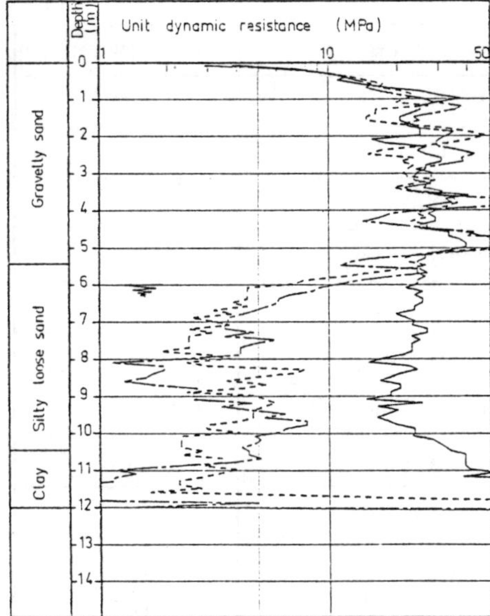

Fig.2 - Interaction mode between dynamic penetrometer and soils :
—— point with the same diameter that rods
------ enlarged point
——- enlarged point and bentonite filling

4. MEASUREMENT OF THE UNIT POINT RESISTANCE

In site investigations, the aim of dynamic probing is to determine unit dynamic point resistances characterizing perfectly a soil layer whatever its depth, its position, its nature, its density or consistency may be.

To obtain such measures, the dynamic probing must be executed with a constant operating equipment and the next specifications :

- a free fall hammer, with a constant height and a driving rate about 30 blows per minute. A reduction of the impact energy leads to an overvaluation of the unit dynamic point resistance.
- a perfectly connected anvil to the extension rods, to impart at best the impact energy.
- straight rods of the same length, allowing a tight screwing to make easier the energie propagation
- an enlarged and lost point, fitting to the end of the rods. Fig.3 shows that a lost point gives as good results as a fixed point.
- hollow rods, to can fill with drilling mud, the annular void between the rod side and the boring wall, to prevent parasitic friction resistance (Fig.4).

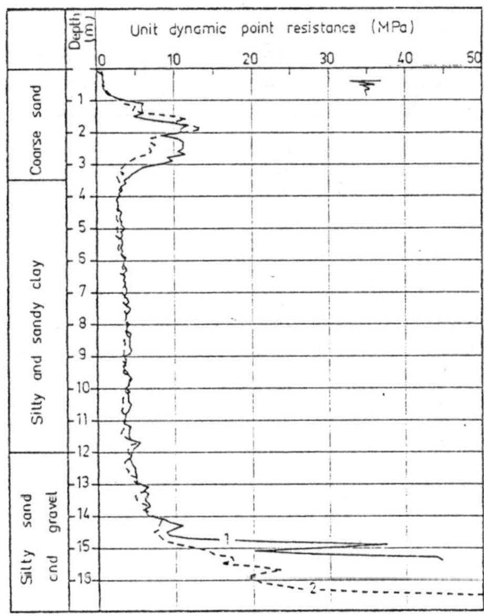

Fig.3 - Dynamic probing diagrams with an enlarged lost (1) or fixed (2) point and bentonite filling (average curves of three tests)

To compute the unit dynamic point resistance, we councel, at the real stage of our researchs, to use the Dutch formula (1) which gives values comparable to static penetrometer ones.

On the other hand, if one considers the reduced formula (2)

$$rd = \frac{M.g.H}{A.e} \qquad (2)$$

the computed values, rd, are higher than static penetrometer ones.

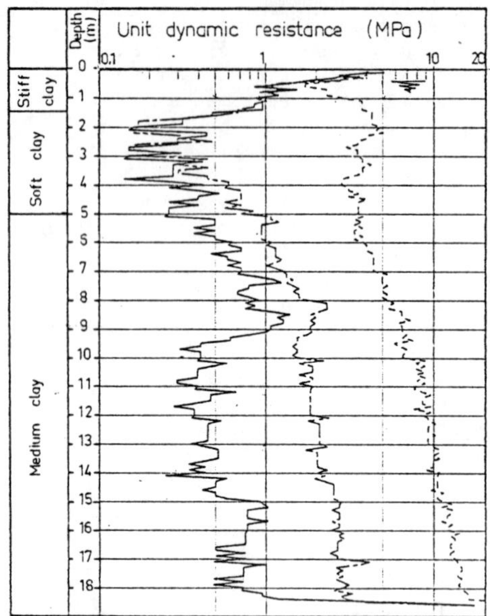

Fig. 4 - Dynamic probing in cohesive soils
.... point with the same diameter that rods
—·—· enlarged point
—— enlarged point and bentonite filling

So for our dynamic penetrometer, in the formula (1), the ratio M/M+M' ranges between 0.8 to 0.4 for an associated depth of 0 and 30 m.

5. DYNAMIC CHARACTERISTICS OF SOILS

Between 1973 and 1981, we have performed many tests, in different soils, with a dynamic penetrometer Sermes, described by Baudrillard (1974).

The main technological data are :
- hammer with variable mass : 30, 60 or 90 kg
- lifting mechanism : pneumatic engine joint to the hammer
- height of free fall : 0,4 m
- cylindroconical point diameter : 61.8m
- hollow rod diameter : 40 mm
- rod length : 1 m
- rod mass : 3.65 kg
- driving rate : 30 blows par minute

All the soils have been tested by three types of procedures :

1- point with the same diameter that rods
2- enlarged lost point
3- enlarged lost point and drilling mud filling

The shapes of the experimental diagrams are all comparable to those given on the Fig.2 and 4. The tests with bentonite filling give the smallest results comparable to the static penetrometer ones.

From all these tests, we have established variational intervals of unit dynamic point resistance, qd, corresponding to the different natures of soils as is shown in table 1.

Table 1. Measured unit dynamic point resistance for various soil natures.

Nature of soils	qd (MPa)
Slime	0.1 to 1
Silt	0.6 to 1.5
Soft clay	0.1 to 1.5
Medium clay	1.5 to 3
Stiff clay	3 to 5
Stiff stony clay	3 to 7
Loose sand	0.2 to 4
Dense sand	5 to 30
Clayey Sand	4 to 7
Loose sand and gravel	0.5 to 4
Dense sand and gravel	7 to 35
Soft chalk	0.7 to 4
Hard chalk	10 to 50
Marl	6 to 15
Stiff or hard marl	20 to 100

The dynamic penetrometer diagrams show that in homogeneous soils, qd increases slightly with depth.

However, contrary to the static cone penetrometer, the plastification state which spreads around the enlarged point of the dynamic penetrometer is not perfectly restrained, owing to the existence of an annular void.

Groundwater does not seem to influence the results obtained with an enlarged point and bentonite filling, but it affects the results given by a point with the same diameter that rods.

Besides, we have performed many tests in which the dynamic skin friction has been mobilized to the maximum. So that, we have tried and calculate the ratio between the unit dynamic point resistance, qd, and the unit dynamic skin friction, sd, as a function of soils, presended in the table 2.

Table 2. Ratio between unit dynamic point resistance, qd, and unit dynamic skin friction, sd, for various soil natures.

Soils	qd/sd
Stiff clay	13 to 20
Sandy clay	49 to 84
Silty sand, sandy silt, fine loose sand	30 to 68
Silty or clayey coarse sand or gravel, fine dense sand	54 to 113
Slime, silt, soft clay	50 to 200
Soft to stiff chalk	350
Clean and dense sand or gravel	≫

The results presented in table 2 show that :

- stiff clays can mobilize a great dynamic skin friction
- in clean and dense sands or gravels the dynamic skin friction is negligible
- slimes and silts give a low dynamic skin friction owing to their sensitivity
- soft or stiff chalks offer a very low dynamic skin friction through their thixotropy.

6. ATTEMPS AT APPLICATION OF STANDARD RECOMMENDATIONS

Having a great number of results from tests performed with an enlarged point and bentonite filling, with a dynamic penetrometer called Sermes, we have to do somes tests according to the dynamic probing standard recommendations which were presented at the Tokyo international conference.
So, the modification of the equipment concerns only the hammer, which must presented a mass of 63,5 kg and a height of fall of 0,75 m.
The diagrams of fig.5 show that the results, obtained with the dynamic penetrometer Sermes, are 10 to 20 % smaller than these corresponding to the recommendations. We think that this difference must be set down to the height of fall which is respectively of 0,40 m. and of 0.75 m. But, Dahlberg (1974) has soon indicated that the variation of the height of fall influences very much the dynamic penetration resistance.

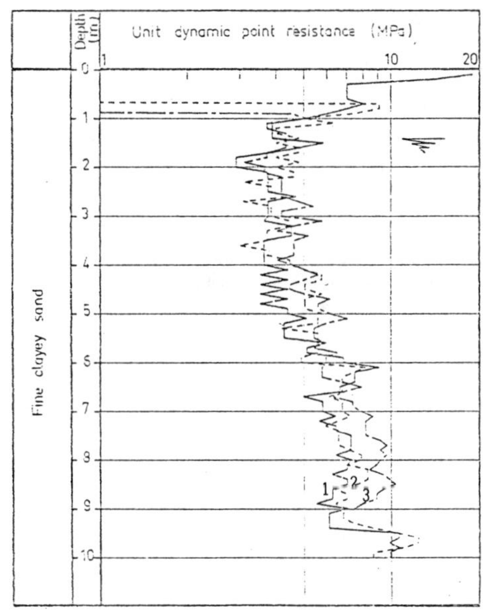

Fig.5 - Dynamic probing diagrams with dynamic penetrometer Sermes (1) tests according to standard recommandations (2)(3)

On the overhand, it seems important to us, that with a constant height of fall, one can dispose of a variable hammer mass which can be adapted to the soil resistance. So, we consider that, in slimes and soft clays the impact energy must not be higher than 200 joules.
We think also that it is necessary to calibrate the impact energy of all dynamic penetrometers.

7. CONCLUSIONS

For site investigation, the dynamic probing must give a unit dynamic point resistance, qd, which have to perfectly characterize a slice of soil at any depth, without any influence of the above soil layers.

It is possible to performe dynamic penetrometer tests in cohesive or granular soils, whatever their consistency or density may be.

To achieve this results, one must use an enlarged lost point and fill the annular void with drilling mud.

So, the dynamic probing, with an adaptation of the hammer mass to dynamic soil resistance, can be conducted in soft soils as well as a static cone penetrometer. But, in gravelly and stony soils, the dynamic penetrometer gives more realistic results than a static penetrometer with a cone diameter lower than 40 mm.

8. REFERENCES

Baudrillard, J., 1974, New development in dynamic penetration testing. Proceedings ESOPT - Stockhlom Vol. 2-2 p. 25-32

Cassan, M., 1969, Les essais in situ en mécanique des sols. Construction 5 : 178 - 181

Dahlberg, R., and V. Bergdahl, 1974, Investigations on the swedish ramsounding method. Proceedings ESOPT - Stockholm Vo.2.2 p. 93 - 101

Meardi, G., then J.W. Gadsby, 1971, The correlation of cone size in the dynamic cone penetration test with the standard penetration test. Discussions Géotechnique 21 n° 2, p. 184 - 188, p. 188 - 189

Mohan, D., V.S. Aggarwall and D.S. Tolia 1970, The correlation of corre size in the dynamic cone penetration test with the standard penetration test. Géotechnique 20 - n° 3 - 315 - 319

Pfister, p.,1973, Utilisation combinée du sondage et de l'essai P.D.S. pour la reconnaissance des sols. Communication au comité français de géologie de l'Ingénieur.

Waschkowski, E., 1979, Etude expérimentale du comportement d'un pénétromètre dynamique. Rapport de recherche L.C.P.C Paris.

Recommended Standard for dynamic probing. Proceeding IX ICSMFE, Tokyo 1977, Vol. 3, p. 110 - 114.

Proceedings of the Second European Symposium on Penetration Testing / Amsterdam / 24-27 May 1982

Correlation between cone penetration resistance, static and dynamic pipe pile response in clays

N.D.WRIGHT
Brunei Shell Petroleum Co. Ltd.

W.R.VAN HOOYDONK & D.J.M.H.PLUIMGRAAFF
Fugro B.V., Leidschendam, Netherlands

INTRODUCTION

The design of large diameter pipe piles depends on semi empirical design rules and engineering judgement. Another potential problem area is the installation of these piles.

Despite the experience gathered in three decades of offshore application, both design- and installation-time are still open for optimisation. A major step in that direction has been taken in Brunei where a long 42 in. diameter pipe pile has been instrumented and load tested. Objective of the test performed in a typical clay profile was to correlate predictions, derived by accepted methods, with the pile behaviour as measured during driving and static load testing.

This paper describes the tests, the results and the conclusions that were drawn.

Special emphasis is given to the correlations between cone penetration resistance on one hand and static and dynamic pile response on the other.

A companion paper (ref. 1) deals with the applications to offshore pile design.

GENERAL

1. The tests were carried out at a site in Kuala Belait, State of Brunei (fig. 1). At this location a large onshore construction project required 42 inch diameter pipe piles to be driven to a penetration of 47 m. A non working pile was statically loaded in compression to establish the ultimate capacity. This pile was instrumented during both the installation and static testing since the size and capacity of the pile are similar to that required for typical offshore installations in the area.

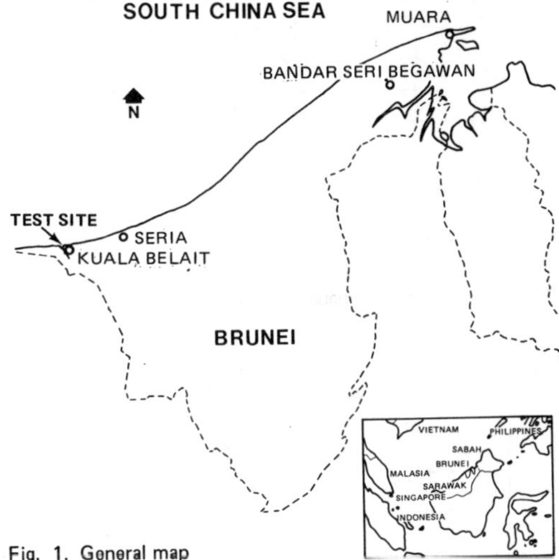

Fig. 1. General map

2. The site investigation for this project consisted of 15 cone penetration tests (c.p.t.'s) and 6 conventional soil borings. The c.p.t.'s were performed with the standarized electric Fugro cone, recording tip resistance and local sleeve friction continuously. The c.p.t. profile nearest the test pile (distance approx. 4 m) and the results of the tests on the samples taken from the borings are presented (figs. 2 and 3).

3. The soil consists predominantly of a normally consolidated clay overlain by 3 to 4 m of fine grained dense sand. The clay is organic and contains various proportions of silt lenses, fine sand pockets and shell fragments. It is very soft near the surface and its undrained shear strength increases linearly with depth.

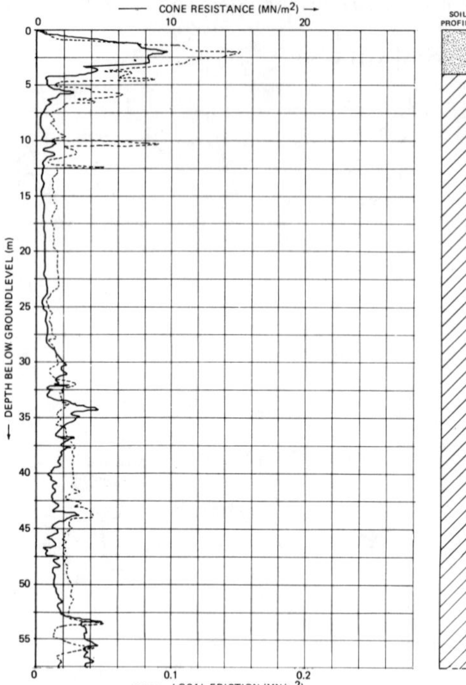

Fig. 2. Electrical cone penetration test

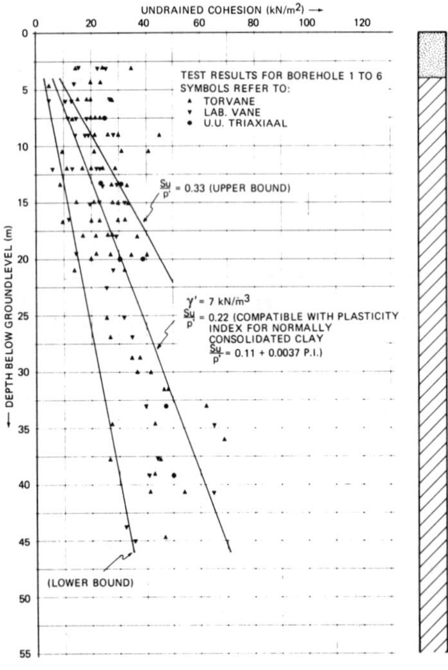

Fig. 3. Summary soil test results borehole 1 to 6

PILE DESIGN

4. Typical design practice requires a soil strength profile, based on the results of the laboratory and/or in-situ tests. The undrained shear strength of clays is used to assess the pile capacity and the soil resistance after a delay in driving, whereas the remoulded shear strength is used to determine the resistance during continuous driving.

5. Three lines representing the undrained shear strength (s_u) based on plasticity index and overburden pressure (p') for normally consolidated clays are presented (fig. 3). The line corresponding with $\frac{s_u}{p'} = 0.22$ has been used in the design because this is standard practice in the area. Another undrained shear strength profile (fig. 4) has been derived from the results of the c.p.t. (fig. 2) using the following formula (ref. 2):

$$s_u = \frac{q_c}{N_k} \qquad (1)$$

where, s_u = undrained shear strength
q_c = cone penetration resistance
N_k = correlation factor

Previous site investigations performed in the South China Sea have produced N_k = 15, which is in line with experience elsewhere.

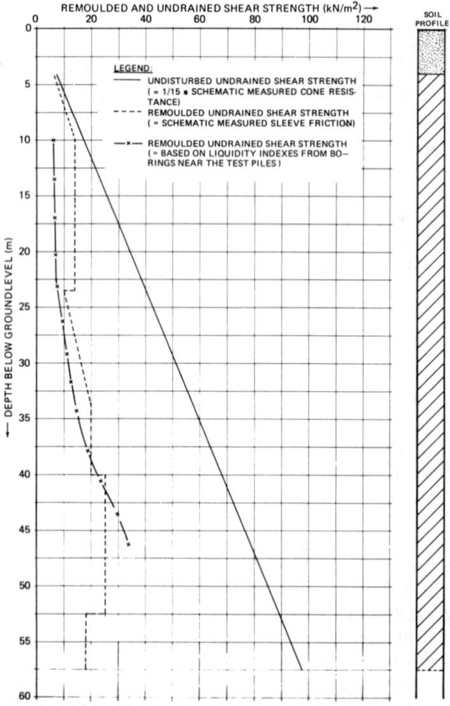

Fig. 4. Remoulded- and undisturbed undrained shear strength vs depth

6. A remoulded shear strength profile has been assessed using liquidity index data from the soil samples (fig. 4). Another approach is to use the sleeve friction measured with the electrical cone (fig. 4). The two curves agree well; the sensitivity of the clay appears to range from 2 to 4 with a typical value of 3 to 3.5.

7. The static capacity of the pile has been computed using traditional methods: API, Lambda and c.p.t. (ref. 2). The capacities are given versus depth (fig. 5), at final penetration the following predictions apply:

Method	Skin friction	End bearing	Total capacity
A.P.I.	5.7 MN	0.7 MN	6.4 MN
Lambda	4.6 MN	0.7 MN	5.3 MN
C.p.t.	6.6 MN	0.8 MN	7.4 MN

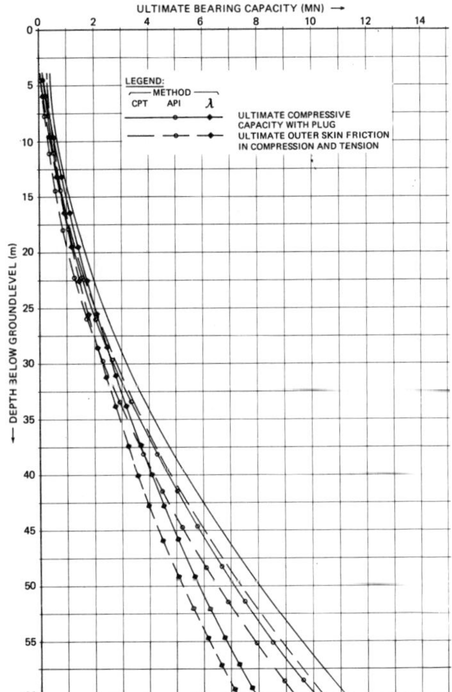

Fig. 5. Ultimate bearing capacity 42″ piles

8. To calculate the Soil Resistance during continuous Driving (SRD) it is general practice in these clays to assume the pile drives unplugged:

$$SRD = \Sigma s_{u,r} \times A_o + K\ \Sigma s_{u,r} \times A_i + L \times A_p \times q_c \quad (2)$$

where: $s_{u,r}$ = remoulded shear strength
A_o = outer pile shaft area
A_i = inner pile shaft area
K = factor to account for reduction of inner friction
L = dimensionless factor
A_p = cross sectional area of the pile annulus
q_c = cone resistance below the pile tip

9. As an absolute upper bound of the SRD the pile is assumed to behave plugged for a while after delays in the installation:

$$SRD_s = \Sigma s_u \times A_o + L \times A_c \times q_c \quad (3)$$

where: SRD_s = SRD after a delay in driving
A_c = cross sectional area of the closed pile tip

10. To predict the driving behaviour, the performance of the hammer, in this case a Kobe 60, has to be quantified as well. Using job specific pile-, soil- and hammermodels, the SRD that a pile can overcome at a certain blowcount can be computed. With this relationship and the SRD profiles described above, predictions of blowcount versus depth can be made (fig. 6).

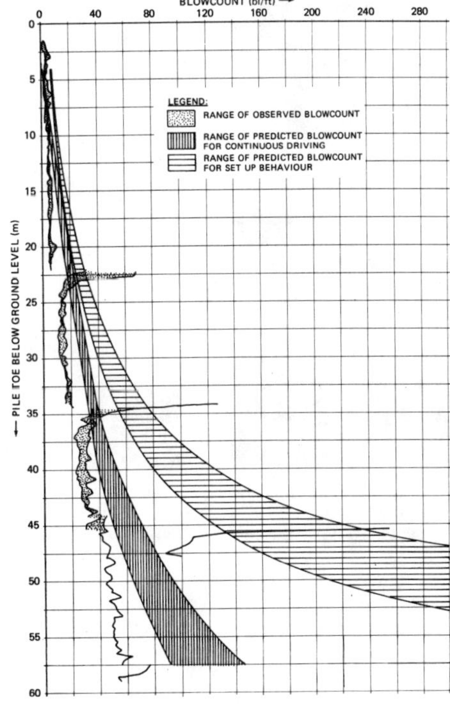

Fig. 6. Comparison of predicted and observed blowcount vs depth

TEST RESULTS

11. During the installation of the test-pile strains and accelerations at pile head level have been measured and recorded (fig. 7), to quantify hammer performance and pile response (ref. 3).
Moreover similar measurements were taken, at the pile tip, purely for research purposes beyond the scope of this paper. On site approx. 100 blows, typical of the installation, were selected and processed. Soil resistances during continuous driving and after delays in driving could thus be computed. This information and the data on sensitivity of the soil allowed an estimate of the static pile capacity

Fig. 7. Measurements during driving.
Test pile (42.0" x 0.75") with Kobe 60. Pile is instrumented with two sets of DPT -transducers.

before the start of the static load test: a total capacity of 6.9 MN, split up into 6.1 MN skin friction and 0.8 MN tip resistance. This procedure is called Dynamic Pile Testing (DPT).

12. The pile was tested in compression using two anchor piles and a reaction beam (fig. 8). During the static load test

Fig. 8. Static load set-up.
Test in compression using two anchor piles and 1000 tonnes loading facility.

applied load and corresponding pile head displacement were recorded. The distribution of load in the pile has been determined using straingauges at four different levels. The displacement of the soil plug inside the pile relative to the pile and the surrounding soil was recorded as well.

13. Load versus pile head displacement and the load distribution for subsequent loadsteps are presented (figs. 9 and 10). The ultimate pile capacity appeared to be 7.65 MN of which 0.9 MN was taken by the pile tip. The soil plug measurements showed that the plug did not move at all relative to the pile, the pile thus behaved as a closed ended one during static loading.

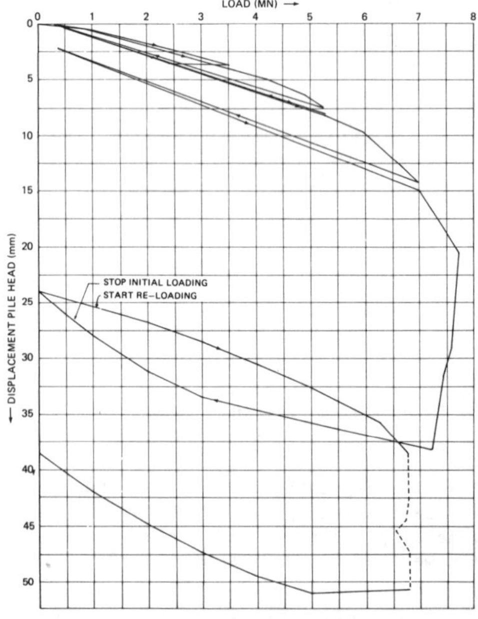

Fig. 9. Pile head displacement versus load

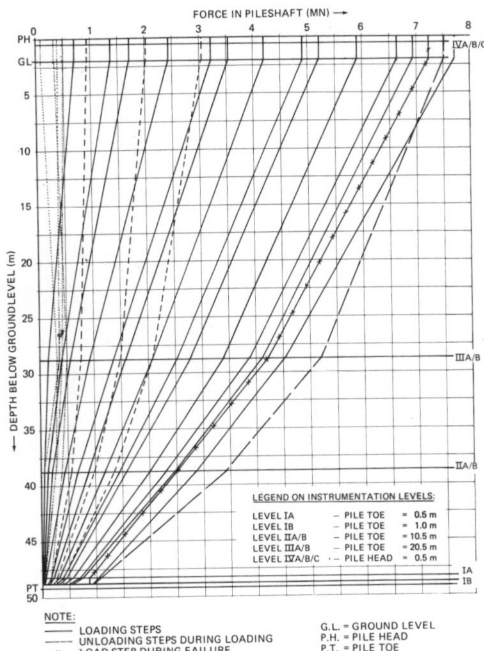

Fig. 10. Load distribution in the pile during loading and unloading (initial loading)

CONCLUSIONS

14. None of the methods used to predict the static capacity overestimated the actual ultimate load. The Lambda- and the API-method appeared to give somewhat conservative results. The c.p.t.- and the DPT-capacity agree remarkably well with the testresult. Both skin friction- and end resistance-predictions were within 10 percent of the actual value:

Method		Predicted Value (MN)	Pred. value / Actual value x 100 % (%)
End bearing	: API	0.7	78
	Lambda	0.7	78
	CPT	0.8	89
	DPT	0.8	89
Skin friction	: API	5.7	85
	Lambda	4.6	68
	CPT	6.6	98
	DPT	6.1	90
Total capacity	: API	6.4	84
	Lambda	5.3	69
	CPT	7.4	97
	DPT	6.9	90

Especially the c.p.t. method has again proven its value. A similar conclusion on the DPT method requires more confirmation by comparable tests.

15. Plotting the inverse of the predicted c.p.t.-load distribution (fig. 5) in the actual load distribution (fig. 10) shows satisfactory agreement as well.

16. The predicted blowcounts for continuous driving appeared to be slightly higher than the observed blowcounts (fig. 6). This type of "conservatism" is considered to be desirable to prevent premature refusal. The soil resistance after delays in driving could be overcome as expected.

ACKNOWLEDGEMENT

The authors wish to thank Brunei Shell Petroleum Co. Ltd. for their permission to present this paper; the cooperation of collegues, IPCO Marine Ltd. and Fugro Singapore Pte. in performing the tests and reducing the data is also highly appreciated.

REFERENCES

Wright,N.D., Tamboezer, A.J., Windle, D., Hooydonk, W.R. van, and Ims, B., (1982). "Pile instrumentation and monitoring during pile driving offshore N.W. Borneo" Proc. offshore Technology Conference, Paper no. 4204, Houston, Texas.

De Ruiter, J. and Beringen F.L. (1979). "Pile foundations for large North Sea structures", Marine Geotechnology, Vol. 3, no. 3.

Beringen, F.L., Hooydonk, W.R. van, and Schaap, L.H.J. (1980). "Dynamic Pile Testing, an aid in analyzing driving behaviour". Proc. Seminar on the application of stress-wave theory, Stockholm.

Proceedings of the Second European Symposium on Penetration Testing / Amsterdam / 24-27 May 1982

Influence of pore water pressure on dynamic and static penetration testings

N.YAGI, M.ENOKI & R.YATABE
Department of Ocean Engineering, Ehime University, Matsuyama, Japan

1 INTRODUCTION

Recently, pore pressure around a tip during penetration has become to being measured, but it has been used for soil identification in most case (Mohsen et al. 1980, and Torstensson 1975). This paper offers some basic consideration of pore pressure on interpretation of static and dynamic penetration tests.

Since most soils have always negative or positive dilating characteristics and since a cone tip is advanced at finite rate, it is inevitable to generate more or less excess pore water pressure around the tip during penetration into saturated soils. Therefore, it is supposed that penetration resistance is affected by pore pressure, because a strength of a soil depends on an effective stress especially for ϕ-material as a sand.

Static and dynamic penetration tests are carried out through sands in a small chamber and volume change or pore pressure are analyzed in relation to penetration resistance. Finally, the similarity of these results to triaxial compression results of sands is indicated.

2 APPARATUS AND PROCEDURE FOR EXPERIMENT

The apparatus is shown in Fig.1. The chamber is 23cm diameter and 50cm height. The upper surface of the soil sample of 40cm thickness is covered with a rubber membrane through which air pressure is applied as overburden load. Diameter of the 60° cone is 2.66cm and the shaft is covered with the rigid pipe so that point resistance can be mainly measured. Lublication is made by putting grease between the inner side of the chamber and the rubber membrane. Four pipes which are led out of the chamber are set in the sand sample in order to measure pore water pressure The drained pipe is attached at the bottom of the chamber and connected to the valve and the burette in order to measure volume change of saturated samples.

Tested soil is Toyoura sand (D_{10}=0.12mm). The tip of the cone is set at 22.5cm depth under the soil surface which corresponds to the position of the pore pressure measurement u_2.

The cone is advanced at 0.12cm/sec in static penetration tests, and is driven by dropping the weight (3.88kg) from 24cm hight in dynamic tests. Tests are conducted over many conditions of humidity of the sample (dry or saturated), void ratio (loose or dense), overburden pressure (1.5, 1.0, 0.5kgf/cm^2) and drainage (drained or undrained from the bottom of the chamber). Saturated condition is firmed by de-airing with vacuum.

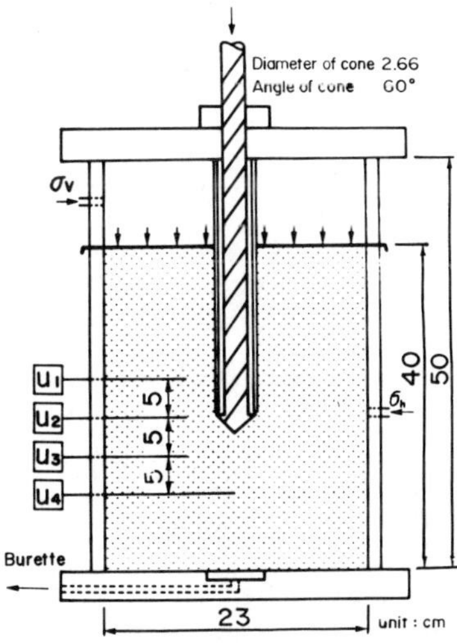

Fig.1. Testing apparatus

3 TEST RESULTS AND ANALYSIS

3-1 STATIC PENETRATION TEST

The drained and undrained test results for the loose and dense samples are shown in Fig.2, Fig.3, Fig.4 and Fig.5, respectively. The volume change and the pore pressure are positive in the loose state and negative in the dense state. This tendency agrees with dilatancy characteristics of each state of sand and means that penetration is mainly due to shear deformation and a little of volume compression.

Influence of pore pressure on penetration resistance is estimated by following tests; 1) test on dry sand, 2) drained test on saturated sand, 3) undrained test on saturated sand.

The change of pore pressure during the penetration test, even if it is positive or negative, may be largest in the test 3), and least in the test 1). Therefore, it may be supposed that the penetration resistance is maximum in the test 1) and minimum in the test 3) for the loose sample, and vice versa in the dense sample. This reason is due to that the pore pressure increases in the loose sample and decreases in the dense sample during penetration, if a compressive pore pressure is taken as positive. This supposition is made sure by the test results

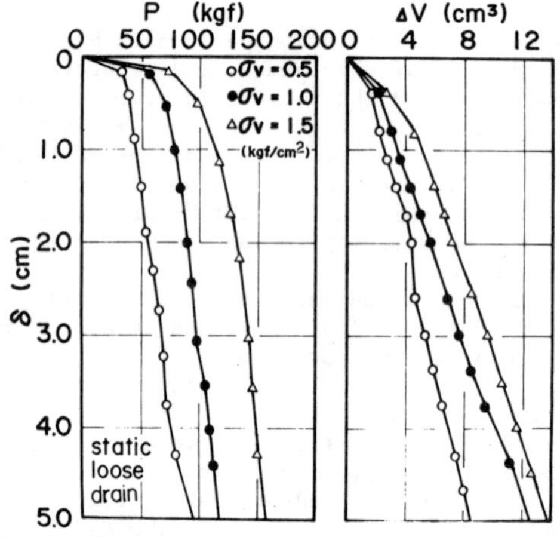

Fig.2. Plots of static penetration resistance P and volume change ΔV against amount of penetration.

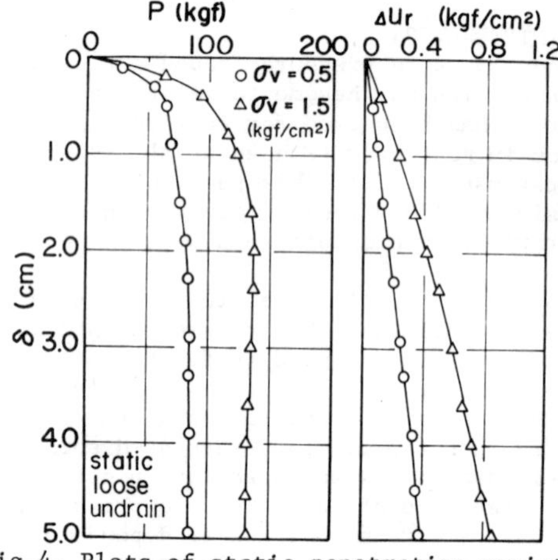

Fig.4. Plots of static penetration resistance P and pore pressure Δu against amount of penetration.

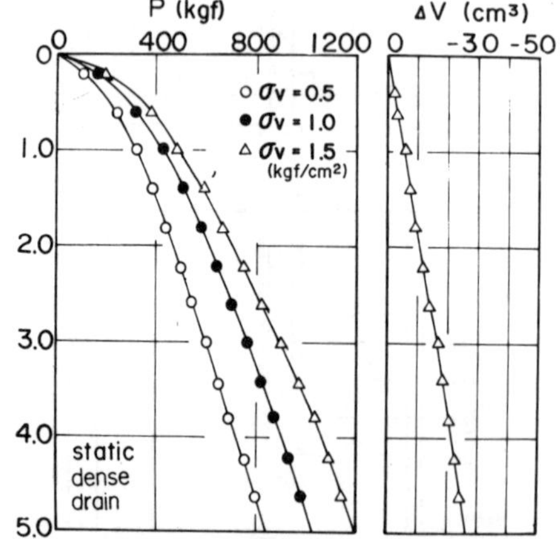

Fig.3. Plots of static penetration resistance P and volume change ΔV against amount of penetration.

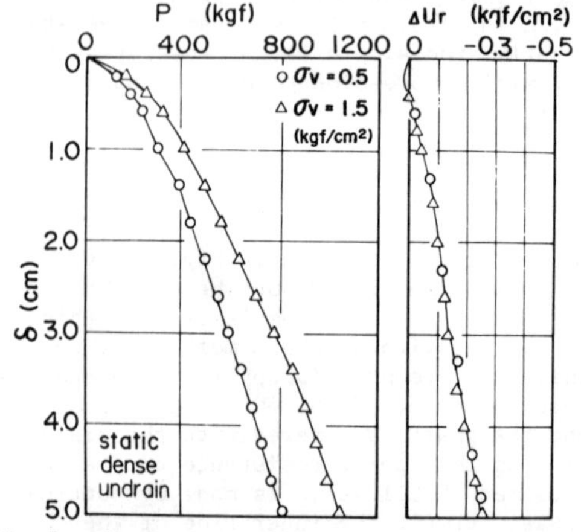

Fig.5. Plots of static penetration resistance P and pore pressure Δu against amount of penetration.

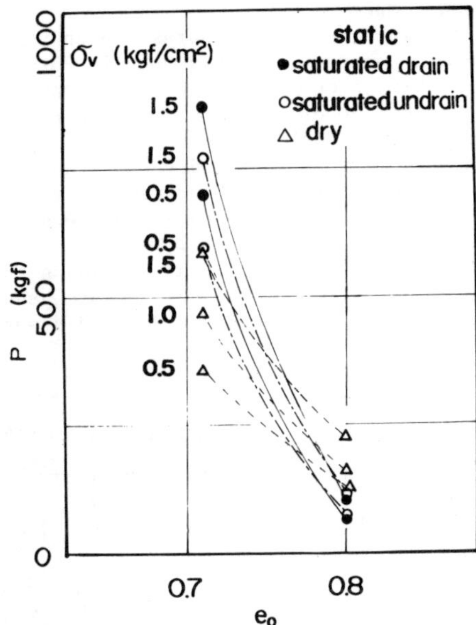

Fig.6. The variation of static penetration resistance P with void ratio e_0.

shown in Fig.6. Although a drainage condition in a field sand layer is not same as in the chamber test, the consideration and test results mentioned above may aid a precise interpretation of a static penetration test.

3-2 DYNAMIC PENETRATION TEST

Dynamic tests are carried out under the same conditions of the sample and drainage as the static tests. The typical vibrograms of the pore pressure around the cone tip, when being applied an impact load in the loose and dense samples, are shown in Fig.7 and Fig.8. The pore pressure is measured under the backpressure of 1.0kgf/cm^2 and the effective overburden pressure of 0.5 kgf/cm^2. Fig.7 and Fig.8 are the vibrograms for the first drop of the weight, and similar vibrograms are obtained for successive blows, though the amplitude of oscillation decreases with a number of blows.

The pore pressure immediately after impact tends to positive at the pore pressure measurement u_4 underneath the cone tip, while it tends to negative at u_2 beside the cone tip in the loose and dense samples. The pore pressure alternates larger in positive side than in negative side in the loose sample. Damping of pore pressure occurs rapidly after the second or third wave. After damping the residual pore pressure which is larger than the value before impact, exists in the undrained test, though it disappear in the drained test. These behaviours of the pore pressure, for example the tendency that it is generated larger in the loose sample than in the dense one, coincide with one obtained in the static test.

The penetration resistance N, the volume change ΔV after dissipation of the pore pressure and the dynamic pore pressure Δu_d immediately after impact are demonstrated in Fig.9 and Fig.10 for the drained test. N is the number of blows needed in 1.0cm penetration of the cone into which the number of blows needed in about 0.2cm penetration of the cone is converted. For example, if the number of blows needed in

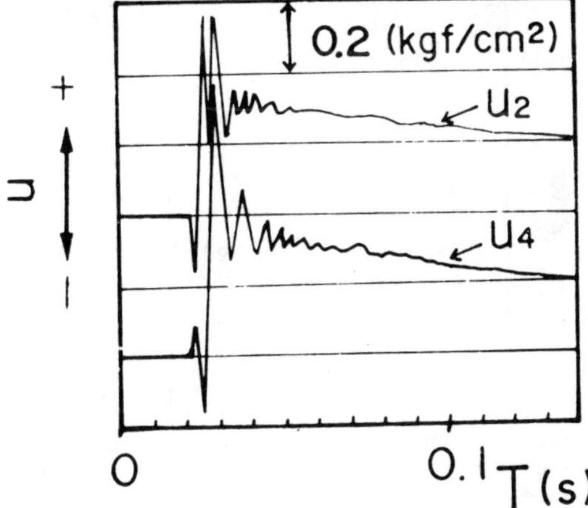

Fig.7. The typical vibrograms of the pore pressure measured at u_2 and u_4 (loose).

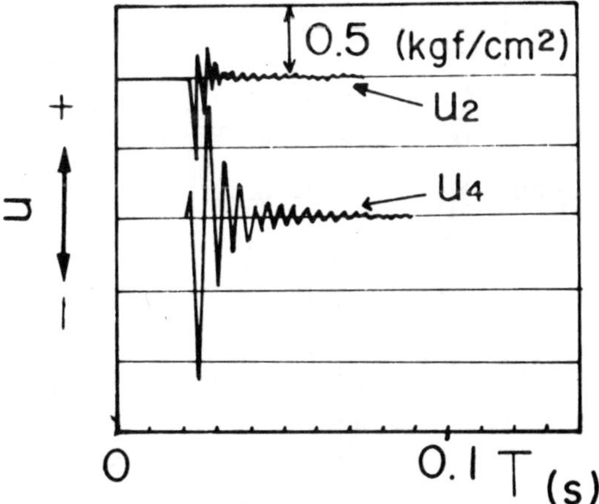

Fig.8. The typical vibrograms of the pore pressure measured at u_2 and u_4 (dense).

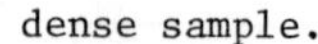

0.22cm penetration is three, N is (10/2.2) × 3 = 13.6. The volume change characteristic in the dynamic test has the same tendency as static one. A little positive volume change is observed even in the dense sample. This means that the sand around the tip subjected to repeating shear deformtion due to impacts.

N, Δu_d and the residual pore pressure Δu_r measured just before next impact are demonstrated in Fig.11 and Fig.12 in the undrained test. The resudial pore pressure which accumulates with the number of blows, is larger in the loose sample than in the dense sample.

Influence of the pore pressure on the dynamic penetration resistance is rather complex than on the static one, because of the existance of two kinds of pore pressure Δu_d and Δu_r. The dynamic penetration resistance is maximum in the dry state and minimum in the undrained saturated state for loose sample, and vice versa for dense sample as shown in Fig.13. This is similar tendency to the static test and means that there is influence of pore pressure generated during penetration on dynamic penetration resistance.

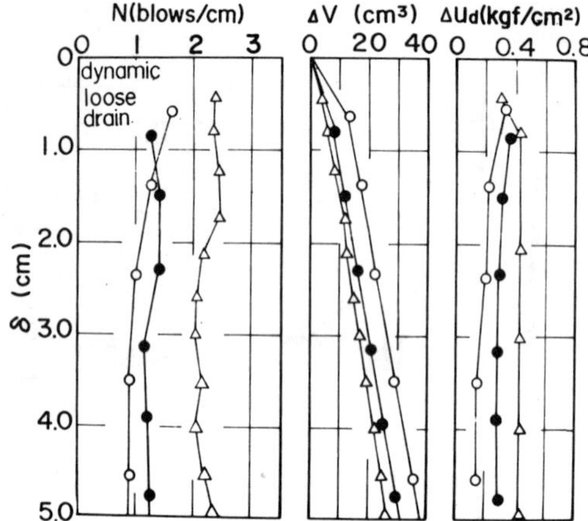

Fig.9. Plots of dynamic penetration resistance N, volume change ΔV and dynamic pore pressure Δu_d against total amount of penetration δ.

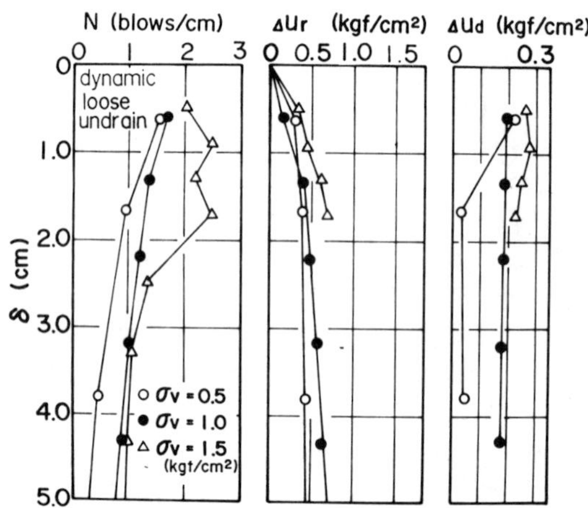

Fig.11. Plots of dynamic penetration resistance N, residual pore pressure Δu_r and dynamic pore pressure Δu_d against total amount of penetration δ.

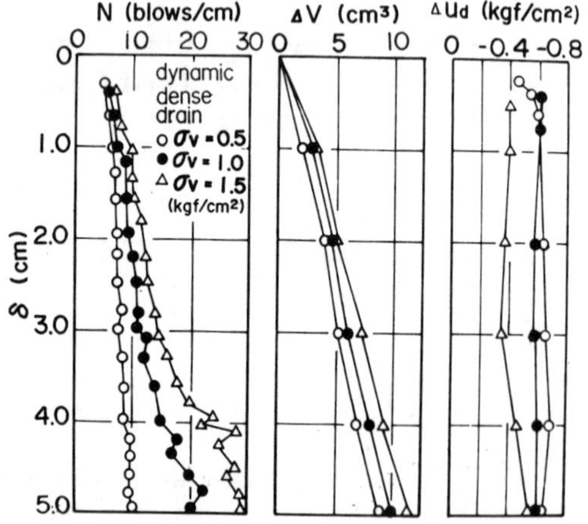

Fig.10. Plots of dynamic penetration resistance N, volume change ΔV and dynamic pore pressure Δu_d against total amount of penetration δ.

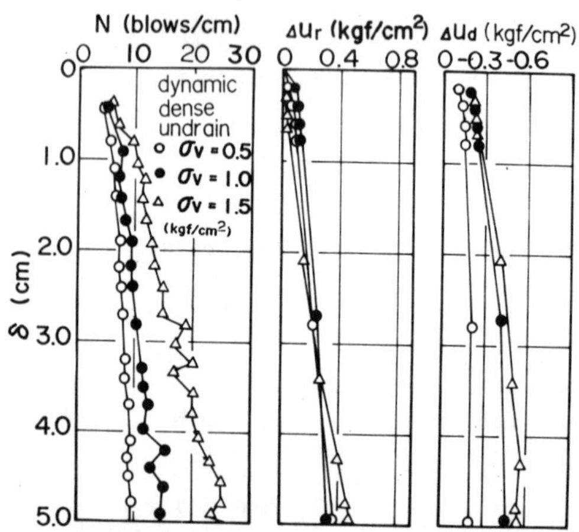

Fig.12. Plots of dynamic penetration resistance N, residual pore pressure Δu_r and dynamic pore pressure Δu_d against total amount of penetration δ.

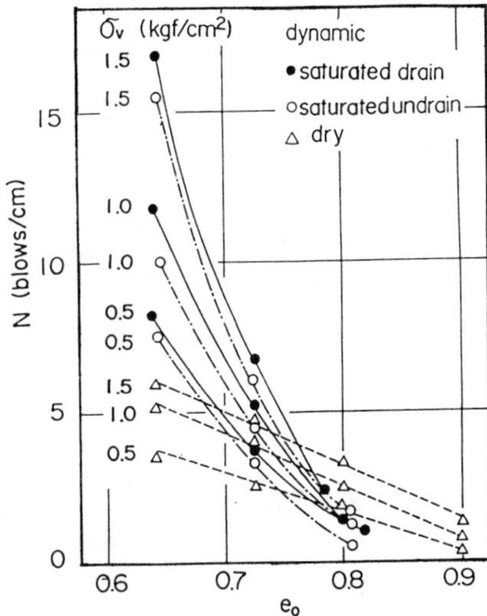

Fig.13. The variation of dynamic penetration resistance N with void ratio e_0.

3-3 SIMILALITY TO TRIAXIAL COMPRESSION TEST

Typical results of triaxial compression tests on loose and dense sands are shown in Fig.14 (a) and (b). Since positive pore pressure for volume decrease occurs with shear deformation for loose sample, undrained strength is less than drained one. On the other hand, since negative pore pressure or volume increase occurs for the dense sample, undrained strength is more than drained one. These characteristics are similar to those obtained by penetration tests. This similality is explained by the consideration that the deformation of sand around the cone tip consists mainly

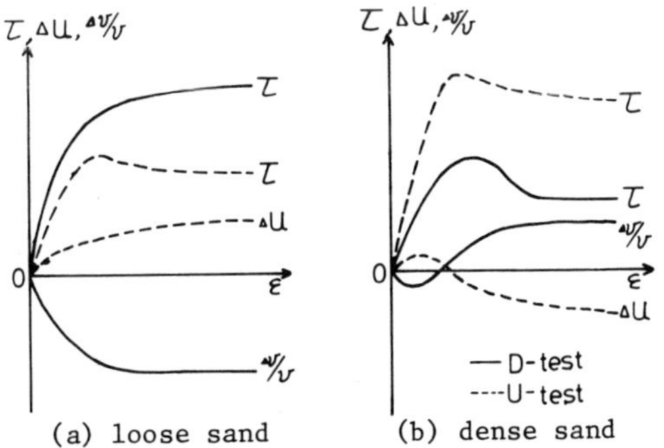

(a) loose sand (b) dense sand
Fig.14. Stress-strain-volume change and stress- strain-pore pressure relation.

of shear deformation and dissipation of the generated pore pressure requires an unnegligible period of time. Further, the positive pore pressure accumulated by the repeating deformation due to the impacts is added for the dynamic penetration.

From the stand point of effective stress, schematic diagrams of the relation between penetration resistance N and total overburden σ_v can be drawn as shown in Fig.15 for loose sand and in Fig.16 for dense sand. The difference of penetration resistance between these three kinds of the state may depend on the effective stress state. The effective overburden pressure $\bar{\sigma}_v$ can be expressed as $\bar{\sigma}_v = \sigma_v - \Delta u_r - \Delta u_d$. There exists no pore pressre in the dry state. In the

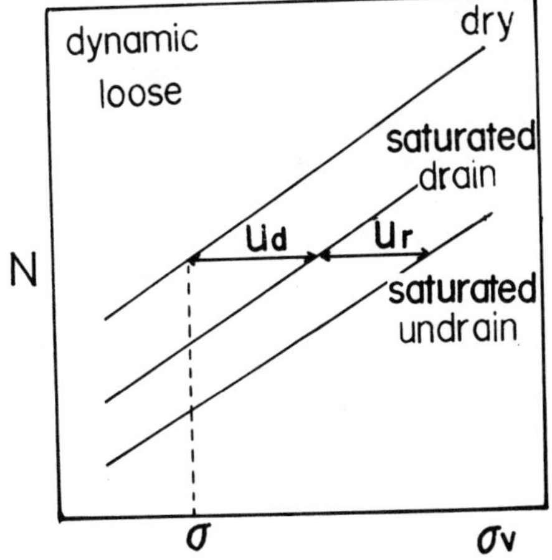

Fig.15. Schematic diagram of relationship of N and total overburden pressure σ_v.

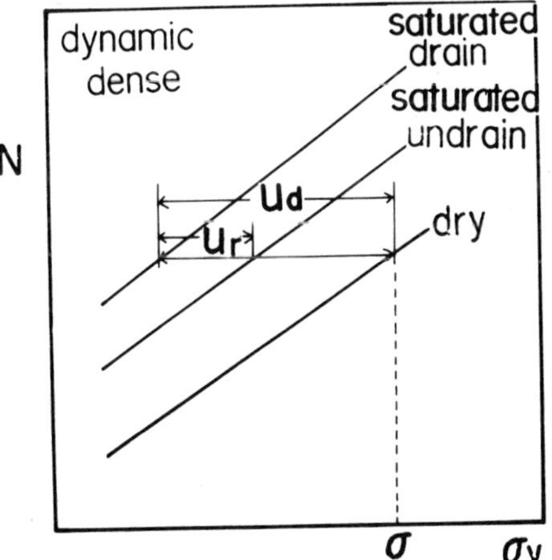

Fig.16. Schematic diagram of relationship of N and total overburden pressure σ_v.

drained state the dynamic pore pressure Δu_d only exists and in the undrained state both dynamic and residual pore pressure Δu_d, Δu_r exist. Further, Δu_r is always positive, and Δu_d is considered to be positive in the loose state and negative in the dense state. Then we must take account of pore pressure to compare penetration resistance of various conditions. The same schematic diagrams can be drawn for the static penetration.

From these diagrams qualitative characteristics can only be explained but not quantitative one, because of the following reason. Firstly, as suggested by Schmertmann (1974), both positive and negative pore pressure can exist simultaneously below and above the cone tip, and a uniform distribution around the tip can not be expected. Secondly, in these chamber tests, the vertical overburden pressure σ_v can not be uniformly distributed within the soil sample along the depth, because of the friction between the soil sample and inner side of the chamber.

4 CONCLUSION

1) In both static and dynamic penetration tests for the loose and dense sample, the same dilative characteristics are obtained as in triaxial compression tests.
2) The dynamic pore pressure generated just after impact is observed in the test of the saturated samples. Especially, in the undrained test, the positive residual pore pressure exists and it accumulates with the number of blows.
3) In both static and dynamic penetration tests for the loose sample, the penetration resistance is most in the dry state and the least in the saturated undrained state. The result is reverse for the dense sample. This fact reflects that the pore pressure due to dilatancy of sand is important for estimation of soil properties from the penetration resistance.

5 ACKNOWLEDGEMENT

The authors wish to thank Mr. Ohnishi and Mr. Takechi for their experiments.

6 REFERENCES

Mosen, M.B., V. Vivatrat, & C.C. Ladd 1980, Cone Penetration in Soil Profiling, Jour. ASCE, GE4, p. 447-461.

Schmertmann, J.H. 1974, Penetration Pore Pressure Effects on Quasi-Static Cone Bearing, q_c, Proc. of ESOPT, Vol.2, p. 345-351.

Torstensson, B.A., 1975, Pore Pressure Sounding Instrument, Proc. of Conf. on In Situ Measurement of Soil Properties, ASCE Vol.2, p. 48-54.

Proceedings of the Second European Symposium on Penetration Testing / Amsterdam / 24-27 May 1982

Scale modelling of static and dynamic pile penetration

AMOS ZELIKSON
Laboratoire de Mécanique des Solides, Ecole Polytechnique, Palaiseau, France

1. METHODOLOGY

The piling problem lacks a guiding theory, because its basic features are grain rearrangements in the plastic zones, which result in either work hardening or softening, residual stresses and pore pressure build up. One has to rely on experiments. Small scale models are tested in the laboratory under fully controled conditions facilitating extensive measurements. In cases like pile groups small scale models are the only solution. Centrifugal models have recently been used to test structures founded on piles and subject to earthquakes (1). The single pile is a very adequate problem for small centrifuges, now quite abundant (2). The Hydraulic Gradient method has been used for pile tests since 1962. It has been recently used for earthquake models of structures (3) and following these an installation built to test piles dynamically. The experimental program aims to correlate penetration tests, driving tests and static and dynamic performance of pile groups.

The mechanical properties of soils follow from the intergranular forces. Thus changing the scale of stresses is equivalent to changing of the material. A correct soil similitude is based on stress conservation. Euler's equation of motion is :

$$(\rho v)_{,t} + \mathrm{div}\,(\sigma + \rho v \otimes v) = f$$

($_{,} \equiv \partial/\partial t$. ρ = density . v = material velocity . σ = stress . f = volume force)

(m designs model . Am/A is the scale A^* of A)

$1 = \sigma^* = (\rho v \otimes v)^*$ if $\rho^*=1$; $v^*=1$ $t^* = \ell^*$; $\dot{v}^*=\ell^{*-1}$. All accelerations and frequencies are increased by ℓ^{*-1} which provides for a correct scale of inertial forces.

According to the Hydraulic Gradient method the condition for the acceleration of gravity $g^* = \ell^{*-1}$ is met by introducing inside the soil an additional volume force, the drag on the grains of a percolating fluid, normally water. Figure 1 shows the installation.

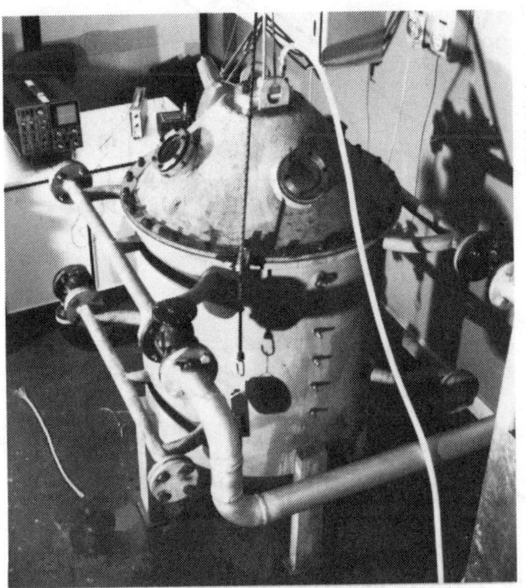

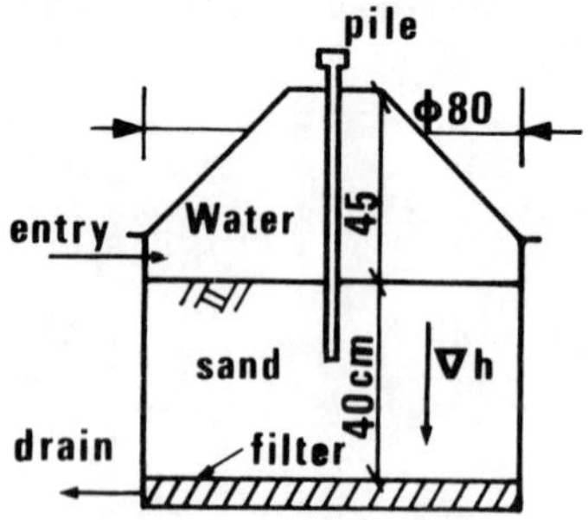

Fig.1 Hydraulic gradient cell

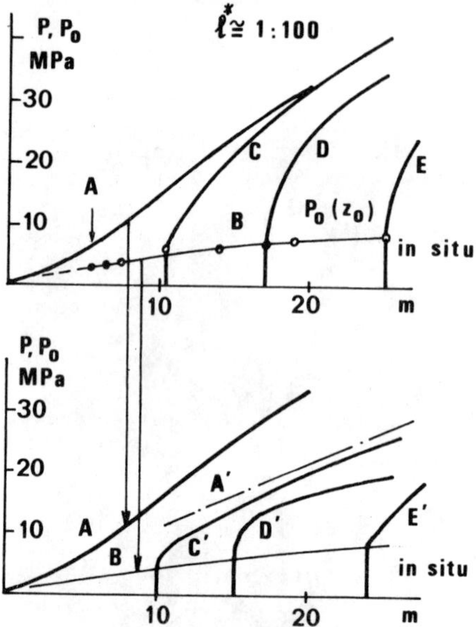

Fig. 2 static penetration tests

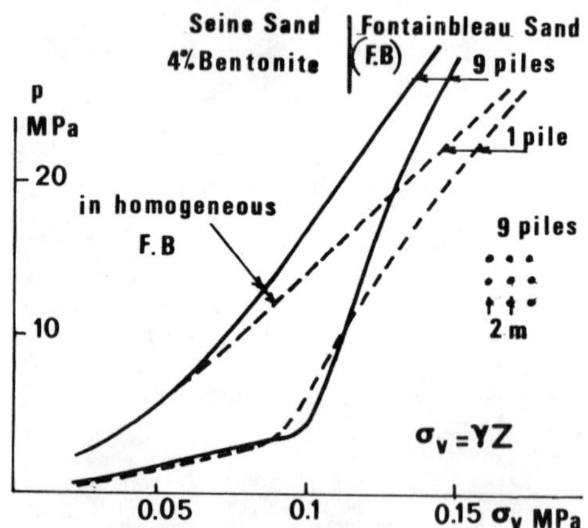

Fig.5 Penetration test on 1 and 9 piles in two-layered and homogeneous soil

If a head of h=40m is maintained between the soil's free surface and the base filter, a constant head gradient of 100 is obtained. As the soil's submerged unit weight relative to water, γ_{sr}, is about 1, $g^* \simeq \nabla h/\gamma_{sr} \simeq \nabla h = 100$. In this exemple a length scale of 1/100 is obtained. (details of the method in (4)(5)). The weight of structures should be accounted for by additional forces according to ℓ^{*2}. This great reduction in forces facilitates testing of model pile groups.

The time scale presents problems : the dissipation of pore pressure follows a diffusion time scale $t^*_d=\ell^{*2}$. By reducing the grain size sand's permeability can be reduced by ℓ^* (down to $\ell^* \simeq 1/100$) without changing the mechanical properties so that $t^*_d=\ell^*$ as needed. The soil's damping in the Elastic range is mostly due to radiation and is correctly scaled. In the Plastic range the main process is Coulomb friction, also correctly scaled. Any strain rate effects are not conserved, as they call for $t^*=1$. By mixing small quantities of clay with the sand and by changing the relative density layers of greatly different mechanical properties and equal permeability are obtained. Water flows through the pile's base. Thus a constant gradient is maintained in sands and buried foundations.

2. STRENGTH AND PENETRATION TESTS

Increasing attention is being paid to earthquakes and shocks. Piles need to be rated by the performance of groups under various cycles of loading. Accordingly information from static or dynamic vertical loading in situ of single piles must have a fragmentary character, while being very expensive. A common attitude is : "the tests cost so much that they must be interpreted at any price". In contrast, static and dynamic penetration tests use light equipment, can be highly instrumented and computerized and furnish much information. Small scale model tests should be able to relate penetration tests directly to pile group performance.

2.1. Some results of Static model tests

In the following p means the total force divided by the pile's (or anchor's) sectional area. Figure 2 shows clearly that the notion of strength as a function of depth does not exist. Curve A shows p versus the pile's penetration depth, Z, starting from the surface. C, D, E are loading tests started at different depths, Z_0, giving the virginal strength, p_0, (curve B) and then p as a function of further penetration. (C, D, E have A as an envelop). C', D', E' started as C, D, E then cycles of pulling-pushing performed every 4m

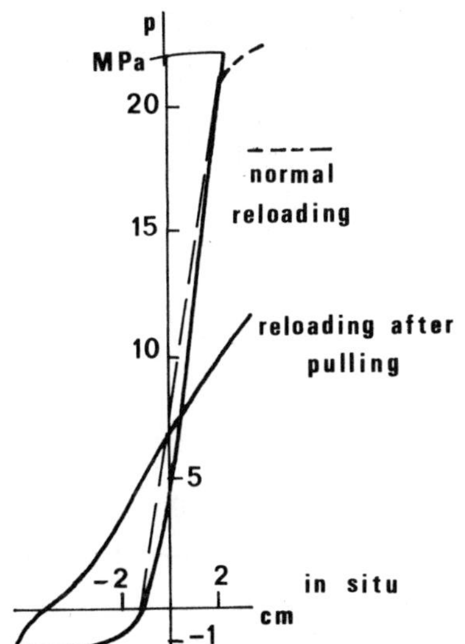

Fig.3 Penetration-pulling-penetration

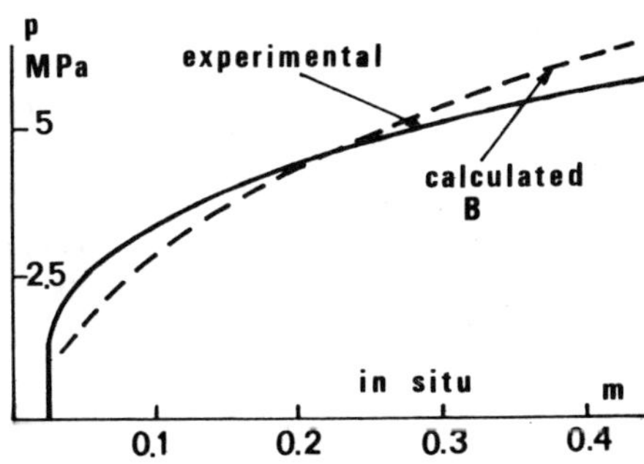

Fig.4 Pulling a cylindrical anchor

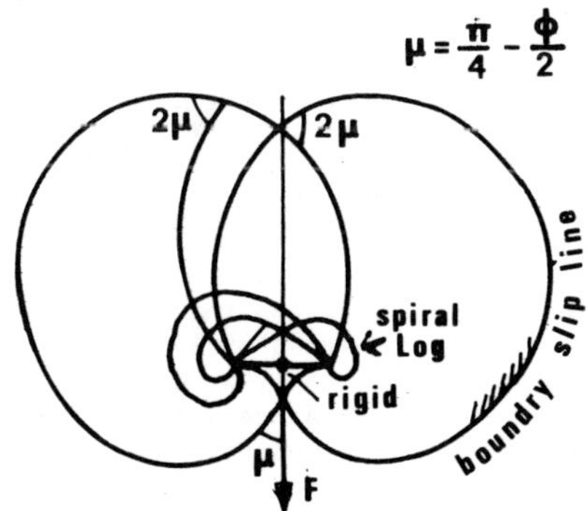

Fig.6 Characteristic lines for an anchor-disc

"in situ". They are enveloped by the curve A'. In fact any curve between B and A could in principle be obtained by a suitable loading history. When a pile is drilled in, jetted in, cast in place, driven, grouted, its placing history is equivalent to a curve between B and A.

The mechanism is work hardening. In fact, unloading at any point on those curves and reloading show an Elastic response up to the starting point (figure 3). A most important technical conclusion is that performance depends above all on preloading during construction. The Hydraulic Gradient method is most suitable for long duration tests which show that preloading is not relaxed, unless changes in surrounding stress occur. Such changes are reproduced by cycling the hydraulic gradient which partially destroys work hardening and reduces the pile's yield point. The implication to liquefaction and set is immediate. During earthquakes a pile might come under pulling which also partly reduces work hardening as is shown in figure 3. Another conclusion is that there is much resemblance between the static penetration test and tests on anchors. An extensive test program on models of anchors showed the similitude between anchors and piles (e.g. figures 2, 4).

The common case of a rigid layer overlaid by soft soil is depicted in figure 5, where single piles and groups are compared. The result of figure 5 showing a negligible influence of the overlaying soft soil are supported by precise full-scale tests (6).

In conclusion the small scale modelling method discussed is a suitable tool comparing static penetration tests to piles' performance.

3. THEORIES

Closed form solutions of a spherical cavity are helpful approximations. The Elastic case forms Lamé's problem while shock implosion was studied by Rayleigh (7). Curve B of figure 4 is based on an Elasto-Plastic calculation from (4) (1967). For a Coulomb material of $\phi = 35°$, by fitting the calculated curve to an anchor's test curve a value of 170 MPa is obtained for the modulus E. According to calculation, upon unloading a plastic zone of active pressure is formed around the cavity together with a field of residual stresses.

A plastic rigid solution by characteristics is given in (4). Any geometry of characteristics which terminate on the pile is incompatible with a flow field, so the problem of an anchor disc was chosen where the soil flows from the front face to the back one (figure 6). The characteristics rotate by $(1.25\,\pi - 0.5\,\phi)$ radians which means a ratio of several hundred between the pressure in front and at back of the anchor. Both solutions leave too much margin to curve fitting.

The Elastic approximation of Sharp (8) gives the expansive displacement u of a cavity of radius R subject to an internal pressure signal in the form of a unit step function :

$u = \sin \alpha t \ \exp\,(- \alpha t / \sqrt{2})$

with $\alpha = \sqrt{E / \rho R^2}$. Comparing to a single degree solution of the form :

$u = \sin \sqrt{(1-D^2)\,K/M}\ \exp\,(- \sqrt{K/M}\ Dt)$

one gets : radiative damping D=0.37 spring constant K=ER and an attached soil mass $M=\rho R^3$.

A plastic shock expansion is treated in (9). A differential equation is constructed the solution of which gives a relation between the velocity v, the acceleration $\dot{v}$ and the pressure σ at the cavity.

$\sigma = A + 2\rho v^2 / (nR^4 - 4R^4) - \rho\dot{v} / (nR-R)$

A is a function of the cohesion and ϕ. In a Viscoplastic calculation a term proportional to v appears.

The wave equation for strings was found by d'Alembert (1747) and during the following discussions by d'Alembert, Euler and Bernoulli, Lagrange, then a young mathematician, proposed a solution in 1759 based on concentrating the mass of the wire at discreet points and then passing to the limiting continuous case (10). An extensive calculation of the impact of rods was given by St. Venant (11). The transmission of energy along the pile, each element of which is subject to the forces mentioned in the above dynamic solutions and attached to some soil mass is in fact the classical problem of the telegraphist treated by Kelvin for the trans-Atlantic cable, and is solved normally by characteristics.

All these theories claim success in terms of curve fitting, while ignoring pore pressure, grain rearrangements and residual stress fields. The question remains whether, using these theories, a static loading test or driving test give enough information for modellize problems like sea wave action or earthquakes.

The problem of the difference between the Plastic zones in the dynamic and static cases presents itself. Static model tests have shown that changing or moderately enlarging the pile's base does not change the penetration curve (figure 7) in contrast with tests of dynamic penetration of rigid bodies (12). In conclusion, theories must be tested in detail by measurements in the soil around the pile, and on the pile itself, to show what actually happens during the different types of loading (figure 8).

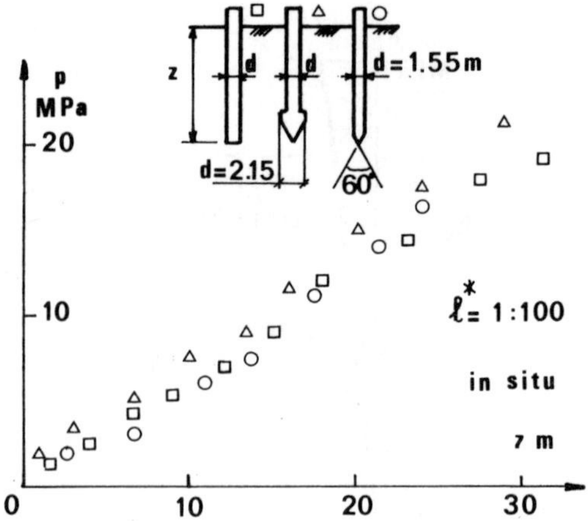

Fig.7 Influence of the form of the base

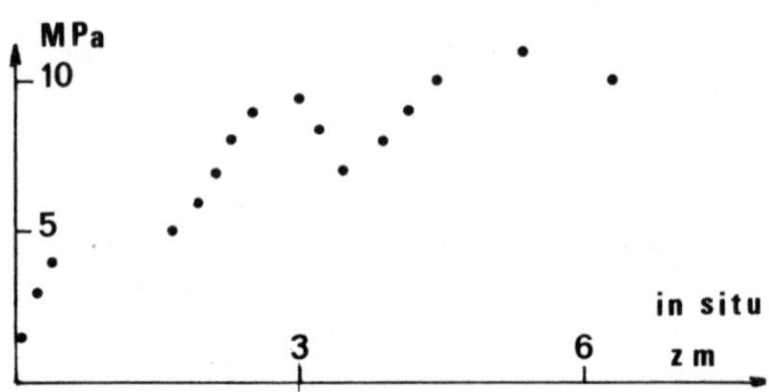

Fig.8 Instability of penetration curve due to collapse of interlocking in a density decseasing sand profile

4. DYNAMIC MODELS (figure 1)

The dynamical set up is intended for comparison between pile driving tests and various cases of impact and dynamical loading of groups of piles and structures on pile foundation. At the central part there is room for a model of 20cm diameter and 40cm high (20m and 40m at $\ell^* = 1/100$). The maximum gradient according to performance tests is about 200 ($\ell^* = 1/200$), modelling a soil mass of 160m diameter and 80m depth.

For pile driving the model is a steel pipe with the wiring lines inside. Stresses on the pile are measured by strain gauges. The pile is prolongated by a light plastic rod of negligible impedance which forms a Hopkinson bar for velocity measurements. Miniature accelerometers, pore pressure and total pressure transducers are fixed on the pile and in the soil. According to similitude with $v^* = 1$, the hammer stroke is conserved while its mass scale equals ℓ^{*3}.

Some results are given as an example.

The soil was homogeneous Fontainebleau sand with 10% ground sand which decreased permeability 40 times as needed for $\ell^*=1/40$. It was of medium compaction, equivalent to that of the static tests. The pile was a smooth stainless steel pipe open at the base. Its diameter was 10mm and the wall thickness was 2mm. It passed through a Teflon guide in the cell's cover and was statically loaded through rubber bands to represent the pile's self-weight. The hammer's model, made of steel, was guided by the plastic rod and hit directly the pile head suitably enlarged. The following tables show the role of the different parameters :

1. $\nabla h = 40$ $\ell^* = 1/40$
   pile's diameter in situ 0.40m
   hammer's mass in situ M = 6.4 ton .
   Stroke = 1m

Z_m cm	14	23	31
N/mm	10	10	10
Z(in situ) m	5.6	9.2	12.4

N is the number of blows . N/10mm equals 75 per 30cm at $\ell^* = 1/40$

2. Zm = 30 cm Mm = 100g

∇h	68	40	28
N/mm	25	10	8.4
N/30cm	110	75	75
Z m	20	12	8.4
M t	31.5	6.4	2.2

3. $\nabla h = 40$ Mm = 100g $\sim$ M = 6.4^t
   Mm = 400g $\sim$ M = 25.6^t

Z m cm	5.5	14	23	31
100g stroke 1m;N/mm	-	10	10	10
400g stroke 0.2m;N/mm	1	5	12.5	12.5

5. CONCLUDING REMARKS

At the end of an either static or dynamic loading cycle the soil near the pile is transformed. Plastic deformations cause residual stresses and grain interlocking which become predominant in the future performance of the pile. Tests consisting of measurements at one or two points of a single pile incorporate the contributions of many soil elements. Like in Plato's similitude of the people in the cave those tests only give the shadows of what happens outside the pile in the soil. The actual picture can only be given by using many transducers in the soil during extensive laboratory tests and correlation can only be made with highly instrumented light test-pile's results in the field obtained by competent people. It is held that objectively useful information is that which is cheap and abundant permitting parts of which to be discarded if doubtful. The normal inclination to attach importance to information from huge single events because it is expensive makes it of little help.

Only modelling methods conserving stresses in the soil can be called true, meaning that they present a reality close to what happens in situ. The centrifuge and the Hydraulic Gradient are true modelling methods which complete each other. The Hydraulic Gradient methode is much cheaper and simpler in application for the cases it can treat. Both supply large quantities of consistent results. It is firmly believed that only extensive measurements in the soil will bring advance.

6. REFERENCES

1 Zelikson, A., P. Leguay and C. Pascal 1982, Centrifuge model comparison of pile and raft foundations subject to earthquakes. Int. Conf. Soil dynamics and earthquake engineering, Southampton, England (to be published)

2 Scott, R.F. 1981, Pile testing in a centrifuge, X ICSMFE, Stockholm, Sweden

3 Zelikson, A. and J. Bergues 1981, Running waves in large sand models for the study of liquefaction utilizing the Hydraulic Gradient similitary method. VII Int. Symp. Military Application of Blast Simulation (MABS), Medicine Hat, Canada

4 Zelikson, A. 1967, Représentation de la pesanteur par gradient hydraulique dans les modèles réduits en Géotechnique. Ann. DE l'Inst. techn. du bâtiment et des T.P. Novembre 1967.

5 Zelikson, A. 1969, Geotechnical models using the Hydraulic Gradient similarity method. Géotechnique 19: n°4

6 Tcheng, Y. 1977, Modèles hydrauliques des fondations. IX ICSMFE Tokyo, Japan

7 Rayleigh, 1917, On the pressure developped in a liquid during the collapse of a spherical cavity. Phil. Mag. 34 (6th series)

8 Sharp, J. 1942, The propagation of elastic waves by explosive pressures. Geophysics 7.

9 Zelikson, A. 1979, Essais sur modèles réduits utilisant une centrifugeuse pour représenter l'action d'une explosion profonde dans une roche sur une cavité voisine. VI MABS, Cahors, France
10 Carslaw, H.S. 1930, Theory of Fourier Series and integrals, Dover
11 Goldsmith, W. 1960, Impact. Edward Arnold Publ., London
12 Yankelevsky, D.Z. and M.A. Adin 1980, A simplified analytical method for soil penetration analysis. Inter. Jour. Num. and Anal. Meth. in Geotech. 4: n°3.